Water Science and Application 6

Calibration of Watershed Models

Qingyun Duan
Hoshin V. Gupta
Soroosh Sorooshian
Alain N. Rousseau
Richard Turcotte
Editors

American Geophysical Union
Washington, DC

Calibration of Watershed Models
Water Science and Application 6

Library of Congress Cataloging-in-Publication Data
Calibration of watershed models / Quingyun Duan ...[et al.], editors.
    p.cm -- (Water science and application ; 6)
    Includes bibliographical references.
    ISBN 0-87590-355-X
        1. Watersheds--Mathematical models. 2. Watershed management--
Mathematical Models. 3. Hyrdrologic models. I. Duan, Quingyun, 1960-. II. Series.
GB980.C35 2002
551.48--dc21   2002038510

ISBN 0-87590-355-X
ISSN 1526-758X

Copyright 2003 by the American Geophysical Union
2000 Florida Avenue, N.W.
Washington, DC 20009

FSC
www.fsc.org
MIX
Paper from
responsible sources
FSC® C013604

CONTENTS

CONTENTS

PREFACE

During the past four decades, computer-based mathematical models of watershed hydrology have been widely used for a variety of applications including hydrologic forecasting, hydrologic design, and water resources management. These models are based on general mathematical descriptions of the watershed processes that transform natural forcing (e.g., rainfall over the landscape) into response (e.g., runoff in the rivers). The user of a watershed hydrology model must specify the model parameters before the model is able to properly simulate the watershed behavior.

In this regard, hydrologic models are no different from mathematical models of other physical or natural systems. If the physical processes governing the system are well understood, the values for model parameters can often be determined to a high degree of precision. In some instances, such models gain acceptance as "physical laws," and their parameters are referred to as "physical constants" (e.g., the gravitational constant in Newton's law of gravity and the gas constant in the ideal gas law). In hydrology, however, the physical processes of concern are complex and not well understood. Although model parameters are conceptually related to characteristic properties of the structure of a landscape, for example, these properties have been found to be highly variable in space, and not easily measurable at the spatial and temporal scales required by the models in use. As a result, model parameters must be estimated for each specific application of the model.

There are two main approaches to estimating the model parameters. The first (*a priori* approach) estimates model parameters by relying on theoretical or empirical relationships that relate such parameters to observable (measurable) characteristics of the watershed, such as soil and vegetation properties, watershed geomorphology, topographical features, and more. The second approach (model calibration) adjusts model parameter values, so that the model input-output (e.g., rainfall-runoff) response closely matches the observed (measured) input-output response of the watershed for some historical period for which data have been collected. When adequate amounts and quality of calibration data are available, hydrologists have typically preferred the model calibration approach, or some combination of the two.

Unfortunately, past experience has shown the profound complexity of estimating values for hydrologic model parameters, either by the *a priori* or model calibration approaches. Because all models are approximations of the real world, model equations and associated parameters are idealized representations which are not directly (unambiguously) related to measurable watershed properties. Furthermore, there is a variety of errors in the model structure and uncertainties in the data used for parameter estimation, which introduce considerable inaccuracy into model behavior. These factors have made it difficult to develop reliable procedures for model parameter estimation, and to provide suitable estimates of uncertainties in the resulting model predictions.

During the past several decades, scientists and practitioners have devoted considerable research effort to the model parameter estimation problem, leading to a variety of different approaches. In a process analogous to the proverbial blind men studying the elephant, different perspectives and opinions have arisen that are energetically debated and which do not enjoy universal acceptance. Nonetheless, each perspective is based on a different approach to the problem and, therefore, points to important "truths" that call for assimilation into a more complete understanding of the "beast" (the watershed modeling problem).

Despite imperfect knowledge and understanding, scientists recognize the progress made, and a degree of consensus has begun to emerge. The study of the problem has been greatly facilitated by advances in computing power, advances in measurement technologies (including remote sensing), and by improved mathematical and statistical theories.

Here, then, is a monograph that broadly reflects the state of the art in the methods and philosophies for model calibration now available from leading researchers worldwide. A broad range of topics are discussed within the context of the following questions: (1) what constitutes the best estimates for the parameters of a watershed model?; (2) what computational procedures are necessary to ensure proper model calibration and meaningful evaluation of model performance?; (3) how are calibration methods developed and applied to watershed models?; (4) what calibration data are needed, and how are these data obtained and analyzed, in order to obtain reliable parameter values?; and (5) how can model parameters be estimated using a combination of expert knowledge of the model physics and *a priori* knowledge of land surface characteristics?

The papers in the monograph are organized into seven sections: (1) Introduction, (2) Advances in Calibration Methodologies, (3) Optimization Algorithms for Parameter Estimation, (4) Interactive Strategies for Parameter Estimation, (5) Automatic Strategies for Parameter Estimation, (6) Methods for Developing *a priori* Parameter Estimates, and (7) Process Representation, Parameter Sensitivity, and Data Informativeness. The introduction by John Schaake presents a historical perspective on watershed model calibration, a brief overview of each chapter, and a discussion of emerging opportunities and future directions in watershed model development and calibration. Section 2 covers a range of perspectives and philosophical approaches to model calibration, arising from different emphases and computational approaches. In some cases, similar arguments are made, and this redundancy has been consciously retained to reflect the consensus now emerging (due in no small measure to improved global communication technologies). Section 3 presents a number of state-of-the-art optimization algorithms that can be used to find optimal estimates (and uncertainty bounds) for model parameters, when the watershed model calibration problems are posed as optimization problems. Section 4 reviews various interactive calibration procedures that incorporate human expert experience and knowledge, while taking advantage of modern computational and graphical tools such as GIS and Graphical-User-Interfaces (GUIs). Section 5 illustrates the applications of automatic calibration to various practical hydrological applications including rainfall-runoff modeling, hydrochemical modeling and land surface modeling. Section 6 discusses how *a priori* parameter estimates can contribute to the fine-tuning of parameters. Finally, Section 7 addresses factors critical to the success of model calibration, including data requirements, process representation and interpretation of model parameters.

Despite more than forty years of experience with digital watershed modeling, a book that focuses on the achievements and advances in calibration of watershed models has yet to be published. This monograph is intended to fill that void. It is suitable for both didactical and reference purposes, and should prove valuable to a wide audience, including university researchers and graduate students, practicing hydrologists, civil and environmental engineers, and water resources managers. We particularly hope that the materials contained herein will motivate generations of students to bring new and broader perspective to the "study of the elephant," leading to resolution of the many problems that still command our attention.

The monograph derives from a special session, "Advances in Calibration of Watershed Models," held at the 2000 Fall Meeting of the American Geophysical Union (AGU) in San Francisco, California. The excitement generated by the presentations, discussions, and posters at this special session, along with strong encouragement from the AGU Books Department, led to the suggestion that the papers be compiled into an AGU Monograph as part of the Water Science and Application Series. It also contains invited papers by a number of leading researchers who were unable to attend the AGU meeting.

The editors wish to acknowledge many of our colleagues who contributed to the volume by reviewing individual chapters, thereby ensuring the scientific integrity of the monograph as a whole. We also gratefully acknowledge support from our respective institutions. In particular, we appreciate the partial support provided for editors Duan, Gupta and Sorooshian by SAHRA (Sustainability of semi?Arid Hydrology and Riparian Areas), under the STC Program of the National Science Foundation, Agreement No. EAR-9876800. Finally, we express our gratitude to our AGU acquisitions editor, Allan Graubard, and AGU production editor, Terence Mulligan, for their editorial guidance, cooperation, and patience in publishing this volume.

Qingyun Duan
Hoshin V. Gupta
Soroosh Sorooshian
Alain N. Rousseau
Richard Turcotte

Introduction

John C. Schaake

NOAA/NWS, Hydrology Laboratory, Office of Hydrologic Development, Silver Spring, Maryland

HISTORICAL PERSPECTIVE

Conceptual hydrologic models that account for the continuous dynamics of hydrologic processes were introduced in the early 1960's. The Stanford Watershed Model (Crawford and Linsley, 1962) was the first integrated attempt to take advantage of the advent of digital computers to describe quantitatively the hydrologic processes that take place in a watershed "within the limitations of current understanding and the limitations of the computer".

The limitations of our hydrologic understanding and the limitations of the computer both have evolved since 1960's. Computing power is at least a billion times what it was then. But computing still may pose practical limits for hydrologic modeling and parameter estimation today. Hydrologic understanding remains limited in several ways that are scale dependent. A great challenge in hydrology is to make predictions and test hypotheses at space and time scales of practical interest. Understanding depends on: knowledge of the physics of hydrologic processes at different scales, knowledge of soils, vegetation and topographic characteristics and knowledge of water and energy forcing that varies in time and space. Knowledge of all of these factors is both limited and imperfect. In the end it is not possible to resolve every detail of every aspect of the hydrology of the "real" world. So, effective hydrologic modeling is both the art and the science of applying limited and imperfect understanding.

These issues were well understood by Crawford and Linsley (1966) who wrote:

A hydrologic model is nothing more than a collection of quantitative hydrologic concepts that are given mathematical representations. If each of these concepts is a well established physical law that has an exact mathematical representation, and if every physical component of the watershed is present in the model, the entire model structure would be unique and all physical processes in the watershed could be accurately simulated. Prohibitive amounts of input data would be required, far beyond practical limitations even for small watershed plots.

Since most of the action in hydrology occurs underground where it cannot be directly observed, this assessment remains valid today. Nevertheless, Crawford and Linsley's Stanford Watershed Model was a remarkably successful attempt to achieve an acceptable level of complexity using physically relevant components and a moderate number of quantitative components to represent a broad range of hydrologic behavior. A variant of the original model, but with the addition of water quality components, is used today as EPA's Hydrological Simulation Program.

Since the advent of the Stanford Watershed Model, a plethora of hydrologic models have been proposed and many are being used for a variety of different applications. Twenty-six of the world's most popular computer models of watershed hydrology were documented recently by Singh (1995). More recently Singh and Frevert (2002a,b) put together a 2-volume book that gives a comprehensive account of 38 mathematical models of large and small watershed hydrology not included in Singh's 1995 book. Some notable models that have been widely used throughout the world include: the Tank model (Sugawara, 1995) that was a contemporary of SWM; the Sacramento model in the National Weather Service River Forecast System (Burnash, 1995); the Precipitation Runoff Modeling System (PRMS) developed by the United States Geological Survey (Leavesley and Stannard, 1995); the HBV model developed in Sweden (Bergstrom, 1995) and the SHE

Calibration of Watershed Models
Water Science and Application Volume 6
Copyright 2003 by the American Geophysical Union
10/1029/006WS01

model developed in Europe (Bathurst, et al, 1995) that also has several widely used variants.

A comprehensive review of what has happened in hydrologic modeling since the 1960's was prepared by Beven (2001). Interest in hydrologic models initially was focused on solving practical engineering problems. Models were often explained in terms of how they could be used. There was much more interest in how they performed than in why. Although engineering interest in hydrologic models continues, there also has emerged a more reflective, scientific interest in hydrologic models. This has been enhanced during the last decade by a rapidly growing collaboration of hydrologists and atmospheric scientists to develop improved representations of the role of the land surface as a lower boundary in weather and climate models. There is renewed interest in hydrologic prediction for ungaged basins, and this is calling attention to the importance of parameter estimation.

The first comprehensive attempt to intercompare different hydrologic models was the World Meteorologic Organization (WMO) hydrologic model intercomparison study (WMO, 1975). Subsequently, WMO led intercomparison studies of snowmelt models (WMO, 1986) and realtime applications of hydrologic models (WMO, 1988). Recently there has been a number of intercomparison studies of models used to represent the land surface in atmospheric models (Henderson-Sellers *et al.*, 1993).

Every conceptual model has parameters that are the coefficients and exponents in the model equations. These parameters must be estimated for a given catchment and for each computational segment of the model. They must be estimated either by some relationship with physical characteristics or by tuning the parameters so that model response approximates observed response, a process known as calibration.

The process of model calibration is quite complex because of limitations of the models, limitations of the input and output data, imperfect knowledge of basin characteristics, mathematical structure of the models and limitations in our ability to express quantitatively our preferences for how best to fit the models to the data. As a result of these limitations, it is not clear that a unique set of values exists for the model parameters for a given watershed. And there is a degree of uncertainty about which parameter values may be "best". When comparing model outputs to observations, a basic question is what causes the differences. Are they because of limitations in the model structure, limitations in the parameter set, errors in the forcing data or errors in the output measurements? Improvements in the calibration process are needed to deal better with this issue.

There are two primary parts to the calibration process. The first is to decide how to judge whether one set of parameter values is preferred over another set. In the case of automatic calibration this means to specify an objective function or a set of objective functions. The second part is to find preferred sets of parameters and possibly select one of these to apply the model. This may be done manually, automatically using an optimization technique or by some combination. It may also involve a hierarchical process or other strategy to estimate different parameters at different stages in the process.

In the early days of hydrologic modeling, the existing optimization techniques would tend to converge on local optimal solutions and would not reliably find the global optimum. Objective functions for hydrologic model calibration are notorious for having many local optima. Therefore it was not clear if the limitations of a given model application to a particular watershed were due to poor calibration or to limitations in the model or the input data. The optimization part of the calibration problem, though yet to be solved completely, was no longer a major limiting factor with the development of the Shuffled Complex Evolution (SCE-UA) optimization method, a highly reliable technique at finding global optimal solutions even for difficult objective functions with many local optima (Duan *et al.*, 1992 & 1994; Duan, this volume).

Now it is possible to focus attention much better on the most important aspects of the model calibration process. Two aspects of this that are receiving increased attention are analysis of uncertainty and consideration of multiple objectives. Uncertainty in models and data leads to uncertainty in model parameters and model predictions. Bevin and Binley (1992) proposed a method called generalized likelihood uncertainty estimation (GLUE) that uses prior distributions of parameter sets and a method for updating these distributions, as new calibration data become available, to make probabilistic estimates of model outputs. Other recent studies of parameter uncertainty include Kuczera (1997), Kuczera and Parent (1998), Thiemann et al. (2001), Bates and Campbell (2001). Bevin (2001) notes that the GLUE methodology provides one way of recognizing the possible "equifinality" of models and parameter sets. Bevin introduced the term equifinality to recognize there may be no single, correct set of parameter values for a given model and that different parameter sets may give acceptable model performance. Multi-objective approaches to calibration that recognize there may be no single, optimal set of model parameters, have been pioneered by Gupta et al (1998) and Yapo et al (1997).

This book presents an interesting view of the state-of-the art in model calibration. The contributions presented herein are organized into the following six sections:

a)   Advances in Calibration Methodologies;
b)   Optimization Algorithms for Parameter Estimation;
c)   Interactive Strategies for Parameter Estimation;

d)   Automatic Strategies for Parameter Estimation;
e)   Methods for Developing *A Priori* Parameter Estimates;
f)   Sensitivities of Model Calibration to Various Factors.

ADVANCES IN CALIBRATION METHODOLOGIES

Gupta *et al.* (this volume, "Advances...") offer an interesting reflection on recent advances in automatic calibration of watershed models. The goal of calibration strategies should be to explicitly account for all of the following - *a priori* model uncertainty, input, state, structure parameter and output uncertainties, and multiple sources and types of information, while allowing recursive processing of data as they become available. They suggest that a multi-objective approach offers a way forward by emulating the ability of manual-expert calibration to employ a number of complementary ways to evaluate model performance and to extract greater amounts of information from the data. They raise several questions for further investigation.

Major sources of uncertainty in the modeling process are a lack of objective approaches to evaluate model structures and the inability of calibration approaches to distinguish between the suitability of different parameter sets. Therefore, Wagener *et al.* (this volume), propose a framework for identification and evaluation of conceptual rainfall-runoff models that is based on multi-objective performance and a novel dynamic identifiability analysis framework (DYNIA). They illustrate their approach with an application to a catchment located in the south of England and propose several areas of possible application to the modeling process.

Understanding the nature of data and model errors should be an essential part of the calibration process. Kavetski, *et al.* (this volume) discuss shortcomings in existing calibration methodologies and outline a Bayesian Approach to Total Error Analysis (BATEA) framework that integrates model and data uncertainty representations into the calibration process. They propose that distinguishing the various sources of error will improve our understanding of uncertainty in both parameter values and model predictions.

Freer *et al.* (this volume) apply the Generalized Likelihood Uncertainty Estimation (GLUE) approach to assess the changing dynamics of a hydrologic model applied to data from the 41-ha Panola Mountain Research Watershed, Georgia. They conclude that there needs to be a more thoughtful approach to specification of performance measures and that further development of the model to better represent effects of seasonality is also required.

Seibert and McDonnell (this volume) suggest that new progress in watershed modeling may be possible by comple-

menting traditional hard data measures used in model calibration with qualitative process understanding that exists for most small research catchments. Their idea is to include soft data in automatic calibration procedures using a multi-criteria approach as a way to mimic hydrologic reasoning that is done implicitly in manual calibration approaches. They present a framework to use soft data from the experimentalist through fuzzy measures of model simulation and parameter value acceptability. They illustrate their ideas for the Maimai research catchment in New Zealand.

OPTIMIZATION ALGORITHMS FOR PARAMETER ESTIMATION

Recent progress in developing robust, global optimization techniques is reviewed by Duan (this volume). Three global optimization methods commonly used in watershed model calibration: simulated annealing; genetic algorithm and shuffled complex evolution are presented in detail. The relationship between these and earlier classical local search methods is discussed.

Improvements to the original SCE optimization algorithm have led to development of the Shuffled Complex Evolution Metropolis (SCEM-UA) algorithm (Vrugt *et al.*, this volume). Two enhancements have been made that prevent the search from becoming mired in a small domain of attraction. It is demonstrated that the new algorithm is more efficient that alternative approaches to accomplish the same objective.

Calibration of basins that only recently are gaged and therefore have only limited historical data or basins where land use may be changing might best be done with a recursive algorithm. Misirli *et al.* (this volume) present a Bayesian Recursive Estimation (BaRE) algorithm that considers uncertainty associated with model structure, parameters, states and the input and output measurements. Comparisons to batch calibration using the SCE-UA algorithm show that BaRE is a powerful on-line, adaptive calibration tool.

Model calibration using a single objective function does not adequately measure the ways in which a model fails to match important characteristics of the observed data. Gupta *et al.* (this volume, "Multiple...") present the MOCOM-UA algorithm as an effective and efficient methodology for solving the multi-objective optimization problem and illustrate this in a simple hydrologic model calibration study.

INTERACTIVE STRATEGIES FOR PARAMETER ESTIMATION

The National Weather Service (NWS) uses hydrologic models as an integral part of its river and flood forecasting system. Experience with calibration of many basins throughout the United States has led to development of a comprehensive interactive calibration strategy now being used by NWS River Forecast Centers presented by Smith *et al.* (this volume). Data analysis techniques, calibration procedures and future enhancements to the calibration process are discussed.

A multi-step automatic scheme (MACS) that emulates the thought processes of expert-manual calibration of the Sacramento model is described by Hogue *et al.* (this volume). Different objective functions are used at different steps in the process. Application to three river basins in different climate regimes demonstrates improved quality calibrations comparable to the existing River Forecast Center and other automatic calibrations. This method offers a reliable, time-saving approach to obtain quality calibrations.

An approach to estimate parameters by assigning each parameter to one of several objectives is discussed by Turcotte *et al.* (this volume). Parameters affecting objectives characterized by long time scales are calibrated first while those characterized by short time scales are calibrated last. Adjustments to parameters estimated earlier are considered by repeating the process until satisfactory performance is attained. Objectives to minimize errors and stratify parameter values pertain to: (i) prolonged summer drought recessions, (ii) annual and monthly flow volumes, (iii) summer and fall high flows, (iv) high flow synchronization, (v) winter recessions and (vi) spring runoff from snowmelt.

A case study of a fully distributed hydrologic model calibrated with a systematic manual adjustment of parameters for the Illinois and Blue river basins in Oklahoma is presented by Vieux and Moreda (this volume). An ordered physics based parameter adjustment (OPPA) procedure is used in which parameters are associated with criteria for their estimation and the order of estimation considers the role played by each parameter and its sensitivity to other parameters in the model.

AUTOMATIC STRATEGIES FOR PARAMETER ESTIMATION

A new hybrid multi-criteria calibration approach that combines the strengths of automatic and manual calibration methods is presented by Boyle *et al.* (this volume). The new approach is used to explore the benefits of different levels of spatial and vertical representation of important watershed hydrologic variables. Suggestions are made for further research using this approach to investigate simultaneously the effects of spatial resolution and vertical structural complexity on model performance and parameter calibration.

An approach allowing calibration of hydrologic models over a range of time scales using wavelet analysis is presented by Parada *et al.* (this volume). The multi-resolution approach can be applied in a similar way as the single-scale approach to different objective functions. It is applied to a sub-humid basin in northern California where it was found that the multi- resolution approach was superior to the single-scale approach and was less sensitive to the representativeness of the period selected for calibration. Suggestions for choice of optimization criteria also are offered.

The MOCOM-UA multi-objective approach was used by Meixner *et al.* (this volume) together with sensitivity analyses to investigate parameter estimates, model structure and natural processes using the Alpine Hydrochemical Model (AHM) of the Emerald Lake watershed. The sensitivity analysis was used to develop four sets of criteria for MOCOM-UA. Improved estimates of several hydrologic and biochemical process parameters were made and a flaw was found in the current representation of mineral weathering in the AHM. Also, some conflicts were found between the kinds of conclusions that might be drawn from sensitivity and calibration analyses.

The relationship between parameter values and the ability of a land surface model to simulate surface heat fluxes as well as water and energy state variables is discussed by Bastidas *et al.* (this volume). The potential is explored for using remotely sensed ground surface temperatures and surface soil moisture to bound the parameters of land surface models and thereby to improve the ability to simulate surface heat fluxes to the atmosphere. Although both the surface state variables and the surface heat fluxes could be simulated accurately, different parameter sets were required to do this, raising questions about the adequacy of the model structure and how to interpret the relationship between observations and model state variables.

METHODS FOR DEVELOPING *A PRIORI* PARAMETER ESTIMATES

When neighboring basins are calibrated independently there may be far more spatial variability in the calibrated parameters than might seem reasonable relative to the variability of basin characteristics. Koren *et al.* (this volume) developed an objective estimation procedure that uses *a priori* parameter estimates to initialize the calibration process, to provide limits to constrain the feasible parameter space for basins being calibrated, and to transfer calibrated parameters to ungaged basins. Tests involving Sacramento model applications to a number of headwater watersheds in the Ohio river basin suggest that soil derived parameters can improve the spatial and physical consistency of estimated parameters while maintaining hydrologic performance of both gaged an ungaged watersheds.

The USGS has been developing an integrated modeling framework that can be used to assess objective parameter estimation methodologies and process conceptualizations. Leavesley *et al.* (this volume) present methods and results from initial testing of the USGS Modular Modeling System (MMS) for three major snowmelt regions of the western United States. The study is concerned with estimation of parameters for distributed models, the application of *a priori* information and the role of calibration in the parameter estimation process. The chapter includes a comprehensive evaluation of the results and the performance of various parts of the MMS. The ability to identify sources of error, such as model, data and parameter are needed to provide an objective assessment of estimation methodologies and model coneptualizations.

SENSITIVITIES OF MODEL CALIBRATION TO VARIOUS FACTORS

An interesting review of issues important in the development of hydrologic models, estimation of model parameters and applications of models is given by Burges (this volume). This review is motivated by the author's extensive hydrologic research experience and its relationship to hydrologic modeling. A wide range of topics needing attention are suggested, including need for more and better measurements and a need to establish "natural laboratories" with nested measurements.

Calibration and validation model results were analyzed for 37 sets of conceptual rainfall runoff model experiments by Gan and Biftu (this volume). These experiments were from five different models applied to five different catchments from wet, semi-wet and dry climates. Generally more dependable results were obtained for wet catchments. Model performance was found to depend more on model structure and on data quality than on model complexity or data length. Because parameter estimates are data dependent, adequate data are needed for estimation. Although hydrologic processes in dry catchments are more complex than in wet, good quality hydrologic data can support hydrologic modeling of dry catchments.

Quantification of nutrient loads from nonpoint sources is investigated by Baginska and Milne-Home (this volume) using the Annualized Agricultural NonPoint Source (AnnAGNPS) model for a small rural watershed in New South Wales, Australia. Even though all of the model inputs can be measured in the field, calibration of model parameters improves the results and helps to understand uncertainties and sensitivities. Interdependence of model parameters was found to complicate the calibration process. Particular attention is needed during the verification process to assure that simulated flow volumes match observed so that flow volume inconsistencies are not transformed and amplified in subsequent water quality simulations.

Land use changes over a period of years may lead to significant changes in flow peaks, shorter times to peak flows, changes in recession characteristics, etc. These hydrologic changes may have important societal effects. Loaiciga (this volume) discusses the relevance of hydrologic model calibration within the context of forensic hydrology, a branch of hydrology that supports legal investigations and that deals with the study of flood events with the objective of determining the probable causes and sources of human-induced contributions to flood damages.

REFLECTIONS ON THE FUTURE OF MODEL CALIBRATION

Calibration methodologies must extract as much information as possible from available data. Often, the only available measured, endogenous, variable is streamflow. Jakeman and Hornberger (1993) argued that only very limited model complexity involving few model parameters is appropriate if the only source of data for model calibration is streamflow. Since this usually is the case, improved diagnostic tools are needed to extract more information from both the input forcing data and streamflow data. This would also improve our understanding of how the climatic variability of the forcing is modified by catchments to control the climatic variability of the streamflow.

Several contributions to this volume use multi-objective techniques, sometimes associating subsets of parameters with different objectives. If we had improved diagnostic tools they could be used to better understand how different model structures and parameter values function together to approximate the behavior of real catchments. Improved diagnostic tools might also be used to develop improved approaches to multi-objective calibration.

There has been an exponentially growing recent interest in distributed hydrologic modeling that has been fueled by growing availability of GIS-related information. The distinction between lumped and distributed hydrologic models is simply whether the catchment is represented by a single, lumped hydrologic element or a set of spatially distributed elements. In any case, the smallest element of all distributed models is a lumped model. If there are enough distributed elements the size of the smallest area may be small enough that models of point physical processes may be reasonable representations of local hydrologic processes. Nevertheless, a great challenge is to estimate the parameters of distributed models. Although there may be a wealth of GIS data to help establish model parameters, there is also a great lack of data about the detailed physical characterization of the sub-surface where most hydrologic processes occur. This means that detailed, distributed, "physically based" models can be improved through calibration of at least some model parameters. Practical application of all hydrologic models require data related to how physical processes work to calibrate key model parameters. While distributed hydrologic models may better represent some aspects of the physical processes in a catchment and offer *a priori* methods to estimate model parameters using GIS data, they also present a great challenge for model calibration procedures and for improved diagnostic tools to use limited streamflow data.

A key theme of a few authors is to find additional sources of endogenous measurements, even qualitative information, and then develop ways to use these in the calibration process. It was illustrated that this can lead not only to improved parameter estimates but to better understanding of limitations of our models because different sets of parameter values may be needed to match different sets of observed data. Although some special measurements may only be available for brief periods during special research projects, others may become available routinely, especially from satellite remote sensing. To use additional measurements, more attention is needed to the relationship between measured variables and related model variables. Such relationships may be very complex and may only be possible to define empirically. If so, should additional parameters in such relationships be calibrated together with the original model parameters?

Several chapters note that measurement errors have significant effects on the calibration process and propose strategies to deal with this. Improved understanding of how measurement errors lead to uncertainty in both parameters and predictions requires improved knowledge of measurement errors and methods to estimate properties of measurement errors. Most existing methods to account for uncertainty do not explicitly distinguish between different sources of uncertainty. More attention is needed to model all of the sources of uncertainty explicitly and to estimate how this leads to uncertainty in model parameters and model predictions. It might be interesting to consider how a model would respond to an ensemble of equally likely traces of forcing variables that might be repeated for an ensemble of equally likely parameter sets. Could such ensemble approaches assist in various aspects of the modeling process?

More attention is needed to deal with the fact that every model is an imperfect representation of a real catchment. The goal of the calibration process is to somehow "fit" the model to the real catchment. The best approach to the fitting process depends on how the model will be used. For example, a model might be used to make probabilistic predictions about the occurrence of one or more endogenous variables. Or it might be used to make an "optimal" estimate of the endogenous variable at various times in the past or future. Modeling approaches designed to produce optimal estimates may not be the preferred approaches to make probabilistic predictions, although they may be related.

Some of the implications of imperfect models are that parameter values do not have exact physical meaning and that calibrated parameter values are partly an artifact of the model structure. Changing only a part of the model structure could lead to changes in all of the resulting calibrated parameter values. This does not mean that parameter values have no physical basis. Indeed there may be some relationship between parameter values and physical or climatological catchment characteristics that could be derived empirically using data from many catchments.

Finally, a key step in improved hydrologic modeling is to have good *a priori* estimates of model parameters and *a priori* estimates of the uncertainty in these parameter estimates. This is important to apply models to ungaged or poorly gaged areas, to apply distributed models to well gaged areas and to constrain the calibration process. An international Model Parameter Estimation Experiment (MOPEX) is being conducted to develop improved *a priori* parameter estimates and procedures for relating model parameters to physical and climatological basin characteristics (Schaake *et al.*, 2001). Data sets from hundreds of basins in the United States have been compiled and hundreds more from throughout the world are being sought. These data sets include model forcing and model output measurements as well as basin characteristics data. Many investigators from the international hydrologic community are beginning to apply the concepts presented in this book to these data, both to improve approaches to model calibration and to develop improved approaches to *a priori* parameter estimation.

REFERENCES

Bates, B.C. and E.P. Campbell, (2001), A Markov chain Monte Carlo scheme for parameter estimation and inference in conceptual rainfall-runoff modeling, *Water Resour. Res.*, 37(4), 937-947.

Bathurst, J. C., J. M. Wicks, and P. E. O'Connell, (1995), The SHE/SHESED basin scale water flow and sediment transport modelling system, in *Computer models of watershed hydrology*, ed V. P. Singh, Water Resour. Pub., p165-214.

Bergstrom, S., (1995), The HBV model, in *Computer models of watershed hydrology*, ed V.P. Singh, Water Resour. Pub.,p165-214.

Bevin, K. J. and A. M. Binley, (1992), The future of distributed models - model calibration and uncertainty prediction, *Hydrological Processes* 6(3): 279-298.

Beven, K. J., (2001), Rainfall-runoff modeling, Wiley, 360pp.

Burnash, R. J. C., (1995), The NWS river forecast system - catchment modeliing, in Computer models of watershed hydrology, ed V. P. Singh, Water Resources Publications, p165-214.

Crawford, N. H. and R. K. Linsley, 1962, The synthesis of continuous streamflow hydrographs on a digital computer, Technical Report 12, Civil Engineering Department, Stanford University

Crawford, N. H. and R. K. Linsley, 1966, Digital Simulation in Hydrology: Stanford Watershed Model IV, Technical Report 39, Civil Engineering Department, Stanford University

Duan, Q., S. Sorooshian, and V.K. Gupta, (1992), Effective and Efficient Global Optimization for Conceptual Rainfall-Runoff Models, *Water Resour. Res.*, 28(4), 1015-1031

Duan, Q., S. Sorooshian, and V.K. Gupta, (1994), Optimal Use of the SCE-UA Global Optimization Method for Calibrating Watershed Models, *J. of Hydro.*, 158, 265-284

Gupta, H. V., S. Sorooshian, and P. O. Yapo, (1998), Toward improved calibration of hydrologic models: multiple and noncommensurable measures of information, *Water Resources Research* 34: 751-763.

Henderson-Sellers, A., Z.L. Yang, R.E. Dickinson, (1993), The project for intercomparison of land-surface parameterization schemes, *Bull. Amer. Meteor. Soc.*, 74(7), 1335-1349

Jakeman, A. and G. Hornberger, (1993), How much complexity is warranted in a rainfall-runoff model? *Water Resources Research* 29(8): 2637-2649

Kuczera, G., (1997), Efficient subspace probabilistic parameter optimization for catchment models, *Water Resources Research* 33(1): 177-185.

Kuczera, G. and E. Parent, (1998), Monte Carlo Assessment of parameter uncertainty in conceptual catchment models: the Metropolis algorithm, *Journal of Hydrology* 211: 69-85.

Leavesely, G. H. andL. G. Stannard, (1995), The precipitation-runoff modeling system, in *Computer models of watershed hydrology*, ed V. P. Singh, Water Resources Publications, p165-214.

Schaake, J., Q. Duan, V. Koren, and A. Hall,(2001) Toward improved parameter estimation of land surface hydrology models through the Model Parameter Estimation Experiment (MOPEX), in *Soil- Vegetation-Atmosphere Transfer Schemes and Large-Scale Hydrological Models*, edited by Dolman A. J., A. J. Hall, M. L. Kavvas, T. Oki, and J. W. Pomeroy, IAHS Publ. No. 270, 91-97.

Singh, V. P., (1995), Computer models of watershed hydrology, Water Resources Publications, 1130pp

Singh, V. P., D. K. Frevert. (2002a), *Mathematical models of large watershed hydrology*, Water Resources Publications, 891pp

Singh, V. P., D. K. Frevert. (2002b), *Mathematical models of small watershed hydrology and applications*, Water Resources Publications, 950pp

Sugarawa, M., (1995), Tank model, in *Computer models of watershed hydrology*, ed V. P. Singh, Water Resources Pub., p165-214.

Thiemann, M.M., M. Trosset, H. Gupta, and S. Sorooshian, 2001: Bayesian recursive parameter estimation for hydrological models, Water Resources Research, 37,10,2521-2535.

WMO, (1975), Intercomparison of conceptual models used in hydrological forecasting, Operational Hydrology Technical Report No. 7, WMO, Geneva

WMO, (1986), Intercomparison of models of snowmelt runoff, Oper. Hydrol. Rep. No. 23, WMO-No. 646, WMO, Geneva.

WMO, (1988), Real-time intercomparison of hydrological models, Report of the Vancouver Workshop, Technical Report to Chy No 23, WMO/TD No. 255, WMO, Geneva

Yapo, P. O., H. Gupta and S. Sorooshian, (1997), Mult- objective global optimization for hydrologic models, *J. of Hydrol.* 204: 83-97.

John C. Schaake, NOAA/NWS, Hydrology Laboratory, Office of Hydrologic Devlopement, 1325 East-West Highway, Silver Spring, MD 20910

Advances in Automatic Calibration of Watershed Models

Hoshin V. Gupta, Soroosh Sorooshian, Terri S. Hogue, and Douglas P. Boyle[1]

SAHRA, NSF STC for Sustainability of semi-Arid Hydrology and Riparian Areas
Department of Hydrology and Water Resources, The University of Arizona, Tucson, Arizona

There is an urgent need to develop reliable automatic methods for identification of watershed models. The goal of such research should be to develop strategies that explicitly account for all of the following— *a priori* model uncertainty, input, state, structure, parameter and output uncertainties, and multiple sources and types of information, while allowing recursive processing of data as they become available, and providing quantified (perhaps probabilistic) estimates of model output uncertainty. The "Turing Test" of such a strategy would be its ability to provide reliable model performance that is indistinguishable from, or demonstrably superior to what can be obtained by an expert hydrologist. Traditional Automatic methods based on techniques of non-linear regression fail in this regard. Major weakness include their underlying assumption that the model structure is correct, inability to handle various sources of uncertainty, dependence on a single aggregate measure of model performance, and emphasis on identifying a unique optimal parameter set. The multiple-criteria approach offers a way forward by emulating the ability of Manual-Expert calibration to employ a number of complementary ways of evaluating model performance, thereby compensating for various kinds of model and data errors, and extracting greater amounts of information from the data. The outcome is a set of models that are constrained (by the data) to be structurally and functionally consistent with available qualitative and quantitative information and which simulate, in an uncertain way, the observed behavior of the watershed. This chapter explores the historical development of current perspectives on calibration and raises questions for further investigation.

1. INTRODUCTION

1.1. Conceptual Watershed Models

A watershed model is a conceptual-numerical representation of the dominant processes controlling the transformation of precipitation over a watershed into streamflow in the river channel. Such models are commonly designed to compute streamflow at the watershed outlet, but a number of models now also attempt to compute the flow at various interior locations. The reasons for wanting to do this vary, ranging from the construction of flood frequency curves for engineering design, to the simulation of the potential impacts of land use change or climate change, to operational real-time flood forecasting. Our own research, conducted primarily at the University of Arizona, has focused on the Sacramento Soil Moisture Accounting Model (SAC-SMA, see Figure 1) developed by Burnash et al. [1973; Burnash, 1995] in the 1970's and extensively used by the US National Weather Service (NWS) for flood forecasting at over 4000 forecast points throughout the United States [Ingram, 1996; Smith et al., this volume] (Note: the model is also widely used by other hydrologic agencies throughout the world).

Since the 1960's a variety of "different" conceptual watershed models have been developed, differing somewhat in the particular details of their design and equations used, but arguably similar to the diagram shown in Figure 1 in terms of

[1]Now at the Desert Research Institute, University and Community College System of Nevada, Reno, NV

their overall structure. Examples of such models are the Stanford series of watershed models [Crawford and Linsley, 1966], the Boughton Model [Boughton, 1965], and the Xinanjiang Model [Zhao et al., 1980], to name just a few. The above-mentioned models are generally considered to be *lumped parameter* models because they were historically designed to represent and be applied at an aggregate (watershed or sub-watershed) scale. Recently, with the advent of more powerful computers and access to distributed data at relatively fine scales, a number of *distributed parameter* watershed models have been developed, including the MIKE-SHE model [Refsgaard and Storm, 1995], TOPMODEL [Beven and Kirby, 1976; 1979], KINEROS [Smith et al., 1995], HBV [Bergstrom, 1995], and IHDM [Calver and Wood, 1995], to name just a few. The latter models attempt (in varying ways) to represent the spatial heterogeneity of the inputs, states, watershed properties and outputs at the sub-watershed scale.

In this paper, we shall focus our attention almost exclusively on the issue of model calibration for *lumped parameter* watershed models, hereafter simply called watershed models. Further, we shall refer mainly to the research conducted by our own group at the University of Arizona over the past two decades with some reference to the work of others; other chapters of this monograph are well representative of the important contributions made by eminent researchers throughout the world.

1.2. The Problem of Model Identification

Watershed models, such as the SAC-SMA, are (for the most part) based on the assumption that the dominant hydrologic processes controlling the transformation from

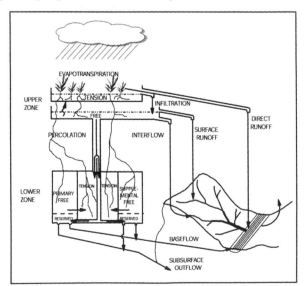

Figure 1. Conceptual diagram of SAC-SMA model.

precipitation (over a watershed) to runoff (at the watershed outlet) are *structurally* similar for all (most) watersheds of interest, and that the same (similar) model structure can be used at a wide variety of locations throughout the world simply by adjusting the values of the model parameters so that the model-simulated precipitation-runoff response is *similar* to the observed response of the watershed in question. As mentioned before, lumped watershed models are designed to represent the aggregate (watershed or sub-watershed) hydrologic response of the watershed (or some sub-region within it), and hence the model structure and parameters are "effective" conceptual representations of spatially and temporally heterogeneous watershed properties.

This assumption of similar watershed structure and behavior works quite well for applications such as engineering design and flood forecasting concerned with streamflow estimates at the watershed outlet. This is because physical watersheds are themselves functional integrators of the sub-watershed hydrologic processes, channeling the precipitation distributed over the basin through a relatively small watershed outlet. However, the *effective* (watershed scale, lumped, and time-invariant) model representation of the watershed structure and parameters (soil and vegetation properties) makes them difficult, if not impossible, to specify by means of direct observations conducted in the field. This is because many of these properties are not easily observable and measurable, and/or because the available measurement technologies are incapable of providing appropriate measurements at the correct (watershed aggregate) scale.

The general problem of model identification [Gupta and Sorooshian, 1985b] therefore involves the selection of appropriate *structures* for the various model components and specification of values for the *parameters* of those model components, such that the resulting model provides a sufficiently accurate (depending on the application) simulation/prediction of watershed response. Historically, due primarily to computational limitations, it has been common to choose a particular model structure (such as the SAC-SMA) based on the recommendations of other people, or one's own experience (e.g., James and Burges, [1982], offer guidance on hydrologic model selection). The specific identification problem, therefore, has been the simpler issue of selecting values for the model parameters, given the fixed model structure [Gupta and Sorooshian, 1985b].

1.3. The Process of Parameter Estimation

Unfortunately, the problem of parameter estimation has not proved to be simple. Many (if not all) of the parameters are effective quantities that cannot, in practice, be measured in the field, and must therefore be estimated by indirect means. The typical way to achieve this is to try and adjust the parameter values by various means (described below) so that the input-output (precipitation-runoff) behavior of the model approximates, *as closely and consistently as possible*, the observed response of the watershed over some historical period of time for which precipitation, streamflow, and other relevant measured data are available (see Figure 2). The process by which parameters are estimated in this way has come to be called model calibration. A model calibrated by such means can be used for the simulation or prediction of events outside the historical record used for model calibration, if it can be reasonably assumed that the physical characteristics of the watershed and the hydrologic/climatic conditions remain similar.

The manual process of model calibration typically proceeds via three steps [Boyle et al., 2000]:

Step 1. The hydrologist examines the data that are available about the watershed characteristics and develops crude a priori estimates (guesses) of the range of likely values for each of the parameters. Boyle et al. [2000] refer to this as Level Zero estimates. For example, the NWS may look at the range of parameter values from a number of similar watersheds in the same region to develop Level Zero estimates. This process involves little or no use of historical precipitation-runoff data. In the absence of any other sources of information, the Level Zero range of parameter estimates can be defined conservatively, based on the maximum plausible ranges for the parameters based on physical reasoning.

Step 2. More refined (Level One) ranges for some of the parameter estimates are computed by identifying and analyzing the characteristics of specific segments of the observed streamflow hydrograph that are thought to be predominantly controlled by a specific parameter (or sub-set of

parameters) in isolation from the effects of other parameters. For example, the estimate of the baseflow rate parameter (that represents the average rate at which the groundwater drains into the river) can be refined by analyzing the mean slopes of the extended recession segments of the hydrograph. The multi-dimensional region of the parameter space bounded by the Level One upper and lower ranges for the parameters is called the feasible parameter space.

Step 3. The model is used to simulate the input-output response of the watershed using a carefully chosen representative period of historical data (called calibration data) and one (or more) representative parameter sets selected from within the feasible parameter space that was estimated via steps one and two. The simulated and observed output responses (streamflow hydrographs) are then compared (as described below) and an incremental, trial-and-error process of parameter adjustments is attempted (within the feasible parameter space) to get the simulated response to approach more closely the observed watershed response. This step has, in practice, proven to be quite difficult to carry out in a reliable and consistent manner, because:

a) there are typically a large number of parameters that can be adjusted (the SAC-SMA has 15),

b) these parameters usually have either similar or compensating (interacting) effects on different portions of the modeled hydrograph,

c) there is no uniquely unambiguous way of evaluating the *closeness* of the simulated and observed streamflow hydrograph time series, and

d) the input data, model conceptualization, and output data are all to some extent imperfect (contain errors or uncertainties).

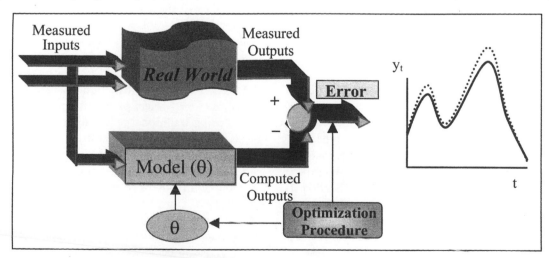

Figure 2. Strategy used for model calibration.

The resulting model, with this further refined region of parameter values, which we call Level Two estimates, is then typically evaluated for consistency (and possible violations of modeling assumptions) by testing it over one or more independent (i.e., not used for calibration) periods of historical data to establish some degree of confidence that the calibrated model will continue to provide consistent and reasonable simulations when used in practice. This final step is often called model verification or validation, although we prefer the more accurate term *model evaluation*. For more details of the subtleties of model verification/evaluation, please refer to the work of authors such as Klemes [1986a; 1986b] and Yapo et al. [1996], among others.

1.4. Level Two Parameter Estimation

Let us state quite explicitly that we do not consider model identification to be a process of "fitting the model to the data". The process of model identification consists of a series of steps in which the initial (large) uncertainty in the model structures and associated parameter estimates is progressively reduced in such a way that the model is constrained to be structurally and functionally (behaviorally) consistent with the available qualitative (descriptive) and quantitative (numerical) information about the watershed. Any selected model will be, at best, a structural and functional approximation to the true (unknown) watershed structure and function. Hence, it will be impossible to reduce the model uncertainty to zero, even if the input and output data were to be perfectly observed. The best we can achieve is some minimal representative set of models (structures & parameter values) that closely and consistently approximates, in an (unavoidably) uncertain way, the observed behavior of the watershed.

The methods for model identification described below will refer primarily to estimation of the model parameters under the assumption that a specific model structural form (set of equations) has been selected. However, with enough computational power, these methods are also in principle generally applicable to the more general problem of simultaneous structure and parameter estimation [see Boyle, 2001; Boyle et al., this volume].

In general, two general approaches to Level Two parameter estimation have evolved since the early 1970s. One, the Manual-Expert approach, relies on the judgement and expertise of a trained and experienced hydrologist (see Section III of this volume). The other, the Automatic approach, employs the power, ability to follow systematic programmed rules, and speed (machine intelligence or computer power) of a digital computer. In either of these two approaches, the process of obtaining Level Two parameter estimates requires:

a) an estimate of the approximate parameter region to be searched (feasible parameter space),

b) a strategy for evaluating the "closeness" between the modeled and historically observed watershed responses, and

c) a strategy (procedure) for making parameter adjustments (within the search region) that bring the simulated responses *closer* (as defined by (b)) to the observed responses.

In both approaches, the Level Zero and Level One procedures can be used to estimate the feasible parameter space. In Manual-Expert calibration the closeness of the simulated and observed hydrographs (i.e. model performance) is evaluated primarily by subjective visual inspection (usually supported by a number of statistical "goodness-of-fit" measures) and the parameter adjustments are based on expert guesses. In contrast, most strategies for Automatic calibration have (till recently) used a single explicitly defined measure of closeness (most commonly an aggregate goodness-of fit statistic such as the mean squared error over the simulation period) and the parameter adjustments are made by an optimization algorithm. Traditional automatic calibration strategies, have therefore, strongly borrowed from the techniques of non-linear regression.

1.5. The Turing Test of Model Calibration

Table 1 compares the advantages and disadvantages of the Manual-Expert and Automatic approaches. For lumped watershed models having 15 or fewer parameters, a carefully executed Manual-Expert approach can give superior results, but at the expense of considerable time and energy. In contrast, the traditional Automatic approach is both fast and relatively simple to apply, but somehow lacks some of the sophistication of the Manual-Expert approach and tends to give relatively "poor" parameter estimates and less "consistent" model performance. We have observed, over the past two decades, that NWS hydrologists responsible for making critical flood forecast decisions are unwilling, for a variety of reasons, to rely on models calibrated using traditional automatic calibration procedures. It has become apparent, therefore, that improvements to the traditional automatic calibration procedures are necessary, and that the test of these improved procedures should be their ability to pass inspection by a team of expert hydrologists. Our goal, therefore, is to develop calibration procedures that result in parameter estimates and model performance that are essentially indistinguishable from (comparable to or better than) those obtained by a highly trained expert. We think of this as the equivalent "Turing

Table 1. Comparative features of Manual-Expert and Automatic calibration.

Manual Calibration	Automatic Calibration
User Knowledge and Expertise	Speed and power of computer
Subjective (realistic)	Objective (statistics)
Complicated and highly labor intensive	Computer Intensive
Time Consuming	Time Saving
Excellent Results	Results may not be acceptable

Test" of model calibration. (The test was originally posed by Turing [1950], in the context of machine intelligence. He classified an artificial system as "intelligent", if its response to questions cannot be distinguished from that of an intelligent human being.). The Turing test analogy was first proposed to us in 2000 at the Federal Interagency Hydrologic Modeling Conference in Tucson, Arizona, by a hydrologist whose name we, unfortunately, cannot remember.

2. ADVANCES IN AUTOMATIC MODEL IDENTIFICATION METHODS

2.1. The Need for Improved Automatic Methods

Although the Manual-Expert approach to watershed model calibration can give very good results when performed by an experienced hydrologist with considerable calibration skill and knowledge about the watershed, there is an urgent need to develop fast and reliable computer-based methods. In particular, the NWS has over 4000 forecast points in the U.S.A. for which the SAC-SMA model (and its future versions) must be calibrated within the next few years [Ingram, 1996]. This number severely taxes the capabilities of the existing limited number of NWS hydrologists and forecasters trained in calibration skills. Further, it may take several hundred hours or more of training time to bring a novice model calibrator up to a reasonable level of skill [Mike Smith, NWS Office of Hydrology, Personal Communication, 1999]. The magnitude of the problem is growing rapidly with the expanding number of forecast points. Another, and perhaps more important reason, is that the availability of spatially distributed information (NEXRAD radar-based precipitation data) is now encouraging the use of semi-lumped and distributed watershed model representations, having much larger numbers of parameters than can be practically handled using the Manual-Expert approach.

2.2. Historical Background

Research conducted during the past two and a half decades has revealed that the traditional Automatic calibration methods suffer from a number of serious conceptual and practical weaknesses. An overarching problem is that the approach is based on classical non-linear regression theory, which operates under the central assumption that *the available model structure is true*, and therefore seeks to identify a unique "optimal" (unbiased, minimum variance) set of parameter estimates. In practice, it has proved difficult to identify, with confidence, unique parameter estimates that optimize any of the wide variety of objective functions that have been tested by numerous researchers [Gupta et al., 1998]. Until the early 1990s the available optimization procedures could not even be relied upon to find the actual global optimum of an objective function. Any parameter estimate obtained in this way was found to be very sensitive to the choice of the objective function and the data set used for calibration [Gupta et al., 1998; Sorooshian et al., 1993].

One early response to these problems was to seek a rigorous statistical footing for the parameter estimation problem. Sorooshian and Dracup [1980] pointed out that the output measurement data (streamflows) have measurement errors that can be considered to be temporally auto-correlated and heteroscedastic (non-constant variance) and demonstrated that the use of objective functions derived using Maximum Likelihood theory reduces the sensitivity of the estimated parameters to such errors. The Heteroscedastic Maximum Likelihood Estimator they proposed (HMLE, see Appendix), based on the form of the rating curve commonly used in the US for deriving streamflow volumes from depth measurements, directly countered the previously held wisdom that objective functions should provide greater weight to peak flow measurements. In parallel work, Kuczera [1988] (see also Kavetski et al., this volume) posed the identification problem in the context of Bayesian statistical theory with similar results, thereby also demonstrating the value of accounting for the measurement error properties of the data, while showing how statistical confidence bounds for the parameter estimates could be estimated. Based on their work, it has since become common to apply a power transformation [Box and Cox, 1964]. We use the following version of the transformation:

$$Q = [(q+1)^\lambda - 1] / \lambda \tag{1}$$

with values of $\lambda \sim 0.3$-0.5 applied to the observed and simulated streamflows q, thereby helping to stabilize the error variance, resulting in more stable and consistent parameter estimates.

A second response was to study how the structural parameterization of the model might contribute to difficulties in parameter identification. Gupta and Sorooshian, [1983] showed that the representation of percolation in the SAC-SMA model could lead to severe interaction among the model parameters, contributing significantly to an ill-posed identification problem. They showed that the problem can be partially alleviated, but not entirely resolved, by a judicious reparameterization of the model. The results serve to highlight the necessary care that must be applied during model design.

A third early response was to explore the role played by the data selected for model calibration. Sorooshian et al. [1983] and Gupta and Sorooshian [1985b] showed that the type and quality of data is more critical than the amount of data used for model calibration. They pointed out that it is more important that the calibration data contain a wide variety of hydrologic behaviors from dry to wet conditions, than that the data focus on historical flooding periods (see also Crawford and Linsley, [1966]). For example, theoretical analysis was used to show that it is the number of times the capacity of a model tank component exceeds and drops below its critical threshold value that controls the identifiability of that component parameter, not how long the tank remains in overflow mode [Gupta and Sorooshian, 1985a]. Numerous students in Hydrologic Modeling classes at the University of Arizona have since empirically verified this fact, which was not previously obvious. Gupta and Sorooshian [1985b] showed theoretically and empirically that the marginal benefit of additional measurement data having similar information context diminishes as the reciprocal of the square root of the length of the data set, suggesting that in the absence of new information content, no more than three years of daily data should be necessary for model calibration. Yapo et al. [1996] however, conducted a more comprehensive empirical study for the SAC-SMA model using several different objective functions and 40 years of data for the Leaf River, Mississippi, concluding that approximately eight years of daily calibration data are necessary to ensure minimal sensitivity to the period of record used in model identification. This number is consistent with the longstanding claim by NWS hydrologists that approximately 11 years of data should be used to calibrate the SAC-SMA.

The fourth area that was extensively investigated was the choice of method for exploration of the objective function response surface in search of the "optimal" parameter values. It was well known in the Systems Theory (ST) and Operations Research (OR) literature that efficient optimization could be carried out using gradient-based optimization methods such as the Gauss-Newton family of algorithms. However, because it was perceived that the threshold structures common to watershed models made the derivation of derivatives difficult, early attempts at Automatic Calibration used direct-local-search methods such as the Pattern Search method of Hooke and Jeeves [1961], the Rotating Directions method of Rosenbrock [1960], and the Downhill Simplex method of Nelder and Mead [1965]. Johnston and Pilgrim [1976] published a seminal paper showing that automatic search of the objective function (using such methods) gave widely differing "optimal" parameter estimates, when initiated at different initial guesses. In more than two years of extensive investigation, they were unable to confidently claim that they had discovered the optimum to their watershed model calibration problem. Ibbitt [1970] tested eight different optimization strategies on the Stanford Watershed Model IV [Crawford and Linsley, 1966] and was unable to find a reliable method for finding the global solution. Various researchers investigated this problem on different models with similar findings. To make possible the use of more powerful optimization strategies, Gupta and Sorooshian [1985a] investigated the question of derivative computations and showed that watershed model derivatives can indeed be easily derived even for threshold structures (for any watershed model). However, tests by Gupta and Sorooshian [1985a] and Hendrickson et al. [1988] achieved no benefit by the application of Gauss-Newton methods to the watershed calibration problem.

The attention then began to shift to the use of global search methods, which were still in the infancy of their development in the fields of ST and OR. Brazil and Krajewski [1987] tested the use of the Adaptive Random Search (ARS) strategy for finding good initial guesses for the calibration of the SAC-SMA model, followed by application of the Pattern-Search direct-local-search method, with encouraging results. However, Armour [1990] and Weinig [1991] conducted an exhaustive investigation of the ARS method and found it to be both inefficient and incapable of identifying the region of the known global optimum of a watershed model problem with a reasonable degree of confidence.

It was not until the arrival of adequate computational resources in the early 1990's that a comprehensive diagnosis of the true nature and difficulty of the watershed model optimization problem could be achieved. Duan et al. [1992] conducted an exhaustive computer based evaluation of the structure of the objective function response surface for a typical watershed model, and reported the existence of numerous small "local optima" nested within the several larger "regions of attraction" (Figure 3). Their research finally explained the reasons for convergence problems reported by previous studies (Table 2). It also made clear that for any optimization strategy to be suitable for calibration of watershed models, it must have the ability to avoid being trapped by unpredictable numbers of minor optima en route to the global solution, while being insensitive to the initial guess. These insights led to the development of the Shuffled Complex Evolution (SCE-UA) optimization algorithm, with global convergence properties [Duan et al., 1992; Duan et al., 1993; Sorooshian et al., 1993].

The strength and reliability (efficiency and effectiveness) of the SCE-UA algorithm have since been independently tested and proven by numerous researchers and the algorithm is now extensively used world-wide [e.g. Sorooshain et al., 1993; Gan and Biftu 1996; Luce and Cundy, 1994; Tanakamaru, 1995; Kuczera, 1997; Franchini et al., 1998, Hogue et al., 2000]. The SCE-UA has also been used in related areas such as subsurface hydrology, soil erosion, remote sensing and land surface modeling [Mahani et al., 2000; Contractor and Jenson, 2000, Scott et al., 2000; Nijssen et al., 2001; Walker et al., 2001].

The SCE-UA has been generally found to be robust, effective and efficient. A number of researchers have explored possible modifications to the algorithm [e.g., Wang et al., 2001; Santos et al., 1999]. Yapo et al. [1997] extended the SCE-UA to a multi-objective framework [see Gupta et al., this volume "Multiple ..."]. For further discussion of the SCE-UA and other effective global search algorithms, see Duan [this volume].

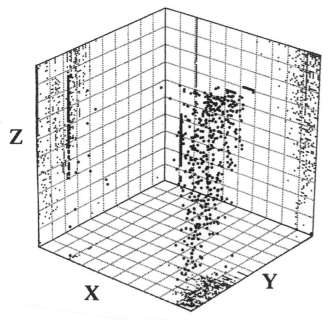

Figure 3. Three-parameter subspace of a simple conceptual watershed model (SIXPAR), showing locations of multiple local optima (dark dots).

2.3. Current Perspectives

The availability of the SCE-UA algorithm helped to reduce the confusion regarding causes of poor calibration performance. One can now have confidence that the global optima of an objective function are found. The analysis that led to the development of the SCE-UA algorithm also revealed that the optimization problem itself is poorly posed. Regardless of the objective function, the response surface contains numerous very similar solutions (in terms of objective function value) at widely differing locations in the parameter space. Therefore, in spite of our confidence in the optimization tools now at our disposal, uncertainty in the calibrated parameter estimates still remains very large. Further, the actual locations of these globally optimal solutions are highly sensitive to the choice of objective function (i.e., to the hypothesized structure of the noise in the input-output data).

There have emerged at least three responses to this situation. One response has been to argue that the phenomenon is evidence of a condition named equifinality [Beven and Binley, 1992] in which the available information is insufficient to distinguish between a number of alternative models (including different parameter sets for a given model structure), and therefore one should retain all such models as being plausible until evidence to the contrary become available. This line of reasoning has been further developed to argue that the concept of calibration is itself suspect and that we should instead focus on strategies that translate the uncertainty in our ability to select a specific model into uncertainty bounds on the model predictions. Beven and Binley [1992] [see also Freer et al., 1996; Franks and Beven, 1997] promote a strategy for this called the Generalized Likelihood Uncertainty Estimation (GLUE) method (see Freer et al., this volume).

A second response has been to suggest that the "equifinality" phenomenon is evidence of models that are too complex in relation to the information content of the data available for model development and calibration. This line of reasoning has encouraged the exploration of various data-based-mechanistic modeling approaches which seek to provide strategies for developing hybrid metric-conceptual watershed models with only as much complexity as can be supported by the available data (using rigorous statistical testing). Examples of these approaches are proposed by Jakeman and Hornberger [1993], Wheater et al. [1993], and Young et al. [1996]. Models developed in this way typically consist of a non-linear component (that partitions the precipitation into precipitation excess, additions to soil moisture, and evapotranspiration), followed by a linear routing component (that allows for both fast and slow rates of drainage from the watershed). Proponents of these

Table 2. Response surface characteristics that complicate the optimization problem in watershed models.

Characteristic	Reason for Complication
Regions of attraction	More than one main convergence region
Minor local optima	Many small "pits" in each region
Roughness	Rough response surface with discontinuous derivatives
Sensitivity	Poor and varying sensitivity of optimum, and nonlinear parameter interaction
Shape	Nonconvex response surface with long curved ridges

approaches have argued that the evidence suggests that lumped watershed input-output data can only support models having approximately five parameters [Jakeman and Hornberger, 1993].

Our own response (at the University of Arizona) has been to suggest that the so-called problems of equifinality and extreme parameter interdependence (leading to claims of model over-complexity) are largely *consequences of a weakness in the design of the model identification problem itself.* In support of this assertion, we note that the traditional automatic calibration strategy relies on the use of a single aggregate measure (such as the RMSE or HMLE objective functions) to evaluate model performance (goodness), and that there can be several *quite different* model simulated hydrographs (associated with different model structures or sets of parameter values) that give essentially equivalent values for the objective function (Figure 4). Therefore similar values for an aggregate objective function are *not* necessarily evidence of similar model behavior (i.e. equifinality). In fact the lack of ability of a single objective function to distinguish between different model behaviors is clear evidence that the traditional calibration strategy is unable to extract all the information available in the data.

In support of our assertion, we note that the hydrologists at the NWS consider a major strength of the Manual-Expert

calibration approach to lie in its use of *a wide variety of subjective ways of evaluating model performance.* In particular, they pay careful attention to a number of specific *visual* (local) characteristics of the hydrograph during storm periods (e.g., the slope of the rising limb, volume of runoff, and magnitude and timing of the peak flow), and during inter-storm periods (e.g., the rates at which the hydrograph recedes during the early quick recession, and the later slow recession). To supplement the visual analysis of local hydrograph characteristics, they also examine a number of overall (global) hydrograph behaviors, summarized for the entire calibration period using a variety of statistical measures (e.g., the total error variance and bias, monthly bias, and flow biases in various flow regimes). During Manual-Expert calibration, the hydrologist tries to get the model to obtain a suitable compromise in matching *all* of the visual hydrograph characteristics as closely as possible, while achieving acceptable values for the summary statistics.

We contend, therefore, that the weaknesses in the model calibration/evaluation procedures at our disposal must be resolved before confident conclusions can be drawn regarding the overly-complex nature of watershed models or the inability to discriminate between alternative model hypotheses. One way to do that is to adopt a multiple-criteria perspective.

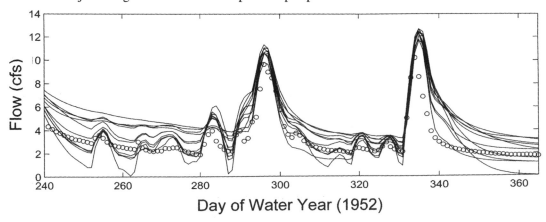

Figure 4. Multiple hydrograph simulations generated using the SAC-SMA model, having similar RMSE values (observed values indicated by circles, and simulated values indicated by solid lines).

3. THE MULTIPLE CRITERIA APPROACH

3.1. The Problem of Model Structural Errors

We mentioned above that during Manual-Expert calibration, the hydrologist tries to adjust the parameters of the model so as to obtain a suitable *compromise* in matching (as closely as possible) several different *local* aspects of visual fit between the simulated and observed hydrographs, while achieving acceptable values for the *global* summary statistics. The reason the hydrologist must seek such a compromise is, that the task of simultaneously reproducing all of these local and global behaviors of the observed watershed response is significantly complicated by inadequacies in the model formulation (model errors) and the errors in the input-output data. However, even if data errors could be ignored (or treated using statistical techniques such as Maximum Likelihood), the model structural errors make it impossible to match the observed hydrograph without having to trade-off the ability to match one or more hydrograph behaviors against the others.

The strength of the Manual-Expert calibration approach lies, therefore, in the ability of the hydrologist to make judicious decisions regarding the relative importance of different kinds of model (and data) errors and to select one or more parameter sets that provide some appropriate compromise among the aforementioned different measures of model performance. On the face of it, this process of balancing objectives might seem to pose a serious difficulty to the hydrologist. In fact, however, each of the competing targets has the effect of constraining the parameter space in different ways so that, although the space of suitable parameter solutions for *each* objective might be large, the "acceptable region" consisting of the intersection of solution spaces is much more tightly constrained (Figure 5). An important consequence is that the "set of acceptable parameters" is less likely to contain solutions that result in unbalanced model performance, so that forecast performance tends to be more reliable. Further, the calibration results are less likely to be overly sensitive to the choice of any individual measure of model performance.

3.2. The Multiple-Criteria Optimization Approach

Referring back to the "Turing Test" of model calibration, it becomes apparent that if an automatic calibration procedure is to pass inspection by a team of Experts, it must be capable of somehow emulating or improving upon the Manual-Expert procedures mentioned above. The Expert is, of course, attempting to optimize a number of subjective and objective "measures" of model performance. Therefore, in a series of papers, Gupta et al. [1998], Yapo et al. [1998] and Boyle et al.

[2000] have proposed that this process be formalized by replacing the subjective visual evaluation of local hydrograph behaviors by objective criteria that measure the goodness-of-fit for each of those behaviors.

Drawing upon multiple-criteria methods from the field of economic analysis, Gupta et al. [1998] proposed that the watershed model parameter estimation problem be reformulated as a multiple-criteria optimization problem which seeks, instead of a single unique solution, a Pareto-set of trade-off solutions (see Gupta et al., this volume, "Multiple ...")

Consider a model having the p-dimensional parameter vector $\theta = \{\theta_1, \ldots, \theta_p\}$ which is to be calibrated using time series observations $(O_j(t_j), t_j = ta_j, \ldots, tb_j, j=1, \ldots k)$ collected from times ta_j through tb_j on k different response variables. The different responses represent the different model outputs, e.g., sensible heat flux, latent heat flux, ground heat flux, runoff, etc. To measure the distance between the model-simulated responses Z_j and the observations O_j, separate criteria $f_j(\theta)$ for each model response are defined. The criteria and their mathematical form depend on the goals of the users. It is common practice to use a measure of residual variance such as the root mean square error. For a discussion of this, see *Gupta et al. [1998]*. The multi-criteria model calibration problem can then be formally stated as the optimization problem:

$$\text{Minimize } F(\theta) = \{f_1(\theta), \ldots f_k(\theta)\} \text{ subject to } \theta \subset \Theta$$

where the goal is to find the values for θ within the feasible set Θ that simultaneously minimize all of the k criteria.

The multi-objective minimization problem does not, in general, have a unique solution. Due to errors in the model structure (and other possible sources), it is not usually possible to find a single point θ at which all the criteria have their minima. It is common to have a set of solutions with the property that moving from one solution to another results in the improvement of one criterion while causing deterioration in another. A case with two parameters (θ_1, θ_2) and two-criteria response functions $\{f_1, f_2\}$ is illustrated in Figure 6. In Figure

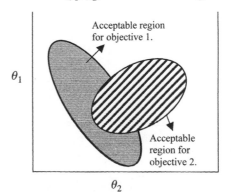

Figure 5. Intersection of solution spaces.

6a the feasible parameter space Θ is shown, and the corresponding projection of the parameter space into the function space (shaded area) is shown in Figure 6b. Criterion f_1 is minimized at point α, and criterion f_2 is minimized at point β. The thick line indicates the set P of multi-criteria minimizing points to the function $\{f_1, f_2\}$. If $\gamma \in P$ and $\delta \notin P$ are points selected arbitrarily, then every point γ is superior to every point δ in a multi-criteria sense because each point has the property that $f_j(\gamma) < f_j(\delta)$, for $j = 1,2$. However, it is not possible to find another point $\gamma^* \in P$ such that γ^* is superior to γ; instead γ^* will be superior to γ for one criterion but inferior for at least one other criterion. The set P of solutions is variously called the *trade-off* set, *non-inferior* set, *non-dominated* set, *efficient* set or *Pareto* set. Here, we call it the *Pareto* set.

Because the solution of the multiple-criteria optimization problem is not unique, but consists of a *Pareto region* of the parameter space, classical optimization algorithms (including the SCE-UA) that seek a single point in the parameter space are not well suited. Yapo et al. [1998] therefore adapted the population-based global search strategies of the SCE-UA to handle multiple-criteria and developed the Multiple-Objective Complex Optimization Method (MOCOM-UA) that converges to an approximation of the Pareto solution set.

To implement the automatic multiple-criteria approach so that it emulates the Manual-Expert calibration of watershed models, one must select a set of objective (mathematical criteria) that formally represent the subjective visual methods

by which various local hydrograph behaviors are evaluated. Boyle et al. [2000; 2001] therefore proposed the use of three "local" criteria (see also Boyle [2001], and Boyle et al., [this volume]), one each to represent the errors in matching of the rising limb of the hydrograph (precipitation driven response), the early recession (quick, non-driven relaxation response), and the late recession (slow, non-driven relaxation response) (Note the correspondence of this proposal and the three model components identified by the data-based-mechanistic approach mentioned earlier — nonlinear precipitation partitioning, quick recession and slow recession). They further proposed the use of two additional global criteria, the overall error variance and overall error bias, to further constrain the solution set.

The result of a watershed model calibration using multiple-criteria and the MOCOM-UA optimization algorithm is a discrete set of possible parameter sets that represent trade-offs between different *optimal* ways of constraining the model to be consistent with the observed data. This computerized approach is (in principle) superior to the Manual-Expert approach, because it quickly searches and rejects "bad regions" of the entire parameter space (as defined by the criteria) and identifies the limited "good regions" of the parameter space for which the model is consistent with the data [Figure 7a]. The hydrologist is then left with the relatively minor task of selecting a final "solution" from the sample of Pareto-optimal solutions. This last step allows for

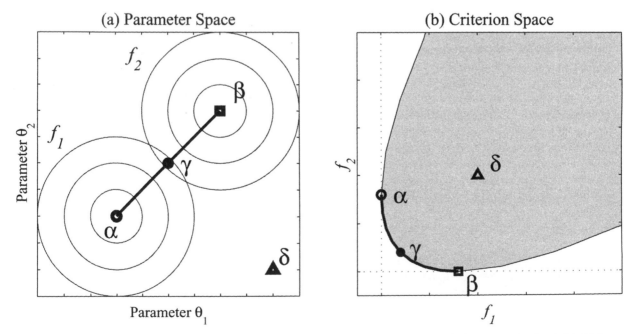

(a) Parameter Space

(b) Criterion Space

Figure 6. Example showing the Pareto Solution set for a two-parameter problem (θ_1, θ_2) and two criteria (f_1, f_2): (a) feasible parameter space and (b) criterion space. Point α minimizes f_1, and point β minimizes f_2. The thick line indicates the Pareto set P of multi-criteria minimizing points to the function $\{f_1, f_2\}$; $\gamma \in P$ is superior to any $\delta \notin P$.

the application of additional criteria that were not included in the automated approach, or simply for the selection of a solution that satisfies one's own personal preference. However, even if a single solution is to be selected, the Pareto-solution set can be used to generate a Pareto-ensemble of simulated hydrographs, displayed as a Trade-off-uncertainty region on the hydrograph plots (Figure 7b), which shows the uncertainty in the model simulations due to different possible ways of trading-off the model (and data) errors.

The use of Pareto parameter sets to represent model structural uncertainty and Pareto-ensemble simulations to represent model prediction uncertainty provides useful new ways for evaluating models and their performance. For example, Figure 7b shows that by allowing trade-offs between different aspects of model performance, the calibrated watershed model tends to provide biased under-estimates of the hydrograph recession suggesting structural problems with the model. Further, because better (more accurate) models would have smaller model errors, they would be expected to result in both smaller Pareto solution sets and smaller values for the objective functions. To illustrate this, Figure 8 (a & b) shows that the more complex SAC-SMA model provides a better (more accurate and less uncertain) representation of the Leaf River hydrograph when compared to the simpler HyMod model. Figure 9 illustrates how the multiple-criteria analysis also has the potential to reveal varied strengths among mod-

els. Models A and B are Pareto-equivalent in the sense that each is better (worse) than the other with regard to one of the criteria, while both are Pareto-superior to model C.

Boyle et al [2000, 2001] have tested their automatic multiple-criteria (AMC) approach on several watershed data sets and compared the results with traditional single-criteria calibration using SCE-UA and with Manual-Expert calibrations conducted by NWS hydrologists. The AMC approach was found to provide better solutions than the traditional automatic approach when applied to a lumped calibration of the Leaf River watershed [Boyle et al., 2000]. In a comparison with NWS Manual-Expert calibrations, the AMC provided superior solutions when applied to a more complex semi-lumped (multiple sub-watershed) calibration for the Blue River watershed [Boyle et al., 2001]. They have also demonstrated that AMC provides a quick and powerful tool for evaluating and comparing alternative model structures and components [Boyle et al., 2001, Boyle et al., this volume].

In the past few years, several other researchers have also developed and tested various formulations of the multiple-criteria approach for watershed calibration and have reported good results [see Beldring, 2002; Franks et al., 1999; Hogue et al., 2000; Madsen, 2000; Madsen et al., 2002; Seibert, 2000; Sen et al., 2001; Wagener et al., 2001c, among others]. Also, a robust multiple-criteria sensitivity analysis procedure has been developed by Bastidas et al.

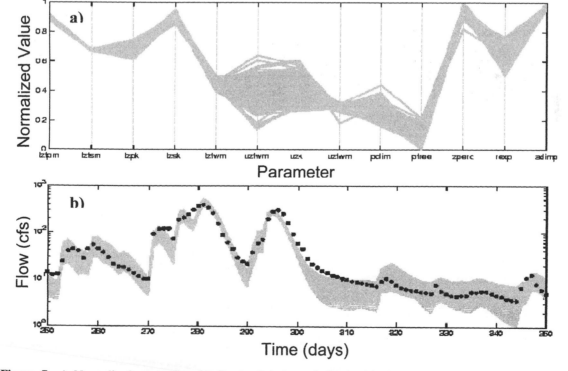

Figure 7. a) Normalized parameters for Pareto Solutions obtained with the automatic multi-criteria approach. b) Hydrograph range associated with the Pareto solution set in logarithmic scale. Observed values indicated by solid dots.

[1999] based on extensions of the Regional Sensitivity Analysis (RSA) procedure of Spear and Hornberger [1980]. Finally, it should be mentioned that applications of the multiple-criteria approach to hydro-chemistry models [Meixner et al., 1999; 2000] and to water-energy-carbon balance land-surface models [Bastidas et al., 1999; Gupta et al., 1999; Leplastrier, 2001; Pitman et al., 2001, Zeng et al., 2001] are also being developed and explored. For further discussion and applications of the multiple-criteria philosophy see Wagener et al. [this volume], Boyle et al. [this volume], Meixner et al. [this volume], Bastidas et al., [this volume], and Hogue et al., [this volume].

4. LOOKING TOWARDS THE FUTURE

4.1. Emerging Model Structures

Two technological developments are now exerting a major influence on the development of watershed modeling and calibration procedures. The first is, of course, the ever-increasing pace of desktop computational power. The second is the availability of distributed data, driven primarily by the boom in radar and satellite based remote sensing. In particular, watershed modeling in the USA is now being strongly influenced by the interest in using NEXRAD Stage IV radar precipitation data that are available at 4x4 km resolution every 60 minutes for much of the country. This has promoted greater interest in the development of semi-lumped- and distributed-parameter watershed models. For example, the NWS is sponsoring the Distributed Model Intercomparison Project (DMIP) to encourage a community wide dialog on this topic, with a view to influencing the next generation of flood forecast models for the USA [Smith et al., 1999]. A related project, the Model Parameter Estimation eXperiment (MOPEX) is seeking to encourage community collaboration on the issue of how to parameterize distributed models, par-

ticularly using distributed soils and vegetation data, for regions where precipitation-runoff data may not be readily available for calibration [Schaake et al., 1998]. In this regard, Koren et al. [2000] (see also Koren et al., this volume) have proposed a procedure, based on the use of soil data, for computing approximate sub-watershed scale estimates of the parameters of the SAC-SMA (thereby allowing the SAC-SMA to be applied in semi-lumped mode).

A number of modeling toolboxes are beginning to become available, with a view to facilitating the use of appropriate (different) model components based on the unique hydrology of a place. For example, the Modular Modeling System [Leavesley et al., 1996; Leavesley et al., this volume] developed by the USGS provides a sophisticated drag-and-drop environment to make model development relatively simple, along with integrated tools for parameter estimation via both a priori methods (using distributed soils and vegetation data sets) and automatic calibration (the most recent version under testing now includes both the SCE-UA and MOCOM-UA algorithms).

A simultaneous counter-move is also underway, in which some researchers (influenced by the databased-mechanistic movement and philosophy of equifinality) are investigating methods for developing simpler watershed models, which are only as functionally complex as can be supported by the hydrologic data available. For example, Young et al. [1997] and Young and Beven [1994], show that accurate, hydrologically consistent, but simply structured conceptual watershed models suitable for streamflow forecasting can be developed using precipitation and runoff data. Wheater et al. [1993] and Wagener et al. [2001a] are investigating the simplest possible conceptual model structures that are able to provide accurate streamflow predictions for UK watersheds, while having well determined (low uncertainty) parameter estimates, so that rules for regionalization (i.e., extrapolating model structures and parameter values to

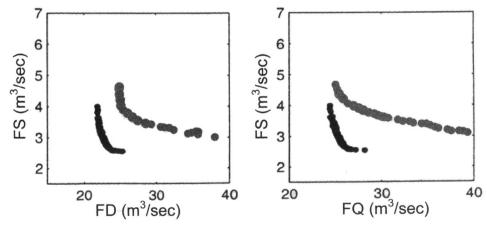

Figure 8. Comparison of SAC-SMA (black points) and simpler HyMOD (gray points) models in objective function space.

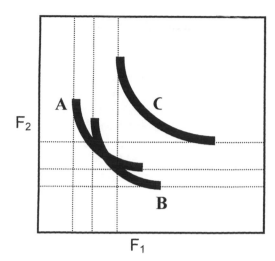

Figure 9. Multi-criteria objective space for three pareto sets (A, B, and C) in two dimensions (F_1 and F_2).

ungaged watersheds) might be developed. The Rainfall Runoff Modeling Toolbox (RRMT) and the Monte Carlo Analysis Toolbox (MCAT) being developed at Imperial College, London, are designed to facilitate *parsimonious* lumped watershed models with a high level of parameter identifiability [Wagener et al. 2001a,b; Wagener et al., this volume].

Finally, there is an increasing interest in the development of models that compute multiple types of interacting fluxes. These include hydro-chemical models such as the Alpine Hydrochemical Model [Bales et al., 1993; Wolford and Bales, 1996; Wolford et al., 1996; Meixner, 1999; Meixner et al., 2000] and the Birkenes Model [De Grosbois et al., 1988; Hooper et al., 1988; etc] and hydro-meteorology models which describe the interdependence of water, energy, and various environmental chemicals (such as carbon) [Dickinson et al., 1998; Liang et al., 1994; Mitchell et al., 1999; Sellers et al., 1996]. These ever more complex descriptions of the hydrologic environment will continue to demand evolutionary developments in model identification technology.

4.2. Methods for Model Identification

On the issue of parameter estimation strategies, there is currently a tension between the school that suggests that if a watershed model is constructed to be structurally consistent with the kinds of (distributed) data now available, then little or no calibration (i.e. parameter adjustments using input-output data) should be necessary [e.g., Leavesley et al., 1996], and the view (which we ourselves subscribe to) that useful models are conceptual simplifications of reality and therefore the need for some degree of calibration will likely still remain for the foreseeable future. Also, the concept of "calibration" has been questioned by proponents of the philosophy of equifinality (in particular the GLUE strategy). They suggest that

we should put the notion of calibration aside, and instead focus on the problem of identifying all models (structures and parameter sets) that cannot be rejected by the input-output and other watershed data at our disposal, and on translating this uncertainty into probabilistic streamflow forecasts.

There is no fundamental inconsistency between the equifinality view and our own notion of model calibration as discussed here. If calibration is viewed as a progressive process of reducing the uncertainty regarding which model structures and parameter sets are consistent with the available data, then the end result is a family of plausible models (not a single one), which can (and should) be used to generate uncertain streamflow forecasts. In support of this, Vrugt et al. [this volume] have recently developed the Shuffled Complex Evolution Metropolis optimization algorithm (SCEM-UA; a modified version of the SCE-UA method) that converges to a stationary approximation of the posterior distribution of the parameter values. The SCEM-UA therefore provides both an estimate of the mode of the posterior parameter distribution (the traditional "best" parameter set) *and* a sample set of parameter values describing the probabilistic representation of remaining parameter uncertainty. This posterior description of parameter (model) uncertainty is used to produce probabilistic model forecasts (most likely forecast and 95% confidence intervals at each time step). The method does not exclude the use of multiple model structures, although this has not yet been explicitly demonstrated. Our view on calibration is also reflected in the BaRE algorithm for Bayesian Recursive Estimation of watershed model parameters which begins with a family of plausible models (in principle various structures and their feasible parameter spaces can be included) and seeks to reduce progressively the model *and* hence the forecast uncertainty through stepwise recursive (in time) processing of the input-output data as they become available [Theimann et al., 2001]. For further discussion of the BaRE approach, see Misirli et al., [this volume].

4.3. Two Important Issues That Need Addressing

While both the GLUE and the BaRE strategies are (in differing ways) rooted in Bayesian mathematics, neither has succeeded in providing a complete description of the model identification problem. To be comprehensive, future strategies for model identification should seek to account explicitly for *all* of the following sources of uncertainty:

a)   input uncertainty (observation error)
b)   state uncertainty
c)   structural uncertainty,

d)   parameter uncertainty, and

e)   output uncertainty (observation error)

which could (we think) be treated within a Bayesian framework (something for ambitious graduate students to think about!).

A related (and perhaps more interesting) issue that also remains to be addressed, is how to resolve the Bayesian uncertainty framework, which is based on the computation of a single aggregate measure of model performance with the multiple-criteria philosophy which argues for *multiple non-commensurable* measures of model performance to enhance identifiability. One strategy that contributes towards ways to address this dilemma is the DYNIA (Dynamic Identifiability Analysis) method [Wagener and Wheater, 2001; Wagener et al., 2001c; and Wagener et al. this volume] which combines the multiple-criteria approach with a recursive strategy for identifying data periods of high information content (A feature of the DYNIA method is its ability to detect violations of the assumption that the parameter values are constant with respect to time). A goal of future model identification research should be to develop a strategy that:

a)   allows for descriptions of *a priori* model uncertainty

b)   explicitly incorporates reasonable descriptions of input, state, structure, parameter and output uncertainties

c)   incorporates multiple sources and types of information

d)   permits recursive processing of data as they become available, and

e)   provides probabilistic estimates of the model outputs (e.g., "most likely" values and 95% confidence intervals).

5. SUMMARY

Watershed models are conceptual-numerical descriptions of the dominant hydrologic processes occurring in a watershed. Since the 1960's there has been a progressive evolution of watershed model structures ranging from lumped to semi-lumped, distributed and now multi-flux descriptions. In each case, application of the model to a specific watershed required that estimates for the model parameters be provided. The general problem of model identification involves the selection of appropriate structures for the various components of the model and specification of the values of the parameters for those components, such that the resulting model provides a sufficiently accurate simulation/prediction of the watershed response.

The problem of model identification (both structure and parameter values) has proved to be difficult. In most cases a single model structure is selected and various strategies are employed to adjust the parameter values so that the modeled input-output response approximates, as closely as possible, the observed response of the watershed for some historical period of data. Parameter estimation proceeds through three stages (Levels Zero, One, and Two) that progressively reduce the initial uncertainty in the parameter values. The resulting model is then evaluated for consistency and accuracy using an independent period of data.

Model identification is a process that seeks to constrain a model to be structurally and functionally consistent with the available qualitative and quantitative information about the watershed. Because there are many sources of uncertainty, including data error and conceptual model error, the best possible outcome of model identification is some minimal set of models (structures and parameter values) that closely and consistently approximate, in an uncertain way, the observed behavior of the watershed. The two main strategies for parameter estimation are the Manual-Expert approach and the Automatic approach. While the Manual-Expert approach relies on the subjective judgement of a trained hydrologist, the Automatic approach has traditionally been based on techniques of non-linear regression.

Hydrologists responsible for calibrating the SAC-SMA model of the NWS have eschewed the traditional Automatic approach in favor of the Manual-Expert approach, because they find the latter to provide more accurate and consistent model calibrations. However, there is an urgent need to develop reliable Automatic calibration methods because of the large number of watersheds that must be calibrated for flood forecasting purposes, and because the Manual-Expert procedures may be unable to handle the increasing complexity of emerging models.

A major problem with traditional Automatic calibration methods is their underlying assumption that the available model structure is correct, leading to the elusive goal of finding a unique optimal parameter set. Although global optimization algorithms, such as the SCE-UA, allow us to find the global optimum of an objective function with a high degree of confidence, research has indicated that there are typically large regions of the feasible parameter space for which the objective function values are very similar. There have been at least three responses to this finding. One is that this phenomenon is evidence of equifinality among models, a condition in which the available data is insufficient to distinguish between competing models. Another is that the phenomenon is evidence of models that are too complex in relation to the information content of the data, leading to the contention that watershed input-output data can typically only support models with approximately five parameters.

While issues of equifinality and over-complexity exist, these phenomena result partly as a consequence of inherent weaknesses in the design of the model identification problem. One major weakness lies in the dependence of the identification process on a single aggregate measure of model performance that is unlikely to constitute a rigorous test from which differences in model performance can be inferred. In contrast, the Manual-Expert calibration approach uses a number of complementary, although subjective, ways of evaluating model performance. This allows the hydrologist to compensate for various kinds of model and data errors, and to extract greater amounts of information from the data.

We contend that it is necessary to resolve the deficiencies of the automatic approach, before jumping to the conclusions of equifinality and/or over-complexity, and that single criteria identification methods are fundamentally inadequate for the identification of hydrologic models. The multiple-criteria approach offers a way to improve the power of the Automatic approach, by emulating the procedures used in Manual-Expert calibration. The approach identifies a Pareto-region of the parameter space, which represents the trade-offs that can be made between different "optimal" ways of constraining the model to be consistent with the data in the presence of model and data error. By using the Pareto-solution set one can generate a Pareto-ensemble of simulated outputs, so that the uncertainty in the model simulations due to different ways of trading-off the model and data errors can be examined. This provides new and useful ways to evaluate competing models and their performance. Numerous researchers have demonstrated the value of the multiple-criteria approach for watershed modeling, and extensions to multiple-flux, hydro-chemical, and hydro-meteorological models have also been explored.

The relentless pace of technological development is promoting increasing interest in models that provide spatially distributed predictions of multiple interdependent fluxes (including water, energy and various chemical substances). These more complex descriptions of the hydrologic environment will, no doubt, continue to demand evolutionary developments in model identification technology. While there have been suggestions that the increasing availability of various kinds of spatially distributed data will soon obviate the need for calibration technologies, it is our view that the need for model calibration methods remains. Further, there is a growing need for procedures that provide reliable estimates of the confidence in the model forecasts.

The goal of future model identification research should be to develop a strategy that allows for descriptions of *a priori* model uncertainty, explicitly incorporates descriptions of input, state, structure, parameter and output uncertainties,

incorporates multiple sources and types of information, permits recursive processing of data as they become available, and provides probabilistic estimates of the model outputs. The outcome of this strategy should be able to pass the "Turing Test" of model identification; i.e., model performance that is either indistinguishable, or demonstrably superior to what can be obtained by a highly trained expert.

6. SOME CLOSING COMMENTS AND CONJECTURES

We close with five comments and conjectures that build on our earlier work [Gupta, 2001].

1. All model identification problems are inherently multi-criteria in nature. Every modeling assumption implies a criterion that could (in principle) be testable. It is self-evident, therefore that we should be careful to acknowledge and carefully examine all assumptions.

2. It is not clear that identification methods based on traditional statistical approaches are powerful enough to facilitate extraction of all relevant information from hydrologic data, particularly as model complexity continues to grow. Is it really useful to strive for model residuals that have the properties of being "homogeneous, uncorrelated, and independently identically distributed", and that "belong to a Gaussian distribution?"

3. We will be unable to establish confidence in the emerging generation of sophisticated watershed models if our model identification procedures are weak. Nor can we make inferences about the "information content of the data", or "amount of supportable model complexity", by using a single aggregate measure of model performance.

4. The observation by some researchers that only model structures having approximately five parameters can be confidently estimated from hydrologic input-output data is an artifact of the single criteria strategy rooted in traditional regression theory. We conjecture that a single aggregate measure of model performance only has the "power" to identify approximately 3 to 5 parameters of a conceptual dynamic model. We suggest that the number of criteria (N_{crit}) necessary to identify a model having N_{par} parameters must satisfy the condition $N_{crit} \leq N_{par}$, and that a rule of thumb may be $N_{crit} \sim N_{par} / 5$ This is supported by Boyle et al., [2000] who obtained good results using three criteria for the 15 parameter SAC-SMA model.

5. There is no "best" way to select the criteria for a multiple-criteria analysis. The selected criteria should measure different, complementary, attributes of system behavior. It is likely that there may be several functionally equivalent ways to define these. When setting up a model identification problem with multiple criteria there can be interdependence among criteria (in a fashion dual to the problem of interdependence among parameters). More insight into this issue is needed.

As always, we invite open discussion and collaboration on these and related issues. It is through the free and unfettered exchange of ideas that rapid progress is accomplished. We have had the good fortune to collaborate with numerous researchers from around the world, and for those opportunities we are extremely grateful.

APPENDIX

The HMLE estimator is the maximum likelihood, minimum variance, asymptotically unbiased estimator when the errors in the output data are Gaussian, zero mean, uncorrelated and have variance related to the magnitude of the output (streamflows). Nonstationary variance errors are believed to be common in streamflow data. The HMLE is related to the SLS (Simple Least Squares): it reduces to the SLS function in the special case when the streamflow measurement errors have equal variance. The HMLE estimator has the form:

$$\min_{\theta, \lambda} HMLE = \frac{\frac{1}{n}\sum_{t=1}^{n} w_t \varepsilon_t^2}{\left[\prod_{t=1}^{n} w_t\right]^{\frac{1}{n}}} \qquad (A.1)$$

Where $\varepsilon_t = q_{t,obs} - q_t(\theta)$ is the model residual at time t; $q_{t,obs}$ and $q_t(\theta)$ are observed and simulated flows, respectively; w_t is the weight assigned to time t, computed as:

$$w_t = f_t^{2(\lambda - 1)} \qquad (A.2)$$

Where $f_t = q_{t,true}$ is the expected true flow at time t, n is the number of data points, and λ is the unknown transformation parameter which stabilizes the variance. The expected flow $q_{t,true}$ is approximated using either $q_{t,obs}$ or $q_t(\theta)$ [Sorooshian et al., 1983]. Fulton [1982] showed that the estimator can become unstable when $q_t(\theta)$ is used to approximate f_t and recommends using $q_{t,obs}$. While this is the recommended pro-

cedure at present, it should be noted that use of $q_{t,obs}$ will cause some degree of bias in the estimate of λ [Gupta, 1984].

The HMLE is solved in two stages. First, given a set of model parameters, the residuals of the model are obtained. Next, an estimate of λ must be obtained; Sorooshian [1978] showed that this could be done by solving the implicit expression A.3 to obtain an estimate of the parameter λ, using an interative numerical procedure:

$$\left[\sum_{t=1}^{n} \ln(f_t) \bullet \sum_{t=1}^{n} w_t \varepsilon^2_t\right] - n \bullet \left[\sum_{t=1}^{n} w_t \ln(f_t) \bullet \varepsilon^2_t\right] = 0 \qquad (A.3)$$

The value of λ is substituted into eqs. A.1 and A.2 to compute the value of the HMLE function. Duan [1991] devised an equivalent, but more stable procedure, for estimating, by rearranging eq. A.3, as follows:

$$R = \frac{R_d}{R_h} - 1 = 0 \qquad (A.4)$$

where
$$R_d = \sum_{t=1}^{n} w_t \varepsilon^2_t \qquad (A.5)$$

$$R_h = \sum_{t=1}^{n} w_t \varepsilon^2_t a_t \qquad (A.6)$$

$$a_t = \frac{\ln(f_t)}{a_d} \qquad (A.7)$$

$$a_d = \frac{1}{n}\sum_{t=1}^{n} \ln(f_t) \qquad (A.8)$$

With this arrangement of terms, the HMLE function value can be computed as:

$$HMLE = \frac{\frac{1}{n} R_d}{\exp[2(\lambda - 1)a_d]} \qquad (A.9)$$

The revised procedure for estimating λ and computing HMLE, given $q_{t,obs}$ and $q_t(\Phi)$, is as follows:

a) Select $f_t = q_{t,obs}$ or $q_t(\Phi)$ or $(\alpha\, q_{t,obs} + \beta\, q_t(\Phi))$, where $\alpha + \beta = 1$; $\alpha, \beta \geq 0$; $t = 1,2,...,n$;

b) Compute a_d, using eq. A.8, and α_t (for t = 1,2,...n) using eq. (A.7)

c) Use an iterative procedure (e.g. Golden Section, or Fibonacci Method) to estimate λ such that the R = 0 in eq. (A.4). If the procedure requires an initial value, use $\lambda = 1$.

d) Compute HMLE using eq. (A.9).

Acknowledgements: We gratefully acknowledge the hard work, dedication and innovative contributions of the numerous graduate students and colleagues who have worked with us on the research projects described in this review paper, including: Qingyun Duan, Patrice Yapo, Luis Bastidas, Kuolin Hsu, Douglas Boyle, John Schaake, Steve Burges, Keith Beven, George Kuczera, Howard Wheater, Thorsten Wagener, Arthur Armour, Walter Weining, Michael Thiemann, Tom Meixner, Feyzan Misirli, Jasper Vrugt, Eylon Shamir, YuQiong Liu, and Newsha Khodatalab. Partial funding to investigate these ideas was provided, over the years, by several grants, including recent awards by the Hydrologic Research Laboratory of the U.S. National Weather Service [grants NA85AA-H-HY088, NA90AA-H-HY505, NA47WH0408, NA57WH0575, NA77WHO425, NA87WH0581, NA87WHO582, NA86GPO324, NA07WH0144], NASA [EOS grant NAGW2425, NAG536405, NAG81531], HyDIS program (NASA funded Hydrological Data and Information System grant NAG58503, NOAA [grant NA86GP0324], NSF [grants ECE-86105487, BCS-8920851, EAR-9415347, EAR-9418147, EAR-9876800], the Salt River Project Doctoral Fellowship in Surface Water Hydrology, and SAHRA [NSF-STC grant EAR-9876800].

REFERENCES

Armour, A.D., 1990: Adaptive Random Search Evaluated as a Method for Calibration of the SMA-NWSRFS Model, M.S. Thesis, Dept. of Hydrology and Water Resources, University of Arizona, Tucson, AZ, 119 p.

Bales, R.C., M.W. Williams, and R.A. Wolford, 1993: Acidification Potential of Snowpack in Sierra-Nevada-Discussion, *Journal of Environmental Engineering-ASCE*, 119(2), 399-401.

Bastidas, L., H.V. Gupta, S. Sorooshian, W.J. Shuttleworth and Z.L. Yang, 1999: Sensitivity Analysis of a Land Surface Scheme using Multi-Criteria Methods, *Journal of Geophysical Research, Atmospheres*, 104(D16), 19481-19490.

Beldring, S., 2002: Multi-Criteria Validation of a Precipitation-Runoff Model, *Journal of Hydrology*, 257, 189-211.

Bergstrom, S., 1995: The HBV Model, V. P. Singh (Editor), *Computer Models of Watershed Hydrology,* Water Resources Publications, Highlands Ranch, CO, 443-476.

Beven, K.J., and A.M. Binley, 1992: The Future of Distributed Models: Model Calibration and Predictive Uncertainty, *Hydrological Processes*, 6, 279-298.

Beven, K. J., and M. J., Kirkby, 1976: Towards a Simple Physically-based Variable Contributing Area Model of Catchment Hydrology, Working Paper 154, School of Geography, University of Leeds, UK.

Beven, K. J., and M. J., Kirkby, 1979: A Physically Based Variable Contributing Area Model of Basin Hydrology, *Hydrological Science Bulletin*, 24(1), 43-69.

Boughton, W. C., 1965, A New Estimation Technique For Estimation Of Catchment Yield, Rep. 78, Water Resources Lab., University of New South Wales, Manly Vale, Australia.

Box, G.E.P., and O.R. Cox, 1964: The Analysis of Transformation, *Journal of the Royal Statistical Society, Series B*, 26(2), 211-252.

Boyle, D.P., 2001: Multicriteria Calibration Of Hydrologic Models, Ph.D. Dissertation, Department of Hydrology and Water Resources, University of Arizona, Tucson, AZ.

Boyle, D.P. H.V. Gupta, and S. Sorooshian, 2000: Toward Improved Calibration of Hydrologic Models: Combining the strengths of Manual and Automatic Methods, *Water Resources Research*, 36(12) 3663-3674.

Boyle, D.P., H.V. Gupta, S. Sorooshian, V. Koren, Z. Zhang, and M. Smith, 2001: Toward Improved Streamflow Forecasts: Value of Semi-distributed Modeling, *Water Resources Research*, 37(11), 2749-2759.

Brazil, L.E., 1988: Multilevel Calibration Strategy for Complex Hydrologic Simulation Models, Ph.D. Dissertation, Dept. of Civil Engineering, Colorado State University, Fort Collins, CO, 217 p.

Brazil, L.E., and W.F. Krajewski, 1987: Optimization of complex hydrologic models using random search methods, paper presented at Conference on Engineering Hydrology, Hydraulics Division American Society of Civil Engineers, Williamsburg, VA, Aug 3-7, 1987.

Burnash, R. J. C., R. L. Ferreal, and R. A. McGuire, 1973: A Generalized Streamflow Simulation System: Conceptual Modeling for Digital Computers, U.S. Department of Commerce, National Weather Service and State of California, Department of Water Resources.

Burnash, R.J.C., 1995: The NWS River Forecast System – Catchment Modeling, V. P. Singh (Editor), *Computer Models of Watershed Hydrology*, Water Resources Publications, Highlands Ranch, CO, 311-366.

Calver, A., and W.L. Wood, 1995: The Institute of Hydrology Distributed Model, V. P. Singh (Editor), *Computer Models of Watershed Hydrology,* Water Resources Publications, Highlands Ranch, CO, 595-626.

Contractor D.N., and J. W. Jenson, 2000, Simulated Effect Of Vadose Infiltration On Water Levels In The Northern Guam Lens Aquifer, *Journal of Hydrology*, 229, 232-245.

Crawford, N.H., and R. K. Linsley, 1966: Digital Simulation in Hydrology – Stanford Watershed Model IV, Technical Report No. 39, Department of Civil Engineering, Stanford University, Stanford, CA.

De Grosbois, E., R.P. Hooper, and N. Christophersen, 1988: A Multisignal Automatic Calibration Methodology for Hydrochemical Models-A Case-Study of the Birkenes Model, *Water Resources Research*, 24(8), 1299-1307.

Dickinson, R.E., M. Shaikh, R. Bryant, and L. Graumlich, Interactive Canopies for a Climate Model, *Journal of Climate*, 11(11), 2823-2836.

Duan, Q., 1991: A Global Optimization Strategy For Efficient And Effective Calibration Of Hydrologic Models, Ph.D. dissertation, University of Arizona, Tucson, AZ.

Duan, Q., V.K. Gupta, and S. Sorooshian, 1992: Effective and Efficient Global Optimization for Conceptual Rainfall-Runoff models, *Water Resources Research*, 28, 1015-1031.

Duan, Q., V.K. Gupta, and S. Sorooshian, 1993: A Shuffled Complex Evolution Approach for Effective and Efficient Global Minimization, *Journal of Optimization Theory and its Applications*, 76, 501-521.

Freer, J., K.J. Beven and B. Ambroise, 1996: Bayesian Estimation Of Uncertainty In Runoff Prediction And The Value Of Data: An Application Of The GLUE Approach, *Water Resources Research*, 32(7), 2161-2173.

Franchini, M., G. Galeati, Giorgio, S. Berra, 1998: Global Optimization Techniques for the Calibration of Conceptual Rainfall-Runoff Models, *Journal of Hydrolologic Science*, 43(3), 443-458.

Franks, S. and K. Beven, 1997: Bayesian Estimation Of Uncertainty In Land Surface-Atmosphere Flux Predictions, *Journal of Geophysical Research*, 102 (D20), 23,991-23,999.

Franks, S.W., K.J. Beven, and J.H.C. Gash, 1999: Multi-objective Conditioning of a Simple SVAT Model, *Hydrology and Earth System Sciences*, 3(4), 477-489.

Fulton, J., 1982: Discussion, Modification, And Extension Of Some Maximum Likelihood Techniques For Model Calibration With Application To Rainfall-Runoff Models, M.S. Thesis, Case Western Reserve University, Cleveland, Ohio.

Gan, T.Y. and G.F. Biftu, 1996: Automatic Calibration of Conceptual Rainfall-Runoff models: Optimization Algorithms, Catchment Conditions, and Model Structure, *Water Resources Research*, 32(12), 3513-3524.

Gupta, V.K., 1984: The Identification Of Conceptual Watershed Models, Ph.D. Dissertation, Case Western Reserve University, Cleveland, Ohio.

Gupta, V.K., and S. Sorooshian, 1983: Uniqueness and Observability of Conceptual Rainfall-Runoff Model Parameters: The Percolation Process Examined, *Water Resources Research*, 19, 269-276.

Gupta, V.K., and S. Sorooshian, 1985a: The Automatic Calibration of Conceptual Catchment Models using Derivative-Based Optimization Algorithms, *Water Resources Research*, 21(4), 473-485.

Gupta, V.K., and S. Sorooshian, 1985b: The Relationship Between Data and the Precision of Parameter Estimates of Hydrologic Models, *Journal of Hydrology*, 81, 57-77.

Gupta, V.K., S. Sorooshian, and P.O. Yapo, 1998: Towards Improved Calibration of Hydrologic Models: Multiple and Non-commensurable Measures of Information, *Water Resources Research*, 34, 751-763.

Gupta, H.V., L. Bastidas, L., S. Sorooshian, W.J. Shuttleworth and Z.L. Yang, 1999: Parameter Estimation of a Land Surface Scheme using Multi-Criteria Methods, *Journal of Geophysical Research, Atmospheres*, 104(D16), 19491-19503.

Gupta, H.V., 2000: The Devilish Dr. M Or Some Comments on the Identification of Hydrologic Models, presented at the Seventh Annual Meeting of the British Hydrological Society, September, 2000, Newcastle-upon-Tyne, UK

Hendrickson, J.D., S. Sorooshian, and L. Brazil, 1988: Comparison of Newton-type and Direct Search Algorithms for Calibration of Conceptual Rainfall-Runoff Models, *Water Resources Research*, 24(5), 691-700.

Hogue, T.S., S. Sorooshian, H. Gupta, A. Holz, and D. Braatz, 2000: A Multistep Automatic Calibration Scheme for River Forecasting Models, *AMS Journal of Hydrometerology*, 1, 524-542.

Hooke, R. and T.A. Jeeves, 1961: Direct Search Solutions of Numerical and Statistical Problems, *Journal Assoc. Computer Mach.*, 8(2), 212-229.

Hooper R.P., A. Stone, N. Christophersen, E. De Grosbois, and H.M. Seip, 1988: Assessing the Birkenes Model of Stream Acidification Using a Multisignal Calibration Methodology, *Water Resources Research*, 24(8), 1308-1316.

Ibbitt, R.P., 1970: Systematic Parameter Fitting for Conceptual Models of Catchment Hydrology, Ph.D. Thesis, University of London, UK.

Ingram J., 1996: Lesson Taught by Floods in the United States of America, Presented at ICSU SS/IDMDR Workshop on River Flood Disasters, Koblenz, Germany, November, 26-28, 1996.

Jakeman, A., and G. Hornberger, 1993: How Much Complexity is Warranted in a Rainfall-runoff Model? *Water Resources Research* 29(8), 2637-2649, 1982.

James, L. D., and Burges, S. J., Selection, Calibration, And Testing Of Hydrologic Models, in *Hydrologic Modeling of Small Watersheds*, edited by C. T. Haan, H. P. Johnson, and D. L. Brakensick, American Society of Agricultural Engineers Monograph, 5, 437-472.

Johnston, P.R. and D. Pilgrim, 1976: Parameter Optimization for Watershed Models, *Water Resources Research*, 12(3), 477-486.

Klemes, V., 1986a: Dilettantism in Hydrology: Transition or Destiny?, *Journal of Hydrology*, 65, 1-23.

Klemes, V., 1986b: Operational Testing of Hydrological Simulation Models, *Hydrological Science Journal*, 31(1-3), 13-24.

Koren, V.I., M. Smith, D. Want, and Z. Zhang, 2000: Use of Soil Property Data in the Derivation of Conceptual Rainfall-Runoff Model Parameters, Preprints, 15th Conference on Hydrology, Long Beach, CA, American Meteorological Society, Paper 2.16, January 10-14 .

Kuczera, G., 1988: On Validity of First-order Prediction Limits for Conceptual Hydrological Models, *Journal of Hydrology*, 103, 229-247.

Kuczera, G., 1997: Efficient Subspace Probabilistic Parameter Optimization for Catchment Models, *Water Resources Research*, 33(1), 177-185.

Leavesley, G.H., S.L. Markstrom, M.S. Brewer, R.J. Viger, 1996: The Modular Modeling System (MMS) – The Physical Process Modeling Component of a Databased-centered Decision Support System for Water and Power Management, *Water Air and Soil Pollution*, 90(1-2), 303-311.

Leplastrier, M. A.J. Pitman, H. Gupta, and Y. Xia, 2001: Exploring the Relationship Between Complexity and Performance in a Land Surface Model using the Multi-Criteria Method, submitted to *Journal of Geophysical Research*.

Liang, X., D.P. Lettenmaier, E.F. Wood, and S.J. Burges, 1994: A Simple Hydrologically Based Model of Land Surface Water and Energy Fluxes for General Circulation Models, *Journal of Geophysical Research*, 99, 14,415-14,428.

Luce, C.H., and T.W. Cundy, 1994: Parameter Identification for a Runoff Model for Forest Roads, *Water Resources Research*, 30, 1057-1069.

Madsen, H., 2000: Automatic Calibration of a Conceptual Rainfall-runoff Model using Multiple Objectives, *Journal of Hydrology*, 235(3-4), 276-288.

Madsen, H., G. Wilson, and H.C. Ammentorp, 2002: Comparison of Different Automated Strategies for Calibration of Rainfall-runoff Models, *Journal of Hydrology*, 261, 48-59.

Mahani, S.E., X. Gao, S. Sorooshian, B. Imam, 2000: Estimating Cloud Top Height And Spatial Displacement From Scan-Synchronous GOES Images Using Simplified IR-Based Stereoscopic Analysis, *Journal of Geophysical Research*, 105(D12): 15,597-15,608.

Meixner T., H.V. Gupta, L.A. Bastidas, and R.C. Bales, 1999: Sensitivity Analysis Using Mass Flux and Concentration, *Hydrological Processes*, 13(14-15), 2233-2244.

Meixner, T, R.C. Bales, M.W. Williams, D.H. Campbell, and J.S. Baron, 2000: Stream Chemistry Modeling of Two Watersheds in the Front Range, Colorado, *Water Resources Research*, 36(1), 77-87.

Mitchell, K., J. Schaake, D. Tarplay, F. Chen, Y. Lin, M. Baldwin, E. Rogers, G. Manikin, A. Betts, Z. Janjic, Q. Duan, and V. Koren, 1999: Recent GCIP Advancements in Coupled Land-surface Modeling and Data Assimilation in the NCEP Mesoscale ETA Model, Preprints, AMS 14th Conference on Hydrology, Dallas, TX.

Nelder, J.A., and R. Mead, 1965: A Simplex Method for Function Minimization, *Computer Journal*, 7(4), 308-313.

Nijssen, B. G. O'Donnell, D.P. Lettenmaier, D. Lohmann and E.F. Wood, 2001: Predicting the Discharge of Global Rivers, *Journal of Climate*, 14(15), 3307-3323.

Pitman, A.J., M. Leplastrier, Y. Xia, H.V. Gupta, and A. Henderson-Sellers, 2001: Multi-criteria Method Calibration of a Land Surface Model Using Multi-data Sets and Multi-time Scales, submitted to *Global and Planetary Change*.

Refsgaard, J. C., and B. Storm, 1995: MIKE SHE, V. P. Singh (Editor), *Computer Models of Watershed Hydrology*, Water Resources Publications, Highlands Ranch, CO, 809-846.

Rosenbrock, H.H., 1960: An Automatic Method of Finding the Greatest or Least Value of a Function, *Computer Journal*, 3, 175-184.

Santos, C.A.G., K. Suzuki, and M. Watanabe, 1999: Modification of SCE-UA Genetic Algorithm for Runoff-Erosion Modeling, Proceedings of the International Symposium, 1999, V21, Nepal Geological Society, 131-138.

Schaake, J., V. Koren, Q.Y. Duan, S. Cong, and A. Hall, 1998: Model Parameter Estimation Experiment (MOPEX): Data Preparation and Some Experimental Results, Presentation at GCIP Mississippi Climate Conference, St. Louis, MO, June 8-12, 1998.

Scott, R.L., W.J. Shuttleworth, T.O. Keefer, and A.W. Warrick, 2000: Modeling Multiyear Observations Of Soil Moisture Recharge In The Semiarid American Southwest, *Water Resources Research*, 36(8), 2233-2248.

Seibert, J., 2000: Multi-criteria Calibration of a Conceptual Runoff Model Using a Genetic Algorithm, *Hydrology and Earth System Science*, 4(2), 215-224.

Sellers, P.J., D.A. Randall, G.J. Collatz, J.A. Berry, C.B. Field, D.A. Dazlich, C. Zhang, G.D. Collelo, and L. Bounoua, 1996: A Revised Land Surface Parameterization (SiB2) for Atmospheric GCMs, I, Model Formulation, *Journal of Climate*, 9, 676-705.

Sen, O.L., L.A. Bastidas, W.J. Shuttleworth, Z.L. Yang, H.V. Gupta, and S. Sorooshian, 2001: Impact of Field-calibrated Vegetation Parameters on GCM Climate Simulations, *Quarterly Journal of the Royal Meteorological Society*, 127(574), 1199-1223.

Smith, M. B, V. Koren, D. Johnson, B. D. Finnerty, and D.-J. Seo, 1999: Distributed Modeling: Phase 1 Results, NOAA Technical Report NWS 44, National Weather Service Hydrologic Research Lab, Silver Spring, MD, 210 pp.

Smith, R.E., D. C. Goodrich, D.A. Woolhiser, and C.L. Unkrich, 1995: KINEROS – A KINematic Runoff and EROSion Model, V. P. Singh (Editor), *Computer Models of Watershed Hydrology*, Water Resources Publications, Highlands Ranch, CO, 697-732.

Sorooshian, S., 1978: Considerations of Stochastic Properties in Parameter Estimation of Hydrologic Rainfall-Runoff Models, Ph.D. Dissertation, University of California-Los Angeles, Los Angeles, California.

Sorooshian, S. and J.A. Dracup, 1980: Stochastic Parameter Estimation Procedures for Hydrologic Rainfall-Runoff Models: Correlated and Heteroscedastic Error Cases, *Water Resources Research*, 16(2), 430-442.

Sorooshian, S., V.K. Gupta, and J.L. Fulton, 1983: Evaluation of Maximum Likelihood Parameter Estimation Techniques for Conceptual Rainfall-Runoff Models: Influence of Calibration Data, Variability and Length on Model Credibility, *Water Resources Research*, 19(1), 251-259.

Sorooshian, S., Q. Duan, and V.K. Gupta, 1993: Calibration of Rainfall-Runoff Models: Application of Global Optimization to the Sacramento Soil Moisture Accounting Model, *Water Resources Research*, 29, 1185-1194.

Spear, R.C., and G.M. Hornberger, 1980: Eutrophication in Peel Inlet, II, Identification of Critical Uncertainties via Generalized Sensitivity Analysis, *Water Resources Research*, 14, 43-49.

Tanakamaru, H., 1995: Parameter Estimation for the Tank Model Using Global Optimization, *Transactions of the Japanese Society of Irrigation, Drainage and Reclamation Engineering*, 178, 103-112.

Thiemann, M., M. Trosset, H. Gupta, and S. Sorooshian, 2001: Bayesian Recursive Parameter Estimation for Hydrologic Models, *Water Resources Research*, 37, 2521-2535.

Turing, A.M., 1950: Computing Machinery and Intelligence, *Mind*, Vol. 59, 433-460.

Wagener, T., and H.S. Wheater, 2001: On The Evaluation Of Conceptual Rainfall-Runoff Models Using Multiple-Objectives And Dynamic Identifiability Analysis, In Littlewood, I. (ed.) *BHS Occasional Papers – Continuous River Flow Simulation: Methods, Applications And Uncertainties*, No. 13, Wallingford, UK, 45-51.

Wagener, T., M.J. Lees, and H.S. Wheater, 2001a: A Toolkit For The Development Of And Application Of Parsimonious Hydrological Models, In *Mathematical Models of Large Watershed Hydrology – Volume I*, Edited by V.P. Singh and D. Frevert, Water Resources Publishers, USA p. 87-136.

Wagener, T., D.P. Boyle, M.J. Lees, H.S. Wheater, H.V. Gupta, and S. Sorooshian, 2001b: A Framework for Development and

Application of Hydrological Models, *Hydrology and Earth System Sciences*, 5(1), 13-26.

Wagener, T., N. McIntyre, M.J. Lees, H.S. Wheater, and H.V. Gupta, 2001c: Towards Reduced Uncertainty In Conceptual Rainfall-Runoff Modeling: Dynamic Identifiability Analysis, *Hydrological Processes*, in press.

Walker, J.P., G.R. Wilgoose, and J.D. Kalma, 2001: One-Dimensional Soil Moisture Profile Retrieval By Assimilation Of Near-Surface Measurements: A Simplified Soil Moisture Model And Field Application, *Journal of Hydrometeorology*, 2, 356-373.

Wang, J., Lu, Z., and Habu, H., 2001: Shuffled Complex Evolution Method for Nonlinear Constrained Optimization, *Hehain University Science Journal*, 29(3), 46-50 (in Chinese).

Weinig, W.T, 1991: Calibration of the Soil Moisture Accounting Model Using The Adaptive Random Search Algorithm, M.S. Thesis, Dept. of Hydrology and Water Resources, University of Arizona, Tucson, AZ, 99 p.

Wheater, H.S., Jakeman, A.J., and Beven, K.J., 1993: Progress And Directions In Rainfall-Runoff Modeling, In: *Modeling Change In Environmental Systems* (Ed. A.J. Jakeman, M.B. Beck, and M.J. McAleer), Wiley, Chichester, UK, 101-132.

Wolford, R.A., and R.C. Bales, 1996: Hydrochemical Modeling of Emerald Lake watershed, Sierra Nevada, California: Sensitivity of Stream Chemistry to Changes in Fluxes and Model Parameters, *Limnology and Oceanography*, 41(5), 947-954.

Wolford, R.A., R.C. Bales, and S. Sorooshian, 1996: Development of a Hydrochemical Model for Seasonally Snow-covered Alpine Watersheds: Application to Emerald Lake Watershed, Sierra Nevada, California, *Water Resources Research*, 32(4), 1061-1074.

Yapo, P.O., H.V. Gupta, and S. Sorooshian, 1996: Calibration of Conceptual Rainfall-Runoff Models: Sensitivity to Calibration Data, *Journal of Hydrology*, 181, 23-48.

Yapo, P.O., H.V. Gupta, and S. Sorooshian, 1998: Multi-objective Global Optimization for Hydrologic Models, *Journal of Hydrology*, 204, 83-97.

Young, P.C., and K.J. Beven, 1994: Data-based Mechanistic Modeling and the Rainfall-flow Nonlinearity, *Environmetrics*, 5(3), 335-363.

Young, P.C., S. Parkinson, and M.J. Lees, 1996: Simplicity out of Complexity in Environmental Modeling: Occam's razor revisited, *Journal of Applied Statistics*, 23, 165-210.

Young, P.C., A.J. Jakeman, and D.A. Post, 1997; Recent Advances in the Data-based Modeling and Analysis of Hydrological Systems, *Water Science & Technology*, 36, 99-116.

Zeng, X., M. Shaikh, Y. Dai, R.E.Dickinson, and R. Myneni, 2001: Coupling of the Common Land Model to the NCAR Community Climate Model, submitted to *Journal of Climate*.

Zhao, R.J., Y. L. Zhuang, L.R. Fang, X.R. Liu, and Q.S. Zhang, 1980, The Zinanjiang Model, *Hydrological Forecasting Proceedings Oxford Symposium*, IAHS 129, p. 351-356.

Hoshin V. Gupta, Soroosh Sorooshian, Terri S. Hogue, SAHRA, NSF STC for the Sustainability of semi-Arid Hydrology and Riparian Areas, Department of Hydrology and Water Resources, Harshbarger, Bldg. 11, University of Arizona, Tucson, AZ 85721, USA.

Douglas P. Boyle, Desert Research Institute, University and Community College System of Nevada, Reno, NV, 89512, USA.

Identification and Evaluation of Watershed Models

Thorsten Wagener[1] and Howard S. Wheater

*Department of Civil and Environmental Engineering, Imperial College of Science,
Technology and Medicine, London, United Kingdom*

Hoshin V. Gupta

*SAHRA, NSF STC for Sustainability of semi-Arid Hydrology and Riparian Areas
Department of Hydrology and Water Resources, University of Arizona, Tucson, Arizona*

Conceptual modeling requires the identification of a suitable model structure and, within a chosen structure, the estimation of parameter values (and, ideally, their uncertainty) through calibration against observed data. A lack of objective approaches to evaluate model structures and the inability of calibration procedures to distinguish between the suitability of different parameter sets are major sources of uncertainty in current modeling procedures. This is further complicated by the increasing awareness of model structural inadequacies. A framework for the identification and evaluation of conceptual rainfall-runoff models is presented, based on multi-objective performance and identifiability approaches, and a novel dynamic identifiability analysis (DYNIA) method which results in an improved use of available information. The multi-objective approach is mainly used to analyze the performance and identifiability of competing models and model structures, while the DYNIA allows periods of high information content for specific parameters to be identified and model structures to be evaluated with respect to failure of individual components. The framework is applied to a watershed located in the South of England.

1. INTRODUCTION

Many if not most rainfall-runoff model structures currently used can be classified as conceptual. This classification is based on two criteria: (1) the structure of these models is specified prior to any modelling being undertaken, and (2) (at least some of) the model parameters do not have a direct physical interpretation, in the sense of being independently measurable, and have to be esti-

mated through calibration against observed data [*Wheater et al.*, 1993]. Calibration is a process of parameter adjustment (automatic or manual), until observed and calculated output time-series show a sufficiently high degree of similarity.

Conceptual rainfall-runoff (CRR) model structures commonly aggregate, in space and time, the hydrological processes occurring in a watershed (also called catchment), into a number of key responses represented by storage components (state variables) and their interactions (fluxes). The model parameters describe aspects such as the size of those storage components, the location of outlets, the distribution of storage volumes etc. Conceptual parameters, therefore, usually refer to a collection of aggregated processes and they may cover a large number of sub-processes that cannot be represented separately or explicitly [*Van Straten and Keesman*, 1991]. The underlying assumption however is that these parameters are, even if not measurable properties,

---

[1] Now at SAHRA, NSF STC for Sustainability of semi-Arid Hydrology and Riparian Areas, Department of Hydrology and Water Resources, University of Arizona, Tucson, Arizona

Calibration of Watershed Models
Water Science and Application Volume 6

at least constants and representative of inherent properties of the natural system [*Bard*, 1974, p.11].

The modeller's task is the identification of an appropriate CRR model (or models) for a specific case, i.e. a given modelling objective, watershed characteristics and data set. A model is defined in this context as a specific parameter set within a selected model structure. Experience shows that this identification is a difficult task. Various parameter sets, often widely distributed within the feasible parameter space [e.g. *Duan et al.*, 1992; *Freer et al.*, 1996], and sometimes even different conceptualisations of the watershed system [e.g. *Piñol et al.*, 1997; *Uhlenbrock et al.*, 1999], may yield equally good results in terms of a predefined objective function. This ambiguity has serious impacts on parameter and predictive uncertainty [e.g. *Beven and Binley*, 1992], and therefore limits the applicability of CRR models, e.g. for the simulation of land-use or climate-change scenarios, or for regionalisation studies [*Wheater et al.*, 1993].

Initially it was thought that this problem would disappear with improved automatic search algorithms, capable of locating the global optimum on the response surface [e.g. *Duan et al.*, 1992]. However, even though powerful global optimisation algorithms are available today, single-objective calibration procedures still fail to completely replace manual calibration. One reason for this is that the resulting hydrographs are often perceived to be inferior to those produced through manual calibration from the hydrologist's point of view [e.g. *Gupta et al.*, 1998; *Boyle et al.*, 2000]. It has been suggested that this is due to the fundamental problem that single-objective automatic calibration is not sophisticated enough to replicate the several performance criteria implicitly or explicitly used by the hydrologist in manual calibration. This problem is increased by indications that, due to structural inadequacies, one parameter set might not be enough to adequately describe all response modes of a hydrological system. Therefore, there is a strong argument that the process of identification of dynamic, conceptual models has to be rethought [*Gupta et al.*, 1998; *Gupta*, 2000].

Three reactions to this problem of ambiguity of system description can be found in the hydrological literature. The first is the increased use of parsimonious model structures [e.g. *Jakeman and Hornberger*, 1993; *Young et al.*, 1996; *Wagener et al.*, 2001b], i.e. structures only containing those parameters, and therefore model components, that can be identified from the observed system output. However, the increase in identifiability is bought at the price of a decrease in the number of processes described separately by the model. There is therefore a danger of building a model (structure) which is too simplistic for the anticipated purpose. Such a model (structure) can be unreliable outside the range of watershed conditions, i.e. climate and land-use, on

which it was calibrated, due to the restriction to 'justifiable' components [*Kuczera and Mroczkowski*, 1998]. It is also particularly important that the data used has a high information content in order to ensure that the main response modes are excited during calibration [*Gupta and Sorooshian*, 1985, *Yapo et al.*, 1996].

The second reaction is the search for calibration methods which make better use of the information contained in the available data time-series, e.g. streamflow and/or groundwater levels. Various research efforts have shown that the amount of information retrieved using a single objective function is sufficient to identify only between three and five parameters [e.g. *Beven*, 1989; *Jakeman and Hornberger*, 1993; *Gupta*, 2000]. Most CRR model structures contain a larger number. More information can become available through the definition of multiple objective functions to increase the discriminative power of the calibration procedure [e.g. *Gupta et al.*, 1998; *Gupta*, 2000]. These measures can either retrieve different types of information from a single time-series, e.g. streamflow [e.g. *Wheater et al.*, 1986; *Gupta et al.*, 1998; *Dunne*, 1999; *Boyle et al.*, 2000; *Wagener et al.*, 2001a], or describe the performance of individual models with respect to different measured variables, e.g. groundwater levels [e.g. *Kuczera and Mroczkowski*, 1998; *Seibert*, 2000], saturated areas [*Franks et al.*, 1998], or measurements of streamflow salinity [*Mroczkowski et al.*, 1997; *Kuczera and Mroczkowski*, 1998]. However, the usefulness of additional data can depend on the adequacy of the model structure investigated. *Lamb et al.* [1998] found that the use of groundwater levels from one or only a few measurement points as additional output variable(s) helped to reduce the parameter uncertainty of Topmodel [*Beven et al.*, 1995]. The use of many (>100) groundwater measurement points however, leads to an increase in prediction uncertainty indicating structural problems in the model. *Seibert and McDonnell* [this volume] show in a different approach how the parameter space can be constrained when soft data, i.e. qualitative knowledge of the watershed behaviour, is included in the calibration process. The soft data in their case included information, derived through experimental work, about the contribution of new water to runoff and the restriction of parameter ranges to a desirable range. The result is a more realistic model, which will however yield sub-optimal performances with respect to many specific objective functions, in their case the Nash-Sutcliffe efficiency measure [*Nash and Sutcliffe*, 1970]. *Chappell et al.* [1998] give another example of how expert knowledge of internal catchment dynamics (e.g. saturated areas) can be used to constrain the parameter space.

Thirdly, some researchers abandoned the idea of a uniquely identifiable model in favour of the identification of a model population [e.g. *van Straten and Keesman*, 1991; *Beven and Binley*, 1992; *Gupta et al.*, 1998]. This can for example be a population of models with varying degrees of (some sort of) likelihood to be representative of the watershed at hand, the idea behind the Generalized Likelihood Uncertainty Estimation (GLUE) approach [*Freer et al.*, this volume]. Or an approach based on the recognition that the calibration of a rainfall-runoff model is inherently a multi-objective problem, resulting in a population of non-dominated parameter sets [*Goldberg*, 1989, p.201] in the presence of model structural inadequacies [*Gupta et al.*, 1998].

Here, we seek to increase the amount of information made available from an output time-series and to guide the identification of parsimonious model structures, consistent with a given model application as explained below. We use multi-objective approaches to performance and identifiability analysis and a novel dynamic identifiability analysis (DYNIA) method for assumption testing. These can be integrated into a framework for model identification and evaluation. An application example at the end of this chapter shows the use of the framework for a specific case.

2. IDENTIFICATION OF CONCEPTUAL RAINFALL-RUNOFF MODELS

The purpose of identifiability analysis in CRR modelling is to find (the) model structure(s) and corresponding parameter set(s) which are representative of the watershed under investigation, while considering aspects such as modelling objectives and available data. This identifiability analysis can be split into two stages: model structure selection and parameter estimation, which can, however, not be treated as completely separate [*Sorooshian and Gupta*, 1985] (in order to evaluate model structures fully, one has to analyse their performance and behaviour which requires some form of parameter estimation).

Traditional modelling procedures commonly contain, amongst others, an additional third step [e.g *Anderson and Burt*, 1985]. This is a validation or verification step often used to show that the selected model really is the correct representation of the watershed under investigation. This results in the following three steps as part of a longer procedure:

(1) Selection or development of a model structure, and subsequently computer code, to represent the conceptualisation of the hydrologic system which the hydrologist has established in his or her mind for the watershed under study.
(2) Calibration of the selected model structure, i.e. estimation of the 'best' parameter set(s) with respect to one or more (often combined) criteria.

(3) Validation or verification of this model by (successfully) applying it to a data set not used in the calibration stage.

It is important to stress that the original meanings of the words validation and verification are different. Verification is the stronger statement, meaning to establish the truth, while validation means to establish legitimacy [*Oreskes et al.*, 1994]. In the context of hydrological modelling, these terms are often used synonymously, describing a step to justify that the chosen model is an acceptable representation of the real system. An in-depth discussion on this topic can be found in *Oreskes et al.* [1994].

These three steps are similar to the logic of induction often used in science. This idea of induction is founded on the underlying assumption that a general statement can be inferred from the results of observations or experiments [*Popper*, 2000, p.27]. It includes the assumption that a hypothesis, e.g. a chosen model structure, can be shown to be correct, i.e. a hypothesis can be validated or verified, through supporting evidence. The steps taken in this traditional scientific method are [for example modified from *Magee*, 1977, p. 56]:

(1) Observation and experiment;
(2) inductive generalization, i.e. a new hypothesis;
(3) attempted verification of hypothesis, i.e. proof or disproof of hypothesis;
(4) knowledge.

However, the logical error in this approach is, (as *Magee* [1977, p. 20] derives from statements by the philosopher Hume), that no number of singular observation statements, however large, could logically entail an unrestrictedly general statement. In rainfall-runoff modelling this is equivalent to the statement that, however often a model is capable of reproducing the response of a particular watershed, it can never be concluded that the true model has been found. It could for example be that future measurements will capture more extreme events, exciting a response not captured by earlier data and therefore not included in the model. Similarly, Popper concluded that no theory or hypothesis could ever be taken as the final truth. It can only be said that it is corroborated by every observation so far, and yields better predictions than any known alternative. It will however, always remain replaceable by a better theory or turn out to be false at a later stage [*Popper*, 2000, p.33].

The idea that a model can be verified (verus, meaning true in Latin [*Oreskes et al.*, 1994]) is therefore ill-founded and alternative modelling frameworks have to be found. One such alternative approach was suggested by *Popper* [2000].

He realised that, while no number of correctly predicted observations can lead to the conclusion that a hypothesis is correct, a single unexplained observation can lead to the falsification of the hypothesis. Hence he replaced the framework of verification with a framework of falsification, allowing the testing of a hypothesis.

This framework of falsification as suggested by Popper can be outlined as follows [modified from *Magee*, 1977, p.56]:

(1) The initial problem or question, often resulting from the fact that an existing hypothesis has failed;
(2) one (or more) proposed new hypothesis(es);
(3) deduction of testable propositions from the new hypothesis;
(4) attempted falsification of the new hypothesis by testing the propositions;
(5) preference established between competing hypotheses.

The procedure is repeated as soon as the new hypothesis fails. It is thus possible to search for the truth, but it is not possible to know when the truth has been found, a problem which has to be reflected in any scientific method.

Additionally, *Beven* [2000, p.304] pointed out that it is very likely, at least with the current generation of CRR models, that every model will fail to reproduce some of the behaviour of a watershed at some stage. However, even if one knows that the model is inadequate, one often has to use it due to the lack of alternatives. And for many cases, the use of this inadequate model will be sufficient for the selected purpose. Or as Wilfried Trotter put it more generally: In science the primary duty of ideas is to be useful and interesting even more than to be 'true' [*Beveridge*, 1957, p. 41].

How this general idea of hypothesis falsification can be put into a framework for CRR modelling is described below.

2.1. Identification of Model Structures

A large number of CRR modelling structures is currently available. These differ, for example, in the degree of detail described, the manner in which processes are conceptualised, requirements for input and output data, and possible spatial and temporal resolution. Despite these differences, a number of model structures may appear equally possible for a specific study, and the selection process usually amounts to a subjective decision by the modeller, since objective decision criteria are often lacking [*Mroczkowski et al.*, 1997]. It is therefore important to deduce testable propositions with respect to the assumptions underlying the model structure, i.e. about the hypothesis of how the watershed works, and to find measures of evaluation that give some objective guidance as to whether a selected structure is suitable or not. However, *Uhlenbrock et al.* [1999] have shown that it is difficult to achieve this using single-objective Monte-Carlo-based calibration approaches. They were able to derive good performances with respect to the prediction of streamflow, from sensible, as well as incorrect conceptualisations of a watershed. *Mroczkowski et al.* [1997] encountered similar problems when trying to falsify one of two possible model structures, including and excluding a groundwater discharge zone respectively, to represent two paired watersheds in Western Australia. This was impossible for both watersheds when only streamflow data was used. The additional use of stream chloride and groundwater level measurements allowed at least for the falsification of one of the model structures in case of the second watershed which had undergone considerable land-use changes.

Testable propositions about a specific model structure can be either related to the performance of the model or its components, or they can be related to its proper functioning.

A test of performance is the assessment whether or not the model structure is capable of sufficiently reproducing the observed behaviour of the natural system, considering the given quality of data. However, an overall measure of performance, aggregating the residuals over the calibration period, and therefore usually a number of response modes, hides information about how well different model components perform. It can be shown that the use of multiple-objectives for single-output models, measuring the model's performance during different response modes, can give more detailed information and allows the modeller to link model performance to individual model components [e.g. *Boyle et al.*, 2001; *Wagener et al.*, 2001a]. Additional information will also be available in cases where the model produces other measurable output variables, e.g. groundwater levels or hydro-chemical variables, as mentioned earlier.

Evaluation of the proper functioning of the model means questioning the assumptions underlying the model's structure, such as: Do the model components really represent the response modes they are intended to represent? And is the model structure capable of reproducing the different dominant modes of behaviour of the watershed with a single parameter set? A model structure is usually a combination of different hypotheses of the working of the natural system. If those hypotheses are to be individually testable, they should be related to individual model components and not just to the model structure as a whole [*Beck*, 1987; *Beck et al.*, 1993].

One, already mentioned, underlying assumption of conceptual modelling is the consideration of model parameters as constant in time, at least as long as for example no changes in the watershed occur that would alter the hydrological response, such as land-use changes. Different researchers [e.g. *Beck*, 1985; 1987; *Gupta et al.*, 1998; *Boyle et al.*, 2000; *Wagener et al.*, 2001a] have shown that this assumption can be tested, and that the failure of a model structure to simulate different response modes with a single parameter set suggests inadequacies in the functioning of the model.

Beck used the Extended Kalman Filter (EKF) extensively to recursively estimate model parameters and to utilize the occurrence of parameter deviation as an indicator of model structural failure [e.g. *Beck*, 1985; 1987; *Stigter et al.*, 1997]. For example, in the identification of a model of organic waste degradation in a river, changes in optimum parameter values in time from one location in the parameter space to another were identified [*Beck*, 1985]. Beck concluded from this observation that the model hypothesis had failed, i.e. the parameters were changing to compensate for one or more missing aspect(s) in the model structure. The subsequent step is to draw inference from the type of failure to develop an improved hypothesis of the model structure. However, there are limitations to the EKF approach. Beck concluded with respect to the use of the EKF for hypothesis testing that the performance of the EKF is not as robust as would be desirable and, inter alia, is heavily compromised by the need to make more or less arbitrary assumptions about the sources of uncertainty affecting the identification problem [*Beck*, 1987].

A trade-off in the capability to simulate different response modes can occur, as shown by *Boyle et al.* [2000] for the example for a popular complex rainfall-runoff model (Sacramento with 13 calibrated parameters [*Smith et al.*, this volume]), thus it was not possible to reproduce (slow) recession periods and the remaining system response modes simultaneously. Their multi-objective analysis suggests that the cause for this problem is mainly an inadequate representation of the upper soil zone processes.

There are therefore ideas to address the problem of model structure identification in a more objective way. However, they are not without weaknesses, as the Beck statement about the use of EKF showed earlier in the text. These need to be addressed to derive more suitable approaches.

2.2. Identification of Parameters

The second stage in the model identification process is the estimation of a suitable parameter set, usually referred to as calibration of the model structure. In this process, the parameters of a model structure are adjusted until the observed system output and the model output show acceptable levels of agreement. Manual calibration does this in a trial-and-error procedure, often using a number of different measures of performance and visual inspection of the hydrograph [e.g. *Gupta et al.*, 1998; *Smith et al.*, this vol-

ume]. It can yield good results and is often a good way to learn about the model, but it can be time consuming, requires extensive experience with a specific model structure and an objective analysis of parameter uncertainty is not possible. Traditional single-objective automatic calibration on the other hand is fast and objective, but will produce results which reflect the choice of objective function and may therefore not be acceptable to hydrologists concerned with a number of aspects of performance [*Boyle et al.*, 2000]. In particular the aggregation of the model residuals into an objective function leads to the neglect and loss of information about individual response modes, and can result in a biased performance, fitting a specific aspect of the hydrograph at the expense of another. It also leads to problems with the identification of those parameters associated with response modes which do not significantly influence the selected objective function [*Wagener et al.*, 2001a]. Selecting, for example, an objective function which puts more emphasis on fitting peak flows, e.g. the Nash-Sutcliffe efficiency value [*Nash and Sutcliffe*, 1970], due to its use of squared residual values [*Legates and McCabe*, 1999], will often not allow for the identification of parameters related to the slow response of a watershed [e.g. *Dunne*, 1999].

An example to demonstrate this problem is briefly presented. It uses a simple model structure consisting of a Penman two-layer soil moisture accounting component [*Penman*, 1949] to produce effective rainfall and a linear routing component using two conceptual reservoirs in parallel to transform it into streamflow. A comparison of hydrographs produced by different parameter sets within the selected structure, which yield similar objective function values, shows that these hydrographs can be visually different. Figure 1 shows a hundred days extract of six years of daily streamflow data, where the observed time-series (black line) is plotted with seven different realisations (grey lines), i.e. using the same model structure, but different parameter sets. The objective function used during calibration is the well known Root Mean Squared Error (RMSE). Each of the models presented yields a RMSE of 0.60mm/d when the complete calibration period (6 years) is considered. However, the hydrographs produced are clearly visually different. The added dotty plots of the two residence times of the (linear) routing component show that while the quick flow residence time, k(quick) is very well identified, the slow flow residence time, k(slow), is not. This is consistent with the observation that the main difference between the hydrographs can be observed during low flow periods. This effect is due to the use of squared residuals when calculating the RMSE.

This result demonstrates that traditional single-objective optimisation methods do not have the ability to distinguish

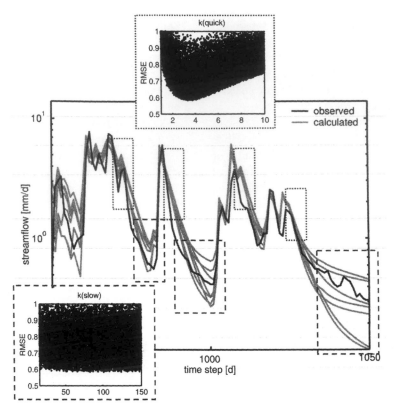

Figure 1. Hundred days extract of six years of daily streamflow data. Observed flow in black, seven different model realizations in gray. Inlets show dotty plots for the time constants k(quick) and k(slow) with respect to the Root Mean Squared Error (RMSE). The model structure used consists of a Penman soil moisture accounting and a parallel routing component of linear reservoirs with fixed flow distribution (see application example for details).

between visually different behaviour [*Gupta*, 2000]. The requirement for a parameter set to be uniquely locatable within the parameter space, i.e. to be globally identifiable, is that it yields a unique response vector [*Kleissen et al.*, 1990; *Mous*, 1993]. The unique response vector, in this case a unique (calculated) hydrograph, might be achievable, but this uniqueness is often lost if the residuals are aggregated into a single objective function. Such problems cannot be solved through improved search algorithms. They are rather inherent in the philosophy of the calibration procedure itself.

Clearly, the complex thought processes which lead to successful manual calibration are very difficult to encapsulate in a single objective function. This is illustrated by the requirements defined by the US National Weather Service (NWS) for the manual calibration of the Sacramento model structure [*NWS*, 2001]:

(1) Proper calibration of a conceptual model should result in parameters that cause model components to mimic processes they are designed to represent. This requires the ability to isolate the effects of each parameter.

(2) Each parameter is designed to represent a specific portion of the hydrograph under certain moisture conditions.

(3) Calibration should concentrate on having each parameter serve its primary function rather than overall goodness of fit.

It can be seen from these requirements that manual calibration is more complex than the optimisation of a single objective function, and that traditional automatic calibration procedures will in general not achieve comparable results. It is for example often not possible to isolate the effects of individual parameters and treat them as independent entities as done in the manual approach described above. Another aspect is that the goal of single-objective optimisation is purely to optimise the model's performance with respect to a selected overall goodness of fit measure which is very different from requirement three. This is not to say that traditional 'single' objective functions are not important parts of any model evaluation. The point is rather that they are not sufficient and should be complemented by a variety of measures.

Gupta et al. [1998] review this problem in more detail and conclude that a multi-objective approach to automatic calibration can be successful. *Boyle et al.*[2000] show how such a procedure can be applied to combine the requirements of manual calibration with the advantages of automatic calibration. A multi-objective algorithm is used to find the model population necessary to fit all aspects of the hydrograph. The user can then, if necessary, manually select a parameter set from this population to fit the hydrograph in the desired way. This will however, in the presence of model structural inadequacies, lead to a sub-optimal performance with respect to at least some of the other measures [*Boyle et al.*, 2000; *Seibert and McDonnell, this volume*]. The resulting trade-off of the ability of different parameter sets to fit different aspects of the hydrograph usually leads to a compromise solution [*Ehrgott*, 2000] in cases where a single parameter set has to be specified. The procedure of *Boyle et al.* [2000] for example, analyses the local behaviour of the model additionally to its global behaviour [*Gupta*, 2000]. The global behaviour is described through objective functions such as overall bias or some measure of the overall variance, e.g. the Root Mean Squared Error (RMSE). The local behaviour is defined by aspects like the timing of the peaks, or the performance during quick and slow response periods [*Boyle et al.*, 2000; 2001].

Recent research into parameter identification has thus moved away from simply trying to improve search algorithms, but has taken a closer look at the assumptions underlying (automatic) calibration approaches [e.g. *Gupta et al.*, 1998]. This has lead to the use of multi-objective (MO) automatic approaches which so far have given promising results [*Boyle et al.*, 2000; *Wagener et al.*, 2001a]. Further investigations are required to make MO optimization a standard method for parameter estimation. For example questions such as the appropriate number and derivation of OFs within a MO approach must be resolved, and will probably depend on model structure and watershed characteristics [*Gupta*, 2000].

3. EVALUATION OF CONCEPTUAL RAINFALL-RUNOFF MODELS

It was established earlier that the idea of calibration and validation of CRR models is in principle ill-founded, i.e. to establish a model as the true representation of a hydrological system. The model identification problem is therefore seen here as a process of model evaluation. Within this process, models and model structures are evaluated with respect to different criteria and those that fail, in whatever way, are rejected as possible representations of the watershed under investigation. This will usually result in a population of feasible models or even model structures which can then be used for a (combined) prediction, which will result in a prediction range, rather than a single value for each time-step.

This evaluation should be at least with respect to three dimensions:

(1) Performance, with respect to reproducing the behaviour of the system.
(2) Uncertainty in the parameters, which is assumed to be inversely related to their identifiability.
(3) Assumptions, i.e. are any assumptions made during the development of the model (structure) violated.

The smaller the population of models (or even model structures) that survives this evaluation, i.e. those that are corroborated by it, the more identifiable is the representation of the natural system in mathematical form. Approaches to test models with respect to these three criteria are described below.

3.1. Evaluation of Competing Model Structures—Multi-objective Performance and Identifiability Analysis

Multi-objective (MO) approaches can be applied to establish preferences between competing model structures or even model components, i.e. competing hypotheses, with respect to their performance and their identifiability. A MO approach is advantageous because the use of multiple objective criteria for parameter estimation permits more of the information contained in the data set to be used and distributes the importance of the parameter estimates among more components of the model. Additionally, the precision of some parameters may be greatly improved without an adverse impact on other parameters [*Yan and Haan*, 1991]. More detailed descriptions of MO model analysis can be found in the chapters by Gupta et al. and Boyle et al. [this volume].

3.1.1. Measures of performance and identifiability. It was already established earlier in the text that it is advantageous to evaluate the global and the local behaviour of models to increase the amount of information retrieved from the residuals in the context of single output rainfall-runoff models. Global behaviour is measured by traditional OFs, e.g. the RMSE or the bias for the whole calibration period, while different OFs have to be defined to measure the local behaviour. One way of implementing local measures is by partitioning the continuous output time series into different response periods. A separate OF can then be specified for each period, thus reducing the amount of information lost through aggregation of the residuals, e.g. by mixing high flow and recession periods.

Partitioning schemes proposed for hydrological time series include those based on: (a) Experience with a specific model structure (e.g. the Birkenes model structure in the case of *Wheater et al.*, 1986), i.e. different periods of the streamflow time series are selected based on the modeller's judgement. The intention of *Wheater et al.* [1986] was to improve the identifiability of insensitive parameters, so called minor parameters, with respect to an overall measure. Individual parameters, or pairs of parameters, are estimated using a simple grid search to find the best values for the individual objective functions. This is done in an iterative and sequential fashion, starting with the minor parameters and finishing with the dominant ones. (b) Hydrological understanding, i.e. the separation of different watershed response modes through a segmentation procedure based on the hydrologist's perception of the hydrological system (e.g. *Harlin*, 1991; *Dunne*, 1999; *Boyle et al.*, 2000; *Wagener et al.*, 2001a). For example, *Boyle et al.* [2000] propose hydrograph segmentation into periods 'driven' by rainfall, and periods of drainage. The drainage period is further subdivided into quick and slow drainage by a simple threshold value. (c) Parameter sensitivity [e.g. *Kleissen*, 1990; *Wagner and Harvey*, 1997; *Harvey and Wagner*, 2000], where it is assumed that informative periods are those time-steps during which the model outputs show a high sensitivity to changes in the model parameters [*Wagner and Harvey*, 1997]. *Kleissen* [1990] for example developed an optimisation procedure whereby only data segments during which the parameter shows a high degree of first order sensitivity are included in the calibration of that parameter (group) utilising a local optimisation algorithm. (d) Similar characteristics in the data derived from techniques like cluster analysis [e.g. *Boogard et al.*, 1998] or wavelet analysis [*Gupta*, 2000] can be used to group data points or periods based on their information content. The different clusters could then be used to define separate objective functions.

While these methods help to retrieve more information, they also show some weaknesses. Approaches (a) and (b) are subjective and based on the hydrologist's experience, and so are not easily applicable to a wide variety of models and watersheds. Approach (c), while objective, does not recognise the effects of parameter dependencies, and may not highlight periods which are most informative about the parameters as independent entities, i.e. periods where the dependency with respect to other parameters is low. The sensitivity of the model performance to changes in the parameter is a necessary requirement, but it is not sufficient for the identifiability of the parameter. Furthermore, if the parameter sensitivity is measured locally [e.g. *Kleissen*, 1990], the result is not guaranteed over the feasible parameter space. However, *Wagner and Harvey* [1997] show that this problem can be reduced by implementing a Monte Carlo procedure where the sensitivity for a large number of

different parameter combinations is assessed using parameter covariance matrices. Approach (d) is independent of any model structure and links between the results and the model parameters still need to be established.

There is therefore scope to improve the objectivity, applicability and robustness of approaches to hydrograph disaggregation, with the goal of improving model structure and parameter identifiability.

The evaluation of the model performance should, if possible, also include objective functions tailored to fit the specific purpose of the model. An example is the use of the model to investigate available quantities for abstraction purposes. Assuming that abstraction can only take place during periods when the water level is above a minimum environmentally acceptable flow and below a maximum water supply abstraction rate allows the definition of a specific objective function. This measure would only aggregate the residuals of the selected period and can give important information about how a model performs with respect to the anticipated task. However, it is important to mention that this should never be the only evaluation criterion.

However, how can one estimate the identifiability of the individual parameters with respect to the different OFs defined? A simple measure of parameter identifiability is defined by *Wagener et al.* [2001a]. It is based on the parameter population conditioned by the selected measure of performance (Figure 2). A uniform random sampling procedure is performed, and the resulting OF values are transformed so that the best performing parameter set is assigned the highest value and all measures sum to unity (these are termed support values in Figure 2). The best performing 10% of all parameter sets are selected and the cumulative marginal distributions for each parameter are plotted. A uniform distribution would plot as a straight line, while a population showing a clear peak will show a curved line. The stronger the conditioning, the stronger the curvature will be. The range of each parameter is subsequently split into M containers and the gradient of the cumulative distribution in each container is calculated. The highest gradient will occur where the conditioning of the distribution is strongest, i.e. at the location of a peak. The amplitude of the gradient is also indicated by the grey shading of the bar, with a darker colour indicating a higher gradient. Other measures of identifiability are possible [e.g. *Wagener et al.*, 1999], but this one has been shown to be robust and easy to calculate.

3.1.2. Multi-objective framework. The above described multi-objective performance and identifiability approaches can be put into an analytical framework to estimate the appropriate level of model complexity for a specific case [Figure 3, adapted from *Wagener et al.*, 2001a].

The hydrologist's perception of a given hydrological system strongly influences the level of conceptualisation that must be translated into the model structure. The importance of different system response modes, i.e. key processes that need to be simulated by the model, however, depends on the intended modelling purpose. Therefore, the level of model structural complexity required must be determined through careful consideration of the key processes included in the model structure and the level of prediction accuracy necessary for the intended modelling purpose.

On the other hand there is the level of structural complexity actually supported by the information contained within the observed data. It is defined here simply as the number of parameters, and therefore separate model components and processes, that can be identified. Other aspects of complexity [e.g. *Kleissen et al.*, 1990] like the number of model

states or interactions between the state variables, or the use of non-linear components instead of linear ones, are not considered here.

An increase in complexity will often increase the performance. However, it will also often increase the uncertainty, for example due to reduction in parameter identifiability caused by increased parameter interaction. What trade-off between performance and identifiability is acceptable depends on the modelling purpose and the hydrologist's preference. In a regionalisation study, a more identifiable model with reduced performance might be adequate, while parameter identifiability might be of minor importance for extension of a single-site record.

It was already established earlier in the text that such a framework has to use a multi-objective approach to allow for an objective analysis. Using various objective functions to represent different system response modes is especially suitable for comparison studies since it allows us to attribute the model performance during different system response modes to different model components, for example either the moisture accounting or the routing components [*Wagener et al.*, 2001a]. Using the segmentation approach by *Boyle et al.* [2000] as described earlier in the text, it is possible to establish that a certain model structure might perform better during "driven" periods because of a superior moisture accounting component, while another model structure containing a more appropriate slow flow routing component could result in higher performance during "non-driven slow" periods. A single-objective framework does not allow the comparison of model components and consequently important information relevant to identifying the most suitable model structure is lost. *Boyle et al.* [2001] use

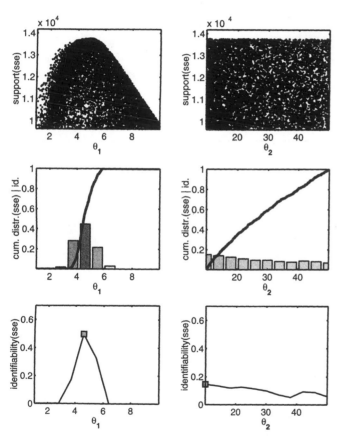

Figure 2. A measure of identifiability can be defined as follows: an initially uniform distribution is conditioned on some OF, the resulting dotty plot is shown in the top plots, selecting the top percentile (*e.g.* 10%) and plotting the cumulative distribution of the transformed measures leads to the middle plots, the gradient distribution of the cumulative distribution is a measure of identifiability, see bottom. The plots in the left column show an identifiable parameter, while the plots in the right column show a non-identifiable one.

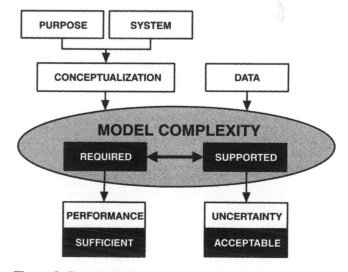

Figure 3. Framework for the evaluation of competing rainfall-runoff model structures.

this to evaluate the benefit of "spatial distribution" of model input (precipitation), structural components (soil moisture and streamflow routing computations) and surface characteristics (parameters) with respect to the reproduction of different response modes of the watershed system.

This framework will also necessarily be comparative, i.e. different models and usually different model structures will have to be compared to identify a suitable model or models. The reason is that the level of performance that can be reached is unknown, due to unknown influences of data error and of natural randomness. Those models and model structures that severely under-perform can be refuted and excluded from further consideration. In cases where all models fail, one has to go back and relax the criteria for under performance [*Beven*, 2000, p. 304].

Model structures producing more than a single output variable, e.g. groundwater levels or water quality parameters, can be tested with respect to all of those variables if measurements are available. One could say that the informative (or empirical) content of these structures is higher and they have therefore a higher degree of testability or falsifiability [*Popper*, 2000, p.113]. However, a hypothesis, or a model structure in our case, which has a higher informative content, is also logically less probable, because the more information a hypothesis contains, the more options there are for it to be false [*Popper*, 2000, p.119; *Magee*, 1977, p. 36]. Multi-output models are beyond the scope of this chapter though.

3.2. Evaluation of Individual Model Structures—Dynamic Identifiability Analysis

There is an apparent lack of objective procedures to evaluate the suitability of a specific conceptual model structure to represent a specific hydrological system. It has been shown earlier how different and competing structures can be compared. However, it is also possible to analyse individual structures with respect to the third criterion mentioned in the beginning of section 3, namely the model assumptions.

3.2.1. Failure, Inference and Improved Hypotheses. Recently, *Gupta et al.* [1998; see also *Boyle et al.*, 2000 and *Wagener et al.*, 2001a] showed how a multi-objective approach can be applied to give an indication of structural inadequacies. The assumption is that a model should be capable of representing all response modes of a hydrological system with a single parameter set. A failure to do so indicates that a specific model hypothesis is not suitable and should be rejected, or preferably, replaced by a new hypothesis which improves on the old one. This idea was already the basis of some of Beck's work [e.g. *Beck*, 1985], as described earlier in the text. *Wagener et al.* [2001c] developed a new approach based on this assumption. Their methodology analyses the identifiability of parameters within a selected model structure in a dynamic and objective manner, which can be used to analyze the consistency of locations of good performing parameter values in (parameter) space and in time.

In cases where the variation of parameter optima can be tracked in time it will sometimes be possible to directly relate changes in a particular parameter to variations in forcing or state variables [examples in *Beven*, 2000, p. 93ff.; and *Bashford and Beven*, 2000]. However, in many cases the development of improved hypotheses will be more complex and depend on the capability of the hydrologist. Unfortunately(?), there is no logical way to create new ideas; the hydrologist therefore has to apply his depth of insight and creative imagination to derive a new hypothesis, which can replace the old one, that has failed.

3.2.2. Dynamic Identifiability Analysis. The DYNamic Identifiability Analysis (DYNIA) is a new approach to locating periods of high identifiably for individual parameters and to detect failures of model structures in an objective manner. The proposed methodology draws from elements of the popular Regional Sensitivity Analysis [RSA; *Spear and Hornberger*, 1980; *Hornberger and Spear*, 1981] and includes aspects of the Generalized Likelihood Uncertainty Estimation [GLUE, *Freer et al.*, this volume] approach, wavelet analysis [e.g. *Gershenfeld*, 1999] and the use of Kalman filtering for hypothesis testing as applied by *Beck* [1985].

In the original RSA approach, a model population is sampled from a uniform distribution. This population is divided into behavioural and non-behavioural models depending on whether a model resulted in a certain response or not [*Spear and Hornberger*, 1980]. *Beven and Binley* [1992] extended the approach by conditioning the model population on a likelihood measure, which in their case, can be a transformation of any measure of performance. These are the building blocks from which a new method of assessing the identifiability of parameters is created [*Wagener et al.*, 2001c].

The steps taken in the procedure can be seen in the flow chart in Figure 4. Monte-Carlo sampling based on a uniform prior distribution is used to examine the feasible parameter space. The objective function associated with each parameter set, i.e. model, is transformed into a support measure, i.e. all support measures have the characteristic that they sum to unity and higher values indicate better performing parameter values. These are shown here in form of a dotty plot (Fig. 4(a)). The best performing parameter values (e.g. top 10 %) are selected and their cumulative distribution is calculated (Fig. 4(b)). A straight line will indicate a poorly identified parameter, i.e. the highest support values are widely distrib-

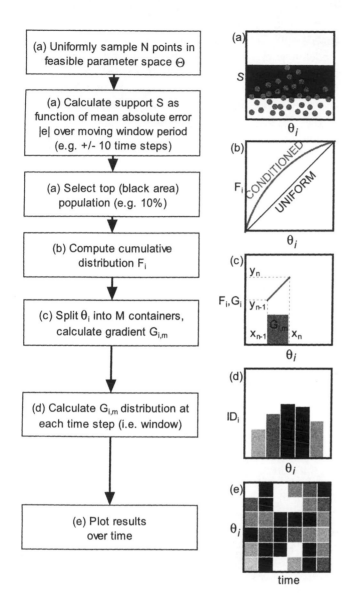

Figure 4. Schematic description of the DYNamic Identifiability Analysis (DYNIA) procedure.

uted over the feasible range. Deviations from this straight line indicate that the parameter is conditioned by the objective function used. The gradient of the cumulative support is the marginal probability distribution of the parameter, and therefore an indicator of the strength of the conditioning, and of the identifiability of the parameter. Segmenting the range of each parameter (e.g. into 20 containers) and calculating the gradient in each container leads to the (schematic) distribution shown in Fig. 4(d). The highest value, additionally indicated by the darkest colour, marks the location (within the chosen resolution) of greatest identifiability of the parameter. *Wagener et al.* [2001a] show how this measure of identifiability can be used to compare different model structures in terms of parameter uncertainty, which is

assumed to be inversely related to identifiability. They calculate the identifiability as a function of measures of performance for the whole calibration period and for specific response modes, derived using the segmentation approach by *Boyle et al.* [2000] described earlier in the text. It can be shown that the identifiability of some parameters, and therefore individual model components, is greatly enhanced by this segmentation [*Wagener et al.*, 2001a].

Calculating the parameter identifiability at every time step using only the residuals for a number of time steps n before and after the point considered, i.e. a moving window or running mean approach, allows the investigation of the identifiability as a function of time (Fig. 4(e)). The gradient distribution plotted at time step t therefore aggregates the residuals between t-n and t+n, with the window size being 2n+1. The number of time steps considered depends upon the length of the period over which the parameter is influential. For example, investigation of a slow response linear store residence time parameter requires a wider moving window than the analysis of a quick response residence time parameter. Different window sizes are commonly tested and the ones most appropriate are used to analyse individual parameters. A very small window size can lead to the result being largely influenced by errors in the data. However, this is not a problem where the data quality is very high, for example in the case of tracer experiments in rivers [*Wagener et al.*, 2001d]. Conversely, if the window size is too big, periods of noise and periods of information will be mixed and the information will be blurred.

The results are plotted for each parameter versus time using a colour coding where a darker colour indicates areas, in parameter space and time, of higher identifiability. Care has to be taken when interpreting the DYNIA results of time steps at the beginning and the end of time-series. Here the full window size cannot be established and the result is distorted. This is an effect similar to the cone of influence in wavelet analysis [*Torrence and Compo*, 1998].

While this approach is not intended to evaluate parameter dependencies in detail, the significance of dependencies to the identifiability is implicit in the univariate marginal distribution which is structurally represented by Figure 4(d). A strong dependency during any period would tend to inhibit the information of a strong univariate peak, i.e. the effect of the involved parameters cannot be singled out. Parameter interdependence can be estimated in detail by the investigation of the response surface or the variance-covariance matrix [e.g. *Wheater et al.*, 1986; *Hornberger et al.*, 1985].

A limitation of the proposed measure of identifiability arises if any near-optimal parameter values are remote from the identified peak of the marginal distribution, as the relevance of such values would be diminished. It is therefore important that a detailed investigation of the dotty plots is used to verify periods of high identifiability. The approach also requires that feasible parameter ranges are defined sensibly and the selected model population (usually the best 10%) represents only the top of the distributions.

DYNIA requires that sensible feasible ranges for each parameter can be defined and that the number of models (i.e. parameter sets) considered reflects the shape of the response surface. The procedure can then be applied to separate periods that do and those that do not contain information about specific parameters, and track parameter variations in time.

The subjective decision for a particular objective function in this procedure is usually not critical for the result and the mean absolute error criterion is usually adopted.

3.3. A Combined Framework of Corroboration and Rejection

The earlier introduced multi-objective framework [*Wagener et al.*, 2001a] can be extended to incorporate the DYNIA approach as an additional step in order to derive a framework of corroboration and rejection (Figure 5). Similar frameworks are for example proposed by *Beven* [2000, p.297ff.], and, more generally, by *Oreskes et al.* [1994].

The initial steps are similar to those in the multi-objective framework described earlier. The hydrologist selects (or develops) model structures that seem suitable for the given modeling purpose, watershed characteristics and data.

One can then apply a multi-objective procedure to establish preferences between the competing model structures, or preferably structural components. Under-performing structures (components) can be rejected at this stage, based on their performance and/or uncertainty.

During the next stage, the DYNIA approach can be used to further analyze the remaining model structures. Further rejections might be possible. The suitability of a model structure not failing is further corroborated. A model structure is (temporarily) accepted when no better performing structure can be found and no underlying assumption is violated.

In the last stage, the parameter space 'within' the remaining model structures can be analyzed to find all those models, i.e. parameter sets that are in line with the behavior of the natural system. It is very likely that such a procedure will result in a range of acceptable or 'behavioral' models or even model structures. The appropriate response is to combine the predictions of all models to derive an ensemble prediction of the systems behavior. A popular approach to do so is the GLUE approach [*Freer et al.*, this volume], however, other methods to combine the predictions of different models are possible [e.g. *Shamseldin et al.*, 1997]. Within the

GLUE approach, a likelihood value is derived for every model. The models are usually drawn from a uniform distribution. Basically any measure of performance which can be transformed so that higher values indicate better models and all measures add up to one, can be used as a likelihood measure in this approach. The likelihoods are then used to weight the prediction of every model at every time step. The cumulative distribution of the weighted streamflow values, even for different models, allows the extraction of percentiles, e.g. 5% and 95%, to derive the, in this case, 90% confidence limits for the predictions. The likelihoods of different models could be combined through simple addition.

4. APPLICATION EXAMPLE

4.1. Modelling Tools and Selected Model Structures

The Rainfall-Runoff Modelling Toolbox (RRMT) and Monte-Carlo Analysis Toolbox (MCAT), developed at Imperial College, are used here for calculation and visualisation of results [*Wagener et al.*, 1999; 2001b].

The RRMT has been developed in order to produce parsimonious, lumped model structures with a high level of

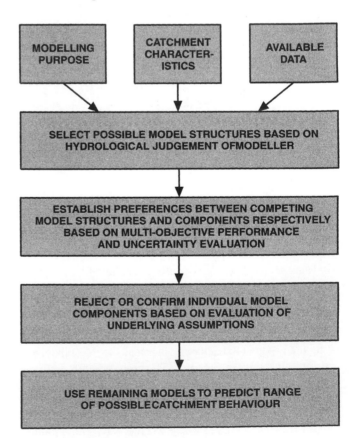

Figure 5. The proposed modeling procedure.

parameter identifiability. It is a generic modelling shell allowing its user to implement different model structures to find a suitable balance between model performance and parameter identifiability. Model structures that can be implemented are spatially lumped, relatively simple (in terms of number of parameters), and of conceptual or hybrid metric conceptual type. Hybrid metric-conceptual models utilise observations to test hypotheses about the model structure at watershed scale and therefore combine the metric and the conceptual paradigm [*Wheater et al.*, 1993]. All structures consist of a moisture accounting and a routing module.

MCAT is a collection of analysis and visualisation functions integrated through a graphical user interface. The toolbox can be used to analyse the results from Monte-Carlo parameter sampling experiments or from model optimisation methods that are based on population evolution techniques, for example, the SCE-UA [*Duan*, this volume] or the MOCOM-UA [*Gupta et al.*, this volume, "Multiple ..."] algorithms. Although this toolbox has been developed within the context of ongoing hydrological research, all functions can be used to investigate any dynamic mathematical model. Functions contained in MCAT include an extension of the Regional Sensitivity Analysis [RSA, *Spear and Hornberger*, 1980] by *Freer et al.* [1996], various components of the Generalised Likelihood Uncertainty Estimation method [GLUE, *Freer et al.*, this volume], options for the use of multiple-objectives for model assessment [*Gupta et al.*, 1998; *Boyle et al.*, 2000], and plots to analyse parameter identifiability and interaction.

Both toolboxes are implemented in the Matlab [*Mathworks*, 1996] programming environment.

A large variety of lumped parsimonious model structures can be found in the literature [e.g. *Singh*, 1995]. However, the range of components on which these structures are based is relatively small. Some of the most commonly found components are selected here in a component library shown in Figure 6. Further details about these components can be found in *Wagener et al.* [2001b; and in the references given here].

The soil moisture accounting components used are:

- The catchment moisture deficit [cmd, *Evans and Jakeman*, 1998]. A conceptual bucket with a bottom outlet to sustain drainage into the summer periods.
- The catchment wetness index [cwi, *Jakeman and Hornberger*, 1993]. A metric approach based on the Antecedent Precipitation Index [API, e.g. *Shaw*, 1994].
- The probability distributed soil moisture stores [pd3 and pd4, *Moore*, 1999]. A probability distribution of conceptual buckets based on a Pareto distribution. Evapotranspiration is either at the potential rate, as

long as soil moisture is available, or at a rate declining linearly with soil moisture content.

- A simple bucket type structure (buc), evaporating at the potential rate as long as soil moisture is available.
- The Penman storage model [*Penman*, 1949]. A layered structure of two conceptual buckets connected by an overflow mechanism. Evapotranspiration occurs at potential rate from the upper layer, similar to the root zone, and at a reduced rate, 12% of PE, from the bottom layer. An additional bypass mechanism diverts a fraction of the rainfall from the SMA component to contribute to the effective rainfall at time-steps where rainfall exceeds PE.

The routing components used are:

- Conceptual reservoirs in various combinations and in linear and non-linear form [e.g. *Wittenberg*, 1999].

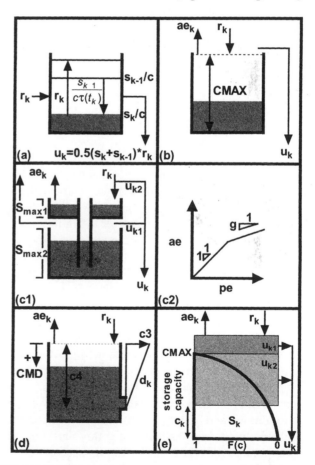

Figure 6. Table showing the soil moisture accounting 'component library' used in the application example. The components are: (a) catchment wetness index (cwi), (b) simple bucket (buc), (c1) and (c2) Penman structure (ic1), (d) catchment moisture deficit (cmd), and probability distribution of soil moisture stores (pdX).

4.2. Data

The river selected for this study is the Lower Medway at Teston (1256.1 km^2) located in South Eastern England. Six years (10/04/1990 – 14/07/1996) of data (daily naturalised flows, precipitation, potential evapotranspiration (PE) and temperature) are available. The Medway watershed is characterised by a mixture of permeable (chalk) and impermeable (clay) geologies subject to a temperate climate with an average annual rainfall of 772 mm and an average annual PE of 663 mm (1990-1996).

4.3. Methodology

Multi-objective (MO) analysis and DYNIA are performed, based on the results of Monte Carlo sampling procedures. For the MO analysis, 20000 parameter sets, i.e. models, are randomly sampled from the feasible parameter space for each individual model structure, based on a uniform distribution.

For each of these models, five OFs are calculated. These are the overall RMSE and four OFs derived for different response modes of the watershed. The segmentation applied is based on an approach by *Wagener and Wheater* [2001] which uses the slope of the hydrograph and an additional threshold as segmentation criteria to split the hydrograph into different response modes. The slope separates periods when the watershed is wetting up or is "driven" [*Boyle et al.*, this volume] by rainfall, i.e. positive slope, and when the watershed is draining, i.e. falling slope. A threshold is used to separate periods of high and low flow, i.e. the mean flow during driven and 50% of the mean flow during drainage periods. Four OFs are therefore derived when the residuals during the different periods are aggregated separately using the RMSE criterion: FDH, "driven" flow during high flow, FDL, "driven" flow during low flow, FQ, quick drainage (high flows), and FS, slow drainage (low flows). This is a modification of the initial approach by *Boyle et al.* [2000], which was based on the analysis of flow and rainfall. However, the approach presented here has been shown to be more suitable for British watersheds as modelled in the example presented here. These OFs are based on the assumption that different processes are dominant during periods of high and low flow, and during periods of watershed wetting-up and drainage. The residuals, i.e. the differences between observed and simulated flows are calculated and summarised in form of the root mean squared error for each period. The performance and identifiability analysis is based on these measures.

The resulting parameter populations are used to rank all models or model structures, with respect to their performance and identifiability, using the measures introduced ear-

lier. The best model structures are retained and a more thorough analysis using DYNIA is performed. DYNIA is based on a random sampling procedures using 2500 parameter sets collected from a uniform distribution. The smaller sample size is due to computational limitations of the current DYNIA application in the Matlab [*Mathworks*, 1996] environment.

4.4. Results and Discussion

The main results of the MO analysis as shown in Figure 7 are as follows:

- At a general level for the SMA modules (Figure 7, top): the probability distributions of storage elements (pd3 and pd4) seem to perform best, followed by the simple bucket (buc), and the cmd and cwi modules.
- The cm1, i.e. a cmd that always evaporates at the potential rate, performs much more poorly than the rest with respect to those objective functions which mainly describe periods of high flow, RMSE(total), FDH and FQ. This is also the case for the cmd module, but not as pronounced. However, the cmd and cm1 modules do very well during low flow periods. This is caused by the bottom outlet of the bucket, which sustains the production of effective rainfall even during periods of severe moisture deficits in the SMA module.
- The overall result of the performance analysis is that the pd3 and pd4 SMA modules in combination with 2pll or 2pln routing modules are superior. The cmd is a useful component when the modelling purpose demands the accurate prediction of low flow periods and periods of high flows are of minor importance.
- A detailed analysis of the routing components shows that the use of a non-linear conceptual reservoir in parallel with a linear one (2pln), performs better at the peaks (RMSE(total) and FDH), see Figure 7(top).
- The uncertainty analysis (Figure 7, bottom) however reveals that the identifiability of the cmd parameters is very low and this module is rejected here on this basis. For some applications, this aspect might be of minor importance, however.

The pd3 and the pd4 SMA components are retained for further analysis with the DYNIA approach. Assuming that our interest is in low flows, e.g. for water resources purposes, only a linear parallel routing structure (2pll) is considered. A non-linear component would be advisable for high flow periods.

The results of the DYNIA are shown in Figures 8 and 9, for the structures pd3-2pll and pd4-2pll. This reveals some problems with the pd3 SMA module.

Figure 8 shows the dynamic identifiability of the five parameters of the pd3-2pll structure. These are: (1) cmax, the

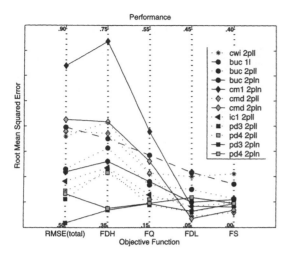

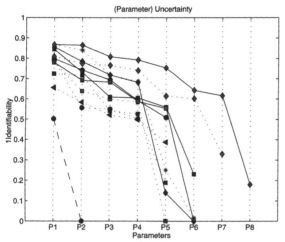

Figure 7. Results of the model structure comparison.

maximum storage capacity, (2) b, the shape parameter of the Pareto distribution of storage capacities, (3) k(quick), the residence time of the quick linear reservoir, (4) alpha, the fraction of flow going through the quick flow component, and (5) k(slow), the residence time of the slow flow linear reservoir.

The plot for the parameter cmax exposes some ambiguity about the optimum values for this parameter. The confidence limits (cfls) narrow into two different parts of the parameter space, towards low values after wet periods and towards high values during periods of wetting up, indicating inadequacies within the model structure. Similarly, but much less pronounced, the parameter b shows a slight shift of optimum after the wet period, i.e. the lower cfls go up. It is mainly identifiable during low flow events (e.g. dark areas just before time step 700). The residence times of the routing component show the expected behaviour, i.e. the cfls of k(quick) narrow down on the quick falling limbs of the hydrograph, while darker areas appear for k(slow) during the long recessions. The cfls for k(slow) hardly narrow during periods of identifiability,

suggesting that the peaks on the response surface are quite small, and that the difference between different values for this parameter is not large. Values for this parameter are therefore still widespread, since the top 10% are selected here. The example of the two residence times also demonstrates the need for different window sizes. A small size (11 time steps) is required for k(quick), whose influence is only very local, while a much larger window (81 time steps) is need to capture the effect of k(slow). Finally, the parameter alpha is most identifiable during periods where the split between quick and slow response is occurring. However, further investigations, which are outside the scope of this example, are required to explain the behaviour of this parameter. In general, this structure is too simplistic to reproduce all aspects of the hydrograph with one parameter set. This is especially reflected in the results for cmax.

The difference between pd3 and pd4 is that, while pd3 always evaporates at the potential rate, pd4 decreases the evapotranspiration with decreasing soil moisture content in a linear manner. However, without adding an additional (scaling) parameter, i.e.

$$AE_t = S_t / S_{max} \cdot PE_t \qquad (1)$$

The effect of this change can be seen in the dynamic results shown in Figure 9. The ambiguity with respect to cmax is removed and the cfls only narrow towards larger values indicating a better structure.

It is interesting to remember that the MO performance analysis had shown that the pd3 component actually performed better. The reason is that the pd4 component puts an additional constraint on the behaviour of the watershed system. The result is that the structure becomes less flexible. The pd3 component can therefore perform better with respect to the different OFs. However, this is due to the expense of a larger variation in parameter values as shown in the dynamic analysis. This indicates that pd4 is actually the better SMA component and should be retained, while pd3 should be rejected. This result supports the statement by *Gupta et al.* [2001] that consistency in a model is more important than optimality.

5. SUMMARY AND CONCLUSIONS

Test everything. Hold on to the good. Avoid every kind of evil.
1 Thessalonians 5, 21:22, New International Version

The identification of suitable conceptual rainfall-runoff (CRR) models is a difficult problem. It has been increased by the recent awareness of the influence of model structural inadequacies.

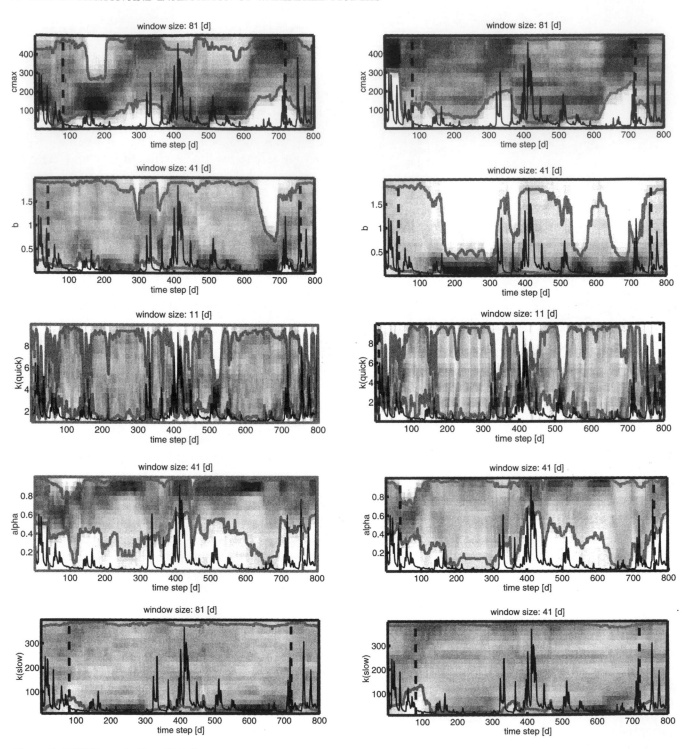

Figure 8. DYNIA results for pd3-2pll.

Figure 9. DYNIA results for pd4-2pll.

A framework of corroboration and rejection is presented to embed the identification problem into a scientific method as outlined by *Popper* [2000]. The framework uses multi-objective and novel dynamic approaches to the evaluation of CRR models and model structures. The theory and methods underlying this framework are described and an application example is presented. It demonstrates that a range of approaches is required for an objective analysis of the suitability of models and model structures.

DYNIA is an attempt to develop an approach to complement traditional calibration methods resulting in increased discriminative power. Advantages of the approach are its simplicity and its general applicability (for example, an application to a solute transport model can be found in *Wagener et al.* [2001d]). Possible areas of application of DYNIA are [see *Wagener et al.*, 2001c for details]: (1) the pure estimation of parameters, (2) the analysis of model structures, (3) relating model parameters and response modes, and (4) to investigate data outliers and anomalies.

Current work is focusing on the extension of this framework to include the identification of CRR models at ungauged sites, using parameter regionalisation approaches.

The RRMT and MCAT toolboxes are available for download free of charge for non-commercial use from the Environmental and Water Resource Engineering Section Web-site on http://ewre.cv.ic.ac.uk/software.

Acknowledgments. This project is funded by NERC under grant GR3/11653. We thank Southern Water for providing the data used in the example application. We also thank Matthew J. Lees and Neil McIntyre for constructive criticism on the presented work.

REFERENCES

Anderson, M. G., and T. P.Burt, Modelling strategies, in *Hydrological forecasting*, edited by M. G. Anderson and T. P. Burt, pp. 1-13, John Wiley and Sons, Chichester, UK, 1985.

Bard, Y., *Non-linear parameter estimation*, Academic Press, 1974.

Bashford, K., and K. J. Beven, Model structures, observational data and predictive uncertainty: explorations using a virtual reality, in Proceedings 7th National Hydrology Symposium, edited by C. Kirby, Newcastle, UK, pp. 3.31-3.37, 2000.

Beck, M. B., Structures, failure, inference and prediction, in *Identification and System Parameter Estimation*, edited by M. A. Barker and P. C. Young, Proceedings IFAC/IFORS 7th Symposium Volume 2, July 1985, York, UK, pp. 1443-1448, 1985.

Beck, M. B., Water quality modelling: a review of the analysis of uncertainty, *Water Resour. Res.*, 23, 1393-1442, 1987.

Beck, M. B., A. J. Jakeman and M. J. McAleer, Construction and evaluation of models in environmental systems, in *Modelling change in environmental systems*, edited by A. J. Jakeman, M. B. Beck and M. J. McAleer, John Wiley and Sons Ltd., USA, pp. 3-35, 1993.

Beven, K. J., Changing ideas in hydrology - The case of physically-based models, *J. Hydrol.*, 105, 157-172, 1989.

Beven, K. J., *Rainfall-runoff modelling – The primer*, John Wiley and Sons Ltd, Chichester, UK, 2000.

Beven, K. J., and A. M. Binley, The future of distributed models: Model calibration and uncertainty in prediction, *Hydrol. Proc.*, 6, 279-298, 1992.

Beven, K. J., R. Lamb, P. Quinn, R. Romanovicz, and J. Freer, Topmodel, in *Computer models of watershed hydrology*, edited by V. P. Singh, Water Resources Publishers, pp. 627-668, 1995.

Beveridge, W. I. B., *The art of scientific investigation*, 3rd edition, William Heinemann Ltd, Melbourne, Australia, 1957.

Boogard, H. F. P. van den, M. S. Ali, and A. E. Mynett, Self organising feature maps for the analysis of hydrological and ecological data sets, in *Hydroinformatics'98*, edited by V. Babovic and L. C. Larsen, Balkema, Rotterdam, NL, pp. 733-740, 1998.

Boyle, D.P., H. V. Gupta, and S. Sorooshian, Towards improved calibration of hydrologic models: Combining the strengths of manual and automatic methods, *Water Resour. Res.*, 36, 3663-3674, 2000.

Boyle, D.P., H. V. Gupta, S. Sorooshian, V. Koren, Z. Zhang, and M. Smith, Towards improved streamflow forecasts: The value of semi-distributed modelling, *Water Resour. Res.*, 37, 2739-2759, 2001.

Chappell, N.A., S.W. Franks and J. Larenus, Multi-scale permeability estimation for a tropical catchment, *Hydrological Processes*, 12, 1507-1523, 1998.

Duan, Q., V. K. Gupta, and S. Sorooshian, Effective and efficient global optimisation for conceptual rainfall-runoff models, *Water Resour. Res.*, 28, 1015-1031, 1992.

Dunne, S. M., Imposing constraints on parameter values of a conceptual hydrological model using baseflow response, *Hydrol. Earth Syst. Sci.*, 3, 271-284, 1999.

Ehrgott, M., *Multicriteria optimization*, Springer-Verlag, Berlin, Germany, 2000.

Evans, J. P., and A. J. Jakeman, Development of a simple, catchment-scale, rainfall-evapotranspiration-runoff model, *Env. Model. Softw.*, 13, 385-393, 1998.

Franks, S. W., P. Gineste, K. J. Beven, and P. Merot. On constraining the predictions of a distributed model: The incorporation of fuzzy estimates of saturated areas in the calibration process, *Water Resour. Res.*, 34, 787-797, 1998.

Freer, J., K. J. Beven, and B. Ambroise, Bayesian estimation of uncertainty in runoff prediction and the value of data: An application of the GLUE approach, *Water Resour. Res.*, 32, 2161-2173, 1996.

Gershenfeld, N., *The nature of mathematical modeling*, Cambridge University Press, Cambridge, UK, 1999.

Goldberg, D. E., *Genetic algorithms in search, optimization, and machine learning*, Addison-Wesley, USA, 1989.

Gupta, H. V., Penman Lecture, paper presented at 7th BHS National Symposium, Newcastle-upon-Tyne, UK, 2000.

Gupta, V. K., and S. Sorooshian, The relationship between data and the precision of parameter estimates of hydrologic models, *J. Hydrol.*, 81, 57-77, 1985.

Gupta, H. V., S. Sorooshian, and P. O. Yapo, Toward improved calibration of hydrologic models: Multiple and noncommensurable measures of information, *Water Resour. Res.*, 34, 751-763, 1998.

Gupta, H. V., S. Sorooshian, and D. P. Boyle, Assimilation of data for the construction and calibration of watershed models, Keynote paper presented at the International Workshop on Catchment Scale Hydrologic Modelling and Data Assimilation, Wageningen, September 2001, NL, 2001.

Harlin, J., Development of a process oriented calibration scheme for the HBV hydrological model, *Nordic Hydrol.*, 22, 15-36, 1991.

Harvey, J. W., and B. J. Wagner, Quantifying hydrologic interactions between streams and their subsurface hyporheic zones, in

Streams and ground waters, edited by J. A. Jones and P. J. Mulholland, Academic Press, San Diego, USA, pp. 3-44. 2000.

Hornberger, G. M., and R. C. Spear, An approach to the preliminary analysis of environmental systems, *J. Env. Management*, 12, 7-18, 1981.

Hornberger, G. M., K. J. Beven, B. J. Cosby, and D. E. Sappington, Shenandoah watershed study: Calibration of the topography-based, variable contributing area hydrological model to a small forested catchment, *Water Resour. Res.*, 21, 1841-1850, 1985.

Jakeman, A. J. and G. M. Hornberger, How much complexity is warranted in a rainfall-runoff model? *Water Resour. Res.*, 29, 2637-2649, 1993.

Kleissen, F. M., Uncertainty and identifiability in conceptual models of surface water acidification, Ph.D. thesis, Imperial College of Science, Technology and Medicine, London, UK, 1990.

Kleissen, F. M., M. B. Beck, and H. S. Wheater, The identifiability of conceptual hydro-chemical models, *Water Resour. Res.*, 26, 2979-2992, 1990.

Kuczera, G., and M. Mroczkowski, Assessment of hydrologic parameter uncertainty and the worth of multiresponse data, *Water Resour. Res.*, 34, 1481-1489, 1998.

Lamb, R., K. J. Beven, and S. Myrabo, Use of spatially distributed water table observations to constrain uncertainty in a rainfall-runoff model. *Adv. Water Resour.*, 22, 305-317, 1998.

Legates, D. R., and G. J. McCabe Jr., Evaluating the use of "goodness-of-fit" measures in hydrologic and hydroclimatic model validation, *Water Resour. Res.*, 35, 233-241, 1999.

Magee, B., *Popper*, 6th Impression, Fontana/Collins, Glasgow, Scotland, 1977.

Mathworks, *Matlab – reference guide*, The Mathworks Inc., Natick, M.A., 1996.

Moore, R. J., Real-time flood forecasting systems: Perspectives and prospects, in *Floods and landslides: Integrated risk assessment*, edited by R. Casale and C. Margottini, Springer, Berlin, pp. 147-189, 1999.

Mous, S. L. J., Identification of the movement of water in unsaturated soils: the problem of identifiability of the model, *J. Hydrol.*, 143, 153-167, 1993.

Mroczkowski, M., G. P. Raper, and G. Kuczera, The quest for more powerful validation of conceptual catchment models, *Water Resour. Res.*, 33, 2325-2335, 1997.

Nash, J. E., and J. V. Sutcliffe, River flow forecasting through conceptual models, I, A discussion of principles, *J. Hydrol.*, 10, 282-290, 1970.

NWS, 2001. Calibration of the Sacramento model structure, http://hsp.nws.noaa.gov/oh/hrl/calb/ workshop/parameter.htm. Accessed 15th February 2001.

Oreskes, N., K. Schrader-Frechette, and K. Belitz, Verification, validation and confirmation of numerical models in the earth sciences, *Sci.*, 263, 641-646, 1994.

Penman, H. L., The dependence of transpiration on weather and soil conditions, *J. Soil Sci.*, 1, 74-89, 1949.

Piñol, J., K. J. Beven, and J. Freer, Modelling the hydrological response of Mediterranean catchments, Prades, Catalonia. The use of distributed models as aid to hypothesis formulation, *Hydrol. Proc.*, 11, 1287-1306, 1997.

Popper, K., *The logic of scientific discovery*, first published 1959 by Hutchinson Education, Routledge, UK, 2000.

Seibert, J., Multi-criteria calibration of a conceptual runoff model using a genetic algorithm, *Hydrol. Earth Syst. Sci.*, 4, 215-224, 2000.

Shamseldin, A. Y., K. M. O'Connor, and G. C. Liang, Methods of combining the outputs of different rainfall-runoff models, *J. Hydrol.*, 197, 203-229, 1997.

Shaw, E. M., *Hydrology in practice*, 3rd Edition, Chapman and Hall, London, UK, 1994.

Singh, V. P., *Computer models of watershed hydrology*, Water Resources Publishers, USA, 1995.

Sorooshian, S., and V. K. Gupta, The analysis of structural identifiability: Theory and application to conceptual rainfall-runoff models, *Water Resour. Res.*, 21, 487-495, 1985.

Spear, R. C., and G. M. Hornberger, Eutrophication in Peel Inlet, II, Identification of critical uncertainties via generalized sensitivity analysis, *Water Res.*, 14, 43-49, 1980.

Stigter, J. D., M. B. Beck, and R. J. Gilbert, Identification of model structure for photosynthesis and respiration of algal populations, *Water Sci. Technol.*, 36, 35-42, 1997.

Torrence, C., and G. P. Compo, A practical guide to wavelet analysis, *Bull. American Meteorol. Soc.*, 79, 61-78, 1998.

Uhlenbrock, S., J. Seibert, C. Leibundgut, and A. Rohde, Prediction uncertainty of conceptual rainfall-runoff models caused by problems in identifying model parameters and structure, *Hydrol. Sci. J.*, 44, 779-797, 1999.

Van Straten, G., and K. J. Keesman, Uncertainty propagation and speculation in projective forecasts of environmental change: a lake-eutrophication example, *J. Forec.*, 10, 163-190, 1991.

Wagener, T. and H. S. Wheater, On the evaluation of conceptual rainfall-runoff models using multiple-objectives and dynamic identifiability analysis, in *Continuous river flow simulation: Methods, applications and uncertainties*, edited by I. Littlewood, and J. Griffin, British Hydrological Society - Occasional Papers, Wallingford, UK, 45-51, 2001.

Wagener, T., M. J. Lees, and H. S. Wheater, A generic rainfall-runoff modelling toolbox, *Eos Trans. AGU*, 80, Fall Meet. Suppl., F203, 1999.

Wagener, T., D. P. Boyle, M. J. Lees, H. S. Wheater, H. V. Gupta, and S. Sorooshian, A framework for the development and application of hydrological models, *Hydrol. Earth Syst. Sci.*, 5(1), 13-26, 2001a.

Wagener, T., M. J. Lees, and H. S. Wheater, A toolkit for the development and application of parsimonious hydrological models, in *Mathematical models of large watershed hydrology – Volume 1*, edited by V. P. Singh and D. Frevert, Water Resources Publishers, USA, pp. 87-136, 2001b.

Wagener, T., N. McIntyre, M. J. Lees, H. S. Wheater, and H. V. Gupta, Towards reduced uncertainty in conceptual rainfall-runoff modelling: Dynamic identifiability analysis. *Hydrol. Proc.*, in press, 2001c.

Wagener, T., L. A. Camacho, M. J. Lees, and H. S. Wheater, Dynamic parameter identifiability of a solute transport model, in *Advances in design sciences and technology*, edited by R. Beheshti, europia, Delft, April 2001, NL, pp. 251-264, 2001d.

Wagner, B. J., and J. W. Harvey, Experimental design for estimating parameters of rate-limited mass transfer: Analysis of stream tracer studies, *Water Resour. Res.*, 33, 1731-1741, 1997.

Wheater, H. S., K. H. Bishop, and M. B. Beck, The identification of conceptual hydrological models for surface water acidification, *Hydrol. Proc.*, 1, 89-109, 1986.

Wheater, H. S., A. J. Jakeman, and K. J. Beven, Progress and directions in rainfall-runoff modelling, in *Modelling change in environmental systems*, edited by A. J. Jakeman, M. B. Beck and M. J. McAleer, Wiley, Chichester, UK, pp. 101-132, 1993.

Wittenberg, H., Baseflow recession and recharge as non-linear storage processes, *Hydrol. Proc.*, 13, 715-726, 1999.

Yan, J., and C. T. Haan, Multiobjective parameter estimation for hydrologic models – Weighting of errors, *Trans. Am. Soc. Agric. Eng.*, 34(1), 135-141, 1991.

Yapo, P. O., H. V. Gupta, and S. Sorooshian, Automatic calibration of conceptual rainfall-runoff models: Sensitivity to calibration data, *J. Hydrol.*, 18, 23-48, 1996.

Young, P. C., S. Parkinson, and M. J. Lees, Simplicity out of complexity in environmental modeling: Occam's razor revisited, *J. Appl. Stat.*, 23, 165-210, 1996.

Hoshin V. Gupta. SAHRA, NSF STC for the Sustainability of semi-Arid Hydrology and Riparian Areas, Department of Hydrology and Water Resources, Harshbarger Bldg. 11, University of Arizona, Tucson, AZ 85721, USA
(hosh_stc@sahra.arizona.edu)

Thorsten Wagener, now at: SAHRA, NSF STC for the Sustainability of semi-Arid Hydrology and Riparian Areas, Department of Hydrology and Water Resources, Harshbarger Bldg., University of Arizona, Tucson, AZ 85721, USA

Howard S. Wheater, Department of Civil and Environmental Engineering, Imperial College of Science, Technology and Medicine, Imperial College Road, SW7 2BU, London, UK
(t.wagener@ic.ac.uk)
(h.wheater@ic.ac.uk)

Confronting Input Uncertainty in Environmental Modelling

Dmitri Kavetski, Stewart W. Franks, and George Kuczera

School of Engineering, University of Newcastle, Callaghan, New South Wales, Australia

The majority of environmental models require calibration of some or all of their parameters before meaningful predictions of catchment behaviour can be made. Despite the importance of reliable parameter estimates, there are growing concerns about the ability of objective-based inference methods to adequately calibrate environmental models. The problem lies with the formulation of the objective or likelihood function, which is currently implemented using essentially ad-hoc methods. We outline limitations of current calibration methodologies, including least squares, multi-objective, GLUE and Kalman filter schemes and introduce a more systematic Bayesian Total Error Analysis (BATEA) framework for environmental model calibration and validation. BATEA imposes a hitherto missing rigour in environmental modelling by requiring the specification of physically realistic uncertainty models with explicit assumptions that can and must be tested against available evidence. Distinguishing between the various sources of errors will reduce the current ambiguity about parameter and predictive uncertainty and enable rational testing of environmental model hypotheses. A synthetic study demonstrates that explicitly accounting for forcing errors leads to immediate advantages over traditional least squares methods that ignore rainfall history corruption and do not directly address the sources of uncertainty in the calibration. We expect that confronting all sources of uncertainty, including data and model errors, will force fundamental shifts in the model calibration/verification philosophy.

INTRODUCTION

Hydrological and environmental modelling has benefited from significant developments over the last decade. The growing understanding of environmental physics, combined with dramatic increases in computing power, has allowed progressively more realistic representation and simulation of catchment dynamics, in many cases solving hitherto intractable analysis and prediction problems.

Paradoxically, these advances have increased the need for improved model calibration and validation methods. Hydrological models invariably require calibration before

predictive use to ensure consistency with observed data. In recognition of the limitations of visual and manual model calibration, several methodologies have been developed for automatic calibration of hydrological models, including classical Bayesian methods (e.g., NLFIT of Kuczera [1994]); multi-objective calibration methods (e.g., MOCOM of Sorooshian et al. [1993] and Gupta et al. [1998]); the GLUE framework of Beven and Binley [1992]; and, less commonly, Kalman filters [*Bras and Rodriguez-Iturbe*, 1985]. As we shall see, these methodologies struggle to properly characterise the fundamental problem of calibration and validation, let alone cope with the ever-increasing number of competing environmental models (see, e.g., Beven and Binley [1992] and Singh [1995]). Due to a lack of widely accepted methods for addressing data uncertainty and model verification (i.e., hypothesis testing) in hydrology, it is difficult to rationally discriminate between competing models and assess trade-offs between model performance and complexity.

Calibration of Watershed Models
Water Science and Application Volume 6
Copyright 2003 by the American Geophysical Union
10/1029/006WS04

This chapter focuses on the calibration paradigms currently in use and on the paradigms that need to be developed in the future. We examine sources of uncertainty in hydrological modelling, survey current calibration methods and then outline a new systematic methodology for hydrological model calibration and validation. We expect that the ideas and concepts presented here are also applicable to other branches of environmental science and engineering, where observations of global system behaviour are inexact and are made on a local scale, and where the system itself is so complicated that its simulation requires considerable approximation.

MATHEMATICS OF ENVIRONMENTAL MODELS

Regardless of whether an environmental model M is conceptual or physical, it has the following functional form

$$\mathbf{Y} \leftarrow f_M(\mathbf{X}, \theta) \tag{1}$$

The term $\mathbf{Y} = \{\mathbf{y}_n; n=1...N\}$ is the response matrix of the catchment. It contains one or more directly observable hydrological quantities at one or more locations within the catchment at a series of times $t = \{t_n; n=1...N\}^T$. In the simplest and most common context, $\mathbf{y}_n$ is a vector of stream flows at several locations within the catchment. However, the definition of $\mathbf{y}_n$ is more general, e.g., it may also contain water table depths at selected locations, saturated areas, etc.

The catchment responds to forcing inputs denoted by the matrix $\mathbf{X} = \{\mathbf{x}_n; n=1...N_x\}$, where the vector $\mathbf{x}_n$ contains one or more directly observable quantities at a series of times. The forcing vector typically comprises rainfall, but evapotranspiration, pumping and injection data can also be included. The dimension N_x need not equal N, but, for time stepping rainfall-runoff models, often $N_x = N$.

The function $f_M(\cdot)$ represents the hydrological model itself and describes the response $\mathbf{Y}$ of the system to the inputs $\mathbf{X}$, e.g., the routing of rainfall into streamflow. The vector $\theta = \{\theta_n; n=1...P\}^T$ contains the conceptual and physical hydrological parameters of the catchment model. Parameters are constants that quantify the hydrological behaviour of the catchment (given a particular mathematical model) and determine the response $\mathbf{Y}$ for a given forcing $\mathbf{X}$. We identify "physical" parameters as those parameters that can be inferred using procedures that are independent of observable catchment responses $\mathbf{Y}$, e.g., local permeability estimates obtained using core samples or slug tests. Conversely, "conceptual" parameters (e.g., discharge coefficients) have no formal physical interpretation and can only be inferred by matching the simulated catchment behaviour $f_M(\tilde{\mathbf{X}}, \theta)$ to the

observed data $\{\tilde{\mathbf{X}}, \tilde{\mathbf{Y}}\}$. The tilde over $\mathbf{X}$ and $\mathbf{Y}$ emphasises that these quantities are estimated and hence subject to sampling and measurement error. The procedure of matching observed data and simulated system behaviour by adjusting the parameters θ is termed calibration and forms the basis for model validation and predictive use in hydrology.

Catchment models are commonly classified into conceptual or physical models. While this distinction is valuable for many purposes, the functional behaviour of successful hydrological models is mathematically similar, since they all simulate the same physical phenomenon (e.g., rainfall-runoff routing). In particular, all hydrological models include quickflow and slowflow simulators. The quickflow is invariably related to the rainfall in the immediately preceding time steps, while slowflow behaviour obeys storage-discharge relationships. The importance of the differentiation of slowflow and quickflow processes will become apparent when the error propagation properties of hydrological models and their impact on parameter estimation are considered.

UNCERTAINTY IN HYDROLOGICAL MODELLING

Although at first glance a simple exercise in optimisation, the calibration of hydrological models is nontrivial and subtle. Most hydrological models are nonlinear and contain multiple parameters. Reliable multi-dimensional nonlinear optimisation is challenging, since it is usually prohibitively difficult to exhaustively analyse the entire parameter space.

Considerable research has been dedicated to the development of robust optimisation methods (stochastic and deterministic, local and global search algorithms). One popular optimisation method in hydrology is the Shuffled Complex Evolution (SCE) algorithm, which makes use of a population of simplexes, the vertices of which are shuffled to improve (although not guarantee) its global convergence properties [Duan et al., 1992].

However, the calibration of hydrological models is profoundly affected by sources of uncertainty completely unrelated to the numerical difficulties of multi-dimensional optimisation:

1. Uncertainty in observed system inputs and responses. For example, rain gauges offer only point estimates of precipitation, while the rating curves used to estimate streamflow are also inexact, particularly when the ratings are extended beyond the data. Observational uncertainty can be further split into two categories: a) uncertainty in forcing inputs (e.g., rainfall, evapotranspiration); and b) uncertainty in output responses (e.g.,

streamflow, piezometer responses). The convenience of separating observation error into two categories is due to the causal structure of environmental models and will become clear later, when parameter estimation methods that take account for uncertainty in observations are considered.

2. Inherent uncertainty in the model hypothesis. Indeed, even the most elaborate model is at best a simplification of the natural environment. Although most models are based on valid physical principles (typically derived at the laboratory scale), they nevertheless remain simplifications of reality, particularly if the grid scale is orders of magnitude larger than the laboratory scale.

Figure 1 shows a schematic of the propagation of errors through environmental models. In general the observations of external forcing (inputs) $\tilde{\mathbf{X}}$, are corrupted by measurement and sampling error, which propagates through the calibrated catchment model to corrupt the simulated responses (output) $\tilde{\mathbf{Y}}$. In addition, $\tilde{\mathbf{Y}}$ will be affected by model and response sampling error.

It is our view that these sources of uncertainty and their propagation characteristics are currently overlooked or misunderstood and a rigorous modelling framework is necessary to provide:

a)   An ability to meaningfully account for observational uncertainty and model errors; and

b)   Parameter estimates with realistic confidence limits, which can then be used for prediction with meaningful uncertainty bounds;

Accurate parameter inference is necessary for meaningful prediction of flows and parameter regionalisation (allowing transfer of parameters from gauged to ungauged catch-

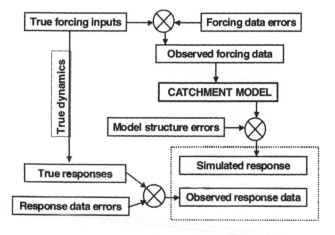

Figure 1. Error propagation in catchment modelling.

ments). We shall see that common calibration methods introduce un-predictable bias into the parameters estimates, confounding regionalisation attempts.

Moreover, it is perhaps ironic that, despite extensive research of environmental physics, relatively little is known about the uncertainty operating in environmental modelling. As a consequence, we often naively combine sophisticated environmental models with simplistic Gaussian error models. Such mismatch weakens the entire modelling process and undermines the validity of its predictions.

In the following sections the significance of data uncertainty in hydrological modelling is examined, with specific references to existing calibration techniques. We then articulate a conceptual framework for model calibration that addresses the two aims described above.

CALIBRATION: A CRITIQUE OF CURRENT PARADIGMS

The calibration of conceptual parameters (e.g., discharge coefficients) is required since, by definition, these parameters cannot be independently measured. In practice, despite advances in instruments and measuring devices, even physical parameters (e.g., soil permeability) often require calibration. The heterogeneity of environmental systems makes even the most accurate probe measurement only a point estimate, perhaps correct for a particular location, but often invalid as a representative average over the model grid cell, let alone the entire domain.

The difference between parameters estimated using various methods is often considerable. For example, Chappell *et al.* [1998] found orders of magnitude differences between permeability estimates based on core samples, inverted model and hillslope transect estimates. In the light of these differences, calibration is an important model verification procedure, since parameter estimates are site-specific and area-effective at the catchment scale. If a model cannot reproduce observed events, it is unlikely to provide accurate predictions of future events.

Least Squares Calibration

Traditionally, calibration is posed as an optimisation problem: obtain estimates of system parameters that, given the model at hand, achieve the best agreement between simulated and observed responses. Many calibration methods have been proposed and used, ranging from visual assessment of hydrographs to sophisticated search algorithms. Before providing a critique of current calibration paradigms, we consider the following case study [*Kavetski et al.*, 2000] based on the well-known hydrological model

TOPMODEL [*Beven and Kirkby*, 1979; *Beven et al.*, 1995].

Figure 2 shows a portion of observed and simulated hydrographs (at this point we do not disclose the origin of the data – the reason for this will become apparent in a moment) obtained by calibrating TOPMODEL using the standard least squares (SLS) objective function.

$$S_{SLS}(\theta) = \sum_{n=1}^{N}\left(\tilde{y}_n - f_n\left(\tilde{\mathbf{X}},\theta\right)\right)^2 \qquad (2)$$

where θ are the parameters, $\tilde{y}_n$ is the observed streamflow at time t_n, $\tilde{\mathbf{X}}$ is the observed rainfall time series and $f(\bullet)$ denotes TOPMODEL. We used several years of data, perhaps more than normally employed in such calibrations, to ensure statistical averaging of parameter estimates.

The calibrated parameters are shown in Table 1. They have been obtained using the SCE algorithm with a tight convergence tolerance. The standard deviations have been estimated using the inverse negative Hessian matrix at the maximum of the objective function. The negative inverse Hessian matrix at the mode of probability density functions (pdfs) converges to the covariance matrix as the probability distribution converges to the Gaussian form [*Gelman et al.*, 1997]. As a large data set has been used, the distribution is sufficiently Gaussian.

Visually, the calibration is successful – the fit is generally good in all areas, the coefficient of determination R^2 of the observed vs predicted responses is close to unity – what else could we hope for? In practical hydrology, far less appealing calibrations are routinely accepted.

Appearances are deceptive, however, and it is now time to reveal the origin of the data. In fact, the streamflow was generated synthetically using TOPMODEL with a pre-determined set of parameters and no streamflow error. The set of "exact" parameters is listed in Table 2, which shows that the fitted parameters are actually quite far from the true values and, more disconcertingly, the uncertainty estimates significantly underestimate the actual errors. So what went wrong?

In fact, we have corrupted the observed rainfall depth of each storm used in the calibration using log-normally distributed error multipliers with an expected value of 1 and a coefficient of variation of 0.2. Note that the mean of the multipliers is 1, i.e., the rainfall estimates are on average unbiased. So why have the parameter estimates become

Table 1. – SLS calibration of TOPMODEL.

Parameter	SLS value ± St.deviation
m-exponent, m	$0.00746 \pm 0.12 \times 10^{-3}$
Log-Transmissivity, ln(m)	$0.879 \quad \pm 0.122$
Root zone storage, m	$0.0555 \pm 0.18 \times 10^{-2}$
Stream velocity, m/hr	$2840 \quad \pm 40.9$

Table 2. – True TOPMODEL parameters and SLS errors, reported as $\mathbf{e}=(\theta_{true}-\theta_{SLS})/\sigma_{SLS}$.

Parameter	True value	SLS Error e
m-exponent, m	0.016	71.2
Log-Transmissivity, ln(m)	1.0	0.99
Root zone storage, m	0.1	24.7
Stream velocity, m/hr	3000	3.94

biased? Could it be because the asymmetric lognormal distribution was used? It turns out that the symmetry of the error corruption is not at fault – using symmetrically distributed multipliers does not remove the bias in parameter estimates. The real culprit is the inadequate error model underlying the standard least squares objective function (2).

Indeed, although least squares fitting dates back to Gauss (who used the method to calculate solar system orbits), it is often misunderstood that the "objectiveness" of the criterion of the sum of squared residuals is not guaranteed. Instead, it corresponds to an assumption that the response series contains additive uncorrelated Gaussian noise,

$$\tilde{\mathbf{Y}} = f\left(\tilde{\mathbf{X}},\theta\right) + \varepsilon \qquad (3)$$

$$\varepsilon \sim N\left(0,\sigma_y^2\right) \qquad (4)$$

where σ_y^2 is a scalar covariance. Bayesian and maximum-likelihood estimators then obtain the most probable value of θ by minimising the sum of squared residuals with respect to θ [*Box and Tiao*, 1973]. Note that the simulated results are obtained using the observed forcing series $\tilde{\mathbf{X}}$.

4.2. Introducing Total Least Squares Calibration

It is not widely recognised that, if error is contained in both the x and y series, the fitting procedure must be modi-

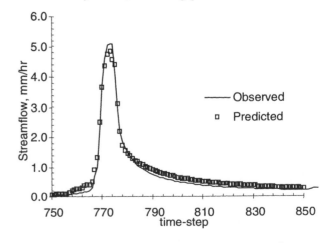

Figure 2. Portion of calibrated TOPMODEL hydrograph.

fied. Fitting functions with errors in both x and y data is not a new research area and dates back decades (see, e.g., Deming [1943] and Macdonald and Thompson [1992]). Indeed, Jefferys [1980] shows that, even when fitting a straight line through the origin, the use of the SLS criterion (2) when both **X** and **Y** series are corrupt yields biased estimates of the slope coefficient:

$$E[b_{SLS}] = b\left(1 - \left(\frac{\sigma_x}{\bar{x}}\right)^2\right) \qquad (5)$$

where $E[\cdot]$ is the expectation, b_{SLS} is the SLS estimate of the true slope b, σ_x is the standard deviation of errors in the x data and $\bar{x}$ is the mean of the observations. The parameter bias is proportional to the error in x.

More importantly, however, equation (5) shows that the bias is independent of the quantity N of data used and, since $\text{var}[b_{SLS}] \sim O(N^{-1})$, the SLS method yields progressively more misleading parameter estimates as more data is included in the analysis. This somewhat surprising result is a serious shortcoming of the SLS method and can be removed by adjusting the objective function, now termed the total least squares (TLS) function

$$S_{TLS}(\boldsymbol{\theta}) = \sum_{n=1}^{N}\left[m_n\left(y_n - bx'_n\right)^2 + \left(x'_n - \tilde{x}_n\right)^2\right] \qquad (6)$$

where x'_n is an estimate of the true value x_n and m_n is a weighting factor dependent on the relative uncertainty in x and y data (typically, $m_n = \sigma^2_{x(n)}/\sigma^2_{y(n)}$). The minimisation of (6) lies is at the heart of TLS methods, also referred to as the "Error in Variables Method" (EVM). The variance of the TLS estimates decays at the same asymptotic rate as the SLS variance, $\text{var}[b_{TLS}] \sim O(N^{-1})$, but unlike the SLS parameter estimates, the TLS estimates converge on the true values, i.e., $E[b_{TLS}] = b$.

Although (6) remains a sum of squared residuals, it is fundamentally different from (2), as it contains N additional unknowns $\mathbf{X}' = \{x'_n; n = 1...N\}^T$, which are referred to as hidden or nuisance variables [*Gull*, 1989], latent variables, or incidental variables [*Zellner*, 1971]. Due to the form of the model $f(\bullet)$, the true values of x are necessary before the simulated values of y can be computed. This requirement has significant implications for the calibration algorithm, increasing the dimensionality of the problem.

We stress that the parameter bias (5) has been derived for the simplest 1-parameter linear model. Although analytical results are unavailable, it is unduly optimistic to expect that, when the model is nonlinear and contains many parameters, the bias in parameter estimates will disappear or decrease – indeed, the opposite is far more likely.

Structure of Hydrologic Models: The Slowflow Blues

Another factor that affects parameter inference is the structure of hydrological models – they represent physical systems with storage components. Baseflow Q_b (slow release of water from storage S) is governed by first-order ODEs, with solutions of a recursive auto-correlated form.

$$Q_b = f(S) \qquad (7)$$

and hence

$$\frac{dS}{dt} = f(S) \qquad (8)$$

A discretised solution of (8) takes the following form

$$Q_b(t_n) = f\left[Q_b(t_{n-1})\right] \qquad (9)$$

All time stepping hydrological models implement some variation of (9) as they step through time from t_{n-1} to t_n. A first-order Taylor series approximation suggests that, due to the recursive form of (9), a perturbation of baseflow at t_n by ε_n will lead, after K time steps Δt, to an accumulated error ε_{n+k} of the form

$$\varepsilon_{n+K} = \varepsilon_n \prod_{k=1}^{K}\left(\frac{df}{dQ_b}\right)^{n+k} + O\left(\varepsilon_n^2\right) \qquad (10)$$

where $n + k$ denotes $t_n + k\Delta t$. The perturbation ε_n could arise, for example, due to observation error in the rainfall, which would propagate into the storage and hence alter the initial conditions for (10). Mathematically, the auto-correlation (10) vanishes at stationary points. In a physical context, this corresponds to the catchment drying out, when the storage is at a minimum and is, in a sense, "reset".

The auto-correlation in the residual error series violates the independence assumption of least squares schemes, decreases the informational content of the data accessible to SLS calibration and leads to additional distortion of the inferred parameters by portions of data dominated by baseflow. In contrast, quickflow processes, which rapidly respond to rainfall, induce little, if any, auto-correlation in the error series. The implication of (10) is that calibration of hydrological phenomena at the time scale of quickflow (within the storm) will not satisfy the independence assumption of least squares methods (2). It is emphasised that auto-correlated residuals present an additional problem (in addition to input errors) – even if there was no auto-correlation in the response series, the presence of input errors would lead to parameter bias similar to (5).

Finally, a third factor must be considered when calibrating environmental models – the environment is not (always) Gaussian. Although central limit theorems indicate convergence of arbitrary distributions to the Gaussian pdf, they remain limit theorems that may have limited applicability in realistic environmental modelling. Due to their relation to Gaussian error models (quadratic forms in (6)), even TLS methods cannot account for this problem. What is needed, therefore, is a general and systematic calibration framework for dealing with various sources of error in modelling that allows scrutiny of its various underlying assumptions. Although we cannot hope to develop an assumption-free framework for system analysis, we can make these assumptions maximally evident.

BAYESIAN METHODS IN HYDROLOGY

We have alluded to statistical methods when discussing least squares methods. Although calibration can be viewed as an optimisation problem, the interpretation of calibration as statistical analysis is arguably more useful, since, as well as identifying the most likely parameters, it is essential to assess the uncertainty associated with these estimates. Error estimates are not easily available within the optimisation paradigm, yet are a natural product of proper statistical inference. In addition, Bayesian statistical analysis allows the combination of both quantitative and qualitative information, i.e., combine rainfall-runoff data with the intuitive knowledge and experience of hydrological practitioners (via prior distributions). Finally, there is a wealth of applications of Bayesian methods in areas ranging from artificial intelligence to pattern recognition and some techniques could migrate to hydrologic analysis.

When applied to rainfall-runoff modelling, Bayes equation yields the posterior pdf $p(\theta \mid \tilde{\mathbf{X}}, \tilde{\mathbf{Y}})$ of the model parameters, conditioned on the observed data

$$p\left(\theta \mid \tilde{\mathbf{X}}, \tilde{\mathbf{Y}}\right) = \frac{p\left(\tilde{\mathbf{X}}, \tilde{\mathbf{Y}} \mid \theta\right) p\left(\theta\right)}{p\left(\tilde{\mathbf{X}}, \tilde{\mathbf{Y}}\right)} \qquad (11)$$

where the denominator $p(\tilde{\mathbf{X}}, \tilde{\mathbf{Y}})$ ensures that the pdf integrates to 1. Typically, it is not necessary to explicitly evaluate $p(\tilde{\mathbf{X}}, \tilde{\mathbf{Y}})$ and any factors independent of θ in the likelihood function $p(\tilde{\mathbf{X}}, \tilde{\mathbf{Y}} \mid \theta)$ can be absorbed into the proportionality constant, giving

$$p\left(\theta \mid \tilde{\mathbf{X}}, \tilde{\mathbf{Y}}\right) \propto L\left(\tilde{\mathbf{X}}, \tilde{\mathbf{Y}} \mid \theta\right) p\left(\theta\right) \qquad (12)$$

or, if a non-informative prior is imposed

$$p\left(\theta \mid \tilde{\mathbf{X}}, \tilde{\mathbf{Y}}\right) \propto L\left(\tilde{\mathbf{X}}, \tilde{\mathbf{Y}} \mid \theta\right) \qquad (13)$$

Although the prior pdf has an important role in classic Bayesian analysis, hydrological applications tend to use non-informative priors, with the justification that this "lets the data speak for itself". Since prior knowledge of model parameters is usually case-specific, we will also use uniform priors, although nothing in the analysis precludes the use of informative prior distributions on θ.

In equations (12) and (13), $L(\cdot)$ is a likelihood function. This function must contain all functional dependencies of $p(\tilde{\mathbf{X}}, \tilde{\mathbf{Y}} \mid \theta)$ on θ. It represents the likelihood of observing the data $\{\tilde{\mathbf{X}}, \tilde{\mathbf{Y}}\}$ given the model parameters θ and the model hypothesis. The form of $L(\cdot)$ must reflect the way error and uncertainty enter and propagate through the system. It is not necessary for $L(\cdot)$ to be a proper pdf and it does not have to integrate to unity – any constant factor can be absorbed into the proportionality relation in (12).

BAYESIAN ANALYSIS OF DATA UNCERTAINTY

Traditional Regression Methods: Hiding the Dirac Delta

Standard Bayesian regression ignores input uncertainty and lumps observed response error and model error into a single white noise term ε according to eqns (3) and (4). The likelihood function is then

$$L_{SLS}\left(\theta, \sigma_y^2\right) = \frac{1}{\sigma_y^N} \exp\left(-\frac{1}{2\sigma_y^2} \sum_{n=1}^{N}\left[\tilde{y}_n - f_n\left(\tilde{\mathbf{X}}, \theta\right)\right]^2\right) \qquad (14)$$

Assuming the non-informative invariant Jeffrey's prior $p(\sigma_y^2) \propto 1/\sigma_y^2$ and integrating σ_y^2 out of the posterior density yields the posterior pdf for the system parameters θ [Box and Tiao, 1973]

$$p_{SLS}\left(\theta \mid \tilde{\mathbf{X}}, \tilde{\mathbf{Y}}\right) \propto \left(\sum_{n=1}^{N}\left[\tilde{y}_n - f_n\left(\tilde{\mathbf{X}}, \theta\right)\right]^2\right)^{-\left(\frac{N-1}{2}\right)} \qquad (15)$$

In practice, it is easier to maximize the logarithm of the posterior density, i.,e,

$$\log p_{SLS}\left(\theta \mid \tilde{\mathbf{X}}, \tilde{\mathbf{Y}}\right) \propto -\left(\frac{N-1}{2}\right) \log\left(\sum_{n=1}^{N}\left[\tilde{y}_n - f_n\left(\tilde{\mathbf{X}}, \theta\right)\right]^2\right) \qquad (16)$$

The variance of Y errors can then be estimated using

$$\sigma_y^2 \approx \frac{1}{N} \sum_{n=1}^{N}\left[\tilde{y}_n - f_n\left(\tilde{\mathbf{X}}, \hat{\theta}\right)\right]^2 \qquad (17)$$

where $\hat{\theta}$ is the most probable parameter set.

Since the extrema of the posterior pdf (16) are identical to those of the SLS objective function (2), maximising (16) is equivalent to minimising the sum of squared errors. In fact, the input error model hidden in the classic least squares regression is the Dirac delta function $\delta(\cdot)$, which assigns all the probability mass to the observations $\tilde{\mathbf{X}}$:

$$p\left(\mathbf{X}\mid\tilde{\mathbf{X}}\right)=p\left(\tilde{\mathbf{X}}\mid\mathbf{X}\right)=\delta\left(\mathbf{X}-\tilde{\mathbf{X}}\right)=\begin{cases}\infty & \mathbf{X}=\tilde{\mathbf{X}} \\ 0 & \mathbf{X}\neq\tilde{\mathbf{X}}\end{cases} \quad (18)$$

or, in other words, $\mathbf{X}=\tilde{\mathbf{X}}$. Indeed, assuming that forcing uncertainty is statistically independent from the hydrological model parameters, i.e., $p(\tilde{\mathbf{X}}\mid\theta)=p(\tilde{\mathbf{X}})$, substituting the Dirac function (18) and integrating over the support $\Omega(\mathbf{X})$ of $\mathbf{X}$ yields

$$\begin{aligned} p\left(\tilde{\mathbf{X}},\tilde{\mathbf{Y}}\mid\theta\right) &= p\left(\tilde{\mathbf{Y}}\mid\tilde{\mathbf{X}},\theta\right)p\left(\tilde{\mathbf{X}}\right) \\ &= C\int_{\Omega(\mathbf{X})}p\left(\tilde{\mathbf{Y}}\mid\mathbf{X},\theta\right)p\left(\mathbf{X}\mid\tilde{\mathbf{X}},\theta\right)d\mathbf{X} \\ &= C\int_{\Omega(\mathbf{X})}p\left(\tilde{\mathbf{Y}}\mid\mathbf{X},\theta\right)\delta\left(\mathbf{X}-\tilde{\mathbf{X}}\right)d\mathbf{X} \\ &= Cp\left(\tilde{\mathbf{Y}}\mid\tilde{\mathbf{X}},\theta\right)=Cp\left(\varepsilon\mid\tilde{\mathbf{X}},\theta\right) \end{aligned} \quad (19)$$

where the fourth line follows from the properties of the Dirac function and $C=p(\mathbf{X})$ is independent of θ.

Since the response noise ε is assumed to be normally distributed according to (4), the likelihood function simplifies to (14). It follows that, since Bayesian SLS regression disregards input uncertainty via (18), it inherits all the weaknesses of SLS optimisation, in particular, parameter bias when error is present in both $\mathbf{X}$ and $\mathbf{Y}$ data, as well as non-robustness with respect to non-Gaussian errors. To a certain extent, however, the auto-correlation in the residuals can be addressed using AR methods.

Auto-Regressive (AR and ARMA) Methods

Auto-regressive (AR) and the more general ARMA models attempt to remove the auto-correlation structure from the residuals and reduce the latter to white Gaussian noise $\phi_n \sim N(0,\sigma_{AR}^2)$ by introducing additional parameters, e.g., for the AR1 model, $\varepsilon_n=a_1\varepsilon_{n-1}+\phi_n$. K-order generalisations are readily obtained, e.g., the AR-K model

$$\varepsilon_n=\sum_{k=1}^{K}a_k\varepsilon_{n-k}+\phi_n \quad (20)$$

The correlation in the residuals can then be addressed by inferring the K parameters $\{a_k;k=1...K\}$ of the AR model in addition to the model parameters θ.

However, AR and ARMA models have the conceptual limitation that they do reflect the physical mechanism that induces the auto-correlation in model residuals in the first place. In rainfall-runoff modelling, the auto-correlation depends strongly on the process dominating the hydrologi-

cal response (quickflow or slowflow) and it is cumbersome to introduce AR models that account for such distinctions (e.g., the AR parameters $\{a_k;k=1...K\}$ are effectively time- and process-dependent). Furthermore, since standard AR and ARMA methods do not explicitly introduce input error models, they suffer the same shortcomings as SLS methods when errors are present in both the $\mathbf{X}$ and $\mathbf{Y}$ data.

Total Least Squares Methods: A Bayesian Perspective

The Bayesian likelihood function corresponding to TLS methods is obtained by assuming Gaussian error models for both $\mathbf{X}$ and $\mathbf{Y}$ data

$$\tilde{\mathbf{X}}=\mathbf{X}+\varepsilon_x \quad \varepsilon_x\sim N\left(0,\sigma_x^2\right) \quad (21)$$

$$\tilde{\mathbf{Y}}=\mathbf{Y}+\varepsilon_y \quad \varepsilon_y\sim N\left(0,\sigma_y^2\right) \quad (22)$$

The limitation of TLS methods in hydrological modelling is that they assume that both input and output uncertainty can be described by independent Gaussian distributions, which is not always appropriate. Since the error models (21) and (22) are embedded in the TLS framework, the latter is not robust against deviations from normality and additive error forms (21) and (22). In addition, since TLS methods introduce one latent variable per each data point, the computational cost of a calibration may become prohibitive even with modern computer power.

However, TLS methods are significant in that they explicitly recognise input uncertainty and its impact on parameter estimation. A modern application of Bayesian TLS methods in pattern recognition [*Nestares et al.*, 2000] shows similarities to hydrological calibration, although the linearity of the models used in that study considerably simplifies the analysis and the approach is hence not readily applicable to non-linear hydrological modelling.

Zellner [1971] offers an interesting Bayesian analysis of TLS methods (which he refers to as the Error-In-Variables Method, EVM). In particular, useful insights can be obtained by considering special cases of the EVM. Zellner [1971] shows that, even for linear models, the attempt to infer both the input variance and output variances leads to an ill-posed problem due to the likelihood function becoming unbounded. This implies that, in the hydrological context, TLS methods cannot estimate the variances of forcing and response errors σ_x^2 and σ_y^2. However, the inference of the ratio of the variances $\lambda=\sigma_x^2/\sigma_y^2$ is well posed given an informative prior on either σ_x^2 or σ_y^2. These results show the subtlety of the apparently straightforward problem of data modelling, even using linear functions.

MULTI-OBJECTIVE CALIBRATION

An alternative approach to model calibration is to employ a composite objective function that contains terms corresponding to several distinct objectives. This approach mimics the manual calibration of hydrologists and is termed multi-objective (or multi-criteria) calibration. The following objectives are normally considered [*Sorooshian et al.*, 1993; *Gupta et al.*, 1998; *Madsen*, 2000]:

a)   Correct flow volumes over the simulation, i.e. correct water balance;

b)   Agreement in shape between the observed and predicted hydrographs;

c)   Agreement of the peak flow characteristics: timing, rate, volume, etc.;

d)   Agreement of recession limbs and low flow periods;

These objectives are expressed in numerical form and can be used to obtain Pareto solutions, reflecting various trade-offs between parameters and calibration criteria. Weighting factors can be used to obtain a composite objective function if a "globally optimal" single parameter set is required.

Multi-objective calibration is intuitive and conceptually simple. It recognises that different parts of the hydrograph can be subject to different error processes; hence it constitutes a conceptual advance compared with crude single-objective calibration methods such as SLS schemes.

However, we feel that multi-objective calibration methods have certain limitations:

1.   The trade-off between various objectives is often unclear and there are no theoretical guidelines for the selection of weighting factors. The use of the entire Pareto front in a way circumvents the need for a single optimal parameter set, yet if predictions are necessary, particular parameter sets still have to be selected. A related pitfall is the possible correlation between the objectives. For example, minimising the discrepancy between observed and simulated responses would generally also improve the mass balance;

2.   Multi-objective calibration is typically response focused and does not explicitly consider the influence of input errors. As a result, if the input history is corrupt and the model imperfect, this approach cannot in principle provide good fits and unbiased parameter estimates;

3.   A fundamental limitation of multi-objective calibration from a Bayesian viewpoint is that it does not articulate an identifiable error model. It is therefore more difficult to appraise the validity of the inference procedure and provide uncertainty bounds on the parameter estimates (as these are strongly related to specific error models).

These shortcomings, in particular, the difficulty in obtaining confidence limits on parameters that can be used to obtain prediction limits on future events, undermine the range of application of multi-objective methods in hydrological modelling and forecasting.

GENERALISED LIKELIHOOD UNCERTAINTY ESTIMATION (GLUE)

The Generalised Likelihood Uncertainty Estimation (GLUE) methodology [*Beven and Binley*, 1992] was developed as a method for calibration and uncertainty estimation using generalised likelihood measures, and is related to the Generalised Sensitivity Analysis of Spear and Hornberger [1980]. The GLUE methodology explicitly recognises the fundamental limitations of simulating rainfall-runoff processes with contemporary hydrological models in data-sparse and data-corrupt applications. Its application so far has been predominantly in rainfall-runoff modelling [*Beven and Binley*, 1992], but GLUE has also been used to assess the uncertainty associated with predictions of land surface to atmosphere fluxes [*Franks and Beven*, 1997], geochemical modelling [*Zak et al.*, 1997] and flood inundation studies [*Romanowicz et al.*, 1994].

GLUE is based on Monte Carlo simulation, generating a large number of model runs with parameter sets sampled from a uniform probability distribution on prior parameter bounds. The likelihood of the parameter sets is then evaluated using a user-defined pseudo-likelihood measure, parameters with a pseudo-likelihood below a threshold are rejected, and the remaining likelihoods normalised to add up to 1. Next, at each time step, the predicted output from the retained runs are likelihood-weighted and ranked to form a cumulative distribution of response variables, from which quantiles can be selected to represent predictive uncertainty.

While GLUE is based on Bayesian conditioning, it does not articulate a specific error model structure - instead it embeds an unknown implicit error model within a suitably lenient pseudo-likelihood measure. As a result, all sources of uncertainty in GLUE manifest themselves as parameter uncertainty, giving rise to the concept of parameter equifinality, which admits multiple disjoint parameter sets that fit the observed data equally well. In the GLUE framework it is difficult to scrutinise and improve the uncertainty model underlying the inference. GLUE therefore lacks the conceptual rigour to address the challenges posed by error structures typified in Figure 1.

KALMAN FILTERS

The Kalman filter is a method widely used in electrical engineering and system analysis, primarily for linear Gaussian dynamics [*West and Harrison*, 1997], and has seen some use in hydrology [*Bras and Rodriguez-Iturbe*, 1985]. Kalman filters, at least in principle, explicitly specify the uncertainty in the system states that arises from imperfect process approximation and from data uncertainty.

The state-space formulation underlying Kalman filters is general and applicable to almost arbitrary models and uncertainty distributions. However, to obtain analytical closed-form solutions to the state estimation equations, the process is assumed to be linear with respect to the state variables and all errors are assumed to have a Gaussian distribution. A Bayesian interpretation of the classic Kalman filter follows.

Consider the discrete time stepping state-space model

$$\mathbf{\Psi}_{n+1} = \mathbf{A}_n \mathbf{\Psi}_n + \mathbf{L}_n \mathbf{u}_n + \mathbf{v}_n \quad (23)$$

where $\mathbf{\Psi}_n$ is the state vector at step n, $\mathbf{A}_n$ is the transfer matrix, $\mathbf{u}_n$ is a control vector, $\mathbf{L}_n$ is the control matrix and $\mathbf{v}_n$ represents model error. When the Kalman filter is used for simulation, $\mathbf{\Psi}$ contains model state variables, e.g., simulated streamflow and internal fluxes. When used for calibration, $\mathbf{\Psi}$ is augmented with the model parameters $\mathbf{\theta}$, giving rise to extended Kalman filters. Often $\mathbf{\Psi} \equiv \mathbf{\theta}$, with the transfer matrix reduced to the identity matrix.

In addition, consider the observation equation

$$\mathbf{z}_{n+1} = \mathbf{H}_{n+1} \mathbf{\Psi}_{n+1} + \mathbf{w}_n \quad (24)$$

where $\mathbf{z}_{n+1}$ contains system observations (e.g., observed forcing inputs and responses), $\mathbf{H}_{n+1}$ is the observation matrix and $\mathbf{w}_{n+1}$ represents observation error.

If the observation history up to and including step n is stored in $\mathbf{Z}_n = \{\mathbf{z}_i; i = 1...n\}$ and a (prior) pdf $p(\mathbf{\Psi}_n \mid \mathbf{Z}_n)$ of the state vector along with the pdf of the noise terms $\mathbf{v}_n$ and $\mathbf{w}_n$ are known at step n, the posterior pdf of the state vector at consequent steps can be constructed in two stages:

Prediction Step

Using total probability, the pdf of the state $\mathbf{\Psi}_{n+1}$ is

$$p(\mathbf{\Psi}_{n+1} \mid \mathbf{Z}_n) = \int_{\Omega(\mathbf{\Psi}_n)} p(\mathbf{\Psi}_{n+1} \mid \mathbf{\Psi}_n, \mathbf{Z}_n) p(\mathbf{\Psi}_n \mid \mathbf{Z}_n) d\mathbf{\Psi}_n \quad (25)$$

which, given the Markovian property of (23), simplifies to

$$p(\mathbf{\Psi}_{n+1} \mid \mathbf{Z}_n) = \int_{\Omega(\mathbf{\Psi}_n)} p(\mathbf{\Psi}_{n+1} \mid \mathbf{\Psi}_n) p(\mathbf{\Psi}_n \mid \mathbf{Z}_n) d\mathbf{\Psi}_n \quad (26)$$

Conditioning Step

The observation can be processed using Bayesian updating

$$p(\mathbf{\Psi}_{n+1} \mid \mathbf{Z}_{n+1}) = \frac{p(\mathbf{z}_{n+1} \mid \mathbf{\Psi}_{n+1}, \mathbf{Z}_n) p(\mathbf{\Psi}_{n+1} \mid \mathbf{Z}_n)}{p(\mathbf{z}_{n+1} \mid \mathbf{Z}_n)} \quad (27)$$

The Markovian property of the observation equation yields

$$p(\mathbf{\Psi}_{n+1} \mid \mathbf{Z}_{n+1}) = \frac{p(\mathbf{z}_{n+1} \mid \mathbf{\Psi}_{n+1}) p(\mathbf{\Psi}_{n+1} \mid \mathbf{Z}_n)}{p(\mathbf{z}_{n+1} \mid \mathbf{Z}_n)} \quad (28)$$

Analytical Solution of State Estimation Equations

In general, (26) and (28) do not possess closed-form analytical solutions. However, if the model and observation equations are linear in the state vector $\mathbf{\Psi}$

$$\frac{\partial \mathbf{A}_n}{\partial \mathbf{\Psi}_{n+1}} = \frac{\partial \mathbf{A}_n}{\partial \mathbf{\Psi}_n} = \frac{\partial \mathbf{H}_n}{\partial \mathbf{\Psi}_{n+1}} = 0 \quad (29)$$

and the random variables follow a multi-Gaussian pdf

$$\begin{bmatrix} \mathbf{\Psi}_n \\ \mathbf{v}_n \\ \mathbf{w}_n \end{bmatrix} \sim N\left(\begin{bmatrix} \hat{\mathbf{\Psi}}_n \\ 0 \\ 0 \end{bmatrix}, \begin{bmatrix} \Sigma_n & 0 & 0 \\ 0 & \mathbf{Q}_n & 0 \\ 0 & 0 & \mathbf{R}_n \end{bmatrix} \right) \quad (30)$$

it is straightforward to derive stepwise exact solutions to (26) and (28) [*West and Harrison*, 1997].

Since data uncertainty in both forcing and response observations can be specified through the observation noise w, while model error can be specified through v, Kalman filters offer a very general framework for the estimation of system parameters accounting for all sources of uncertainty. Moreover, the Kalman filter requires only matrix multiplications and inversions and is consequently computationally fast. However, the derivation of the classic Kalman filter based on assumptions (29) and (30) highlights the limitations of this estimation scheme in rainfall-runoff model calibration.

Limitation 1 – Hydrological Models are Nonlinear

Selecting sufficiently small time steps in principle reduces the effect of model nonlinearity with respect to time and forcing – this is the approach used to control discretisation error in numerical DE solvers. However, it is the nonlinearity of the model with respect to the parameters $\mathbf{\theta}$ within the state vector $\mathbf{\Psi}$ that undermines the validity of Kalman filtering in hydrological calibration. Although statistical and Taylor series linearisations of Kalman filters have been proposed, they are susceptible to nonlinear divergence, when the state estimates converge to incorrect values [*Bras and Rodriguez-Iturbe*, 1985].

Limitation 2 – Modelling Errors may not be Gaussian

The Kalman filter equations are derived by substituting Gaussian kernels into the forecast-update equations. The use of alternative error models requires the re-derivation of the Kalman filter equations and analytical solutions are unlikely in most cases.

Limitation 3 – Implementation of Kalman Filters

Casting hydrological models into the stepwise Kalman matrix forms is not always trivial, complicating the development of general model-independent analysis software. When using extended linearised Kalman filters, the derivatives are typically unavailable in closed form and numerical differentiation must be used.

Due to these and other problems, classic Kalman filters do not offer a complete solution to the calibration problem in hydrological modelling. However, recent developments in Monte Carlo and particle filters [*Carter and Kohn*, 1994; *Fruhwirth-Schnatter*, 1994; *Cargnoni et al.*, 1997] suggest that there is scope for the application of these recent methods in hydrological parameter estimation. Indeed, the Monte Carlo computational tools allowing the generalisation of Kalman filters to nonlinear non-Gaussian models are related to the Monte Carlo Markov Chain tools that we will use to implement the generalised parameter inference with arbitrary error models latter in this chapter (albeit not in the stepwise iterated form characteristic of Kalman filters).

BAYESIAN TOTAL ERROR ANALYSIS (Ba TEA)

We have now surveyed all the major current calibration philosophies and see that, although inherent uncertainty in environmental systems is partially recognised (explicitly by Bayesian SLS and TLS methods and Kalman filters, implicitly by GLUE and MOCOM), limited attempt has been made to rigorously define this uncertainty using realistic error models. The (Bayesian) SLS regression avoids the specification of input error and suffers a parameter bias, TLS schemes account for special cases of input error but do not allow for non-Gaussian error models. GLUE offers a Monte Carlo algorithm for sampling from posterior parameter probability distributions, but does not make explicit the assumptions used to derive the likelihood function. Multi-objective calibration methods recognise the multi-criteria nature of calibration, yet lack the ability to provide confidence limits on parameter values and model predictions. Kalman filters can, at least in principle, explicitly

specify error structure, yet linearisations and Gaussian assumptions undermine their stability properties and hence their suitability to environmental modelling. However, armed with an understanding of current limitations and insights provided by the analysis of TLS and Kalman methods, we can use the very general equation (11) and its simplified variant (13) to develop a family of methods that explicitly account for the various sources of system uncertainty. Whilst we present these methods from a Bayesian perspective, the parameter estimation equations become identical to those obtained from general maximum-likelihood theory when non-informative priors on the hydrologic parameters are employed.

Crude Input Error-Sensitive Approaches

One possible approach in rainfall-runoff modelling to extend the traditional regression framework is to augment the fitted parameter vector with rainfall depth multipliers. In the early approach of Kavetski et al. [2000], no likelihood function was specified for the fitted multipliers m, equivalent to setting

$$m \sim U(a,b) \tag{31}$$

where $U(\cdot)$ denotes the uniform distribution, and a and b are positive bounds, e.g., 0.1 and 10.0, reflecting uncertainty in the magnitude of precipitation depth errors.

Error models such as (31) are not new; a similar approach is implicitly used by the PDM model [*Lamb*, 1999] under the guise of a hydrological parameter (a single rainfall multiplier r_f for all time steps). Equation (31) implies that the informational content of rain gauges is limited to hyetograph shape only and that depth measurements have little or no influence on the hydrological parameter estimates, which is clearly an extreme statement. However, by making error structure assumptions such as (31) explicit, it becomes possible (and necessary) to scrutinise them, motivating our understanding of data corruption mechanisms.

Expected Likelihood Approach

When input uncertainty is included into the likelihood function, the following expression can be obtained

$$
\begin{aligned}
p(\tilde{\mathbf{X}},\tilde{\mathbf{Y}}|\boldsymbol{\theta}) &\propto L(\tilde{\mathbf{Y}}|\tilde{\mathbf{X}},\boldsymbol{\theta})\,p(\tilde{\mathbf{X}}|\boldsymbol{\theta}) \\
&= \int_{\Omega(\mathbf{X})} L_p(\tilde{\mathbf{Y}}|\mathbf{X},\boldsymbol{\theta})\,p(\mathbf{X}|\tilde{\mathbf{X}})\,d\mathbf{X} \cdot p(\tilde{\mathbf{X}}|\boldsymbol{\theta})
\end{aligned}
\tag{32}
$$

where $\mathbf{X}$, the unknown true input, has been integrated out. As shown in (18) and (19), omitting the input error model is a special case of (32), equivalent to assigning $p(\mathbf{X} \mid \tilde{\mathbf{X}}) = \delta(\mathbf{X} - \tilde{\mathbf{X}})$. It leads to the SLS scheme and biased parameter estimates.

The numerical cost of evaluating (32) can be considerable, due to the high dimensionality of the integral. In Bayesian image analysis [*Nestares et al.*, 2000], a similar integral is evaluated by taking advantage of the multi-normal kernel in the equivalent of $p(\mathbf{X} \mid \tilde{\mathbf{X}})$ and a linear likelihood function $L_p(\cdot)$. However, since hydrological models are highly nonlinear, there is little hope in obtaining a closed-form solution of the integral (32), certainly not for arbitrary models. Although classical quadrature schemes based on hyper-lattices (e.g., the trapezoidal rule) could be employed to obtain approximate solutions of (32), these suffer an exponential deterioration of convergence rates in high-dimensional spaces and become computationally infeasible [*Evans and Swartz*, 2000]. Equi-distributed (quasi-random) numbers such as Sobol and Halton sequences raise the rate of convergence to $O([\ln K]^D/K)$, where K is the number of function evaluations and D is the number of dimensions of integration. Although asymptotically this rate is almost as fast as $O(K^{-1})$, it remains impractical in high dimensions [*Geweke*, 1996].

Instead, Monte Carlo (MC) schemes could be employed, since their asymptotic convergence rate is only weakly dependent on the dimensionality of the problem, $O(K^{-1/2})$. A simple MC integration algorithm for (32) is

$$L\left(\tilde{\mathbf{Y}} \mid \boldsymbol{\theta}, \tilde{\mathbf{X}}\right) = \frac{1}{K} \sum_k^K L_p\left(\tilde{\mathbf{Y}} \mid \boldsymbol{\theta}, \mathbf{X}_k\right) + O\left(K^{-1/2}\right) \quad (33)$$

where $\mathbf{X}_k$ is the kth sample from $p(\mathbf{X} \mid \tilde{\mathbf{X}})$. The effective dimension D of integration (33) depends on the input error model and is not necessarily equal to the number of data points. For example, if a single multiplier is used for each storm, then the dimension of integration will be equal to the number of storms in the calibration dataset.

As vividly illustrated by Hammersley and Handscomb [1964], it is possible to reduce the leading constant of the Monte Carlo error term by factors of 100,000's through modifications such as antithetic variates and sampling from orthonormal basis functions. Unfortunately these acceleration techniques exploit special features in the integrand and are cumbersome, if not impossible, in high dimensions. Further, there are hybrids of Monte Carlo schemes with quadrature methods that achieve the highest possible order of convergence, $O(K^{-G/D})$, where G is the highest bounded derivative of the integrand on the domain of integration [*Bahvalov*, 1959], but their efficacy depends on a set of continuity and variational constraints on the integrand that are hard to verify *a priori*. In spite of these challenges, adaptive

Monte Carlo algorithms are available (e.g., the Vegas scheme of Lepage [*Press et al.*, 1992] and the stratification scheme of Press and Farrar [1990]) that could be more efficient than (33). Evans and Swartz [2000] and Fishman [1996] offer good summaries of stochastic integration methods.

Finally, although (32) allows the use of explicit input error models, it has the disadvantage that, after integrating the true input history out of the posterior pdf, it remains difficult to assess the suitability of the selected input error model $p(\mathbf{X} \mid \tilde{\mathbf{X}})$, since estimates of the true input history are not be explicitly available for inspection. As rigorous calibration demands *a posteriori* verification of its assumptions, we are led to reject the use of (32) in favour of a different method of incorporating input uncertainty.

Latent Variables as Subjects of Inference

An alternative to the integration of the true input history via is its Bayesian estimation with the same logical status as the hydrological parameters. Consider the Bayesian inference equation, closely related to

$$p\left(\boldsymbol{\theta}, \mathbf{X} \mid \tilde{\mathbf{X}}, \tilde{\mathbf{Y}}\right) \propto L\left(\tilde{\mathbf{X}}, \tilde{\mathbf{Y}} \mid \boldsymbol{\theta}, \mathbf{X}\right) p\left(\boldsymbol{\theta}, \mathbf{X}\right) \quad (34)$$

where $L(\tilde{\mathbf{X}}, \tilde{\mathbf{Y}} \mid \boldsymbol{\theta} \mathbf{X})$ is the joint likelihood of observing $\tilde{\mathbf{X}}$ and $\tilde{\mathbf{Y}}$ given a hydrological parameter vector $\boldsymbol{\theta}$ and the true input history $\mathbf{X}$. The advantage of using instead of is that the posterior pdf of the true forcing data (e.g., rainfall) becomes available and can be used to verify the data and the assumed probability models.

The joint likelihood $L(\tilde{\mathbf{X}}, \tilde{\mathbf{Y}} \mid \boldsymbol{\theta} \mathbf{X})$ can be re-formulated to maximally separate the probability models for input and response error. Conditional probability yields

$$L\left(\tilde{\mathbf{X}}, \tilde{\mathbf{Y}} \mid \boldsymbol{\theta}, \mathbf{X}\right) \propto L\left(\tilde{\mathbf{Y}} \mid \tilde{\mathbf{X}}, \boldsymbol{\theta}, \mathbf{X}\right) p\left(\tilde{\mathbf{X}} \mid \boldsymbol{\theta}, \mathbf{X}\right) \quad (35)$$

Two further simplifications can be made, based on the following assumptions:

1. $\tilde{\mathbf{X}}$ and $\tilde{\mathbf{Y}}$ are statistically independent, i.e., $\tilde{\mathbf{Y}}$ depends only on the observation error affecting the response and on the true forcing (see Figure 1). The errors affecting the forcing do not causally affect the true response and its associated observation error;
2. $\tilde{\mathbf{X}}$ is statistically independent of $\boldsymbol{\theta}$, e.g., rainfall sampling errors are uncorrelated with the hydrological model parameters.

Subject to the above assumptions and allowing for un-normalised probability models of forcing uncertainty, simplifies to

$$L(\tilde{\mathbf{X}}, \tilde{\mathbf{Y}} \mid \theta, \mathbf{X}) \propto L(\tilde{\mathbf{Y}} \mid \theta, \mathbf{X}) L(\tilde{\mathbf{X}} \mid \mathbf{X}) \qquad (36)$$

Substituting into the posterior pdf yields

$$p(\theta, \mathbf{X} \mid \tilde{\mathbf{X}}, \tilde{\mathbf{Y}}) \propto L(\tilde{\mathbf{Y}} \mid \theta, \mathbf{X}) L(\tilde{\mathbf{X}} \mid \mathbf{X}) p(\theta, \mathbf{X}) \qquad (37)$$

Pdf (37) contains the inference of the model hydrological parameters and the true rainfall history X. It requires the specification of the input error model $L(\tilde{\mathbf{X}}, \mid \tilde{\mathbf{Y}})$ and the response error model $L(\tilde{\mathbf{X}}, \tilde{\mathbf{Y}} \mid \theta, \mathbf{X})$. The analysis of pdf (37) is computationally intensive due to the large dimensionality of the parameter space (now augmented with the latent variables of the forcing error model).

Computational Implementation—The Metropolis Algorithm

A couple of decades ago the formulation of a posterior pdf such as (32) or (37) would have been satisfying from a theoretical viewpoint yet practically useless – there were no effective approaches to examine such distributions, let alone determine their moments and other characteristics. We suspect that it is precisely the formidable computational aspect of (32) and (37) that has hindered the development of a rigorous parameter estimation framework with realistic uncertainty models. Indeed, classic sampling methods such as acceptance-rejection and importance schemes are difficult and inefficient for complicated and high-dimensional pdfs [*Fishman*, 1996].

Fortunately, the development of Monte Carlo Markov Chain (MCMC) methods offers practical ways to sample from probability distributions of considerable complexity and dimensionality. Fishman [1996], Gelman et al. [1997] and Evans and Swartz [2000] provide a good overview of MCMC methods, which originated in nuclear physics (see Metropolis et al. [1953] for historical background) and, in the 1990's, enjoyed an explosive growth in areas as diverse as econometrics [*Geweke*, 1996], biology [*Gelman et al.*, 1997] and hydrology [*Kuczera and Parent*, 1998; *Bates and Campbell*, 2001].

In this work, we implement the Metropolis algorithm [*Chib and Greenberg*, 1995; *Gilks et al.*, 1996; *Gelman et al.*, 1997] in order to generate samples from pdf and summarise the posterior parameter pdf using moments and histograms. It must be emphasised, however, that any other sampler can be used, e.g., importance schemes.

The theory behind MCMC methods is complex and still evolving. It has been described in hydrological context by Kuczera and Parent [1998] and Bates and Campbell [2001]. Our implementation of the Metropolis algorithm follows

1. Sample $\{\theta, X\}$ from a pre-specified symmetric jump distribution, here the multi-normal pdf centred on the current sample location;
2. Evaluate $p(\{\theta \mathbf{X}\}_i \mid \tilde{\mathbf{X}}, \tilde{\mathbf{Y}})$ up to a constant;
3. Accept the new sample with probability given by $r = \min[p(\theta \mathbf{X})_i \mid \tilde{\mathbf{X}}, \tilde{\mathbf{Y}}) / p(\{\theta \mathbf{X}\}_{i-1} \mid \tilde{\mathbf{X}}, \tilde{\mathbf{Y}}), 1$.

Multiple Markov chains are started at the posterior mode of (37) found using the SCE algorithm. After sufficient samples have been collected (the termination criteria of Gelman et al. [1997] were used), the posterior distribution of the hydrological parameters θ and the true forcing parameters $\mathbf{X}$ can be examined using histograms and scatter plots, or summarised using moments and quantiles.

We stress that although the entire input history appears as the random variable in the posterior pdf, the actual number of additional latent variables will depend on the particular error model used. For example, if storm-wise multipliers are assumed, the number of latent variables will equal the number of storms in the calibration data. It follows that the dimensionality of the pdf support is not necessarily prohibitive. The ability to control, to a certain extent, the number of latent variables and hence the dimensionality of the state space, is an important advantage of the generalised BATEA framework over classical TLS schemes, which always introduce N additional latent variables.

Block updating of the sampled variables is advantageous when strong correlation is suspected between these variables [*Gelman et al.*, 1997; *Bates and Campbell*, 2001]. We suspect that considerable computational optimisation can be carried out by a judicious selection of jump distributions and updating sequences. Given enough sampling, however, all these algorithmic variations converge to the same target distribution (37).

CASE STUDY: THE ABC MODEL

The theoretical arguments calling for the more systematic calibration formalism BATEA can be illustrated by numerical experimentation.

Consider the ABC model, a simple time stepping hydrological model with three parameters {a, b, c}, two state variables (discharge Q and storage S, both per unit catchment area) and one forcing term (rainfall r). For any time step i, the ABC equations are given as

$$Q_i = (1 - a - b) r_i + c S_i \qquad (38)$$

$$S_{i+1} = (1 - c) S_i + a r_i \qquad (39)$$

Parameter a represents the fraction of rainfall entering the groundwater storage, b is the evaporation fraction and c is a constant of linear proportionality between storage and discharge. In addition, the initial value S1 is required.

While the ABC model is simplistic and certainly uncompetitive with more complex models, it does have an elementary groundwater store and a quickflow component. We employ ABC to illustrate some of the issues common to all hydrological models. In addition, we use synthetic data to establish the "bona fides" of the calibration methods. If real data were used, there would be no way of checking whether the parameter estimates converge to the true values and whether the confidence limits reflect the actual errors.

The "true" rainfall data $\mathbf{X}$ was generated using the DRIP algorithm [Heneker et al., 2001] using parameters corresponding to the Sydney region in Australia. The "true" streamflow $\mathbf{Y}$ was generated using the ABC model with the "true" parameter set $\theta = \{0.6\ 0.15\ 0.2\}$. The $\mathbf{X}$, $\mathbf{Y}$ and θ data represent the "truth" that will be used to gauge the success of the parameter estimation scheme.

The "observed" rainfall $\mathbf{X}$ was obtained by corrupting each storm depth r within $\mathbf{X}$ using normally distributed storm multipliers, yielding the corrupted depth $\tilde{r}$

$$\tilde{r}_j = m_j r_j; \quad m \sim N\left(0, \sigma_m^2\right) \forall j \tag{40}$$

where j indexes the storms within the rainfall series.

The "observed" streamflow was obtained as

$$\tilde{y}_i = y_i + \varepsilon_i; \quad \varepsilon_i \sim N\left(0, \sigma_y^2\right) \forall i \tag{41}$$

where i indexes the time steps within the simulation.

The additive response error model (41) is the same as that used in the SLS and TLS schemes. The multiplicative input error model (40), although related to the additive TLS input error model, is nonetheless different. It offers an attractive way to parsimoniously parameterise rainfall errors. For example, if the storm largely misses a gauge, its overall temporal pattern will register on the pluviograph, but the true strength will be underestimated. Conversely, if the core of the storm passed directly over the rain gauge, the effective catchment-averaged precipitation would be overestimated.

It is relatively straightforward to accommodate various probabilistic error models within the BATEA model analysis formalism. In contrast SLS and TLS methods, as well as classic Kalman filters, have embedded error models that can not be easily modified.

Unless stated otherwise, $N = 1000$ time steps with $\Delta t = 1$ hr were used (42-day runs with 5 storms). Quantitative results for $\sigma_m^2 = 0.05$ and $\sigma_y^2 = 0.01$ are presented; in general, qualitatively similar behaviour occurred for other range of

σ_m^2 and σ_y^2. In addition, we found no major qualitative differences in calibration behaviour when using alternative input error distributions, e.g., log-normal, instead of normal, multipliers.

The ABC model is calibrated to the "observed" data using a) the SLS scheme; and b) the BATEA formalism. Comparisons are then made, focusing on i) the accuracy of the parameter estimates; and ii) the relation between the estimated uncertainty and the actual parameter errors. Unless synthetic data were used, these assessments, which in our opinion must be applied to any parameter estimation scheme, would have been impossible.

Although typical implementations of SLS schemes are limited to fitting the model parameters using an optimisation scheme, we have performed the additional step of sampling from the posterior SLS parameter distribution (15) using the Metropolis algorithm. This allows a more comprehensive comparison of the posterior parameter distributions inferred by the SLS and BATEA schemes, since the mode of the posterior pdf corresponds to the best-fit parameters, while the shape and spread of the distribution quantify the parameter uncertainty.

Five Metropolis chains with 40,000 samples in each were generated and the first 10,000 samples were discarded. The r-statistic of Gelman et al. [1997] was monitored to ascertain the chains' convergence to the target posterior pdf. Unless stated otherwise, the posterior covariance matrices of the parameters were approximated from the Metropolis samples, since inverse finite difference Hessian approximations at the posterior mode are more "local" and hence less informative.

We also define a dimensionless error measure (θ) for calibrated parameters.

$$\eta(\theta) = \frac{\theta_{true} - E[\theta]}{\sigma[\theta]} \tag{42}$$

where $E[\theta]$ and $\sigma[\theta]$ are the mean and standard deviation of the posterior parameter pdf.

Bayesian SLS Calibration

The SLS-fitted response for a 50-hr segment of the data is shown in Figure 3. Visually, the fit is good in practically all sections of the hydrograph. Undoubtedly the parameter estimates must be good! However, the reader, by now perhaps sceptical of the suitability of SLS schemes in the presence of input error, will not be surprised to learn that, while parameter c has been fitted quite well, parameters a and especially b are poorly identified. Figures 4-5 and Table 3 summarise the SLS calibration results, showing that the

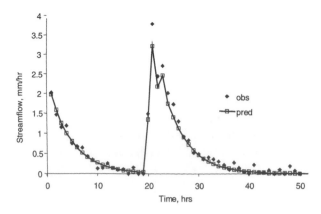

Figure 3. Portion of SLS-calibrated ABC hydrograph.

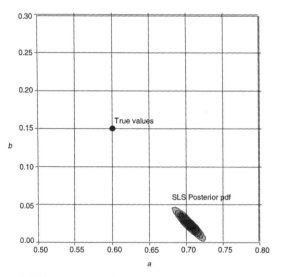

Figure 4. SLS posterior pdf of paramaters *a* and *b*.

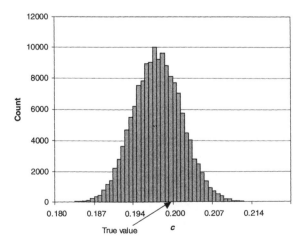

Figure 5. SLS posterior pdf of paramater *c*.

SLS estimates of a and b are ten standard deviations away from the true values, well outside any meaningful confidence limits. The convergence of the SCE search to the global optimum of the SLS objective (likelihood) function was verified by enumerating the parameter space using a fine grid. This confirms that the parameter bias is a result of the inappropriate likelihood function and is not an artefact of the numerical optimisation scheme used.

We stress that in this study no model error was introduced and only moderate unbiased data corruption was used (e.g., the actual response noise corresponded exactly to the additive Gaussian noise in the SLS response error model). The fact that poor estimates were obtained for a and b, while good results were obtained for c, illustrates the unpredictability of parameter inference based on inappropriate models. We are forced to conclude that, given 42 days of data (1000 data points), the SLS calibration fails to identify the correct parameters even when model error is not present.

Another aspect of the ABC model can be seen in Figure 4, which shows a moderately strong correlation between the fitted parameters a and b. In general, correlation indicates that the model is over-determined. A reasonable model validation question then is: *is the correlation between a and b an intrinsic feature of the model or a consequence of data errors?* Due to the simple functional form of the ABC model, this question can be answered analytically. However, for a more complicated model, the solution to this question must be found using some approximate numerical or experimental means – i.e., calibration. But if the SLS scheme does not even admit the possibility of input errors, how can it meaningfully tell the modeller whether parameter correlation is due to model structure or data uncertainty?

SLS Calibration – More Data for Accuracy?

Parameter errors are often blamed on the limited amount of data available for calibration. Indeed, the large parameter errors in Table 3 could have been explained by the fact that only one month of rainfall-runoff data was used. In this synthetic case study, however, we are free to test this statement, since the "truth" is completely known. Extending the data series to 7 months (5000 time steps) and then to 10 yrs (100,000 time steps) and calibrating ABC using the SLS scheme yield the posterior distributions of parameters a and b shown in Figures 6 and 7.

Table 3. – SLS calibration, N=1000 (42 days).

Parameter	Mean ± St. deviation	Error $\eta(\theta)$
a	0.702 ± 0.0106	-9.62
b	0.0250 ± 0.0105	11.9
c	0.198 ± 0.00413	0.484

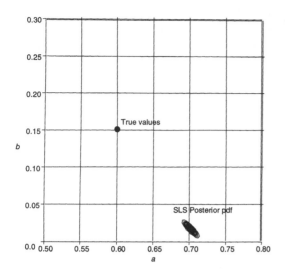

Figure 6. SLS posterior pdf of a and b with N=5000.

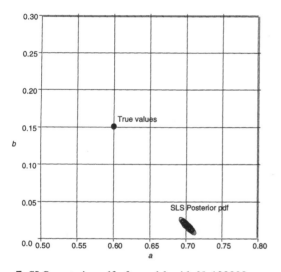

Figure 7. SLS posterior pdf of a and b with N=100000.

Figures 6-7 suggest that large data sets are unlikely to resolve the input error problem in SLS calibration. While the variance of all three parameters has contracted by two orders of magnitude (following the asymptotic $O(N^{-1})$ variance decay rate of SLS schemes), the convergence is to incorrect values. Although the optimal parameter values calibrated using 10 years of data are closer to the true values than those obtained using 1 and 7 months of data, the uncertainty estimates are extremely misleading and the parameter estimates converge to biased values. This bias is a serious shortcoming of the SLS scheme – if the approach cannot handle simple synthetic data error in such a simple hydrologic model (with N = 105 data points), we must be wary of applying it in the real world. However, the non-robustness of SLS schemes with respect to input error is not a mystery

– analytic results dating back to the 1940's show parameter bias even for a 1-parameter straight line fit!

BATEA Calibration Using Diffuse Multiplier Error Model

In this part of the case study, we deliberately mis-specify the input error model, assuming a diffuse uniform likelihood for all values of storm multipliers between 0.1 and 10, as in . The results of the parameter inference are shown in Table 4. The Metropolis analysis of the posterior distribution failed to converge and the standard deviation reported in Table 4 was computed using the inverse finite difference Hessian approximation at the posterior mode.

It can be seen that the introduction of a different incorrect input error model led to little if any reduction of the actual parameter error for all three parameters of the ABC model (although the posterior mode has shifted). These results illustrate potential limitations of ad-hoc approaches that introduce additional model parameters in an attempt to account for rainfall data errors (e.g., the PDM model [*Lamb*, 1999]).

The poor performance of the diffuse multiplier model can be understood by noting that statistical inference extracts information (parameter values) from data (here, rainfall-runoff series) using some set of rules (likelihood functions and prior distributions). Specifying a diffuse uniform likelihood on the storm multipliers instructs the inference scheme to disregard all information on precipitation depth contained in $\tilde{\mathbf{X}}$ and limits the use of the rainfall series to relative hyetograph shape only. The multipliers then become completely unconstrained degrees of freedom that at best merely intensify the computational effort and at worst destabilise the inference algorithm.

The insight of Zellner [1971] into the inability of TLS schemes to operate with non-informative prior distributions on the parameters of the error models is also valuable. Setting a diffuse uniform likelihood on the storm multipliers is equivalent to using the (correct) error model (40) with $\sigma_m^2 = \infty$ and is similar to using a non-informative prior on σ_x^2 in the Bayesian TLS scheme. Since the prior on response noise variance σ_y^2 is also non-informative, it is little surprise that the parameter accuracy is poor and the Metropolis scheme fails to converge – the inference is ill-conditioned.

Table 4. – BATEA calibration, diffuse multiplier model, N=1000.

Parameter	Mean ± St. deviation	Error $\eta(\boldsymbol{\theta})$
a	0.636 ± 0.00872	-4.13
b	0.094 ± 0.00981	5.71
c	0.199 ± 0.00300	0.153

Power of BATEA: Correct Uncertainty Characterisation

The preceding empirical analysis confirms that simplistic treatment of data uncertainty fails to produce accurate and reliable parameter estimates. The Bayesian SLS schemes assumed input uncertainty is described by the Dirac function and the diffuse multiplier model was also incorrect. Both these schemes are specific cases of BATEA, but with incorrect error models for the particular problem. It was this error mis-characterisation that was responsible for the poor calibration results – not the data errors themselves, nor the quantity of data used, nor the model (which we know is exact in this synthetic case study).

To demonstrate the performance of the BATEA formalism, we specify the correct input error model – the Gaussian multiplicative depth error at each storm with the correct value of σ_m^2 in the input error model (some additional comments on selecting and verifying σ_m^2 are made later). The results are shown in Figures 8 and 9, with a summary in

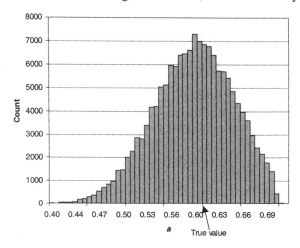

Figure 8. BATEA posterior pdf of a with N=1000.

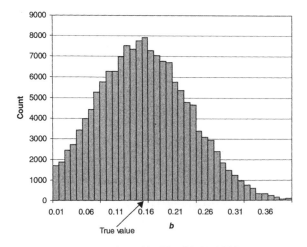

Figure 9. BATEA posterior pdf of b with N=1000.

Table 5.– BATEA calibration, correct error model, N=1000.

Parameter	Mean ± St. deviation	Error $\eta(\theta)$
a	0.593 ± 0.0528	0.125
b	0.155 ± 0.0751	-0.0625
c	0.199±0.00305	0.141

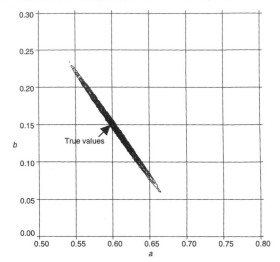

Figure 10. BATEA posterior pdf of a and b with N=5000.

Table 5. Figure 10 and Table 6 show the results of applying BATEA to the 7-month data set.

Figures 8-10 confirm that errors and biases in parameter estimates obtained earlier were not a product of poor data or model inadequacy – they were direct results of applying calibration schemes that did not correctly represent the way errors enter and propagate through the system. As soon as the correct error model is specified, the bias in the parameter estimates disappears. In fact, Table 5 indicates that the parameter estimates inferred by BATEA with 40 days of data are considerably closer to the true values than the parameters calibrated with the SLS scheme using 10 years of data. Although the BATEA posterior parameter uncertainty has increased compared to the SLS case, the mean and mode of the posterior pdf are close to the true values and the error statistic is of the order of 1.0, much below the SLS case. BATEA is hence more "honest" than SLS in reporting parameter uncertainty. At least for this test case, the trade-off between SLS vs BATEA could be described as "precisely wrong vs probably correct"!

As discussed previously, the shift in calibration philosophy from ignoring data errors to rigorously defining and

Table 6. – BATEA calibration, correct error model, N=5000.

Parameter	Mean ± St. deviation	Error $\eta(\theta)$
a	0.604 ± 0.0267	-0.135
b	0.146 ± 0.0376	0.120
c	0.199 ± 0.00150	1.38

treating them incurs a substantial computational cost – the necessity to estimate the latent variables of the error model (in this case, the storm multipliers). This computational effort is not at all worthless, however, since we chose to implement rather than . Following the calibration using , the posterior distribution of the latent variables is available for explicit inspection. In particular, fitted distribution of the latent variables should resemble the error model that was used to describe the corruption process. In our synthetic data case, we can go further and compare the posterior distribution of the multipliers to the known "true" values.

Figure 11 shows a 2D histogram of two of the five storm depth multipliers. The true values were m4 = 1.34 and m5 = 1.11. Figure 11 shows that the multipliers have been estimated with a notable degree of accuracy. The remaining multipliers were also estimated with small errors and with realistic uncertainty bounds. In a practical context, the values of the multipliers (and all other latent variables) must be examined to determine whether they are realistic. If the posterior pdf of latent variables is unreasonable, then either the calibration scheme has failed, or the rainfall-sampling network has produced data inadequate for the calibration of rainfall-runoff models, or the model itself is inadequate.

Another interesting point arises by considering that rainfall depth corruption must necessarily disturb the water balance of the catchment model. Input error insensitive methods, including SLS, GLUE and multi-objective schemes, will attempt to restore the mass balance by adjusting (biasing) the parameters. By including the depth error into the analysis, it becomes possible to directly address the cause of such mass balance errors, in a sense implicitly giving the BATEA calibration methodology multi-objective characteristics.

We also stress that good fits obtained using the SLS scheme (e.g., Figure 3) do not imply that the model predictions in response to future forcing will be correct. Since rainfall data errors were ignored by the SLS scheme, the SLS parameter estimates have been adjusted to cancel out (to the largest extent possible) the input errors in the particular calibration data set. However, it would be naïve to expect that future predictions in response to different forcing would benefit from such "cancellation". It is preferable to explicitly address (in a probabilistic sense) all sources of data uncertainty to maximally remove the influence of data errors from the parameter estimates. The reason for biased SLS parameter estimates is not the model itself, but rather unaccounted rainfall errors distorting the mass balance of the calibrated model.

We are also now in a position to empirically answer one of the earlier questions on the ABC model – is the correlation between a and b an intrinsic feature of the model or an artefact of data error? SLS schemes could not reliably answer that question, as they do not admit input errors. Observing that the correlation between a and b does not vanish when the correct error models are specified (Figure 10), we conclude that the correlation is an intrinsic feature of the model. Analytic assessment confirms this empirical observation.

Validation of Data Uncertainty Models

The reader by now might be asking: the results in Tables 5 and 6 are promising, but how do we tell when the chosen data error model is appropriate? Indeed, a posteriori verification of modelling assumptions is an essential step in model calibration: omitting it throws us back to ad-hoc modelling techniques. Since there are criteria that can be used to reject hypotheses, data uncertainty models can and must be scrutinised through seeking invalidating evidence, not necessarily used in the inference itself. For example, if the rainfall data was collected in a densely gauged catchment, but the inference suggests large rainfall errors, then a flawed data error model must be suspected.

A powerful rejection criterion is the compatibility between the posterior pdf of the latent variables and their corresponding likelihood functions. For example, in the final BATEA case study we assumed that the multipliers can be described by the Gaussian pdf with mean of 1.0 and variance of 0.05. We should therefore inspect the posterior distribution of the multipliers. Although it is hard to confirm trends from 5 and 28 storm multiplier samples, it is reason-

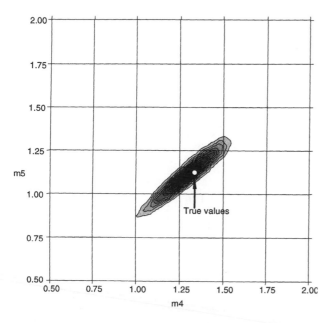

Figure 11. BATEA posterior pdf of multipliers, *N*=1000.

able to expect some basic compatibility. The posterior mean and variance of the fitted multipliers are 1.01 and 0.052 for the 1000-step 5-storm BATEA simulation and 0.99 and 0.057 for the 5000-step 28-storm BATEA calibration. These results are consistent with the rainfall error model used in the calibration. A more rigorous test is to examine the distribution of the multipliers, e.g., using probability plots. Care must be exercised, however, since probability plots based on small sample sizes can depart from linearity even if the data indeed came from that distribution. Figure 12 presents a normal probability plot for the posterior means of the multipliers in the 28-storm calibration. Its comparison with the assumed Gaussian multiplier likelihood function is satisfactory.

The error models do not exist in a vacuum – if the rainfall data came from a densely gauged catchment, we would reduce the variance of the multipliers and inspect the results. In the limit as the rainfall data becomes more exact, the multiplier likelihood would approach the Dirac function and we would arrive at the original SLS scheme. It is not the SLS scheme itself that is faulty; rather, it is its use to calibrate models to corrupt input data that is inappropriate.

Finally, we note that the input error model is not the only model that needs to be assessed. The response error model embedded in SLS and also employed in our ABC case study using BATEA is additive white noise, which is not necessarily appropriate. The verification of the response uncertainty model is also essential in a holistic calibration and can be carried out, e.g., via graphic analysis of residuals. The convenience of the additive Gaussian noise for response is largely mathematical, as it allows analytical integration of the output error variance σ_y^2 from the posterior pdf. Since BATEA employs MCMC methods to infer latent variables, the analytical convenience of the Gaussian residual assumption is useful yet no longer essential. However, since the literature describes many response error models (e.g., AR and ARMA models), but largely overlooks input uncertainty (with the exception of the TLS scheme, which is not used in hydrology and, at least in principle, Kalman filters, which are rarely used), we have focused on the latter to restore the balance.

A VIEW INTO THE FUTURE

We recognise that BATEA is still in its nascency and a range of outstanding issues remain:

a) The current ability to specify probabilistic data and model error models is limited: uncertainty functions must be chosen to reflect our physical understanding of processes contributing to input, response and model corruption. Such functions do not yet exist.

b) Parametric approaches for model error quantification are currently lacking. Nonetheless, the recent work by Gaganis and Smith [2001] evidences the growing recognition that explicit analysis of model error can and should be incorporated into environmental studies.

c) The numerical performance of the inference algorithm is of evident practical importance. Whilst the increased number of latent variables can be managed using Monte Carlo methods, the theory and practice of these algorithms are themselves still evolving. Much work remains to be done to refine sampling convergence criteria and develop more efficient jump rules and state update algorithms.

However, these issues must be kept in perspective – we argue that fundamental and overriding benefits arise from developing a framework that explicitly recognises data and model uncertainty and that produces greater insight about how uncertainty affects our worldview.

A common concern with hydrological modelling is that there appears to be a limit to the model complexity supported by rainfall-runoff data. However, the informational insufficiency of data series to support specific model hypotheses should be demonstrated via excessively wide confidence limits on calibrated parameters, preferably when data uncertainty and its effect on the inference is minimised. Although the parameter confidence limits in Bayesian SLS methods and GLUE are also indirectly related to model and data uncertainty, the lack of explicit error models in these approaches makes that link ambiguous, since confidence limits obtained using incorrect error models will have erroneous significance levels. BATEA, however, is in no way limited to calibration using rainfall-runoff series alone. The data matrices **X** and **Y** contain all the observations of the system – saturated areas, evapotranspiration, piezometer heads and so on. The inclusion of this data into the inference tends to constrain parameter uncertainty as it provides independent information on the model's internal state variables [Franks et al., 1998]. The inclusion of these variables is, in fact, an avenue for more stringent model assessment [Beven, 1993; Kuczera and Franks, 2002].

CONCLUSIONS

Reliable parameter inference is critical for meaningful prediction using environmental models. Yet the calibration of these models is currently accomplished using essentially ad-hoc approaches. For example, standard least squares calibration, although computationally convenient, disregards input uncertainty and suffers from a bias in the estimated parameters. GLUE recognises the inherent uncertainty in modelling, yet does not treat it explicitly, complicating the verification of its underlying assumptions. Multi-objective methods attempt to mimic the intuitive multi-criteria approach of hydrologists, yet, without specifying explicit error models, suffer from subjectivity in the selection of objective functions and cannot produce meaningful confidence limits on their predictions. Currently, only Kalman filters allow specification of model and data error, yet their reliance on system linearisation and Gaussian error models undermines their suitability in hydrological modelling.

Recognising the necessity to account for all sources of uncertainty in environmental modelling, a general Bayesian approach for total error analysis (BATEA) was introduced. It makes explicit the probabilistic error models used to describe the uncertainty in the observed data, notably, in forcing inputs such as rainfall (overlooked by current calibration schemes), as well as in the observed responses. The fundamental benefit of the BATEA formalism is that it forces the modeller to explicitly specify uncertainty models that can and must be verified against available evidence. Ignoring data and model uncertainty during calibration not only subtracts from the veracity of parameter estimates and predictions, but also may lead to stagnation in rainfall-runoff hydrology if hypotheses of catchment dynamics are only weakly challenged, with any discrepancies between observations and predictions attributed to the currently nebulous concept of "data and model error". Conversely, accounting for all sources of error leads to an honest assessment of parameter and predictive uncertainty of hydrologic models and paves the way for rational discrimination between competing models on the basis of explicit criteria, e.g., model error and susceptibility to data uncertainty.

REFERENCES

Bahvalov, N. S., On approximate calculation of multiple integrals, *Vestnik Moscov. Univ. Ser. Mat. Meh. Astr. Fiz. Him.*, 4, 3-18, 1959.

Bates, B. C. and E. P. Campbell, A Markov chain Monte Carlo scheme for parameter estimation and inference in conceptual rainfall-runoff modeling, *Water Resources Research*, 37(4), 937-947, 2001.

Beven, K. J., Prophecy, reality and uncertainty in distributed hydrological modelling, *Advances in Water Resources*, 16, 41-51, 1993.

Beven, K. J. and A. M. Binley, The future of distributed models: model calibration and uncertainty prediction, *Hydrological Processes*, 6, 279-298, 1992.

Beven, K. J. and M. J. Kirkby, A physically based variable contributing area model of basin hydrology, *Hydrological Science Bulletin*, 24, 43-69, 1979.

Beven, K. J., R. Lamb, P. F. Quinn, R. Romanowicz and J. Freer, TOPMODEL, *in Computer Models of Watershed Hydrology*, edited by V. P. Singh, pp. 627-628, Water Resources Publication, 1995.

Box, G. E. P. and G. C. Tiao, *Bayesian Inference in Statistical Analysis.* Wiley, New York, 1973.

Bras, R. L. and I. Rodriguez-Iturbe, *Random Functions in Hydrology.* Addison-Wesley Publishing Co, 1985.

Cargnoni, C., P. Muller and M. West, Bayesian forecasting of multinomial time series through conditionally Gaussian dynamic models, *Journal of the American Statistical Association*, 92, 587-606, 1997.

Carter, C. K. and R. Kohn, On Gibbs sampling for state-space models, *Biometrika*, 81, 541-553, 1994.

Chappell, N. A., S. W. Franks and J. Larenus, Multi-scale permeability estimation for a tropical catchment, *Hydrological Processes*, 12, 1507-1523, 1998.

Chib, S. and E. Greenberg, Understanding the Metropolis-Hastings Algorithm, *The American Statistician*, 49, (4), 327-335, 1995.

Deming, W. E., *Statistical Adjustment of Data.* Wiley, NY, 1943.

Duan, Q., S. Sorooshian and V. K. Gupta, Effective and efficient global optimization for conceptual rainfall-runoff models, *Water Resources Research*, 28(4), 1015-1031, 1992.

Evans, M. and T. Swartz, *Approximating Integrals Via Monte Carlo and Deterministic Methods.* Oxford University Press, Oxford, 2000.

Fishman, G. S., Monte Carlo: Concepts, *Algorithms, and Applications.* Springer-Verlag, 1996.

Franks, S. W. and K. J. Beven, Bayesian estimation of uncertainty in land surface-atmosphere flux predictions, *Journal of Geophysical Research*, 102(D20), 23, 23991-23999, 1997.

Franks, S. W., P. Gineste, K. J. Beven and P. Merot, On constraining the predictions of a distributed model: The incorporation of fuzzy estimates of saturated areas into the calibration process, *Water Resources Research*, 34(4), 787-797, 1998.

Fruhwirth-Schnatter, S., Data augmentation and dynamic linear models, *Journal of Time Series Analysis*, 15, 183-202, 1994.

Gaganis, P. and L. Smith, A Bayesian approach to the quantification of the effect of model error on the predictions of groundwater models, *Water Resources Research*, 37(9), 2309-2322, 2001.

Gelman, A., J. B. Carlin, H. S. Stern and D. B. Rubin, Bayesian *Data Analysis.* Chapman and Hall, London, 1997.

Geweke, J., Monte Carlo simulation and numerical integration, pp. in *Handbook of Computational Economics*, edited by H. Amman, D. Kendrick and J. Rust, North-Holland, Amsterdam, 1996.

Gilks, W. R., S. Richardson and D. J. Spiegelhalter (ed.), *Markov Chain Monte Carlo in Practice*, Chapman and Hall, London, 1996.

Gull, S. F., Bayesian data analysis: straight-line fitting, in *Maximum entropy and Bayesian methods*, edited by J. Skilling, pp. 511-518, Kluwer Academic Publishers, Dordrecht, 1989.

Gupta, H. V., S. Sorooshian and P. O. Yapo, Towards improved calibration of hydrologic models: multiple and non-commensurable measures of information, *Water Resources Research*, 34(4), 751-763, 1998.

Hammersley, J. M. and D. C. Handscomb, *Monte Carlo Methods*, New York, 1964.

Heneker, T. M., M. F. Lambert and G. Kuczera, A point rainfall model for risk-base design, *Journal of Hydrology*, 247, 54-71, 2001.

Jefferys, W. H., On the method of least squares, *The Astronomical Journal*, 85(2), 177-181, 1980.

Kavetski, D., S. W. Franks and G. Kuczera, Calibration of hydrologic models: The role of input errors, *in XIII International Conference on Computational Methods in Water Resources*, edited by L. R. Bentley, J. F. Sykes, C. A. Brebbia, W. G. Gray and G. F. Pinder, pp. 503-510, A. A. Balkema, Calgary, Canada, 2000.

Kuczera, G. and E. Parent, Monte Carlo assessment of parameter uncertainty in conceptual catchment models: The Metropolis algorithm, *Journal of Hydrology*, 211, 69-85, 1998.

Kuczera, G. A. 1994. NLFIT: A Bayesian nonlinear regression program suite. Dept. of Civ. Eng. and Surv, University of Newcastle, Australia.

Kuczera, G. A. and S. W. Franks, Testing hydrologic models: Fortification or falsification?, *in Mathematical Modelling of Large Watershed Hydrology*, edited by V. P. Singh and D. K. Frevert, pp. 141-185, Water Resources Publications, Littleton, Co., 2002.

Lamb, R., Calibration of a conceptual rainfall-runoff model for flood frequency estimation by continuous simulation, *Water Resources Research*, 35(10), 3103-3114, 1999.

Macdonald, J. R. and W. J. Thompson, Least-squares fitting when both variables contain errors: Pitfalls and possibilities, *American Journal of Physics*, 60(1), 66-73, 1992.

Madsen, H., Automatic calibration of a conceptual rainfall-runoff model using multiple objectives, *Journal of Hydrology*, 235, 276-288, 2000.

Metropolis, N., A. W. Rosenbluth, A. H. Teller and E. Teller, Equations of state calculations by fast computing machines, *Journal of Chemical Physics*, 21, 1087-1091, 1953.

Nestares, O., D. J. Fleet and D. J. Heeger, Likelihood functions and confidence bounds for total-least-squares-problems, in IEEE Conference on Computer Vision and Pattern Recognition, pp. 523-530, 2000.

Press, W. H. and G. R. Farrar, Recursive stratified sampling for multidimensional Monte Carlo integration, *Computers in Physics*, 4, 190-195, 1990.

Press, W. H., B. P. Flannery, S. A. Teukolsky and W. T. Vetterling, *Numerical Recipes in Fortran-77: The Art of Scientific Computing*. Cambridge University Press, 1992.

Romanowicz, R., K. J. Beven and J. Tawn, Evaluation of predictive uncertainty in non-linear hydrological models using a Bayesian approach, *in Statistics for the Environment, II, Water Related Issues*, edited by V. Burnett and K. F. Turkman, pp. 297-317, Wiley, New York, 1994.

Singh, V. P. (ed.), *Computer Models of Watershed Hydrology*, Water Resources Publication, 1995.

Sorooshian, S., Q. Duan and V. K. Gupta, Calibration of rainfall-runoff models: Application of global optimization to the Sacramento Soil Moisture Accounting Model, *Water Resources Research*, 29(4), 1185-1194, 1993.

Spear, R. C. and G. M. Hornberger, Eutrophication in Peel Inlet, II: Identification of critical uncertainties via Generalised Sensitivity Analysis, *Water Resources Research*, 14, 43-49, 1980.

West, M. and J. Harrison, *Bayesian Forecasting and Dynamic Models, 2nd Edition*. Springer-Verlag, New York, 1997.

Zak, S., K. J. Beven and B. Reynolds, Uncertainty in the estimation of critical loads: A practical methodology, *Soil, Water and Air Pollution*, 98, 297-316, 1997.

Zellner, A., An Introduction to *Bayesian Inference in Econometrics*. John Wiley and Sons, New York, 1971.

Dmitri Kavetski, Stewart W. Franks, and George Kuczera. School of Engineering, University of Newcastle, Callaghan, NSW 2308, Australia.

Multivariate Seasonal Period Model Rejection Within the Generalised Likelihood Uncertainty Estimation Procedure

Jim Freer and Keith Beven

Lancaster University, Department of Environmental Sciences, IENS, Lancaster, Lancaster, United Kingdom

Norman Peters

U.S. Geological Survey, 3039 Amwiler Road, Suite 130, Atlanta, Georgia

The evaluation of model performance within the Generalised Likelihood Uncertainty Estimation (GLUE) framework allows the use of multiple qualitative and quantitative rejection criteria to identify behavioural model structures and parameter sets from a sample of all possible models. A fundamental question within the GLUE approach (and indeed other calibration / validation methodologies) remains the choice of appropriate rejection criteria to assess model performance in a way that is consistent with our perception of the catchment dynamics. This paper tests the use of seasonal and sub-event performance criteria to assess the changing dynamics of a hydrological model, dynamic TOPMODEL, using data from the 41ha Panola Mountain Research Watershed (PMRW), Georgia, USA. The paper explores commonly used objective functions (performance measures) as rejection criteria within the GLUE procedure. Furthermore the paper shows that it may be difficult to propose a consistently parameterised model structure at PMRW due to significant variability in the observed seasonal responses (i.e. changes in recession dynamics and runoff coefficients). It is possible, given the additional model evaluations used here, to reject all the sampled parameter sets for this application of dynamic TOPMODEL, even though many were apparently acceptable when using 'global' performance measures (PMs) to characterise model performance. There appears to be a need for a more thoughtful approach to our use of PMs when using purely numerical assessments of rainfall-runoff models.

INTRODUCTION

The scientific study of hydrology necessarily requires the characterisation of a natural system that is, to a greater or lesser extent, unknown. Furthermore our perceptual understanding of the hydrological system far outweighs our ability to develop a rigorous theoretical treatment of that system [*Hornberger et al.*, 1985; *Beven*, 2000, 2001a; d]. This is

exemplified by our representation of hydrological flowpaths at the hillslope scale, where we do not have adequate formulations of processes that are known to involve multiple flowpaths with highly non-linear responses and dynamic variations in spatial connectivities upslope. Furthermore better descriptions of these processes and fluxes at the scale required by the model are unlikely to advance significantly in the near future due to limitations in our measurement technologies. Perhaps we should not be surprised that, given the possible range of dynamic output behaviour and the constraints of the water balance in applications of typical conceptual rainfall-runoff models, we can simulate a rainfall-runoff record, but that does not imply that our models are physically correct.

Calibration of Watershed Models
Water Science and Application Volume 6
Copyright 2003 by the American Geophysical Union
10/1029/006WS05

There is a growing realisation in the literature that perhaps we need to be more thoughtful about the measure/s we use to test model performance. The development of modified objective formulae has been reported since computer simulations of hydrology began and automated routines for calibration used [notably *Dawdy and O'Donnell*, 1965; *Nash and Sutcliffe*, 1970; *Garrick et al.*, 1978; *Sorooshian and Dracup*, 1980; *Willmott*, 1981; *Kuczera*, 1982; *Williams and Yeh*, 1983]. Interest is increasing in multi-criteria or multi-objective methods that attempt to make best use of our limited observations or observations other than rainfall-runoff data that may be more fuzzy or even qualitative, and which often involve less integrated catchment responses [e.g. *Moore and Thompson*, 1996; *Lamb et al.*, 1997; *Molicova et al.*, 1997; *Seibert et al.*, 1997; *Franks et al.*, 1998; *Gupta et al.*, 1998; *Yapo et al.*, 1998; *Guntner et al.*, 1999; *Boyle et al.*, 2000]. Not only do multi-criteria (or multi-objective) methods potentially allow more robust analyses of models, they also have the potential to aid hypothesis testing of competing model structures [*Beven*, 2001 a, b, c]. In this paper we analyse the similarities and differences among different global and seasonal and sub-event *PMs* within a multi-criteria framework for model rejection. We propose that a combination of *PMs* is more effective than any global *PM* in discriminating between behavioural and non-behavioural simulations. This type of model rejection framework is consistent with the Generalised Likelihood Uncertainty Estimation (GLUE) methodology of *Beven and Binley* [1992]. Section 2 gives a discussion of the rationale and philosophy of the GLUE methodology, concentrating on aspects of the methodology that relate to multi-criteria model rejection. Sections 3-4 describe and discuss the methods used to analyse seasonal and sub-event model performance from an application of Dynamic TOPMODEL [*Beven and Freer*, 2001a] applied to the Panola Mountain Research Watershed (PMRW).

EQUIFINALITY, MODEL REJECTION, AND GENERALIZED LIKELIHOOD UNCERTAINTY ESTIMATION (GLUE) – A DISCUSSION

In Section 3 below we will outline a conceptualisation of PMRW as three Landscape Units (LUs) within the Dynamic TOPMODEL structure as a reflection of the perceptual understanding of PMRW gained in the field. The penalty of such a distributed conceptualisation (albeit still simple) is an increase in the number of parameters that must be specified. Paradoxically our results may show that we are unable to make definitive statements regarding our 'improved spatial representations', or that we can only make statements associated with significant uncertainty about the nature of the distributed responses in the catchment. Accepting that

our model structures are to some extent in error, suggests that multiple possibilities of good (or behavioural) simulations will exist given the limitations or uncertainties in the data, in our perception of important hydrological processes and in the mathematical formulation of the conceptual model. This belief in multiple possibilities, rejecting the concept of searching for a single optimal representation, has been termed *equifinality* [*Beven*, 1993, 2001a, d; *Beven and Freer*, 2001b] to emphasise that this is not just a problem of identifying the "correct" (or even optimal) parameter set within a model structure.

Generalized Likelihood Uncertainty Estimation (GLUE)

If the concept of equifinality is accepted as a working paradigm, then some way of analysing possibilities for model structures and parameter sets must be developed. One such technique, and the method of choice for this study, is the GLUE methodology of Beven and Binley [1992], which is an extension of the Hornberger/Spear/Young Generalised Sensitivity Analysis [see *Spear and Hornberger*, 1980; *Ratto et al.*, 2001]. The GLUE procedure is a Monte Carlo (MC) based technique that uses different likelihood measures to evaluate multiple simulations resulting from different realisations of parameter sets within a given model structure. Different model structures can also be evaluated, providing that the same likelihood measures can be calculated. The likelihood measures are used to define a set of acceptable or *behavioural* models. All other model realisations are rejected (given a likelihood of zero).

In prediction, the predicted variables over the whole set of behavioural models are calculated. For each variable a likelihood weighted cumulative distribution is formed as:

$$P\left(\hat{Z}_t < z\right) = \sum_{i=1}^{i=N} L\left[M(\Theta_i) \mid \hat{Z}_{t,i} < z\right] \quad (1)$$

where $\hat{Z}_{t,i}$ is the value of variable Z at time t simulated by model $M(\Theta_i)$ with associated likelihood $L[M(\Theta_i)]$. Prediction quantiles can then be evaluated from this distribution. Accuracy in estimating such prediction quantiles will depend on having an adequate sample of models to represent the behavioural part of the model space.

Different likelihood measures can be combined within this framework by a variety of methods including Bayesian multiplication [see for example *Beven et al.*, 2000; *Beven and Freer*, 2001b]. It should be noted that the resulting prediction quantiles will reflect a belief in the model predictions but will be conditional on the model structures chosen, the ranges (and any prior distribution) of parameter values chosen, the sample of models chosen and the choice of likelihood measure(s) and rejection criterion used in evaluation.

Any interaction amongst parameter values in producing behavioural simulations should be reflected implicitly in the likelihood value associated with a model. The errors associated with that model are also treated implicitly and are effectively also assumed to be weighted by the likelihood value.

The methodology is very flexible in terms of a definition of a likelihood measure. The performance measures used in this paper can be treated in this way. Formal statistical likelihood functions, based on strong assumptions about the error structure, can also be used but are not often justified in the face of model structure error.

GLUE has been shown to be an effective method for determining the predictive uncertainty in several environmental model applications, including simulations of stream discharge [*Beven and Binley*, 1992; *Romanowicz et al.*, 1994; *Fisher and Beven*, 1996; *Freer et al.*, 1996; *Franks et al.*, 1998; *Beven and Freer*, 2001b; *Mwakalila et al.*, 2001]; ground water dynamics [*Feyen et al.*, 2001]; SVAT modelling [*Franks and Beven*, 1997; *Franks et al.*, 1997; *Beven and Franks*, 1999]; geochemistry [*Zak et al.*, 1997; *Zak and Beven*, 1999b], flood frequency [*Cameron et al.*, 1999; *Cameron et al.*, 2000]; computational fluid dynamics [*Aronica et al.*, 1998; *Romanowicz and Beven*, 1998; *Hankin et al.*, 2001]; and soil erosion [*Brazier et al.*, 2000; *Brazier et al.*, 2001].

Equifinality and Hypothesis Testing

The concept of equifinality is the basis of the GLUE methodology. It comes from a realisation that due to interactions between physical processes and parameter values defined in the model structure, and limitations of the input data used to drive the model, multiple and competing representations may be able to give simulations that are consistent with the observations that are available (i.e. are *behavioural*) [*Beven and Freer*, 2001b; *Beven*, 2002]. This is not an unreasonable concept given the limits that current measurement technologies place on our ability to characterise the surface / subsurface continuum and to define effective parameter values. As *Naef* [1981] noted, for all hydrological models discharge increases with increasing rainfall but '*an algorithm is not necessarily physically correct if a conceptual model produces reasonable results*'. That sentiment still holds today. Although advances have been made in process understanding in hydrology, hydrologists today still use descriptions of processes that are wrong and are known to be wrong (e.g. we still do not yet have a satisfactory description of flow in a macroporous soil at the hillslope scale?). This comes about because our models are not complete representations of reality and there is danger in assuming that they are [see discussions in *Grayson et al.*, 1992; *Beven*,

1993; *Morton*, 1993; *Beven*, 2001a, 2002]. To embrace the concept that our models are not necessarily 'true' representations but essentially a set of hypotheses [*Beven*, 1992; *Addiscott et al.*, 1995; *Beven*, 2001a, 2002; *Beven and Feyen*, 2002] that can be falsified if there is appropriate data with which to do so is to challenge many of the calibration / validation methodologies currently employed. The GLUE methodology provides a framework for hypothesis testing and model rejection that allows a pragmatic approach to many calibration / validation problems [*Beven*, 2001a; *Beven*, 2001b].

Subjectivity in the Specification of an Appropriate Measure in Rejecting Non-behavioural Simulations

GLUE allows the use of multiple performance measures (*PM*) (qualitative or quantitative), in combination or sequentially through the application of Bayesian updating. The aim is to allow a sensible and justifiable rejection of non-behavioural models and to apply a likelihood weight to the predictions of the retained behavioural models (a 'model' in this context is used to define a model structure and the associated parameter values). Choice of rejection criteria and *PMs* clearly plays an important role in this formulation. The methodology can make use of more traditional likelihood functions where appropriate [see *Romanowicz et al.*, 1994, 1996; *Beven and Freer*, 2001b], but only when the error structure can be formally defined. GLUE has been criticised for the level of subjectivity in the specification of the likelihood measure and behavioural acceptance thresholds [*Melching*, 1995; *Thiemann et al.*, 2001] but some subjectivity would appear to be inevitable for cases where it is not possible to make strong assumptions about the error structure (and where the error structure might vary for different behavioural models). In essence, for nearly all likelihood measures and *PM* there will be a full range of model performance from good (or at least the best that is achievable given the data and model structures available) to poor. There will be no clear boundary between what is a behavioural model and what is non-behavioural. All that is recommended within GLUE is that the *PMs* and acceptance thresholds used should be explicitly defined so that they can be evaluated, discussed and criticised if necessary. In fact, *Freer et al.* [1996], in an application of GLUE to a rainfall-runoff modelling problem, concluded (from an evaluation of the resultant prediction uncertainties for several likelihood measures) that the choice of the behavioural acceptance threshold was not as critical as previously thought.

One approach that tries to include a more objective evaluation of model performance is the multi-criteria calibration of *Boyle et al.* [2000] in which the hydrograph record is separated into driven and non-driven (quick, slow) flows with individual calculations of performance made for each part within a Pareto optimal calibration framework [see also *Wagener et al.*, 2001]. Pareto optimisation attempts to find the set of models that are optimal in the sense of being non-dominated by any other model for all of the criteria considered. While resulting in an objectively chosen set of models, the results of *Boyle et al.* [2000] showed that the 500 Pareto optimal solutions resulted in prediction limits that did not bound the observations for significant periods of the calibration period (in the sample shown). This approach also leads to the rejection of many models that would be considered behavioural and might, indeed, be Pareto optimal given a different calibration period.

In addition, the recent paper of *Thiemann et al.* [2001] gives a good demonstration of the results of making strong assumptions about the nature of the errors within a formal recursive Bayesian methodology. While recognising the multiple sources of error in the modelling process, their procedure follows standard statistical practice in treating the model as if it was the "true" model, lumping all the errors together into an additive model error treated as "measurement error". In their application, this results in a rapid convergence to a single parameter set, with all the uncertainties represented as a combination of a parameter error (negligible in this application) and a (large) "measurement" error. The GLUE methodology, in contrast, implicitly allows for the fact that errors in the data used by the model, the model structure and the available observed responses might yield a population of behavioural models. The use of two such contrasting approaches to the same environmental modelling problem stems from the difficulty of separating these different sources of error in any real application without making very strong (and difficult to justify) assumptions about them.

Previous Use of Measures for the Evaluation Model Performance and Model Rejection?

Past studies have shown that global numerical indices of model performance in predicting discharge are not very good in discriminating between feasible models. Many papers have shown this to be the case in environmental modelling for standard calculations of coefficient of determination (R^2) and the *Nash and Sutcliffe* [1970] Efficiency (E) criteria [*Willmott*, 1981; *Gupta et al.*, 1998; *Legates et al.*, 1999], although these measures are still in general use today. *Legates et al.* [1999] concluded "... *it is clear that correlation-based measures are inappropriate and should not be used to evaluate the goodness-of-fit of model simulations*", and recommended a combination of statistical measures, similar to that suggested by *Gupta et al.* [1998] and *Dunn* [1999]. It is worth mentioning that *Legates et al.*

[1999] report improved sensitivity of the efficiency E criterion by including seasonal variations in the mean discharge (called E') to account for 'baseline adjustments' as suggested by *Garrick et al.* [1978]. What seems clear is that the application of a single numerical expression that adequately summarises model performance for multiple storm sequences is not easily achievable and that more robust methods should now be sought.

Attempts to apply more rigorous assessments of hydrological and coupled hydrogeochemical models have been developed with the use of multi-objective (also termed multi-signal or multi-response) evaluations of additional simulated variables, including contributing area, groundwater storage, water table, isotopic and geochemical signals [*Kuczera*, 1983; *de Grosbois et al.*, 1988; *Hooper et al.*, 1988; *Mroczkowski et al.*, 1997; *Franks et al.*, 1998; *Lamb et al.*, 1998; *Blazkova et al.*, 2002]. Such studies have had mixed results however, with either little improvement in the identifiability of parameter estimates [i.e. *Kuczera and Mroczkowski*, 1998] or that defining appropriate weights for data with varying information content was a non-trivial task [i.e. *Hooper et al.*, 1988]. Furthermore the use of multi-response data will inherently involve observations with different levels of uncertainty in either the measurement techniques or in the characterisation of the underlying hydrological processes and some observations may be incommensurable with the simulated model responses because of scale or model representation problems. The predicted quantity and the observed quantity may not mean the same thing, even if they have the same name.

There seems to be little comment in the literature on how to deal with differing levels of uncertainty in observations in the specification of appropriately formulated likelihood measures, except in terms of correlated and heteroscedastic model errors within a Gaussian framework [*Romanowicz et al.*, 1994]. In the general case for rainfall-runoff modelling, the form (skewness) of the distribution of uncertain predictions varies markedly over the range of streamflow [*Freer et al.*, 1996], and that the appropriate error structure might vary with both the type of data and the model parameter set. GLUE provides a flexible methodology for handling this type of multi-criteria model evaluation but still requires the specification of suitable criteria, especially for model rejection.

Since *Beven and Binley* [1992] first suggested using both quantitative and qualitative (in effect soft or fuzzy) information in the GLUE procedure, only a few studies have used such information within a MC calibration methodology. *Franks et al.* [1998] found that using fuzzy saturated area estimates from remotely sensed data helped significantly constrain parameter estimates [see also *Guntner et al.*, 1999; *Seibert*, 2000]. More recently *Seibert and*

McDonnell (this issue) have called such fuzzy data 'soft information'. This terminology used by *Seibert and McDonnell* (this issue) is somewhat ambiguous. In reality there is no clear threshold between what is 'hard' information and what is poorly defined or 'soft' information in terms of utility in evaluating the predictions of a model. In reality different types of data have varying degrees of uncertainty associated with them including runoff and groundwater information. Furthermore variations in the certainty of values for a single data series will be time variant, for example that reported by *Ambroise et al.* [1996] in relation to the timing of a significant snowmelt event that caused prediction errors for all the models considered throughout the whole summer recession period. The effect of the snowmelt event on the models would not normally be a sufficient reason to reject all the models, which until that event had been considered behavioural (because no models would then be left).

Sampling the Parameter Hyperspace

Environmental models are often highly non-linear and demonstrate complex interactions between the effects of different model parameters in producing acceptable simulation results. These complexities are often revealed in the shape of the response surface for any likelihood or *PM* in the parameter hyperspace. Ideally, we want to examine this surface completely for several different measures, but for high dimensional spaces this can be computationally demanding and difficult to visualise.

For surfaces that show relatively simple forms, density dependent sampling such as Monte Carlo Markov Chain (MCMC) methods can be used [e.g. *Tarantola*, 1987; *Kuczera and Parent*, 1998]. For parameter spaces for which there is strong prior information available, Latin Hypercube sampling can be used to increase the sampling efficiency. For highly complex surfaces, techniques such as Regression Trees [e.g. *Spear et al.*, 1994] can be used to concentrate sampling densities in areas where behavioural model responses are expected.

Generally, however, our experience has been that response surfaces for the practical application of models are complex, and often exhibit an upper limit of model performance that is, presumably, related to errors in the input data as well as errors in the model structure. Models that reach this upper limit may be scattered widely through the model space (it is the parameter set that gives a behavioural simulation rather than values of individual parameters). In this situation, it may be sufficient to obtain an "adequate sample" of behavioural models. Thus, a uniform sampling strategy is often used in GLUE for sampling parameter values from reasonably wide parameter distributions, particularly where there is a lack of strong prior information about effective parameter values and their covariation. The strategy is easy to implement and makes minimal prior assumptions regarding the nature of the response surface, but will be inefficient in identifying behavioural models for some problems. However, increasing availability of cheap powerful parallel computing resources has meant that many sample model realisations can be run for an increasingly wide variety of model types.

Parameters good at simulating one set of data will not necessarily be good at simulating another (note we are extending this here to say that measures that show a model to be good at simulating an entire record are not necessarily compatible with the same measure calculated for a seasonal response). We should also not forget that we could generally reject all models on the basis of our perceptual model of the flow processes. Thus, criteria for rejection will always be, somewhat relaxed so that one or more behavioural models will be retained. Within an optimisation framework this was not an issue, i.e. the best model found was retained since it represented the "best estimate". This is unrealistic, however, if we wish to evaluate the risk of different possible outcomes. However, there remains the issue of what should then constitute an acceptable or behavioural model. The question of when to reject a model, given the problems that we have with our input forcing data and that some data types are likely to be more erroneous than others, is not an easy one.

THE STUDY SITE

The Panola Mountain Research Watershed (PMRW) is a 41-ha forested catchment, 25-km south east of Atlanta, Georgia, USA [*Peters et al.*, 2000]. The catchment, which is in the piedmont physiographic region, is 90% forested, dominated by hickory, oak, tulip poplar, and loblolly pine, and 10% partially vegetated (lichens and mosses) bedrock outcrops (Figure 1). The basin relief is 56 m and slopes average 18%. The forested area varies from 100% deciduous to 100% coniferous [*Cappellato and Peters*, 1995]. The bedrock is predominantly Panola Granite (granodiorite composition). The bedrock contains pods of amphibolitic gneiss particularly at lower elevations. Soils are predominantly ultisols developed in colluvium and residuum intergrading to inceptisols developed in colluvium, recent alluvium, or in highly eroded landscape positions. Typical soil profiles are 0.6 to 1.6 m thick grading into saprolite of variable thickness. The saprolite typically is 0 to 5 m thick over granodiorite and 5 to 20 m over the amphibolite; the valley bottom generally contains the deepest regolith and is the primary surficial aquifer.

The climate is humid continental to sub-tropical. For the period of record, i.e., water years (WY: October through September) 1986-2001, annual precipitation averaged 1,220 mm and ranged from 760 to 1,580 mm. The annual runoff averaged 380 mm and ranged from 150 to 700 mm. The annual water yield averaged 30% and ranged from 16% to 50% with most of the runoff occurring during the winter and early spring from December through April (Figure 2). A breakdown of the WY based on the hydrological state of the catchment by season (Dry, Wetting, Wet and Drying) used for the Monte Carlo simulations, is also summarised in Table 1. A long growing season, warm temperatures, and many sunny days result in a high evapotranspiration demand, particularly during the summer. During the spring and summer from April through September, rainstorms are convective (high intensity and short duration), whereas the remainder of the year precipitation is dominated by synoptic weather systems (low intensity and long duration). During 1951-75, the average number of days in the year with thunderstorms was 47, of which 38 were during the summer [*Court and Griffiths*, 1985]. Less than 1% of the precipitation falls as snow or sleet. High soil moisture deficits can limit runoff from summer rainstorms to a few percent of rainfall. During WY1993, which was chosen for the study herein, the annual precipitation was slightly less than average at 1,120 mm, but runoff was well above average at 560 mm and the annual water yield was the highest of the entire period of record. Most of the precipitation fell during the dormant season (750 mm fell during November through March), during which the evapotranspiration demand is low and the runoff and water yield are high. WY1993 was chosen for modeling because annual precipitation was similar to the long-term average (the monthly precipitation and runoff generally bracket the range for the period of record), but seasonality in precipitation and runoff was pronounced causing the watershed to be both dry and wet for extended periods with distinct wetting up and dry out periods.

Stormflow is flashy and is attributable to runoff generated from bedrock outcrops in the headwaters (Figure 1). Although streamflow decreases rapidly during recession, baseflow is sustained throughout the year, even during droughts. When sufficient runoff is generated by rainfall on the bedrock outcrop, which typically occurs during convective rainstorms with greater than 15 mm of rainfall, a flood wave develops, which propagates rapidly downstream from the base of the 3.6-ha bedrock outcrop. Discharge from the flood wave typically peaks at the upper streamflow gage (250 m downstream from the outcrop and draining the 10-ha sub-catchment, Figure 1) 15 to 20 min after initiation of flow in the channel at the base of the outcrop, and at the lower gauge (200 m downstream from the upper gage) about 20 min later. The time to maximum flow is less when the watershed is wet than when it is dry.

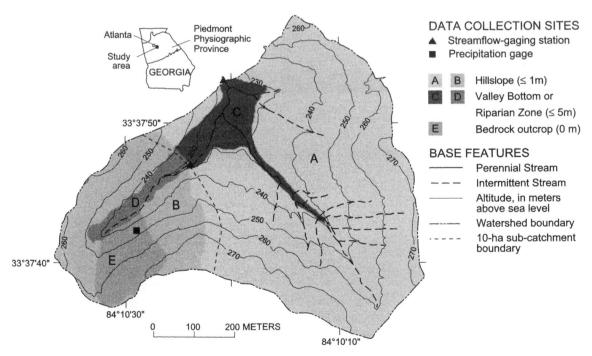

Figure 1. Map of the forested 41-ha Panola Mountain Research Watershed, Georgia showing the spatial connectivity of the three main landscape-units (LU) as used in Dynamic TOPMODEL. LU areas have distinct hydrological functional forms. For PMRW, the LUs were grouped by regolith depth, 0 m for the bedrock outcrop, <1 m for the hillslope, and <5 m for the riparian zone.

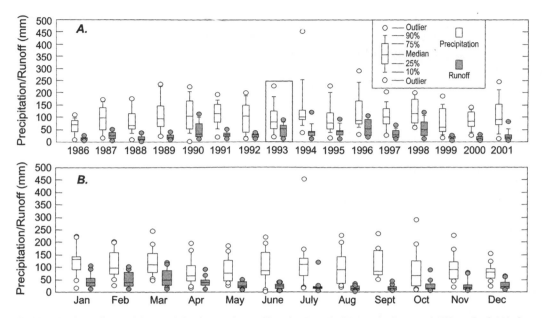

Figure 2. Distribution of monthly precipitation and runoff at the Panola Mountain Research Watershed (A) for each water year (October through September) from 1986 to 2001. Note the data for water year 1993, with the box around them, was used for the analysis herein and (B) for each month for 1986-2001.

Table 1. Panola 1993 Observed Data Set – The different calibration periods used in evaluating different model realisations.

Period	Start Date	End Date	Sum P (mm)	Sum Q (mm)	Q/P	Peak Q (mm/hr)	Mean Q (mm/hr)	POT (mm)
All Year	01/10/92	01/10/93	1116.4	561.8	0.503	3.608	0.064	1091.0
Dry	01/10/92	30/10/92	49.0	22.6	0.461	0.541	0.033	70.2
Wetting	30/10/92	03/01/93	362.7	142.8	0.394	3.608	0.092	62.5
Wet	03/01/93	31/03/93	370.6	207.1	0.559	0.926	0.099	130.0
Drying	31/03/93	30/05/93	122.7	116.2	0.947	0.622	0.081	249.2

METHODS

The Hydrological Model

Dynamic TOPMODEL [*Beven and Freer*, 2001a] is a new version of TOPMODEL that relaxes some of the assumptions of the original form [*Beven and Kirkby*, 1979] following critiques of TOPMODEL by [*Barling et al.*, 1994], *Beven* [1997], and [*Wigmosta and Lettenmaier*, 1999]. This new formulation allows for local accounting of hydrological fluxes and storages, relaxing the quasi steady state assumption of a water table parallel to the local surface slope expressed through the derivation of the $ln(a/tan\beta)$ index of *Kirkby* [1975]. Therefore the seasonal dynamics of the subsurface saturated zone (expanding and contracting) can be simulated. Furthermore the spatial organisation and

connectivity of different landscape units (LUs), each having potentially different functional forms of hydrological (and/or other) responses, are readily accommodated. It is only necessary to know the appropriate conceptual form!

In a comparison of TOPMODEL structures, Dynamic TOPMODEL simulations of streamflow for several years at PMRW improved over that obtained using the Original TOPMODEL form [*Peters et al.*, 2001]. Dynamic TOPMODEL is briefly described below to allow the reader to understand the spatial context of the model structure and associated parameters applied to PMRW. The reader is referred to the paper by [*Peters et al.*, 2001] and to the original paper on Dynamic TOPMODEL by *Beven and Freer* [2001a] for a more detailed treatment of the model application and model theory.

Peters et al. [2001] conceptualised Dynamic TOPMODEL to include the spatial variability of distinct landscape units

(LUs), primarily though the distribution of regolith depths (depth of unconsolidated material including the soil). These LUs were assumed to have different hydrological / physical characteristics that were controlling hydrological response and therefore required the definition of different parameter ranges and/or model structure. The distribution of bedrock depths for the catchment was developed from a detailed (10-m grid) survey of the 10-ha sub-catchment, which contains the 3.6-ha bedrock outcrop [*Zumbuhl*, 1998], together with seismic-refraction transects and well installations. Knowledge of the deep riparian zone gained from the detailed data of the 10-ha sub catchment was transferred to the whole 41ha using depths from other wells and from considering geomorphologically similar landscape positions. This analysis resulted in the identification of three distinct LUs perceived as having notable differences in hydrological responses (see Figure 1). For the hydrological extremes, the relatively deep (> 5m) riparian zone is important for sustaining baseflow (Figure 1, areas C & D) and the bedrock outcrops, i.e., with little or no regolith, are important for generating stormflow (Figure 1, area E). Furthermore, stormflow from the largest bedrock outcrop (3.6 ha) recharges the riparian storage through the channel bottom while the flood wave propagates down the channel. Finally the hillslopes, the majority of the catchment, and the area thought to have the most potential for a seasonally expanding and contracting sub surface saturated zone, and have an average regolith depth of 0.6 m (Figure 1). The variations in regolith depth and the downslope connectivity of flows over a variable bedrock surface might be more important in controlling local hydrological function at PMRW than the soil classification alone [*McDonnell et al.*, 1996; *Freer et al.*, 1997]. In what follows the spatial variability of regolith depths is treated in the simplest possible way (i.e. one depth from a distribution of possible depths that are different for each LU for each simulation), but it would be possible to allow for the change in cross-sectional area for downslope flow arising from a distribution of depths in each LU. Local accounting of fluxes allows for dynamically varying upslope contributing areas 'a' (as previously noted) by the introduction of the additional parameter S_{max} (the maximum effective deficit of subsurface saturated zone), which in a simple form, as in this example, restricts down slope flow only to areas where the local deficit $s_i \geq S_{max}$ Areas with shallow regolith depths (small S_{max}) and areas near the catchment divide, would be more likely to 'disconnect' upslope areas during recession periods. The introduction of a dynamically varying 'a' relaxes the much criticised quasi-steady state assumption in the derivation of a relationship between mean storage deficit or water table depth and local water table depth [*Barling et al.*, 1994; *Franchini et al.*, 1996]. Previously *Beven and Freer*

[2001a] found the best behavioural simulations of discharge at Slapton Wood catchment occurred with a dynamically varying upslope contributing area (i.e. when S_{max} became active). However, good / acceptable (behavioural) simulations also were obtained for simulations where no change in the upslope contributing areas was predicted.

The functional differences in the LUs are here expressed by the differences in the parameter ranges for each unit (see Table 2), i.e., the same functional form is retained for each LU, including the assumption of an exponential decline in transmissivity with depth. The 3-LU model has 17 parameters (*CHV* and SR_{init} are sampled once for each simulation, then assumed constant for all LUs), and, for the bedrock-outcrop LU, the parameter values were either fixed or the ranges narrowed to reflect the absence of regolith (Table 2). The bedrock outcrop is covered with lichens and mosses and contains some solution pits and small vegetation islands. As a result, more than 1 mm of rainfall is needed to generate runoff on the outcrop [*Peters*, 1989], which accounts for the minimal values of SR_{max} and S_{max} for this LU.

The catchment was divided using digital terrain analyses (2 m^2 DEM) into 130 Hydrologically Similar Units (HSUs) for the model simulations. HSUs are sub divisions within each of the 3 LUs using classifier matrices of 'a' and $T_o tan\beta$ defining hydrological similarity between points (upslope contributing area 'a' is included to ensure a general continuity of downslope fluxes between HSUs but is not assumed constant in the predictions of the new model). Transfers between HSUs are calculated using a kinematic wave approximation, where both the upslope (for inputs) and local (for outputs) storages are required. Flux volumes are a function of the storages and the $T_o tan\beta$ values in each case [*Beven and Freer*, 2001a]. Additional subdivisions into different vegetation types were not considered at this stage to avoid adding even more parameters and when the runoff response characteristics were not thought to be so sensitive to variations in the vegetation for this catchment.

GLUE Simulations – Multi-Criteria Likelihood Measures Using Seasonal Periods

In the recent paper by *Peters et al.* [2001] simulations of PMRW using the new Dynamic TOPMODEL were analysed for several water years (WY93, WY94, and WY98) using global likelihood measures. This paper extends this analysis by assessing the multi-criteria seasonal variations in model performance expressed using several performance definitions for WY93 within the GLUE methodology. For each simulation run all parameters listed in Table 2 were randomly assigned a value appropriate to the ranges specified for each LU (where appropriate). The

Table 2. Monte Carlo Sampling - Parameter Ranges

| Parameter | Description | Landscape Unit Parameter Ranges | | |
		Hillslope	Valley Bottom	Bedrock Outcrop
SZM [m]	Form of the exponential decline in transmisivity	0.01 - 0.08	same	same
$ln(T_0)$ [m^2 h^{-1}]	Effective lateral saturated transmissivity	-7 - 1	same	-2 - 3
SR_{max} [m]	Maximum root zone storage	0.005 - 0.05	same	0.002 - 0.012
SR_{init} [m]	Initial root zone deficit	0.0 - 0.05	same	same
CHV [m h^{-1}]	Channel routing velocity	250 - 1500*	n/a	n/a
T_d [m h^{-1}]	Unsaturated zone time delay per unit deficit	1 - 40	same	1 - 10
S_{max} [m]	Maximum effective deficit of subsurface saturated zone	0.2 - 0.7	0.5 - 1.5	0.0 - 0.03

* One sampled *CHV* value is applied to the whole river channel network.

model streamflow predictions for the entire WY93 (known as the global *PMs*) were compared to the observed streamflow using the six *PMs* defined in Table 3. *PMs* identified in Table 3 were calculated separately for each seasonal period identified in Table 1, which were based on the general moisture status of the catchment (*Dry*, *Wetting*, *Wet* and *Drying*). Differences among behavioural parameter sets were evaluated for each seasonal period, using the rejection criteria or behavioural thresholds shown in Table 3.

The GLUE simulations were conducted on the Hydrology and Fluid Dynamics Group parallel LINUX PC system at Lancaster University. The system consists of 33 nodes having a combination of AMD 800MHz and 1500MHz processors. The topology used was a simple master slave combination via 100Mbps Ethernet using basic batch processing scripts for job submissions (one job per slave unit). The 500,000 simulations took 3 days to complete for WY93 (17,520 time steps).

RESULTS AND DISCUSSION

Relationships Between Global PMs

The different definitions of *PMs* shown in Table 3 were chosen to reflect their sensitivity to different hydrologic characteristics of the simulated period. M_{EFF} is biased towards reflecting large errors associated with peak discharges, M_{LOG} is biased towards recession flows, M_{SAE} is a compromise between M_{EFF} and M_{LOG}, M_{BIAS} is the bias for the simulation period and M_{RISE} and M_{FALL} are the biases during the rising and falling limb of the hydrograph, respectively. Relationships among these *PMs* for the behavioural

simulations calculated for the entire WY93 (global set) are shown in Figure 3 as dotty plots, where each point signifies a model simulation having a randomly chosen set of parameters from within the set parameter ranges in Table 2. Figure 3 shows that correlations between *PMs* of the behavioural simulations are quite variable, often having a lot of scatter for one *PM*, when compared with another *PM*. Furthermore the behavioural simulations identified by one *PM* are not necessarily behavioural for another Characteristics of the total compatible sets between *PMs* are summarised in Table 4 (Table 5 also lists the total number of behavioural parameter sets for each *PM*). A comparison of the results for M_{BIAS} with the other *PMs* suggests that the model as currently formulated generally over-predicts total streamflow (negative M_{BIAS}) whilst obtaining good simulations of the larger storm events. Figure 3(a3) clearly identifies this with the top simulations defined using M_{EFF} giving large negative M_{BIAS} values (M_{RISE} and M_{FALL} show similar results as presented in Figures 3(e4) and 3(f4)) that are outside the range of the behavioural simulations for this *PM* shown in Figure 3(d1-5). In contrast, the distributions of behavioural simulations for M_{SAE} and M_{LOG} are symmetrical around $M_{BIAS} = 0$, tending towards a positive M_{BIAS} in each case (see Figures 3(b3) and 3(c3)). Only 585 (5%) of the parameter sets for behavioural simulations were compatible between the M_{EFF} and M_{BIAS} global *PMs* (Table 4). However behavioural simulation parameter sets of M_{BIAS} differ markedly from those for the other global *PMs*. The significant scatter between the M_{EFF}, M_{RISE} and M_{FALL} *PMs* are also interesting, in that some of M_{FALL} behavioural results are clearly not compatible with the behavioural results from M_{EFF}, or M_{RISE}.

Table 3. GLUE - Performance Measures Analysed

Performance Measure	Description	Formulation*	Behavioural Threshold
M_{EFF}	Nash r^2 likelihood measure	$M_{EFF}\left[M(\Theta,\Phi)\right] = \left(1 - \sigma_\varepsilon^2 / \sigma_o^2\right)^N$	0.6
M_{LOG}	Nash r^2 likelihood on log transformed flows	$M_{LOG}\left[M(\Theta,\Phi)\right] = \left(1 - \sigma_{\log\varepsilon}^2 / \sigma_{\log o}^2\right)^N$	0.6
M_{SAE}	Likelihood based on the sum of the absolute errors	$M_{SAE}\left[M(\Theta,\Phi)\right] = \sum \left|Q_{obs} - Q_{sim}\right|$	30% Total Period Discharge**
M_{BIAS}	Likelihood based on the sum of the errors (BIAS)	$M_{BIAS}\left[M(\Theta,\Phi)\right] = \sum Q_{obs} - Q_{sim}$	+/-10% Total Period Discharge***
M_{RISE}	Nash r^2 likelihood measure calculated during the rising limb of the hydrograph	$M_{RISE}\left[M(\Theta,\Phi)\right] = \left(1 - \sigma_\varepsilon^2 / \sigma_o^2\right)^N$ for timesteps where $Q_{(t)} \geq Q_{(t-1)}$	0.6
M_{FALL}	Nash r^2 likelihood measure calculated during the falling limb of the hydrograph	$M_{FALL}\left[M(\Theta,\Phi)\right] = \left(1 - \sigma_\varepsilon^2 / \sigma_o^2\right)^N$ for timesteps where $Q_{(t)} < Q_{(t-1)}$	0.6

* Where $M(\Theta,\Phi)$ is the model output given a set of parameter values Θ and a given error model with parameters Φ [see *Beven and Freer*, 2001], σ_ε^2 is the error variance (log transformed $\sigma_{\log\varepsilon}^2$), σ_o^2 is the variance of the observations (log transformed $\sigma_{\log o}^2$), Q_{obs} is the observed streamflow, Q_{sim} the simulated streamflow, $Q_{(t)}$ is the observed streamflow at timestep t.

The best results for M_{SAE} were the lowest values, * The best results for M_{BIAS} were those sets identified as being closest to zero bias

Relationships Between PMs for Different Seasonal Periods

Analysis of the multi-criteria seasonal *PMs*, rather than just using the global measures can be rewarding for two reasons. One reason was to assess the sensitivity of global measures in characterising model performance for the entire period of record. The second reason was that the understanding the inter-relationships of measures during these periods should give a greater understanding of the model dynamics and potentially help in the future development of the model structure. The use of seasonal *PMs* increases our ability to perform a test on various model structure hypotheses. Behavioural model parameter sets for each global *PM* are compared for the same measure for the four individual seasonal periods (*Dry, Wetting, Wet* and *Drying* in Table 1) in Figure 4. In general, these plots show that there is a considerable amount of scatter in the relationship between the seasonal periods and the global period for each type of *PM* calculated. Relationships become more scattered and less correlated for most measures (except M_{BIAS}) during the drier periods (*Dry* and *Drying*). The most scatter between *PMs* are found for M_{LOG} for the *Dry* and *Drying* periods (Figure 4(e, h)) and M_{RISE} also for the *Dry* period (Figure 4(q)). The global M_{EFF} *PM* is most highly correlated with the *PM* during the *Wetting* period which was expected due to the high maximum flow during this period (Figure 4(b) and Table 1). However given that the *Wet* period overall has 31% more

discharge and the runoff coefficient is significantly higher than the *Wetting* period, the high sensitivity of M_{EFF} to this more extreme event during the *Wetting* period is perhaps undesirable and further questions the use of the M_{EFF} as a global *PM*. Interestingly M_{RISE} and M_{FALL} show more scatter between the seasonal period relationships and have less compatible behavioural parameter sets than M_{EFF} for most seasonal periods (Figure 4(q-x), Table 5).

M_{LOG} and M_{SAE} show similar relationships among the seasonal periods (except for the *Dry* period). Generally M_{BIAS} global has the lowest correlation with M_{BIAS} for each seasonal period, reflecting the potential for the model dynamics to compensate for over- and under-predictions. The range of the seasonal M_{BIAS} when compared with the global M_{BIAS} varies markedly among the different seasonal periods. The behavioural simulations generally under-predict (positive bias) discharge during the Dry and Drying periods (Figure 4(m,p)), and over-predict discharge during the *Wet* period (Figure 4(o)). The transition from under- to over-prediction occurs during the *Wetting* period (Figure 4(n)).

To examine the results of the global and seasonal periods for each *PM* further, the behavioural sets of results for each seasonal period were evaluated for their compatibility with the global set. The compatibility of the behavioural sets were determined by sequentially rejecting the sets for each *PM* starting with the global set and working through the

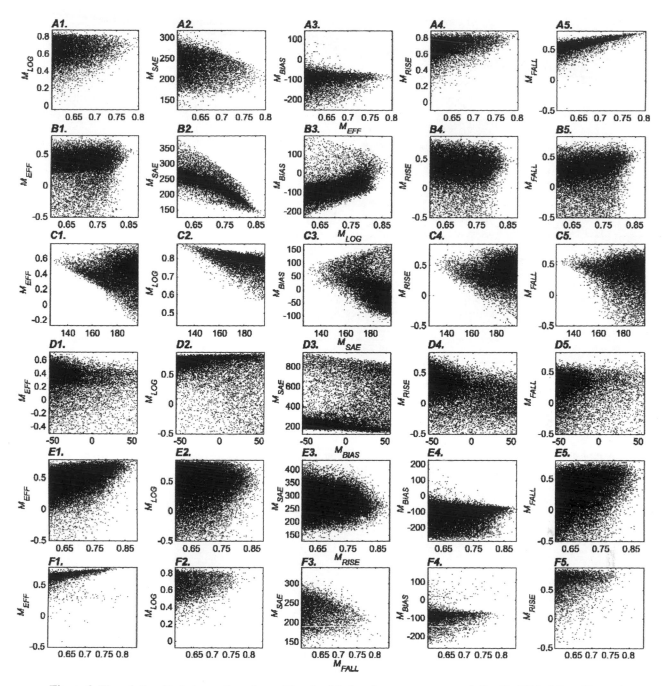

Figure 3. The relationship between the values of the six global performance measures defined in Table 3 from Dynamic TOPMODEL simulations of WY93. For all behavioural simulations identified using each global performance measure (x axes) corresponding values of the other global performance measures for the same simulated parameter sets are shown (y axes) whether they themselves are classed as behavioural or non-behavioural.

individual periods, and also by identifying the union between the four seasonal periods and the global set individually. Results of each type of assessment are summarised in Table 5. By sequentially updating the behavioural sets, the number of compatible parameter sets for each *PM* (Table 5) decrease markedly. The extreme cases are the *PM*s M_{SAE},

M_{BIAS} and M_{LOG} where there are no parameter sets compatible with the behavioural global sets after updating through two (or three) periods. For M_{BIAS} this further reflects the poor correlations shown for this *PM* in Figure 4(m-p). Surprisingly M_{EFF} has the most compatible sets (besides M_{LOG}) after updating on the *Dry* period, and the largest

Table 4. The number of compatible behavioural parameter sets between each performance measure for the 'All Year' global period. Results are listed as both the total number of compatible parameter sets (*italics*) and the equivalent percentage of possible parameter set combinations (i.e. a percentage of the lowest population of behavioural parameter sets being compared for each case).

		Performance Measures					
		M_{EFF}	M_{LOG}	M_{SAE}	M_{BIAS}	M_{RISE}	M_{FALL}
Performance Measures	M_{EFF}		*7088*	*1278*	*585*	*9243*	*4385*
	M_{LOG}	65%		*16129*	*40304*	*15673*	*3723*
	M_{SAE}	12%	100%		*11752*	*1528*	*852*
	M_{BIAS}	5%	66%	73%		*1597*	*288*
	M_{RISE}	85%	32%	10%	3%		*3299*
	M_{FALL}	85%	72%	16%	6%	64%	

Table 5. The reduction in the number of behavioural parameter sets for each performance measure by conditioning sequentially through the seasonal periods. The initial population of behavioural parameter sets is that calculated for the '*All Year*' global period in each case.

Likelihood Measure	Conditioning Period				
	All Year	**Dry**	**Wetting**	**Wet**	**Drying**
M_{EFF}	10858 (2)	3939 (36) {36}	3370 (85) {82}	295 (8) {10}	222 (75) {51}
M_{LOG}	155572 (31)	14390 (9) {9}	7 (0.05) {2}	0 {1}	0 {17}
M_{SAE}	16159 (3)	1493 (9) {9}	0 {0}	0 {71}	0 {94}
M_{BIAS}	61110 (12)	1448 (2) {2}	61 (4) {36}	0 {11}	0 {40}
M_{RISE}	49352 (10)	1033 (2) {2}	723 (70) {63}	456 (63) {11}	433 (95) {31}
M_{FALL}	5186 (1)	2544 (49) {49}	1956 (77) {68}	35 (2) {3}	31 (86) {30}

{}'s Denote the percentage of parameter sets for each seasonal period compatible with the 'All Year' behavioural parameter sets without conditioning on other seasonal periods, ()'s give the percentage retention in the number of parameter sets after conditioning on each seasonal period.

reduction of sets during the *Wet* period (the total number of behavioural sets for M_{EFF}>0.6 during this period was only 2898). The increased sensitivity of the M_{EFF} criterion by taking into account the seasonal variations of the baseline flow is comparable to that noted by *Legates et al.* [1999]. For the union of all seasonal periods with the global period generally the best correlation between *PMs* for all seasonal periods is given by M_{EFF} (Table 5). However M_{SAE} has the highest percentage of compatible sets during the *Wet* and *Drying* periods (Table 5). The *Wet* and *Drying* periods have both the highest percentage of total discharge (72%) and constitute 61% of the total time period considered for the seasonal analysis. The later result is consistent with other GLUE papers, which report that the 90% prediction limits for this measure span the observations for more of the time when compared to those of other *PMs*. However, M_{SAE} highlights the difficulties of applying a globally applied threshold for all seasonal periods having no parameter set combinations that are classed as behavioural for the *Wetting* period.

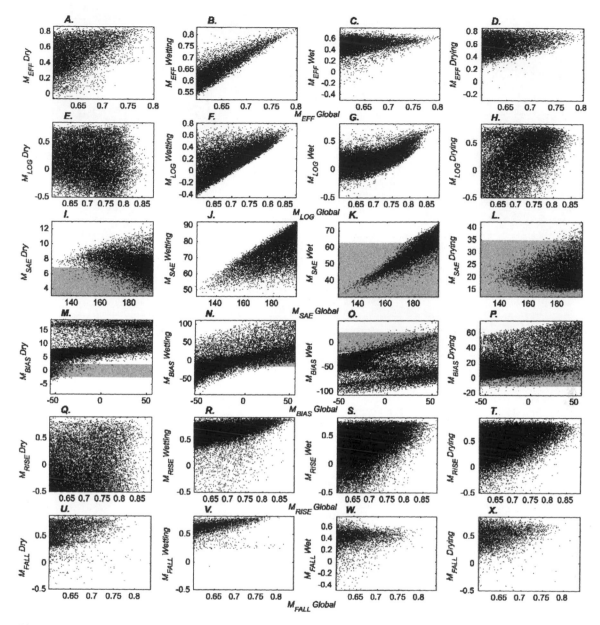

Figure 4. The relationship between the values of the global and seasonal periods for the six performance measures defined in Table 3. For all behavioural simulations identified using each global performance measure (x axes) corresponding values of the same performance measures for each season for the same set of parameters are shown (y axes) whether they themselves are classed as behavioural or non-behavioural. For objective functions M_{SAE} and M_{BIAS} a shaded box has been added to show the range of behavioural simulations for each seasonal period.

Compatibility of Parameter Ranges for PMs and Seasonal Periods

The multi-criteria seasonal results also show changes to the behavioural model parameter distributions for different periods of the year and for different *PMs*. Results showing the variability of parameter distributions for the *Hillslope* LU are shown in Figure 5 and results for the *Valley Bottom* LU in Figure 6 (as noted above, there are no behavioural

simulations for M_{SAE} for the *Wetting* period, see Table 5). Importantly most parameters of the behavioural simulations shown in Figure 5 are highly variable for each *PM* and for each seasonal period. The low overlap among distributions for the inter quartile range (the range of the box in Figure 5) for most parameters demonstrates the need to introduce a greater range of seasonal dynamic behaviour into the model to reduce the incompatibility of parameter sets. This variability between *PMs* and seasonal periods is not well

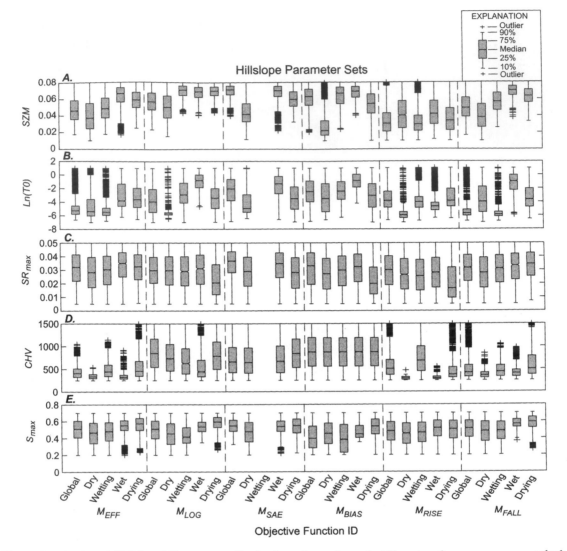

Figure 5. Behavioural *Hillslope* LU parameter distributions shown for each different performance measure calculated for each different seasonal period.

defined for the *Valley Bottom* LU (Figure 6). Furthermore the variability of parameter distributions for the *Bedrock Outcrop* LU (not plotted) showed almost no sensitivity. The area associated with both these LUs is significantly smaller than that of the *Hillslope* LU (*Hillslope* 85.5%, *Valley bottom* 9% and Bedrock Outcrop 5.5% of the total catchment area, see Figure 1), but the low sensitivity is still a surprising result because these LUs were perceived to have important hydrological response characteristics.

For the *Hillslope* LU parameter results, *SZM* varies the most seasonally, highlighting the previously reported sensitivity of this parameter. The variability in *SZM* characterises the changing shape of the observed recession form during the year requiring higher values (less steep recessions) during wetter periods (consistent with observations of changes in recession characteristics). *ln(T0)* also shows a

general sensitivity for each season for all measures, with the largest range of *ln(T0)* values occurring during the wettest periods. The smallest inter-quartile range of *ln(T0)* for both M_{LOG} and M_{SAE} occurs during the *Dry* period. *CHV* is sensitive for the M_{EFF} PM (Figure 5), perhaps compensating for the LUs rapidly responding to rainfall inputs that result from timing errors on the rising limb of the hydrograph (note this is not important for the M_{BIAS} results). S_{max} has the lowest values for M_{BIAS} for the global period, corresponding to the most potential for spatial dynamic variability in the sub-surface saturated zone contributions. Although SR_{max} generally is the least sensitive of the parameters with respect to the different *PMs*, the distribution of this parameter is the most constrained for the M_{BIAS} during the *Drying* period. Finally the variability between parameter distributions for M_{RISE} and M_{FALL} deserve some comment. Generally the rising limb of

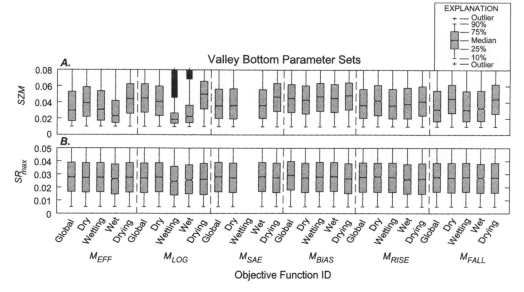

Figure 6. Behavioural *Valley Bottom* and *Bedrock Outcrop* LU parameter distributions shown for each different performance measure calculated for each different seasonal period.

the hydrograph requires steeper recessions, or lower *SZM*; reduced *ln(T0)* values (except for the *Wetting* period) and a similar pattern of *CHV*, with the latter two highlighting changes in the model dynamics associated with time to peak responses (and the difficulties involved in simulating the *Wetting* period in catchments with marked seasonality using models with time invariant parameter sets).

Constraining Model Simulation Responses

As well as assessing the variability in the model parameter distributions for the *PMs* and the seasonal periods similar analyses can be conducted for the distributions and statistics of the model output responses. Figure 7 shows the minimum, maximum, 5th and 95th percentile and mean statistics of three summary model simulation outputs including total simulated discharge (Q_{sum}), maximum discharge (Q_{peak}) and maximum surface saturation (SAT_{max}). The statistics were calculated using the results of the behavioural simulations for each *PM* and for each seasonal period.

These results show that there is significant variability in model responses between measures (especially for maximum simulated discharge). Furthermore, the variability in the model responses for each behavioural set of different *PMs* are high, i.e., total discharge simulated for the entire water year ranges from ~400 to 800mm for most *PMs*). M_{BIAS} is poor at discriminating between model dynamics which would be rejected qualitatively as inconsistent with our perception of the catchment hydrological dynamics at Panola, i.e., a maximum saturated area response of 90% of the catchment area. The results of Figure 7 demonstrate the

potential of using multi-criteria methods in further discriminating between behavioural simulations, but here clearly shows that the improved model formulation still has deficiencies, despite providing optimal simulations for the complete WY93 period that would normally be considered acceptable.

CONCLUSIONS

This paper demonstrates that several global performance measures are poor at discriminating between seasonal behavioural simulations (especially M_{FFF}), especially, as in this case, where there is substantial variability in the observed seasonal discharge responses. However, all model simulations are rejected by a multi-criteria combination of all seasonal *PMs*, even though the rejection criteria are not too strict (indeed all models are rejected for a combination of globally applied *PMs*). Effectively, the successful parameter sets for different *PMs* and different seasons are not consistent. This is not the only time that all models have been rejected within the GLUE framework using multi-criteria evaluation [see also *Zak and Beven*, 1999a]. This is not a problem, of course in optimisation or in multi-criteria Pareto optimisation since there will always be a "best" model, or a set of models at the Pareto front. It does, however, raise the question of whether these "optimal" models will always be acceptable in all aspects.

Perhaps this research has shown that we require a combination of *PMs* (or different *PMs*) to highlight both the seasonal variability in performance and different aspects of the forcing responses for hydrographs [i.e. *Boyle et al.*, 2000].

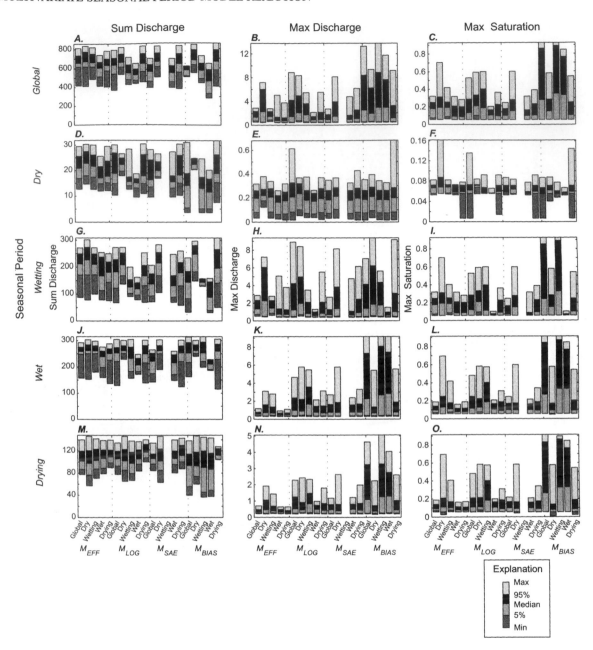

Figure 7. Distributions of summary model responses (cumulative discharge, maximum discharge and maximum saturated area) calculated for each different seasonal period (each row of the plot). Each plot shows for behavioural simulations the variations in the minimum, 5th percentile, mean, 95th percentile and maximum points in the respective model response distribution obtained for each different performance measure calculated for each different seasonal period.

Initially the multi-criteria rejection of models identified acceptable *PMs* for the global period (e.g. maximum M_{EFF}>0.8), which raises the question of what definition should be applied to calibrate dynamic TOPMODEL. This criteria also would need to reflect the knowledge that that there is variable error in the input data, error in the observations that are used to evaluate the models, and error in the model structure. Furthermore, the complexity of the model

dynamics and the effectiveness of observed data to represent processes at the model scale should properly result in rejection criteria that reflect the characteristics of different data or different periods of data. It may be that by imposing even reasonably strict performance criteria, most if not all hydrological models might be rejected. Another way of looking at the results presented here is that the seasonal variation in parameter distributions and model responses suggest that Dynamic

TOPMODEL as applied to PMRW, needs further development to improve the model structure (e.g. to introduce more seasonality into the responses and/or develop better spatial representations of processes for different LUs).

This is not a pessimistic conclusion. It is necessary to have reasons to reject concepts in order for the science to progress. The interesting research question that follows, however, is how to learn from the relative success or failure of the model for different *PMs* in structuring the process of making improvements to the model structure. Does model failure add to understanding of the system? Or is it simply better to revisit the perceptual model of PMRW in trying to refine the model structure. We will be trying to do both in future papers.

Acknowledgements. The study was conducted in cooperation with the Georgia Department of Natural Resources. The authors are grateful for the support provided by the staff of Panola Mountain State Park. The authors also are grateful to the USGS staff for the Panola Mountain Research Project, and in particular, for the database support by B.T. Aulenbach. The research was partly funded by a U.S. National Science Foundation grant (EAR 9743311), United Kingdom National Environmental Research Council (grants GR3/11450 and NER/L/S/2001/00658), and the U.S. Geological Survey's Water, Energy and Biogeochemical Budgets Program.

REFERENCES

Addiscott, T., J. Smith, and N. Bradbury. Critical-evaluation of models and their parameters. *Journal of Environmental Quality, 24:*803-807, 1995.

Ambroise, B., J. E. Freer, and K. J. Beven. Application of a generalized TOPMODEL to the small Ringelbach catchment, Vosges, France. *Water Resour. Res., 32:*2147-2159, 1996.

Aronica, G., B. Hankin, and K. Beven. Uncertainty and equifinality in calibrating distributed roughness coefficients in a flood propagation model with limited data. *Adv. in Water Resour., 22:*349-365, 1998.

Barling, R. D., I. D. Moore, and R. B. Grayson. A quasi-dynamic wetness index for characterizing the spatial-distribution of zones of surface saturation and soil-water content. *Water Resour. Res., 30:*1029-1044, 1994.

Beven, K. J. Future of Distributed Modelling - Special Issue of Hydrol. Processes. *Hydrol. Processes, 6:*253-254, 1992.

Beven, K. J. Prophecy, reality and uncertainty in distributed hydrological modeling. *Adv. in Water Resour., 16:*41-51, 1993.

Beven, K. J. TOPMODEL: A critique. *Hydrol. Processes, 11:*1069-1085, 1997.

Beven, K. J. Uniqueness of place and process representations in hydrological modelling. *Hydrol. and Earth Sys. Sci., 4:*203-213, 2000.

Beven, K. J. How far can we go in distributed hydrological modelling? *Hydrol. and Earth Sys. Sci., 5:*1-12, 2001a.

Beven, K. J. On an acceptable criterion for model rejection. *Circulation, 68:*9-11, 2001b.

Beven, K. J. On hypothesis testing in hydrology. *Hydrol. Processes, 15:*1655-1657, 2001c.

Beven, K. J. *Rainfall-Runoff Modelling: The Primer.* Wiley, Chichester, 2001d.

Beven, K. J. Towards an alternative blueprint for a physically-based digitally simulated hydrologic response modelling system. *Hydrol. Processes, 16:*189-206, 2002.

Beven, K. J., and A. Binley. The Future of Distributed Models - Model Calibration and Uncertainty Prediction. *Hydrol. Processes, 6:*279-298, 1992.

Beven, K. J., and J. Feyen. The future of distributed hydrological modelling. *Hydrol. Processes, 16:*169-172, 2002.

Beven, K. J., and S. W. Franks. Functional similarity in landscape scale SVAT modelling. *Hydrol. and Earth Sys. Sci., 3:*85-93, 1999.

Beven, K. J., and J. Freer. A dynamic TOPMODEL. *Hydrol. Processes, 15:*1993-2011, 2001a.

Beven, K. J., and J. Freer. Equifinality, data assimilation, and uncertainty estimation in mechanistic modelling of complex environmental systems using the GLUE methodology. *J. of Hydrol., 249:*11-29, 2001b.

Beven, K. J., J. E. Freer, B. Hankin, and K. Schulz. The use of generalised likelihood measures for uncertainty estimation in high order models of environmental systems. *in Nonlinear and Nonstationary Signal Processing,* edited by R. L. S. W.J. Fitzgerald, A.T. Walden and Peter Young, pp. 144-183, 2000.

Beven, K. J., and M. J. Kirkby. A physically based, variable contributing area model of basin hydrology. *Hydrological Sciences Bulletin-Bulletin Des Sciences Hydrologiques, 24:*43-69, 1979.

Blazkova, S., K. J. Beven, and A. Kulasova. On constraining TOPMODEL hydrograph simulations using partial saturated area information. *Hydrol. Processes, 16:*441-458, 2002.

Boyle, D. P., H. V. Gupta, and S. Sorooshian. Toward improved calibration of hydrologic models: Combining the strengths of manual and automatic methods. *Water Resour. Res., 36:*3663-3674, 2000.

Brazier, R. E., K. J. Beven, S. G. Anthony, and J. S. Rowan. Implications of model uncertainty for the mapping of hillslope-scale soil erosion predictions. *Earth Surface Processes and Landforms, 26:*1333-1352, 2001.

Brazier, R. E., K. J. Beven, J. Freer, and J. S. Rowan. Equifinality and uncertainty in physically based soil erosion models: Application of the glue methodology to WEPP-the water erosion prediction project-for sites in the UK and USA. *Earth Surface Processes and Landforms, 25:*825-845, 2000.

Cameron, D., K. Beven, and J. Tawn. Modelling extreme rainfalls using a modified random pulse Bartlett-Lewis stochastic rainfall model (with uncertainty). *Adv. in Water Resour., 24:*203-211, 2000.

Cameron, D. S., K. J. Beven, J. Tawn, S. Blazkova, and P. Naden. Flood frequency estimation by continuous simulation for a gauged upland catchment (with uncertainty). *J. of Hydrol., 219:*169-187, 1999.

Cappellato, R., and N. E. Peters. Dry deposition and canopy leaching rates in deciduous and coniferous forests of the Georgia Piedmont: an assessment of a regression model. *J. of Hydrol., 169:*131-150, 1995.

Court, A., and J. F. Griffiths. Chapter 2. Thunderstorm climatology. *in Thunderstorm Morphology and Dynamics 2nd ed.*, edited by E. Kessler, pp. 9-39, University of Okalahoma Press, Norman, Okalahoma, 1985.

Dawdy, D. R., and T. O'Donnell. Mathematical models of catchment behaviour. *Journal of the Hydraulic Division of the American Society of Civil Engineers*, 91:113-137, 1965.

de Grosbois, E., R. P. Hooper, and N. Christophersen. A multisignal automatic calibration methodology for hydrochemical models - a case-study of the birkenes model. *Water Resour. Res.*, 24:1299-1307, 1988.

Dunn, S. M. Imposing constraints on parameter values of a conceptual hydrological model using baseflow response. *Hydrol. and Earth Sys. Sci.*, 3:271-284, 1999.

Feyen, L., K. J. Beven, F. De Smedt, and J. Freer. Stochastic capture zone delineation within the generalized likelihood uncertainty estimation methodology: Conditioning on head observations. *Water Resour. Res.*, 37:625-638, 2001.

Fisher, J., and K. J. Beven. Modelling of stream flow at Slapton Wood using TOPMODEL within an uncertainty estimation framework. *Field Studies*, 8:577-584, 1996.

Franchini, M., J. Wendling, C. Obled, and E. Todini. Physical interpretation and sensitivity analysis of the TOPMODEL. *J. of Hydrol.*, 175:293-338, 1996.

Franks, S., and K. J. Beven. Estimation of evapotranspiration at the landscape scale: A fuzzy disaggregation approach. *Water Resour. Res.*, 33:2929-2938, 1997.

Franks, S., K. J. Beven, P. F. Quinn, and I. R. Wright. On the sensitivity of soil-vegetation-atmosphere transfer (SVAT) schemes: equifinality and the problem of robust calibration. *Agricultural and Forest Meteorology*, 86:63-75, 1997.

Franks, S. W., P. Gineste, K. J. Beven, and P. Merot. On constraining the predictions of a distributed moder: The incorporation of fuzzy estimates of saturated areas into the calibration process. *Water Resour. Res.*, 34:787-797, 1998.

Freer, J., K. Beven, and B. Ambroise. Bayesian estimation of uncertainty in runoff prediction and the value of data: An application of the GLUE approach. *Water Resour. Res.*, 32:2161-2173, 1996.

Freer, J., J. McDonnell, K. J. Beven, D. Brammer, D. Burns, R. P. Hooper, and C. Kendal. Topographic controls on subsurface storm flow at the hillslope scale for two hydrologically distinct small catchments. *Hydrol. Processes*, 11:1347-1352, 1997.

Garrick, M., C. Cunnane, and J. E. Nash. A criterion of efficiency for rainfall-runoff models. *J. of Hydrol.*, 36:375-381, 1978.

Grayson, R. B., I. D. Moore, and T. A. McMahon. Physically Based Hydrologic Modelling .2. Is the Concept Realistic. *Water Resour. Res.*, 28:2659-2666, 1992.

Guntner, A., S. Uhlenbrook, J. Seibert, and C. Leibundgut. Multicriterial validation of TOPMODEL in a mountainous catchment. *Hydrol. Processes*, 13:1603-1620, 1999.

Gupta, H. V., S. Sorooshian, and P. O. Yapo. Toward improved calibration of hydrologic models: Multiple and noncommensurable measures of information. *Water Resour. Res.*, 34:751-763, 1998.

Hankin, B. G., R. Hardy, H. Kettle, and K. J. Beven. Using CFD in a GLUE framework to model the flow and dispersion character-

istics of a natural fluvial dead zone. *Earth Surface Processes and Landforms*, 26:667-687, 2001.

Hooper, R. P., A. Stone, N. Christophersen, E. de Grosbois, and H. M. Seip. Assessing the Birkenes Model of Stream Acidification Using a Multisignal Calibration Methodology. *Water Resour. Res.*, 24:1308-1316, 1988.

Hornberger, G. M., K. J. Beven, B. J. Cosby, and D. E. Sappington. Shenandoah Watershed Study - Calibration of a Topography-Based, Variable Contributing Area Hydrological Model to a Small Forested Catchment. *Water Resour. Res.*, 21:1841-1850, 1985.

Kirkby, M. Hydrograph Modelling Strategies. *in Processes in Physical and Human Geography*, edited by R. Peel, M. Chisholm, and P. Haggett, pp. 69-90, Heinemann, London, 1975.

Kuczera, G. On the Relationship between the Reliability of Parameter Estimates and Hydrologic Time-Series Data Used in Calibration. *Water Resour. Res.*, 18:146-154, 1982.

Kuczera, G. Improved Parameter Inference in Catchment Models .2. Combining Different Kinds of Hydrologic Data and Testing Their Compatibility. *Water Resour. Res.*, 19:1163-1172, 1983.

Kuczera, G., and M. Mroczkowski. Assessment of hydrologic parameter uncertainty and the worth of multiresponse data. *Water Resour. Res.*, 34:1481-1489, 1998.

Kuczera, G., and E. Parent. Monte Carlo assessment of parameter uncertainty in conceptual catchment models: the Metropolis algorithm. *J. of Hydrol.*, 211:69-85, 1998.

Lamb, R., K. J. Beven, and S. Myrabo. Discharge and water table predictions using a generalised TOPMODEL formulation. *Hydrol. Processes*, 11:1145-1168, 1997.

Lamb, R., K. J. Beven, and S. Myrabo. Use of spatially distributed water table observations to constrain uncertainty in a rainfall-runoff model. *Adv. in Water Resour.*, 22:305-317, 1998.

Legates, D. R., M. Jr., and G.J. Evaluating the use of goodness-of-fit measures in hydrologic and hydroclimatic model validation. *Water Resour. Res.*, 35:233-241, 1999.

McDonnell, J. J., J. Freer, R. Hooper, C. Kendall, D. Burns, K. J. Beven, and J. Peters. New method developed for studying flow on hillslopes. *EOS, Trans. AGU*, 77:465/472, 1996.

Melching, C. S. Reliability estimation. *in Computer Models of Watershed Hydrology*, edited by V. P. Singh, pp. 69-118, Water Res. Publ., Colo., USA, 1995.

Molicova, H., M. Grimaldi, M. Bonell, and P. Hubert. Using TOPMODEL towards identifying and modelling the hydrological patterns within a headwater, humid, tropical catchment. *Hydrol. Processes*, 11:1169-1196, 1997.

Moore, R. D., and J. C. Thompson. Are water table variations in a shallow forest soil consistent with the TOPMODEL concept? *Water Resour. Res.*, 32:663-669, 1996.

Morton, A. Mathematical models: questions of trustworthiness. *Brit. J. Phil. Sci.*, 44:659-674, 1993.

Mroczkowski, M., G. P. Raper, and G. Kuczera. The quest for more powerful validation of conceptual catchment models. *Water Resour. Res.*, 33:2325-2335, 1997.

Mwakalila, S., P. Campling, J. Feyen, G. Wyseure, and K. Beven. Application of a data-based mechanistic modelling (DBM) approach for predicting runoff generation in semi-arid regions. *Hydrol. Processes*, 15:2281-2295, 2001.

Naef, F. Can We Model the Rainfall-Runoff Process Today. *Hydrological Sciences Bulletin-Bulletin Des Sciences Hydrologiques,* 26:281-289, 1981.

Nash, J. E., and J. V. Sutcliffe. River flow forecasting through conceptual models, Part 1: A discussion of principles. *J. of Hydrol.,* 10:282-290, 1970.

Peters, N. E. Atmospheric deposition of sulfur to a granite outcrop in the piedmont of Georgia, U.S.A. *in Atmospheric Deposition,* edited by D. J. Delleur, pp. 173-181, IAHS Publication, 1989.

Peters, N. E., J. E. Freer, and K. J. Beven. Modeling hydrologic responses in a small forested watershed by a new dynamic TOP-MODEL (Panola Mountain, Georgia, USA). *in Runoff Generation and Implications for River Basin Modeling,* edited by S. Uhlenbrook, C. Leibundgut, and J. J. McDonnell, pp. 318-325, Freiburger Schriften zur Hydrologie, Band 13, Frieburg, 2001.

Peters, N. E., R. P. Hooper, T. G. Huntington, and B. T. Aulenbach. Panola Mountain Research Watershed - Water, Energy, and Biogeochemical Budgets Program. Fact Sheet 162-99, U.S. Geological Survey, Atlanta, 2000.

Ratto, M., S. Tarantola, and A. Saltelli. Sensitivity analysis in model calibration: GSA-GLUE approach. *Computer Physics Communications,* 136:212-224, 2001.

Romanowicz, R., and K. Beven. Dynamic real-time prediction of flood inundation probabilities. *Hydrological Sciences Journal-Journal Des Sciences Hydrologiques,* 43:181-196, 1998.

Romanowicz, R., K. J. Beven, and J. Tawn. Evaluation of predictive uncertainty in nonlinear hydrological models using a Bayesian approach. *in Statistics for the Environment. II. Water Reslated Issues,* edited by V. T. Barnett, K.F., pp.??-??, John Wiley, Chichester, 1994.

Romanowicz, R., K. J. Beven, and J. Tawn. Bayesian calibration of flood inundation models. *in Flood Plain Processes,* edited by A. M.G., W. D.E., and P. D. Bates, pp. 181-196, Wiley, Chicester, 1996.

Seibert, J. Multi-criteria calibration of a conceptual runoff model using a genetic algorithm. *Hydrol. and Earth Sys. Sci.,* 4:215-224, 2000.

Seibert, J., K. H. Bishop, and L. Nyberg. A test of TOPMODEL's ability to predict spatially distributed groundwater levels. *Hydrol. Processes,* 11:1131-1144, 1997.

Sorooshian, S., and J. A. Dracup. Stochastic parameter estimation procedures for hydrologic rainfall-runoff models: Correlated & heteroscedastic error cases. *Water Resour. Res.,* 16:430-442, 1980.

Spear, R. C., T. M. Grieb, and N. Shang. Parameter Uncertainty and Interaction in Complex Environmental- Models. *Water Resour. Res.,* 30:3159-3169, 1994.

Spear, R. C., and G. M. Hornberger. Eutrophication in Peel Inlet 2: Identification of crucial uncertainties via generalised sensitivity analysis. *Water Research,* 14:43-49, 1980.

Tarantola, A. *Inverse Problems Theory, Methods for Data Fitting and Model Parameter Estimation.* Elsevier, Netherlands, 1987.

Thiemann, M., M. Trosset, H. Gupta, and S. Sorooshian. Bayesian recursive parameter estimation for hydrologic models. *Water Resour. Res.,* 37:2521-2535, 2001.

Wagener, T., D. P. Boyle, M. J. Lees, H. S. Wheater, H. V. Gupta, and S. Sorooshian. A framework for development and application of hydrological models. *Hydrol. and Earth Sys. Sci.,* 5:13-26, 2001.

Wigmosta, M. S., and D. P. Lettenmaier. A comparison of simplified methods for routing topographically driven subsurface flow. *Water Resour. Res.,* 35:255-264, 1999.

Williams, B. J., and W. W. G. Yeh. Parameter-Estimation in Rainfall Runoff Models. *J. of Hydrol.,* 63:373-393, 1983.

Willmott, C. J. On the validation of models. *Physical Geography,* 2:184-194, 1981.

Yapo, P. O., H. V. Gupta, and S. Sorooshian. Multi-objective global optimization for hydrologic models. *J. of Hydrol.,* 204:83-97, 1998.

Zak, S., and K. J. Beven. Equifinality, sensitivity and predictive uncertainty in the estimation of critical loads. *Journal of Environmental Quality,* 236:191-214, 1999a.

Zak, S. K., K. Beven, and B. Reynolds. Uncertainty in the estimation of critical loads: A practical methodology. *Water Air and Soil Pollution,* 98:297-316, 1997.

Zak, S. K., and K. J. Beven. Equifinality, sensitivity and predictive uncertainty in the estimation of critical loads. *Science of the Total Environment,* 236:191-214, 1999b.

Zumbuhl, A. T. Spatial modeling of soil depth and landscape variability in a small, forested catchment. unpublished M.S. Thesis. College of Environmental Science and Forestry, State University of New York, Syracuse, NY, 1998.

Keith Beven and Jim Freer, Lancaster University, Department of Environmental Sciences, IENS, Lancaster, LA1 4YQ, UK

Norman Peters, US Geological Survey, 3039 Amwiler Road, Suite 130, Atlanta, Georgia 30360-2824, USA

Global Optimization for Watershed Model Calibration

Qingyun Duan

NOAA/NWS, Hydrology Laboratory, Office of Hydrologic Development, Silver Spring, Maryland

Optimization methods have been used widely to calibrate the parameters of watershed models since the very beginning of the digital watershed modeling era. Over the years, much progress has been made in both optimization theory and practice, especially in the area of global optimization methods for use in watershed model calibration. This chapter looks back at the past experience of watershed modelers in their endeavor to estimate the proper parameters of watershed models by relying on optimization methods. The many trials and tribulations with classical local search methods are discussed. Recent progress in utilizing the more powerful and robust global optimization methods is reviewed. A survey of the state-of-the-art global optimization methods is provided. Three global optimization methods commonly used in watershed model calibration: Simulated Annealing, Genetic Algorithm and Shuffled Complex Evolution, are described in detail.

1. INTRODUCTION

The need to find the optimal solution to a problem is encountered in virtually every area of human endeavor. In areas such as mathematics, engineering designs, economics, medicine, telecommunications, postal delivery, river forecasting, manufacturing and control, among others, the problem may be represented in the form of a mathematical function, known as the objective function. Solution strategies based on the theory of function optimization can be used to find the optimal solution, typically one that minimizes (or maximizes) the objective function. If the objective function is unimodal (i.e., it has a single minimum (or maximum)), the optimization problem is very well understood, and many successful solution strategies have been developed (see Luenberger, 1984; Fletcher, 2000; *etc.*).

In practice, a great number of the optimization problems have more than one local solution, of which only one may represent the global or "true" optimum. One class of problems involves finding the parameter values of a dynamical model that optimally represents the time-varying output behavior of some physical system. When the model is significantly nonlinear in its input-to-state and/or state-to-output representations, the problem of finding the global solution to the param-

eter optimization problem can be quite difficult. Such is the case for the calibration of many conceptual watershed models.

Conceptual watershed models are formulated using empirical relationships between hydrological variables observed in nature or field experiments or derived based on abstract conceptualization of physical processes. Conceptual watershed models (or simply watershed models) generally have many constants and exponents called model parameters. The performance of a watershed model depends to a great degree on how its parameters are estimated. Even though there is some physical basis for these parameters, they are usually not directly measurable, especially at the scales of our interest (from 10's to 1000's of km^2). To obtain the best match between simulated outputs from the model and observed outputs from the watershed, these parameters need to be tuned. The process of tuning model parameters is called model calibration.

There are two broad approaches to watershed model calibration: manual and automatic. In manual approach, a trial-and-error procedure is used to estimate model parameters. Model knowledge and a multitude of model performance measures (i.e., objective functions), along with human judgment and visual aids, combine to determine the best guesses for model parameters. This process is less prone to the effect of noises in calibration data. But it demands a high level of understanding of the model physics and the inner exchanges among model components. The advent of the interactive graphic-user-interface (GUI) software has made

Calibration of Watershed Models
Water Science and Application Volume 6
Copyright 2003 by the American Geophysical Union
10/1029/006WS06

this process much easier. Still it is tedious and labor-intensive and a novice user needs extensive hands-on training to be proficient. Further, manual calibration procedures take years to develop and are model-specific. For example, the Interactive Calibration Program (ICP) of the National Weather Service (NWS) River Forecast System (NWSRFS) was designed specifically to calibrate the Sacramento Soil Moisture Accounting (SAC-SMA) model (see Burnash et al., 1973; Smith et al., this volume) and it is not easily transferrable for use on another model.

In automatic approach, model calibration problem is formulated as an optimization problem so computer-based optimization methods can be employed to locate the optimal model parameters. This process takes advantages of a myriad of optimization methods available and relies on computer speed and power to perform the mundane task of finding the optimal parameters with respect to a given objective function(s). Automatic calibration procedures can be generalized for use on different models and can be easily grasped by novice model users. However, automatic calibration is by no means a trivial exercise at all. The success of automatic calibration depends heavily on four factors: model structure, calibration data, calibration criteria and optimization methods. Much research has been done to study one or more of these factors (Ibbitt, 1972; Sorooshian et al., 1983; Kuczera, 1983a&b; Gupta and Sorooshian, 1983 &1985; Duan et al., 1992; Yapo et al., 1997; Gan and Biftu, 1996; Kuczera and Mroczkowski, 1998; among others). The chapters throughout this book examine the effects of one or more of these factors on watershed model calibration. This chapter focuses only on the optimization methods.

The main purpose of this chapter is to review the use of optimization methods for watershed model calibration purpose. Special attention is paid to the use of global optimization methods to estimate the parameters of conceptual watershed models. The chapter is organized as follows. Section 2 discusses the use of local search optimization methods for watershed model calibration. Section 3 takes a historical perspective on the use of global optimization methods in watershed model calibration. Section 4 surveys the state-of-the-art methods in global optimization. Section 5 presents three popular global optimization methods that have been used widely in watershed model calibration. Section 6 summarizes this chapter. It also issues a few words of cautions on the limitations of relying on optimization methods to calibrate watershed models.

2. LOCAL OPTIMIZATION METHODS

Sophisticated optimization methods have been used widely to calibrate the parameters of watershed modes since the

very beginning of the digital watershed modeling era. Most early attempts to calibrate watershed models have been based on local-search optimization methods (see Dawdy and O'Donnell, 1965; Nash and Sutcliff, 1970; Chapman, 1970; Ibbitt, 1970; Monro, 1971; Johnston and Pilgrim, 1976; Pickup, 1977; Sorooshian et al., 1983; Gupta and Sorooshian, 1985; Hendrickson et al., 1988; etc.). The popularity of local-search methods is mostly due to the fact that the computer capability then was very limited and local-search methods required relatively small computer processing units (CPU). In contrast, global optimization methods from that time placed relatively high demand on computational resources. Besides, early global optimization theory was not advanced enough to be of practical utility.

There are two broad categories of local search methods: direct-type and gradient-type. Direct type methods (e.g., the Axis-Rotating method of Rosenbrock, 1960, the Pattern Search (PS) method of Hooke and Jeeves, 1961, and the Simplex method of Nelder and Mead, 1965) place few limitations on the form of model equations, and require only that knowledge of the objective function values be available over the feasible parameter space. Gradient type methods require that model equations be continuous to second order, and that knowledge of the values of the objective function as well as the gradient (and sometimes the hessian matrix) be available. The large family of Newton and Quasi-Newton methods belongs to gradient type approach. Gradient type methods usually have faster convergence rate than the direct type methods, but may not perform well when the assumptions of function and derivative continuity are violated.

Ibbitt (1970) conducted the first comprehensive comparative study of different optimization methods for calibration of the Stanford Watershed Model (SWM) (Crawford and Linsley, 1966) and the O'Donnell Model (Dawdy and O'Donnell, 1965). Eight local-search optimization methods and one global-search method were included in the study. The local search methods included direct type methods such as the Rosenbrock Method (Rosenbrock, 1960) and gradient type methods such as Powell's conjugate gradient direct search method (Powell, 1964) and the Levenberg-Marquardt method (Levenberg, 1944; Marquardt, 1963). The global-search method was a simple random search method (Karnopp, 1963). He reported that the effectiveness of local search methods was highly dependent on the choice of initial search points. With reasonable starting points (i.e., within 30% range of the "true" optimum), the Rosenbrock method was the most effective among the different local search methods he tested. He pointed out that Karnopp's random search method was unable to obtain good estimates of the global optimum, even though it might be helpful in finding good starting points for a subsequent local search.

Johnston and Pilgrim (1976) used the Simplex method of Nelder and Mead and a gradient method known as the Davidon method (Fletcher and Powell, 1963) to calibrate the Boughton model (Boughton, 1965). They reported that both methods failed to locate a "true" set of optimal parameters after more than two years of full-time work concentrated on a single watershed. Many other calibration studies echoed the findings cited above (Monro, 1971; Sorooshian and Dracup, 1980; Sorooshian et al.1983; and Hendrickson et al.,1988)

The experience with gradient-type approach for watershed model calibration has been mostly unfavorable. This is due to the difficulties and tedium involved in evaluating the derivatives of model equations, caused by the implicit nature of model equations and the existence of threshold parameters. Some researchers have tried to deal with this problem by approximating the derivatives with finite differences (Ibbitt, 1970; Johnston and Pilgrim, 1976; Pickup, 1977). They reported poor algorithmic performance due to numerical inaccuracies. Goldstein and Larimore (1980) investigated a modified version of the SAC-SMA model, in which the thresholds were replaced by smoothing functions, and the derivatives were explicitly computed. They employed the gradient-type Levenberg-Marquardt Method for estimation of the parameters and reported that good convergence could be achieved if the search was started in the region close to the optimum. However, they also pointed out that the Levenberg- Marquardt Method would be ineffective in cases where the response surface is highly non-quadratic (as is often the case with many watershed models (Sorooshian and Gupta, 1983)). Using a different approach, Gupta and Sorooshian (1985) developed a procedure for explicitly evaluating the derivatives of watershed models with threshold parameters and tested a Newton-Raphson method on a simple watershed model known as the SIXPAR model. Hendrickson et al. (1988) implemented this explicit derivative procedure on the SAC-SMA Model and then compared the calibration performance of two methods: the gradient-type Levenberg-Marquardt method and the direct- type PS method. The gradient-type approach performed poorly in comparison to the direct-type approach, and the evidence presented suggests that this was due to discontinuities in the derivatives of the objective function response surface.

Duan et al. (1992) conducted a detailed investigation into the problems associated with optimizing watershed model parameters. They employed an exhaustive gridding method to examine the objective function and derivative surfaces of the SIXPAR model. Their findings are summarized as follows:

i.   The parameter space contains several major regions of attraction into which a search strategy may converge;

ii. Each major region of attraction contain numerous local minima (stationary points where the first derivatives are zero and the Hessian matrices are positive definite or positive semi-definite);

iii. The objective function surface in the multi-parameter space is not smooth and may not even be continuous. The derivatives are discontinuous and may vary in an unpredictable manner throughout the parameter space;

iv. The parameters exhibit varying degree of sensitivity and a great deal of nonlinear interaction and compensation near the region of global optimum.

The combination of these features makes local-search methods inherently incapable of finding the global optimal parameters for watershed models such as SWM and SAC-SMA. Since the performance of watershed models is highly sensitive to how model parameters are estimated (Duan et al., 2001), the need to have methods capable of obtaining optimal model parameters is real and pressing. Recent research has therefore been directed towards evaluating the suitability of global-search optimization procedures for the calibration of watershed models. Indeed much progress has been made over the last fifteen years in the theory and practice of global optimization for watershed model calibration. This progress has been further facilitated by the fact that computer technologies have rapidly improved, making computationally intensive methods much more practical and affordable. The next section discusses the use of global optimization methods in watershed model calibration.

3. GLOBAL OPTIMIZATION METHODS

Previously it was pointed out that local search procedures are not designed to handle the presence of multiple regions of attraction, multi-local optima, discontinuous derivatives, insensitivities and parameter interdependency, and other problems encountered in the calibration of watershed models. It is therefore imperative that global optimization procedures that are capable of dealing with these various difficulties be employed. To deal with multiple regions of attraction, a search procedure must necessarily possess global convergence properties. It must be able to avoid being trapped by the minor optima. It must not require the availability of explicit analytic expressions for the objective function in terms of its parameters or for the derivatives. It must be robust in the presence of parameter interaction and non-convexity of the objective function surface. Finally, because watershed models usually have a large number of parameters, the algorithm must be efficient in the presence of high dimensionality.

Ibbitt (1970) was probably the first to examine the use of a global search strategy for watershed model calibration. He used a simple, brute-force random search method that can be applied repeatedly to different portions of the feasible space (Karnopp, 1963). This approach is neither efficient nor effective and provides little confidence that the global solution can be found this way. Brazil (1988) proposed the use of the Adaptive Random Search (ARS) method (Masri et. al., 1978, 1980; Pronzato et. al., 1984) to calibrate the SAC-SMA model. The ARS method improves the simple random search method by focusing attention on the promising regions. He reported that ARS method, in conjunction with some heuristic approaches for identifying initial parameter estimates and ranges, was capable of producing promising results. Armour (1990) and Weinig (1991) tested the ARS method extensively on SAC-SMA with both synthetic and real data. They found that, even with synthetic data where the "true" parameter set for the model was known, the ARS method was unable to find the true parameter values if the search space was not confined to a narrow range around the optimum. Their results suggest that ARS method is not well suited to solving multi-optimum problem encountered in watershed model calibration. Those findings were later supported by Duan et al. (1992).

Wang (1991) was the first to use Genetic Algorithm (GA), a random search procedure based on evolutionary principles (Holland, 1975), to calibrate watershed models. He reported that GA was able to consistently locate global optimal parameters of the Xinganjiang Watershed Model (XWM) in 10 random trials. He also indicated that subsequent tuning with a Simplex procedure produced only marginal improvement.

Many other researchers have also used GA to calibrate watershed models (Franchini, 1996; Franchini and Galeati, 1997; Seibert, 2000). Franchini (1996) employed GA, in conjunction with a local search procedure called Sequential Quadratic Programming (SQP), to calibrate the ADM (acronym for A Distributed Model). He reported that GA-SQP achieved an 100% success rate in identifying the exact global optimum when synthetic data were used. In real data study, however, he found that several parameters converged to consistent values, while other parameters scattered over the feasible ranges. He attributed the non-convergence of those parameters to imperfect model structure and data error. In a later study, Franchini and Galeati (1997) compared the performance of a few variants of GAs, along with the a local search method - the PS method, in optimizing the parameters of ADM. They reported a surprising finding that showed the PS method outperforming all GA schemes they tested.

Duan et al. (1992, 1993, 1994) developed a globally based search method known as the Shuffled Complex Evolution (SCE-UA) method. SCE-UA was designed to combine the strengths of existing global and local search methods such as GA and the Simplex method with newly conceived concepts of complex partition and complex shuffling. SCE-UA was compared against the Multi-start Simplex (MSX) method and the ARS method on numerous standard test problems as well as on watershed calibration problems. SCE-UA was shown to be a much superior method than MSX and ARS methods and ARS method was found to be totally ineffective in locating the global optimum in multiple random trials (Duan et al., 1992; Sorooshian et al., 1993).

Numerous researchers have investigated the use of SCE-UA for watershed model calibration purpose (see Luce and Cundy, 1994; Gan and Biftu, 1996; Tanakamura and Burges, 1996; Abdulla et al., 1996; Kuczera, 1997; Franchini et al., 1998, Eckardt and Arnold, 2001; Hogue et al., 2001). SCE-UA was found to be consistently more efficient and robust when it was compared against a variety of search methods. For example, Gan and Biftu (1996) compared SCE-UA, Simplex and MSX methods. They concluded that both MSX and SCE-UA were able to locate global optima, but MSX could not compete against SCE-UA in computational efficiency. They pointed out that for the Simplex method to be effective, model parameters must be divided into groups and be estimated in multiple stages. Kuczera (1997) compared SCE-UA to GA and a few multi-start local search methods. He also found SCE to be more efficient and robust than other methods. Franchini et al. (1998) investigated SCE-UA, GA and PS methods, with the latter two methods coupled with SQP (i.e., GA-SQP and PS-SQP). They reported that SCE-UA was the most reliable of the three methods.

Thyer et al. (1999) and Abdulla and Al-Badranih (2000) studied the use of Simulated Annealing (SA) method in watershed model calibration. SA is based on the analogy to crystallization process of metal in thermodynamics. Annealing refers to the process in which molten metal is cooled at a slow, deliberate pace, mixed with occasional brief re-heating, to attain the most stable crystal state. Thyer et al. (1999) coupled SA with Simplex method (SA-SX) and compared SA-SX to SCE-UA. They found that both SA-SX and SCE-UA were able to identify the optimal parameters for a modified version of the Boughton model, but SCE-UA was six times faster than SA-SX.

The next sections survey the state-of-the-art methods in global optimization and describe in details some of the commonly used global optimization methods in watershed model calibration, including Simulated Annealing, Genetic Algorithm and Shuffled Complex Evolution methods.

4. REVIEW OF GLOBAL OPTIMIZATION METHODS

Work on the global optimization problem has been reported since the 1950s and 1960s (Brooks, 1958; Bocharov and Feldbaum, 1962; Karnopp, 1963; Mockus, 1963; Hill, 1969; and others). The first books which were fully devoted to global optimization methods are by Dixon and Szegö (1975; 1978a). Over the last 15 years, numerous books on global optimization have been published (see Ratschek and Rokne, 1988; Törn and Zilinskas, 1989; Floudas and Pardalos, 1997; Horst and Pardalos, 1995; Horst et al., 1995; Bomze et al., 1996; among others). Some of the books focus on specific approaches. For example, books by Goldberg (1989), Davis (1991), and Michalewicz (1996) focus on the theory and applications of the Genetic Algorithm (GA). Books by Ratchek and Rokne (1988) and Hansen (1992) concentrate on the interval method approach to global optimization. Books by van Laarhoven and Aarts (1987) and Aarts and Korst (1989) describe the Simulated Annealing (SA) method. Mockus' book (1989) discusses the Bayesian approach to global optimization. Various survey papers addressing global optimization methods have been published (Dixon and Szegö, 1978b; Archetti and Schoen, 1984; Rinnooy Kan and Timmer, 1984&1989; Pinter, 1996; Törn and Viitanen, 1999). Some web sites are excellent resources on various topics in global optimization (see http://solon.cma.univie.ac.at/~neum/glopt.html by Neumaier and http://www.cs.sandia.gov/opt/survey by Gray et al.).

A wide variety of global optimization methods have emerged during the last three decades. Many researchers have attempted to classify these methods based on a variety of criteria, but none of the classification schemes have received universal acceptance (see Törn and Zilinskas, 1989). So rather than following a strict classification scheme, paragraphs below provide a survey of the global optimization methods commonly available in the literature. The methods are presented in no particular order with respect to their classifications and origins. All of them are applicable to continuous optimization problems. The methods used exclusively for integer or combinatorial optimization are not covered. Three commonly used global optimization methods in watershed model calibration: Simulated Annealing, Genetic Algorithm, and Shuffled Complex Evolution are described more comprehensively in Section 5.

Generalized Gridding Methods

Generalized Gridding (GG) methods are brute-force methods that sample the entire feasible space exhaustively at pre-specified grids. The most basic approach is to use a rectangular shaped, regularly spaced grid. But the grid spacing need not be uniform. For instance, the grid spacing can be related to the local rate of change of the objective function. Without mathematical verification, it is easy to see that if the density of gridding is high enough, the global optimum can be uncovered with a pre-specified accuracy. Duan et al. (1992) used a uniform gridding method to examine the distribution of local optima in the sub-spaces of a simple watershed model. A local search method may be coupled with GG method and be applied at each grid point to refine the optimal solution. GG is extremely computationally inefficient and is impractical for high-dimensional problems even with today's computational resources.

Interval Methods

Interval methods are based on the idea of finding subregions which contain, or do not contain, the global minimum. For example, through interval mathematics, those regions where the function value is larger than current estimate of global minimum, or where the gradient has non-zero value, or where the second derivative is negative (concave), can be removed from consideration. Pijavskij (1972) and Shubert (1972) independently developed methods using an approximation of the Lipschitz constant to eliminate subregions which do not contain global minimum. Their approach produces a piecewise linear approximation of the lower bound to the objective function. The approximating function is then used to eliminate non-promising regions. Brent (1973) used a quadratic approximation of the lower bound on objective function to eliminate non-optimal regions. This approach apparently depends on accurate second derivative information. Basso (1982) proposed the use of an adaptive bound instead of a global Lipschitz constant over the entire interval of interest. Other algorithms have been developed using similar approaches to those described above (Evtushenko, 1973; Wood, 1985). Interval methods are amenable to methods of classical analysis, and some interesting theoretical properties have been developed (*e.g.*, convergence guarantees, optimal algorithms, *etc.*) (Törn and Zilinskas, 1989; Archetti and Schoen, 1984; and Horst and Tuy, 1987). However, they are generally computationally inefficient and are not suitable for solving high dimensional problems. See Ratchek and Rokne (1988) and Hansen (1992) to learn more on Interval methods.

Trajectory Methods

Trajectory methods are based on modifications to the system equations describing the local descent trajectory. One method is to search for the global minimum by switching between descent (to minima) and ascent (to maxima) trajectories so that the trajectories pass through saddle points (Fiodorova, 1978). Another method can be best described by using an analogy to classical mechanics:

$$m(t)\,\ddot{x}(t) - n(t)\,\dot{x}(t) = -\nabla f(y(t)) \qquad (1)$$

which represents a moving mass $m(t)$ pushed by a field of forces (a potential f and a dissipative force $n(t)\,x(t)$). By a proper choice of $m(t)$ and $n(t)$, the trajectory can be made to escape from local minima and converge to the global minimum under inertial momentum (Griewank, 1981). Branin and Hoo (1972) proposed a method which uses trajectories stemming from the gradient of the objective function. Their method attempts to find the global minimum by searching for the roots of the gradient functions. Trajectory methods are handicapped by the fact that there exist regions of non-convergence. They become impractical when the function has a large number of local minima. Further, their applicability to problems whose analytical derivatives are unavailable is questionable.

Penalty Methods

Penalty methods attempt to find successively lower minima by applying a penalty to the objective function in the region of each local minimum already found. Goldstein and Price (1971) proposed a method using successive polynomial functions to find progressively lower minima. Levy and Gomez (1985) developed the so-called Tunneling Method in which each minimum found thus far becomes a pole in the modified objective function. Ge (1983) proposed the Filled Function Method which tries to fill the regions of attraction of the local minima found so far. Penalty methods are easy to implement because they basically use standard local search methods applied to penalty functions. A major problem with the Penalty methods is that it can be difficult to control the extent and severity of penalty so that false minima will not be introduced or the global minimum will not be missed. Penalty methods are ineffective when the number of local optima is large.

Random Search Methods

Random Search (RS) methods have been the most widely used global optimization procedure for three reasons. First, they are easy to implement on a computer and easy to modify. Second, they are robust, *i.e.*, they are insensitive to discontinuities and irregularities in the objective function. Third, RS methods can theoretically guarantee convergence to global optimum with a probability of 1. The simplest RS method is a Pure Random Search, which randomly samples the parameter space, choosing the best point found as an estimate of the global minimum (Brooks, 1958). A slight modification to this procedure, known as the Single-Start Random Search, adds a local search procedure starting from the best point found. In contrast, the Multi-Start Random Search method employs a local search from each random point (Hartman, 1973). This, of course, usually leads to detection of the same local minimum many times. Duan et al. (1992) and Gan and Biftu (1996) have tested Multi-Start Random Search methods on watershed model calibration problems. RS methods are generally crude and computationally inefficient.

Adaptive Random Search Methods

Adaptive Random Search (ARS) methods utilizes various heuristic strategies to distribute search points non-uniformly in the feasible space, with greater density in promising regions. This approach includes three phases: *Exploration, Decision,* and *Adaptation.* One ARS procedure was presented by Masri et al. (1978; 1980) and was slightly modified by Pronzato et al. (1984). The procedure basically consists of repeated random sampling in different ranges of parameter space. The first round of sampling is conducted over several successively smaller ranges of the parameter space, centered on the initial range. The best point found is assumed to be in the region of the global minimum. Another round of random sampling is then carried out centered on this best point. This procedure is repeated a user specified number of times. Brazil (1988) employed Pronzato's ARS algorithm in an attempt to calibrate the SAC-SMA model, and reported that ARS method was capable of producing promising results. His results were disputed by Armour (1990), Weinig (1991) and Duan et al. (1992) (see Section 3). There are other ARS methods which are based on heuristic and adaptive use of various algorithms in different stages of search process (see Törn and Zilinskas, 1989; Resende and Ribeiro, 2001).

Methods Based on a Stochastic Model of the Objective Function

In this method, the values of the objective function are treated as random variables. The method attempts to find the expected location and value of the global minimum. Several different methods have been proposed. Archetti (1975) proposed an approach which approximates the probability distribution $P(\xi)$ of the objective function $f(x)$ (*i.e.*, $P(\xi)=Prob(f(x)<\xi)$) using an n-th order polynomial. The estimate of the root of $P(\xi) = 0$ is chosen as an approximation of global minimum. A more precise estimate of global minimum is then obtained by using a local search procedure to refine the best point obtained so far. According to Gomulka (1978) this method is able to identify the region of global minimum well, provided the minimum region is not relatively flat. However, Dixon and Szegö (1978b) state that this approach requires a very large number of function evaluations to succeed.

Kushner (1964) treated the objective function as a Wiener process. His method was developed for one-dimensional problems. Using the theory of Wiener processes, the expectation of the minimum value of the objective function over an interval (x_1, x_2) can be computed conditioned on the function values $f(x_1)$ and $f(x_2)$. By eliminating the intervals whose expected minima are greater than the current estimate of global minimum, the location of the interval containing the global minimum can be detected. The procedure stops when the probability of finding a better minimum is significantly small. Rinnooy Kan and Timmer (1989) stated that this method is analytically attractive but requires very cumbersome computations even in the case of a one-dimensional problem. The extension of this method to multi-dimensional problem is thought to be very difficult, if not impossible.

Bayesian Methods

Mockus proposed a Bayesian procedure to determine the expected estimate of global minimum (see Mockus, 1989). In this approach, the *a-priori* distributions of the parameters are pre-specified. These are updated to posterior distributions based on the outcome of sampling procedure. The sampled points are chosen such that the expected value of objective function is minimized. This method strives for best expected results under a limited number of function evaluations. Rinnooy Kan and Timmer (1984) commented that this method yields an estimate of the global minimum which may be too crude for practical purposes. Törn and Zilinskas (1989) pointed out that this method is attractive theoretically but is too complicated for algorithmic realization.

Clustering Methods

A clustering method is one that attempts to group a sample of points into clusters around local minima. Once the clusters are constructed, the local minima can be identified by converging one point from each cluster by means of a local search algorithm. There are three main steps in clustering methods: (i) sample the feasible search space, (ii) group the points around local optima, and (iii) perform clustering analysis to find clusters in the neighborhoods of local optima. If this procedure successfully identifies groups that represent neighborhoods of local minima, then redundant

local searches can be avoided by simply starting a local search for some point within each cluster. Clustering methods are an improvement over the Multi-Start Random Search method because it finds each local minimum only once. Clustering methods, like other random methods, guarantee that the location of global minimum will be found with probability 1 as the sample size is increased. Clustering methods are most effective for low- dimensional problems only. To learn more Clustering methods, refers to Törn and Zilinskas (1989) and Törn and Viitanen (1994).

Complex Evolution Methods

Box (1965) introduced the term "*Complex*" to describe a geometric polyhedron with k vertices in R^n where k must be greater than or equal to $n+1$, where n is the dimension of the problem. The k points constituting the complex may be selected randomly or so as to construct a geometric figure of particular structure. Complex Evolution methods interactively adjust the positions of the individual points so as to move the entire group in the direction of global improvement. Price (1978, 1983, 1987) introduced a Complex Evolution Method which he called the Controlled Random Search (CRS). There are two main steps in the CRS: (i) exploration of the space, and (ii) replacement of the worst point in the sample with a better point. The exploration phase is implemented by randomly (or deterministically) sampling a predetermined number of points from the entire parameter space. The second phase is implemented by randomly selecting "simplexes" of $n+1$ points from the complex, and using the Nelder and Mead (1965) strategy to evolve each simplex by one step in an improvement direction. Each new point found in this manner is used to replace the current worst point of the complex. The second phase is continued until all the points in the complex have converged to within a pre-specified distance of each other. Price (1983) modified the algorithm to always include the best point in the complex in each randomly selected simplex. Later he included a local search phase (Price, 1987). Ali and Storey (1994) made further modifications to CRS method and reported improved performance over its predecessor. CRS was sometimes classified as Clustering method (Törn and Zilinskas, 1989). An advantage of CRS methods over other Clustering methods is that it combine random search and mode-seeking into a continuous process.

5. GLOBAL OPTIMIZATION METHODS COMMONLY USED IN WATERSHED MODEL CALIBRATION

In the previous section, brief capsules on some of the global optimization methods commonly available in the literature are presented. This section describes three of the commonly used global optimization methods in watershed model calibration: Simulated Annealing, Genetic Algorithm and Shuffled Complex Evolution. In presentation below, all optimization problems are assumed to be minimization problems.

Simulated Annealing Methods

Simulated Annealing (SA) method was introduced by Metropolis (1953). The name is drawn from an analogy to the cooling process employed in metallurgy. Molten metal is cooled slowly with intermittent reheating to allow a stable crystal structure to develop. Eventually a thermal equilibrium state is reached. SA method resembles this process in that it accepts both beneficial and detrimental steps along the way towards global minimum. The detrimental steps are accepted probabilistically according to Boltzmann distribution $exp(-\nabla f/T)$, where ∇f is the relative potential of the steps and T is a parameter analogous to the temperature. Thus, SA can basically be regarded as a form of biased random walk that migrates through a sequence of local minima and eventually converges to the global minimum. Many applications of SA have been in the field of combinatorial optimization (Kirkpatrick et al., 1983; Bonomi and Lutton, 1965; Lundy, 1985; *etc.*). Generalizations of SA to continuous problems have been provided by Vanderbilt and Louie (1984), Bohachevsky et al. (1986) and Lucidi and Piccioni (1989). There are many variants of SA algorithms. A simple SA algorithm for continuous optimization is presented below.

i. Set i=0. Initialize the maximum number of function evaluations per equilibrium cycle, N, the minimum number of successful trials required to continue optimization search, κ_{min}, the maximum step size, D, initial temperature, T_i, and temperature scaling factor, α $(0<\alpha<1)$ and minimum temperature, T_{min};

ii. Randomly sample a point, x_i, in the feasible space and compute the corresponding objective function value, $f(x_i)$. Set $\kappa=0$, $x_{best}=x_i$ and $f_{best}=f(x_i)$;

iii. Set i=i+1. Generate a random vector, γ $(-1 \leq \gamma \leq 1)$. Set $x_i = x_{best} + D\cdot\gamma$. Compute $f(x_i)$;

iv. If $f(x_i) < f(x_{best})$, set $\kappa=\kappa+1$, $x_{best}=x_i$ and $f(x_{best})=f(x_i)$. Go to step vi. Else go to step v;

v. Compute $\nabla f=f(x_i)-f(x_{best})$ and Boltzmann probability, $P=exp(\nabla f/T_{i-1})$. Generate a random number, χ $(0<\chi<1)$. If $P>\chi$, set $x_{best}=x_i$ and $f_{best}=f(x_i)$. Continue to step vi;

vi. If i=N, go to step vii. Else go to step iii;

vii. If $T_i<T_{min}$ and $\kappa<\kappa_{min}$, stop. Else, set $T_i= \alpha \cdot T_{i-1}$ and i=0. Go to step iii.

Obviously the selection of algorithmic parameters N, κ_{min}, D, T_0, T_{min}, and α impacts on the effectiveness and convergency speed of SA. Many studies have been done to find the best ways to determine these parameters (see van Laarhoven and Aarts, 1987; Aarts and Korst, 1989). By manipulating the Boltzmann probability function, the convergency speed can be controlled (Ingber, 1993). How to choose step size D and to generate new points have been investigated by many (Vanderbilt and Louie, 1984; Parks, 1990). Obviously initial temperature T_0 should be high enough to ensure a "molten" state at the start. Or in optimization phrase, T_0 should be high enough to allow a sustained search in parameter space. T_i should be decreased at a rate so a stable crystallization process can take place. The algorithm presented above reduces T_i linearly. Alternative methods have been proposed by many researchers (Kirkpatrick et al., 1983; Randelman and Grest, 1986). Random number generators are used frequently throughout the search process and attention should be paid to ensure that they have proper probabilistic properties.

SA methods are very easy to implement algorithmically and are suitable for parallel computer programming.. This method has been shown to be able to converge to global minimum with probability 1 (Faigle and Schrader, 1988). Like many random search algorithms, however, SA methods have been criticized as being slow. Therefore, they are most effective when they are coupled with local search methods (Desai and Patil, 1996, Thyer et al., 1999). For learn more on SA methods, see van Laarhoven and Aarts (1987) and Aarts and Korst (1989).

Genetic Algorithm

Holland (1975) presented a procedure named Genetic Algorithm (GA) which is based on analogies to the principles of genetics and natural selection. Evolution is viewed as a process of reproduction. First, parents are selected that have high probability of generating "fit" offspring. The offspring are then generated by means of information taken from each parent in a process analogous to the sharing of genetic information. Below the steps in a basic implementation of GA are presented:

i. Generate a population (N points) randomly in feasible space, $x^1, x^2, ..., x^N$. Compute corresponding objective function values, $f(x^i)$, $i =1,2,...,N$;

ii. Rank the N points in order of increasing objective function values, $f^*(x^i)$, i=1,2, ... , N, where $f^*(x^1)$ has the lowest objective value and $f^*(x_N)$ the highest;

iii. Assign a probability value to each point using the following probability function:

$$p_i = \frac{2(N+1-i)}{N(N+1)}, \quad i = 1,...,N \quad (2)$$

Here p_1 corresponds to f*(x^1) and p_N to f*(x^N).

iv. Select two "parents", $\mathbf{x^a}$ and $\mathbf{x^b}$, from the population according to the probability function, p_i , i =1,2,...,N;

v. Generate the "offspring", $\mathbf{x^{a*}}$ and $\mathbf{x^{b*}}$, randomly using the genetic operators (to be described later);

vi. Replace the worst points in the population by the two newly generated "offspring";

vii. Repeat steps iv-vi until one of the convergency criteria is satisfied.

GA can be implemented numerically in binary-coding or in real-coding. In binary-coding, each sample point is represented by a binary string (analogous to a chain of chromosomes). For continuous problems, real variables must be discretized and be approximated by integers. The precision of the optimal solution by GA is predicated by the encoded bit length of the integers. The binary representation of an n-dimensional real variable is illustrated below (from Wang, 1991).

Consider a n-dimensional parameter $\mathbf{x}$ =(x_1, x_2, ... , x_{n-1}, x_n), where $a_i \leq x_i \leq b_i$, a_i and b_i are the lower and upper bounds for x_i , i = 1, 2, ... , n. Let an l-bit string to represent x_i. This string ranges from 0 to 2^l - 1 and can be mapped linearly to parameter range, [a_i, b_i]. The parameter range is discretized into 2^l points and the discretization interval is

$$\Delta x_i = \frac{b_i - a_i}{2^l - 1}$$

Table 1 illustrates how any value for x_i can be approximated by a 7-bit binary string. Parameter $\mathbf{x}$ = {x_1, x_2, ... , x_{n-1}, x_n} can thus be represented by connecting the strings of all parameters:

1000101	0010100	...	0101001	1101001
x_1	x_2		x_{n-1}	x_n

Table 1. Example of Parameter Coding for l=7

Binary Code	Integer Value	Parameter Value
0000000	0	a_i
0000001	1	$a_i + \Delta x_i$
0000010	2	$a_i + 2\Delta x_i$
1111110	126	$a_i + 126\Delta x_i$
1111111	127	$a_i + 127\Delta x_i = b_i$

To select parents, a trapezoidal probability function (Eq. 2) is used to favor better points (*i.e.*, points with lower function values) in the reproductive process. Reproduction is realized by using a variety of genetic operators. The most basic genetic operator is the *crossover* operator. *Crossover* is accomplished by exchanging all the bits following a randomly selected location on the strings. For example, if *crossover* occurs after position 5 between two strings:

$$1\ 1\ 1\ 1\ 1\ 1\ 1\ 1$$

and

$$0\ 0\ 0\ 0\ 0\ 0\ 0\ 0,$$

the resulting offspring are

$$1\ 1\ 1\ 1\ 1\ 0\ 0\ 0$$

and

$$0\ 0\ 0\ 0\ 0\ 1\ 1\ 1$$

A slightly more complex *crossover* operator has two random *crossover* location. De Jong (1989) reported that too many crossover location may result in loss of genetic information, even though more diversity is introduced. How to select the crossover location and how to recombine string segments have been the subject of extensive research (Davis, 1991; and Franchini and Galeati, 1997).

Another genetic operator is the *mutation* operator. *Mutation* is realized at a given probability through bit changes on the strings (i.e., 0 to 1 or 1 to 0). *Crossover* and *mutation* operators are the pillars of all GAs. There are other less used genetic operators which allow more than two pairs of parents in reproduction process, or introduce multi-communities (or sub- population) and allow interbreeding between communities (see Goldberg, 1989; Duan et al., 1992, 1993 & 1994).

Binary-coded GAs are best suited for combinatorial or integer optimization problems. It has been criticized for having only limited precision for continuous problems and for having redundance in the coding of parameters (Herrera et al., 1998). An alternative approach to implement GA is to use real-coding (Wright, 1991; Michalewicz, 1996; Herrera et al., 1998). In real-coding, a sample point or a chromosome is represented by a vector of floating point numbers, $\mathbf{x}$ = (x_1, x_2, ... , x_n), where $a_i \leq x_i \leq b_i$, i = 1, 2, ... , n. The precision of a real-coded point is limited only by that of the computer.

As in binary-coding, *crossover* and *mutation* are the main genetic operators in real-coded GA. Let $\mathbf{x^a}$ = (x_1^a, x_2^a, ... ,

x_n^a) and $x^b = (x_1^b, x_2^b, ..., x_n^b)$ be the 2 parents. A simple *crossover* similar to the binary coded implementation described above is illustrated as follows: (1) randomly select a location, $0 \leq l \leq n$; and (2) form the 2 offspring x^{a*} and x^{b*} as follows:

$$x^{a*} = (x_1^a, x_2^a, ..., x_i^a, x_{i+1}^b, ..., x_n^b)$$

and

$$x^{b*} = (x_1^b, x_2^b, ..., x_l^b, x_{l+1}^a, ..., x_n^a)$$

There are many other ways to perform *crossover* operations. For example, the discrete *crossover* operator employed by Seibert (2000) and Seibert and McDonnell (this volume) generates offspring $x^* = (x_1^*, x_2^*, ..., x_n^*)$ by assigning x_i^* to a value randomly chosen from the set $\{x_i^a, x_i^b\}$. A large class of *crossover* operators generate offspring $x^* = (x_1^*, x_2^*, ..., x_n^*)$ by calculating x_i^* based on random or linear combination of x_i^a and x_i^b. To learn more about real-coded *crossover* operations, see Herrera et al. (1998), who have described 11 different *crossover* operators.

Like *crossover* operators, there are numerous ways to implement real-coded *mutation* operators. The random *mutation* operation is executed with a given probability by assigning x_i^* to a value randomly selected from the feasible domain $[a_i, b_i]$. A *mutation* operator known as the *Real Number Creep* is realized at a given probability by randomly assigning x_i^* to a value located in the close neighborhood of the local optimum found so far (Davis, 1991). Readers are referred to Herrera et al. (1998) for more discussion on real-coded *mutation* operators.

There are several important algorithmic parameters in GA: population size N, *crossover* probability p_C and *mutation* probability p_M. Many studies have appeared in the literature on selection of these parameters (see Grefenstte, 1986; Franchini and Galeati, 1997).

The offspring resulted from genetic operations such as *crossover* and *mutation* retain the gene characteristics of their parents. Because "fit" parents are favored in the reproduction process, the offspring tend to be healthier than general population. After healthier offspring displace the "unfit" members in the population, the whole population thus evolves to a healthier state.

GA is a widely popular method among many disciplines, including hydrology. It has been reported to have excellent initial convergency toward the neighborhood of global optimum. However, it has been found to have difficulties converging onto the global solution itself and to be computationally inefficient (Kuczera, 1997; Franchini et al., 1998). Nevertheless, Goldberg (1989) remarked that "*genetic algo-*

rithm method is theoretically and empirically proven to provide robust search in complex spaces". The idea of competitive evolution has motivated the development of other algorithms (Schwefel, 1981; Jarvis, 1975). One of them is the Shuffled Complex Evolution method, described below.

Shuffled Complex Evolution Method

Shuffled Complex Evolution (SCE-UA) method was developed by Duan et al. at the University of Arizona (Duan, 1991; Duan et al., 1992). SCE-UA was originally designed to deal with the peculiarities encountered in calibration of conceptual watershed models. The method is based on a synthesis of four concepts: a) combination of deterministic and probabilistic approaches; b) systematic evolution of a "complex" of points spanning the parameter space, in the direction of global improvement; c) competitive evolution; and d) complex shuffling. The first three concepts are drawn from existing methodologies that have been proven successful in the past including GA, Simplex and CRS methods (Price, 1978, 1983; Nelder and Mead, 1965; Holland, 1975), while the last concept was newly introduced (Duan et al., 1992,1993 & 1994; Sorooshian et al., 1993). A general description of the steps of the SCE-UA method is given below (for a more detailed presentation of the theory underlying the SCE-UA algorithm, refer to Duan et al., 1992; 1993 and 1994):

a. Generate sample: Sample s points randomly in the feasible parameter space and compute the criterion value at each point. In the absence of prior information on the approximate location of the global optimum, use a uniform probability distribution to generate a sample.

b. Rank points: Sort the s points in order of increasing criterion value so that the first point represents the smallest criterion value and the last point represents the largest criterion value.

c. Partition into complexes: Partition the s points into p complexes, each containing m points. The complexes are partitioned such that the first complex contains every $p \times (k-1)+1$ ranked point, the second complex contains every $p \times (k-1)+2$ ranked point, and so on, where $k=1,2,...,m$.

d. Evolve each complex: Evolve each complex according to the Competitive Complex Evolution (CCE) algorithm (The CCE algorithm is elaborated below).

e. Shuffle complexes: Combine the points in the evolved complexes into a single sample population; sort the sample population in order of increasing criterion value; shuffle (i.e. re-partition) the sample population into p complexes according to the procedure specified in step c.

f. Check convergence: If any of the pre-specified convergence criteria is satisfied, stop; else, continue.

g. Check the reduction in the number of complexes: If the minimum number of complexes required in the population, p_{min}, is less than p, remove the complex with the lowest ranked points (or randomly remove a complex); set $p=p-1$ and $s=p \times m$; and return to step d. If $p_{min}=p$, return to step d.

One key component of SCE-UA method is the Competitive Complex Evolution (CCE) algorithm referenced in step d. The CCE algorithm, based on the Nelder and Mead (1965) Simplex downhill search scheme, is briefly presented as follows:

i.   Construct a sub-complex by randomly selecting q points from the complex (community) according to a trapezoidal probability distribution. The probability distribution is specified such that the best point (i.e., the point with the best function value) has the highest chance of being chosen to form the sub-complex while the worst point has the least.

ii. Identify the worst point of the sub-complex and compute the centroid of the sub-complex without including the worst point.

iii. Attempt a reflection step by reflecting the worst point through the centroid. If the reflected point is within the feasible space, go to step iv. Else go to step vi.

iv. If the reflection point is better than the worst point, replace the worst point by the reflection point. Go to step vii. Else go to step v.

v.   Attempt a contraction step by computing a point half way between the centroid and the worst point. If the contraction point is better than the worst point, replace the worst point by the contraction point and go to step vii. Else go to step vi.

vi. Randomly generate a point within the feasible space. Replace the worst point by the randomly generated point in the feasible space.

vii. Repeat step ii through step vi α times, where $\alpha \geq 1$ is the number of consecutive offspring generated by the same sub-complex.

viii. Repeat step i through Step vii β times, where $\beta \geq 1$ is the number of evolution steps taken by each complex before complexes are shuffled.

What differs SCE-UA from traditional GAs is the partition of the population into several communities to facilitate a freer and more extensive exploration of the feasible space in different directions, thereby allowing for the possibility that the problem has more than one region of attraction. The shuffling of communities enhances the survivability by a sharing of the information (about the search space) gained independently by each community.

In the CCE algorithm, each point of a complex is a potential "parent" with the ability to participate in the process of reproducing offspring. A sub-complex functions like pairs of parents, except that it may comprise more than two members. Like GAs, a trapezoidal probability distribution is used to favor better points over worse points in the reproduction process. The Simplex method is utilized to generate offspring because it is insensitive to non-smoothness of the response surface and enables the algorithm to make use of response surface information to guide the search toward the improvement direction. Step vi in CCE generate an offspring at random location under certain conditions to ensure that the evolution process is not interrupted due to some unusual conditions encountered in the search space. This is somewhat analogous to *mutation* operator in GAs in response to stress in biological evolution.

SCE-UA method contains many probabilistic and deterministic components that are controlled by some algorithmic parameters. For the method to perform optimally, these parameters must be carefully chosen. These parameters are:

$m =$ number of points in a complex

$q =$ number of points in a sub-complex

$p =$ number of complexes

$p_{min}=$ minimum number of complexes required in the population

$\alpha =$ number of consecutive offspring generated by each sub-complex

$\beta =$ number of evolution steps taken by each complex

Duan et al. (1994) conducted a detailed study on the selection of SCE-UA algorithmic parameters suggested default values for parameters m, q, α, β and p_{min}. They reported that the most important parameter in SCE-UA is the number of complexes, p, which is dependent on the complexity of the problem. For a 13 parameter optimization problem, they recommended a value of 4 for p. The recommended values for SCE-UA algorithmic parameters were derived based on the experience with the calibration of the 13-parameter SAC-SMA model. Therefore the users of SCE-UA should experiment with the selection of the algorithmic parameters on their own problems.

SCE-UA method has been widely used in watershed model calibration (Sorooshain et al., 1993; Havnø et al.,

1995; Kuczera, 1997; Gan and Biftu, 1996; Franchini et al., 1998; Madsen, 2000; Hogue et al., 2001; among others). It has also been used in other areas of hydrology such as soil erosion, subsurface hydrology, remote sensing and land surface modeling (Mahani et al., 2000; Contractor and Jenson, 2000, Scott et al., 2000; Nijssen et al., 2001; Walker et al., 2001). SCE-UA has been generally found to be robust, effective and efficient. A number of researchers have explored modifications and enhancements to the original SCE-UA (Wang et al., 2001; Santos et al., 1999). Vrugt et al. (this volume) develop the Shuffled Complex Evolution Metropolis (SCEM) algorithm, which combines the elements from both SCE-UA and SA. A major advantage of SCEM is that it provides uncertainty information about the optimal solution. Yapo et al. (1997) extended SAC-UA to multi-objective framework (see Gupta et al., this volume, "Multiple ...").

6. SUMMARY AND A FEW WORDS OF CAUTION

This chapter reviewed the use of optimization methods as a tool to calibrate conceptual type watershed models. It traced back the historical trails on which watershed modelers have struggled with local search methods and started to favor the more powerful global optimization methods. A survey of the state-of-the-art methods in global optimization was presented. Three of the more popular methods commonly used by watershed modelers (i.e., Simulated Annealing, Genetic Algorithm and Shuffled Complex Evolution) are discussed comprehensively. The manner in which the presentation was made was quite informal. No attempts were made to include stringent theoretical treatment. This approach was intentional since most watershed modelers are only interested in the application aspect of the global optimization methods. For those interested in more vigorous treatment, many references were given. It is worthy to point out that as the theory and practice of global optimization continue to evolve, new theories and improved methods are bound to emerge. This trend is made inevitable because of continued investment in research and development of global optimization methods and because of the rapid progress in computational technology (both hardware and algorithms). The watershed modeling community, being always on the cutting edge in developing and applying optimization methods, should stand to benefit from these new developments.

It has to be emphasized that optimization methods serve only as a tool to facilitate the search of the optimal model parameters. It should not be regarded as a panacea which can solve the model calibration problem all by itself. This is because any optimization method, no matter how advanced it is, gives only the optimal solution with respect to the objective function used in a specific case. Any objective function is only a single measure of the difference between the model and the real world aggregated over a long period of time. It can not possibly capture all phases of the hydrograph equally well and is highly impacted by errors, systematic or otherwise, in calibration data. Further, poorly defined model structure leads to insensitive parameters and parameter interdependence. Unless proper care is taken to reduce the effects of data errors on objective functions and to enhance the idenfiability of model parameters, unrealistic or even unphysical model parameters may be resulted from use of automated optimization methods. Another related issue is how many parameters can one possibly estimate by relying on optimization methods for a given set of calibration data. This issue is especially relevant if watershed model calibration is based solely on fitting the simulated streamflow discharge to the observed streamflow discharge (Jakeman and Hornberger, 1993). Much advantage can be gained by studying the type, quantity and quality of calibration data used for model calibration (Sorooshian et al., 1983, Burges, this volume, Gan and Biftu, this volume). A hierarchical strategy that isolates hydrologic processes and associated sub-sets of parameters with different processes may be useful (Harlin, 1991, Turcotte et al, this volume). Use of multiple criteria to calibrate watershed models has been gaining favor in the last few years and many of the chapters in this volume have taken such an approach (Yapo et al., 1997; Boyle et al., Freer et al., Gupta et al., "Multiple .. .", Meixner et al., Parada et al., Seibert and McDonnell, among others, this volume). Finally, given the uncertainties associated with model structure, calibration data and model parameters, any optimal solution is more meaningful if the uncertainties in the parameter estimates and in the model behavior can be quantified (see Kavetski et al., Freer et al, Misirli et al., and Vrugt et al., this volume).

How to deal with various issues related to the use of optimization methods in watershed model calibration has received much attention in the watershed modeling community. A detailed discussion of those topics is beyond the scope of this chapter. Interested readers should refer to the related literature and to other chapters throughout this volume for more in-depth discussion and analysis.

Acknowledgment. The author wishes to acknowledge Hoshin Gupta, Marco Franchini and Henrik Madsen for their constructive comments of the original manuscript.

REFERENCES

Aarts, E., and J. Korst, 1989, *Simulated Annealing and Boltzman Machines,* John Wiley & Sons, Chichester.

Abdulla, F.A, D.P. Lettenmaier, E.F. Wood, and J.A. Smith, 1996. Application of a macroscale hydrologic model to estimate the water balance of the Arkansas-red river basin, *J. Geophys. Res.,* 101(D3), 7,449-7459

Abdulla, F.,and L.Al-Badranih,,2000, Application of a rainfall-runoff model to three catchments in Iraq , *Hydrol. Sci. J.,* 45(1): 13-25

Ali, M.M. and C. Storey, 1994, "Modified Controlled Random Search Algorithms", *Intl. Journal of Computer Mathematics,* 53, 229-235.

Archetti, F., 1975, A sampling technique for global optimization, in: Dixon, C.W.L., and G.P. Szegö, (eds), *Towards global optimization,* North-Holland, Amsterdam, 158-165.

Archetti, F., and F., Schoen, 1984, A survey on the global optimization problem: general theory and computational approaches, *Annals of Operations Research,* 1, 87-110.

Armour, A., 1990, Adaptive random search evaluated as a method for calibration of the SMA-NWSRFS model, MS Thesis, Dept. of Hydro. and Water Resour., Univ. of Arizona, Tucson, AZ 85721.

Basso, P., 1982, Iterative methods for localization of the global maximum, *SIAM Journal on Numerical Analysis,* 19, 781-792.

Bocharov, N., and A.A. Feldbaum , 1962, An automatic optimizer for search of the smallest of several minima, *Auto. & Remote Control,* 23(3).

Bohachevsky, I.O., M.E. Johnson, and M.L. Stein, 1986, Generalized simulated annealing for function optimization, *Technometrics,* 28(3), 209-217.

Bomze,I.M., T. Csendes, R. Horst and P.M. Pardalos, 1996, *Developments in global optimzation,* Kluwer, Dordrecht.

Bonomi, E., and J.-L. Lutton, 1984, The N-city travelling salesman problem: statistical mechanics and the Metropolis algorithm, *SIAM Review,* 26, 551-568.

Boughton, W. C., 1965, A new estimation technique for estimation of catchment yield, Rep. 78, Water Resources Lab., University of New South Wales, Manly Vale, August 1965.

Box, M.J., 1965, A new method of constrained optimization and a comparison with other methods, *Computer Jo.,* 8, 42-52.

Branin, F.H., and S.K. Hoo, 1972, A method for finding multiple extrema of a function of n variables, In: Lootsma, F.A., (ed), *Num. methods of nonlinear optim.,* Academic Press, London, 231-237.

Brent, R.P., 1973, *Algorithms for minimization without derivatives,* Prentice-Hall, New Jersey.

Brazil, L.E., 1988, Multilevel Calibration Strategy for Complex Hydrologic Simulation Models, Ph.D. Dissertation, Dept. of Civil Engineering, Colorado State University, Fort Collins, Colorado.

Brooks, S.H., 1958, A discussion of random methods for seeking maxima, *Operations Research,* 6, 244-251.

Burnash, R.J.C., R.L. Ferral, and R.A. McGuire, 1973, A generalized streamflow simulation system: conceptual models for digital computers, Joint Fed.-State River Forecast Center., Sacramento, CA

Chapman, T. G., 1970, Optimization of a rainfall-runoff model for an arid zone catchment, *UNESCO Pub.96,* 126-143, UNESCO, Wallington, New Zealand.

Contractor D.N., and J. W. Jenson, 2000, Simulated effect of vadose infiltration on water levels in the Northern Guam Lens Aquifer, *J. of Hydrol.,* 229, 232-245

Crawford, N.H., and R.K. Linsley, 1966, Digital Simulation in Hydrology: Stanford Watershed Model IV, *Stanford University Dept. Civil Engr. Tech. Report 39.*

Davis, L., 1991, *Handbook of genetic algorithms,* Van Norstrand Reinhold, N.Y.

Dawdy, D. R., and T. O'Donnell, 1965, Mathematical models of catchment behavior, *J. Hydrau. Div.,* ASCE, 91(HY4), 113-137.

De Jong, K.A., 1975, An analysis of the behavior of a class of genetic adaptive systems, Ph.D. dis., Univ. of Michigan, Ann Arbor, MI.

Desai, R. and R. Patil, 1996, SALO: Combining simulated annealing and local optimization for efficient global optimization, in *Proc. of the 9th Florida AI Research Symposium,* Key West, FL., 233-237

Dixon, C.W.L., and G.P. Szegö (eds), 1975, *Towards global optimization,* North-Holland, Amsterdam.

Dixon, C.W.L., and G.P. Szegö (eds), 1978a, *Towards global optimization 2,* North-Holland, Armsterdam.

Dixon, C.W.L., and G.P. Szego, 1978b, The global optimization problem: an introduction, in Dixon, C.W.L., and G.P. Szegö, (eds), *Towards global optimization 2,* North-Holland, Amsterdam..

Duan, Q., 1991, A global optimization strategy for efficient and effective calibration of hydrologic models, Ph.D. Dis., Univ. of Arizona, Tucson, AZ.

Duan, Q., V.K. Gupta, and S. Sorooshian, 1993, A Shuffled Complex Evolution Approach for Effective and Efficient Global Optimization, *J. Optim. Theo. and Its Appl.,* 76(3), 501- 521

Duan, Q., S. Sorooshian, and V.K. Gupta, 1992, Effective and Efficient Global Optimization for Conceptual Rainfall-Runoff Models, *Water Resour. Res.,* 28(4), 1015-1031

Duan, Q., S. Sorooshian, and V.K. Gupta, 1994, Optimal Use of the SCE-UA Global Optimization Method for Calibrating Watershed Models, *J. of Hydrol.,* 158, 265-284

Duan, Q., J. Schaake, and V. Koren, 2001, A Priori Estimation of Land Surface Model Parameters, in Lakshmi, V. et al. (Eds), Land Surface Hydrology, Meteorology, and Climate: Observations and Modeling, *Water Sci. and Appl. 3,* AGU, Washington, DC, 77-94.

Eckardt, K., and J.G. Arnold, 2001, Automatic calibration of a distributed catchment model, *J. of Hydrol.,* 251, 103-109.

Evtushenko, Y.G., 1973, Numerical algorithms for global extremum search (case of a non-uniform mesh), *USSR Comp. Math. and Math. Phys.,* 11, 1390-1403.

Faigle, U., and R. Schrader, 1988, On the convergence of stationary distributions in simulated annealing algorithms, *Information Processing Letters,* 27, 189-194.

Fiodorova, I., Search for the global optimum of multiextremal problems, Optim.. Dec. Theo., 4, Inst. Of Math. & Cybern., Lithuanian SSR Acad. Of Sci., 93-100.

Fletcher, R., and M.J. Powell, 1963, A Rapidly Convergent Descent Method for Minimization, *The Computer Jo.,* 6, 163-168.

Fletcher, R., 2000, *Practical Methods of Optimization,* Wiley and Sons, New York, New York, 450p

Floudas C.A. and P.M. Pardalos, (Eds)., 1996, *State of the Art In Global Optimization: Computational Methods and Applications,* Kluwer Academic Publishers.

Franchini, M., 1996, Use of a genetic algorithm combined with a local search method for the automatic calibration of conceptual rainfall-runoff models, *Hydrol. Sci. Jo.,* 41(1), 21-37.

Franchini, M. and G. Galeati, 1997, Comparing several genetic algorithm schemes for the calibration of conceptual rainfall-runoff models, *Hydro. Sci. Jo.,* 42(3), 357-379.

Franchini, M., G. Galeati, Giorgio, S. Berra, 1998, Global optimization techniques for the calibration of conceptual rainfall-runoff models, *Hydrol. Sci. J.*, 43(3), 443-458

Gan, T.Y., and Biftu, G. F., 1996, Automatic calibration of conceptual rainfall-runoff models: optimization algorithms, catchment conditions, and model structure, *Water Resour. Res.*, 32(12), 3513-3524

Ge, R.P., 1983, A filled function method for finding a global minimizer, Dundee Biennial Conference on Numerical Analysis.

Goldberg., D.E., 1989, *Genetic algorithms in search, optimization, and machine learning*, Addison-Wesley.

Goldstein, A.A., and J.F. Price 1971, On descent fromlocal minima, *Mathematics of Computation*, 25, 569-574,.

Goldstein, J.D., and W.E. Larimore, 1980, Applications of Kalman filtering and Maximum-Likelihood parameter identification to hydrologic forecasting, The Analytic Sci. Corp., Tech Rep, TR-1480-1, Jacob Way, Reading, MA.

Gomulka, J., 1978, Two implementation of Branin's method: numerical experience, In: Dixon, C.W.L., and G.P. Szegö, (eds), *Towards global optimization 2*, North-Holland, Amsterdam, 19-29.

Grefenstette, J.J.,1986, Optimization of control parameters for genetic algorithms, *IEEE Trans. Syst., Man, Cyber.* SMC-16, 122-128.

Griewank, A.O., 1981, Generalized descent for global optimization, Journal of Optimization Theory and Applications, 34, 11-39.

Gupta, V.K., and S. Sorooshian, 1983, Uniqueness and Observability of Conceptual Rainfall-Runoff Model Parameters: The Percolation Process Examined, *Water Resour. Res.*, 19(1), 269-276.

Gupta, V.K., and S. Sorooshian, 1985, The Automatic Calibration of Conceptual Catchment Models Using Derivative-Based Optimization Algorithms, *Water Resour. Res.*, 21(4),473-486.

Hansen, E.R., 1992, *Global optimzation using interval analysis*, Dekker, N.Y.

Harlin, J., 1991, Development of a process oriented calibration scheme for the HBV hydrological model, *Nord. Hydrol.*, 22, 15-36.

Hartman, J.K., 1973, Some experiments in global optimization, *Naval Research Logistics Quarterly*, 20, 569-576.

Havnø, K., M.N. Madsen, J. Dørge, 1995, MIKE 11 - a generalized river modelling package, in V.P. Singh (Ed), *Computer Models of Watershed Hydrology*, Water Resour. Pub., Colorado, 773-782.

Hendrickson, J.D., S. Sorooshian, and L. Brazil, 1988, Comparison of Newton-type and Direct Search Algorithms for Calibration of Conceptual Rainfall-Runoff Models, *Water Resour. Res.*, 24(5),. 691-700.

Herrera, F., M. Lozano, J.L. Verdegay, 1998, Tackling real-coded Genetic Algorithms: Operators and tools for the behaviour analysis, *Artificial Intelligence Review*, 12, 265-319.

Hill, J.D., 1969, A search technique for multimodal surfaces, *IEEE Trans. Systems Sci. and Cybernetics*, 5, 2-8.

Hogue, T.S., S. Sorooshian, H. Gupta,A. Holz and D. Braatz, 2001, A multistep automatic calibration scheme for river forecasting models, *J. Hydrometeol.*, 1, 524-542.

Holland, J.H., 1975, Adaptation in natural and artificial systems, University of Michigan Press, Ann Arbor, Michigan.

Hooke,R., and T.A. Jeeves,1961, Direct search solutions of numerical and statistical problems, *J. Ass. Comp. Mach.*, 8(2), 212-229.

Horst R. and P.M. Pardalos, 1995, *Handbook of global optimization*, Kluwer, Dordrecht.

Horst, R., P. M. Pardalos and N. V. Thoai, 1995, *Introduction to Global Optimization*, Kluwer Academic Publishers, Dordrecht.

Horst, R., and H. Tuy, 1987, On the convergence of global methods in multiextremal optimization, *J. Optim. Theo.&Appl.*, 54, 253-271.

Ibbitt, R.P, 1970, Systematic parameter fitting for conceptual models of catchment hydrology, Ph.D. Dis., U.of London, London, England.

Ibbitt, R.P., 1972, Effects of random data errors on the parameter values for a conceptual model, *Water Resour. Res.*, 8(1),70-78

Ingber, L., 1993, Simulated annealing: practice versus theory, J. Math. Comp. Model., 18(11), 29-57

Jarvis, R.A., 1975, Adaptive global search by process of competitive evolution, *IEEE Trans. on Syst., Man and Cyb.*, 75, 297-311.

Jakeman, A. and G.M Hornberger, 1993, How much complexity is warranted in a rainfall-runoff model, *Water Resourc. Res.*, 29(8), 2637-2649.

Johnston, P.R., and D. Pilgrim, 1976, Parameter Optimization for Watershed Models, *Water Resour. Res.*, 12(3), 477-486.

Karnopp, D.C., 1963, Random search techniques for optimization problems, *Automatica*, 1, 111-121

Kirkpatrick, S., C. D. Gelatt Jr., M. P. Vecchi, 1983, Optimization by Simulated Annealing, *Science*, 220, 4598, 671-680.

Kuczera, G., 1983a, Improved parameter inference in catchment models 1.Evaluating parameter uncertainty, *Water Resour. Res.*, 19(5), 1151-1162.

Kuczera, G., 1983b, Improved parameter inference in catchment models 2. Combining different kinds of hydrologic data and testing their compatibility, *Water Resour. Res.*, 19(5),1163-1172.

Kuczera, G., 1997, Efficient subspace probabilistic parameter optimization for catchment models, *Water Resour. Res.*, 33(1), 177-185

Kuczera, G., and M. Mroczkowski, 1998, Assessment of hydrologic parameter uncertainty and the worth of multiresponse data, *Water Resour. Res.*, 34(6), 1481-1489

Kushner, M.J., 1964, A new method of locating the maximum point of an arbitrary multipeak curve in the presence of noise, *J. of Basic Engr.*, 86, 97-106.

Levenberg, K., 1944, A method for the solution of certain non-linear problems in least squares, *Quart. of Appl. Math.*, 2, 164-168.

Levy, A.V., and S. Gomez, 1985, The tunneling method applied to global optimization, In: Boggs, P.T., and R.B. Schnabel, (eds), *Numerical optimization 1984*, SIAM, Philadelphia.

Luce, C.H., and T.W. Cundy, 1994, Parameter identification for a runoff model for forest roads, *Water Resour. Res.*,30(4),1057-1069

Lucidi, S., and M. Piccioni, 1989, Random tunneling by means of acceptance-rejection sampling for global optimization, *J. Optim. Theo. & Appl.*, 62(2), 255-277.

Luenberger, D.G., 1984, *Introduction to linear and nonlinear programming*, Addison-Wesley, Menlo Park, California.

Lundy, M., 1985, Applications of the annealing algorithm to combinatorial problems in statistics, *Biometrics*, 72(1), 191-198.

Madsen, H., 2000, Automatic calibration of a conceptual rainfall-runoff model using multiple objectives, *J. Hydrol.*, 235, 276-288.

Mahani, S.E., X. Gao, S. Sorooshian, B. Imam, 2000, Estimating cloud top height and spatial displacement from scan-synchronous GOES images using simplified IR-based stereoscopic analysis, *J. of Geophys. Res.*, 105(D12): 15,597-15,608

Marquardt, D.W., 1963, An algorithm for least-squares estimation of nonlinear parameters, SIAM, *J. Appl. Math.*, 11(2), 431-441.

Masri, S.F., G.A. Bekey, and F.B. Safford, 1978, An adaptive random search method for identification of large scale nonlinear systems, *Proceedings of 4th IFAC Symposium on Identification and System Parameter Estimation 1976*, North-Holland, Amsterdam.

Masri, S.F., G.A. Bekey, and F.B. Safford, 1980, A global optimization algorithm using adaptive random search, *Appl. Math. & Comp.*, 7, 353-375.

Metropolis, N., A.W. Rosenbluth, M.N. Rosenbluth, A.H. Teller, and E. Teller, 1953, Equation of state calculation by fast computing machines, *J. of Chem. Phys.*, 21,1087-1092.

Michalewicz, Z., 1996, *Genetic algorithm + data structures=evolution programs*, 3rd edition, Springer, N.Y.

Mockus, J., 1963, On the application of the Monte-Carlo method in solving multiextreme and combinatorial problems, *Conference Proceedings, Lectures, Vol IV, General problems of application of probabilistic and statistical methods*, State Publ. House of Tech. Lit., Ukrainian S.S.R., Kiev, 30-41.

Mockus, J., 1989, *Bayesian approach to global optimization*, Kluwer Academic Publishers, Dordrecht, Netherlands.

Monro, J.C., 1971, Direct search optimization in mathematical modeling and a watershed model application, NOAA Tech Memo NWS HYDRO-12, U.S. Dept. of Commerce, Silver Spring, MD.

Nash, J.E., and J.V. Sutcliffe, 1970, River flow forecasting through conceptual models, Part I - - A discussion of principles, *J. of Hydrol.*, 10(3), 282-290.

Nelder, J.A., and R. Mead, 1965, A simplex method for function minimization, *Computer J.*, 7, 308-313.

Nijssen, B. G. O'Donnell, D.P. Lettenmaier, D. Lohmann and E.F. Wood, 2001, Predicting the discharge of global rivers, *J. of Clim.*, 14(15), 3307-3323.

Parks, G.T., 1990, Am intelligent stochastic optimization routine for nuclear fuel cycle design, *Nucl. Technol.*, 89, 233-246.

Pickup, G., 1977, Testing the Efficiencies of Algorithms and Strategies for Automatic Calibration of Rainfall-Runoff Models, *Hydrogeological Science Bulletin*, 22(2), 257-274.

Pijavskij, S.A., 1972, An algorithm for finding the absolute extremum of a function, *USSR Comp. Math. & Math. Phys.*, 57-67 (2).

Pinter, J.D., 1996, Continuous global optimization software: a brief review, *Optima*, 52, 1-8.

Price, W.L., 1978, A controlled random search procedure for global optimization, in: Dixon, C.W.L, and G.P. Szegö, (eds), *Towards global optimization 2*, North-Holland, Armsterdam, 71-84.

Price, W.L., 1983, Global optimization by controlled random search, *J. of Optim. Theo. and Appl.*, 40(3), 33-348.

Price, W.L., 1987, Global optimization algorithms for a CAD workstation, *Journal of Optimization Theory and Applications*, 55(1), 133-146.

Powell, M.J.D., 1964, An efficient method for finding the minimum of a function of several variables without calculating derivatives, *Computer Journal &*, 155-162.

Pronzato, L., E. Walter, A. Venot and J.F. Lebruchec, 1984, A general purpose global optimizer: implementation and applications, *Math. and Comp. in Simul.*, 26, 412-422.

Randelman, R.E., and C.S. Grest, 1986, N-city salesman problem-optimization by simulated annealings, *J. Stat. Phys.*, 45, 885-890.

Ratchek, H., and J. Rokne, 1988, *New computer methods for global optimization*, Ellis Horwood Limited, Chichester, England.

Resende, M.G.C. and C.C. Ribeiro, 2001,Greedy randomized adaptive search procedures, To appear in F. Glover and G. Kochenberger (eds), *State-of-the-Art Handbook in Meta-heuristics*, Kluwer Academic Publishers

Rinnooy Kan, A.H.G., and G.T. Timmer, 1984, Stochastic methods for global optimization, *American Journal of Mathematics and Management Sciences*, 4, 7-40.

Rinnooy Kan, A.H.G., and G.T. Timmer, 1989, Global optimization: a survey, in: Penot, J-P., (eds), *New methods in optimization and their industrial uses*, Birkhauser Verlag, Basel, 133-155.

Rosenbrock, H.H., 1960, An automatic method for finding the greatest or least value of a function, *Computer Journal*, 3, 175-184.

Santos, C.A.G., K. Suzuki, and M. Watanabe, 1999, Modification of SCE-UA genetic algorithm for runoff-erosion modelling, *Proceedings of the International Symposium, 1999, V21, Nepal Geo.l Soc.*, 131-138

Schwefel, H.-P.,1981, *Numerical optimization of computer models*, John Wiley and Sons, New York.

Scott, R.L., W.J. Shuttleworth, T.O. Keefer, and A.W. Warrick, 2000, Modeling multiyear observations of soil moisture recharge in the semiarid American Southwest, *Water Resour. Res.*, 36(8), 2233-2248.

Seibert, J., 2000, Multi-criteria calibration of a conceptual runoff model using a genetic algorithm, *Hydro. & Earth Sys. Sci.*, 4, 215-224.

Shubert, B.O., 1972, A sequential method seeking the global maximum of a function, *Siam J. on Num. Anal.*, 9, 379-388.

Sorooshian, S., and J.A. Dracup,1980, Stochastic Parameter Estimation Procedures for Hydrologic Rainfall-Runoff Models: Correlated and Heteroscedastic Error Cases, *Water Resour. Res.*, 16(2), 430-442.

Sorooshian, S., Q. Duan, and V.K. Gupta, 1993, Calibration of Rainfall-Runoff Models: Application of Global Optimization to the Sacramento Soil Moisture Accounting Model, *Water Resour. Res.*, 29(4), 1185-1194

Sorooshian, S., and V.K. Gupta, 1983, Automatic Calibration of Conceptual Rainfall-Runoff Models: The Question of Parameter Observability and Uniqueness, *Water Resour. Res.*, 19(1), 251-259.

Sorooshian, S., and V.K. Gupta, 1985, The Analysis of Structural Identifiability: Theory and Application to Conceptual Rainfall-Runoff Models, *Water Resour. Res.*, 21(4), 487-495.

Sorooshian, S., V.K. Gupta, and J.L. Fulton,1983, Evaluation of Maximum Likelihood Parameter Estimation Techniques for Conceptual Rainfall-Runoff Models—Influence of Calibration Data Variability and Length on Model Credibility, *Water Resour. Res*, 19(1), 251-259.

Tanakamaru, H. and S.J. Burges, 1996, Application of Global Optimization to Parameter Estimation of the Tank Model, *Proceedings, Intl. Conf. on Water Resour. and Env. Res., Vol II,* Kyoto, Japan, 39-46, October 29-31, 1996.

Thyer, M., G. Kuczera, B.C. Bates, 1999, Probabilistic optimization for conceptual rainfall-runoff models: a comparison of the shuffled complex evolution and simulated annealing algorithms, *Water Resour. Res.,* 35(3): 767-773

Törn, A., and A. Zilinskas, 1989, *Global optimization,* Springer-Verlag, Berlin.

Törn, A., and S. Viitanen, 1994, Topographical global optimization using pre-sampled points, *J. Global Optim.,* 5, 267-276.

Törn,A.,and S.Viitanen,1999,Stochastic global optimization: problem classes and solution techniques, *J. Global Optim.,* 14, 437-447

van Laarhoven, P.J.M., and E. Aarts, 1987, *Simulate Annealing: Theory and Applications,* D. Reidel Publishing Company

Vanderbilt, D., and S.G. Louie, 1984, A Monte Carlo Simulated Annealing approach to optimization over continuous variables, *J. Comput. Phys.,* 56, 259-271

Walker, J.P., G.R. Wilgoose, and J.D. Kalma, 2001, One-dimensional soil moisture profile retrieval by assimilation of near-surface measurements: a simplified soil moisture model and field application, *J. Hydrometeo.,* 2, 356-373

Wang, Q.J., 1991, The Genetic Algorithm and its application to calibrating conceptual rainfall-runoff models., *Water Resour. Res.,* 27(9), 2467-2471.

Wang, J., Lu. Z., Habu, H., 2001, Shuffled Complex Evolution method for nonlinear constrained optimization, *Hehain University Science Journal,* 29(3), 46-50 (in Chinese)

Weinig, W., 1991, Calibration of the SMA-NWSRFS Model using the Adaptive Random Search Algorithm, M.S. Thesis, Dept. Hydrol. and Water Resour., University of Arizona, Tucson, AZ

Wood, G.R., 1985, Multidimensional bisection and global minimization, Tech. Rep., University of Canterburry.

Wright, A.H.., 1991, Genetic algorithms for real parameter optimization, in G.J.E. Rawlings (Ed): *Foundations of Genetic Algorithms,* Morgan Kaufman, 205-218

Yapo, P.O., H.V. Gupta, and S. Sorooshian, 1997, Multi-objective global optimization for hydrologic models, *J. Hydrol.,* 204, 83-97.

Qingyun Duan, NOAA/NWS, Hydrology Laboratory, Office of Hydrologic Development,1325 East-West Highway, Silver Spring, MD 20910, USA.

A Shuffled Complex Evolution Metropolis Algorithm for Estimating Posterior Distribution of Watershed Model Parameters

Jasper A. Vrugt

Institute for Biodiversity and Ecosystem Dynamics, The University of Amsterdam, The Netherlands

Hoshin V. Gupta

SAHRA, NSF STC for Sustainability of semi-Arid Hydrology and Riparian Areas
Department of Hydrology and Water Resources, The University of Arizona, Tucson, Arizona

Willem Bouten

Institute for Biodiversity and Ecosystem Dynamics, The University of Amsterdam, The Netherlands

Soroosh Sorooshian

SAHRA, NSF STC for Sustainability of semi-Arid Hydrology and Riparian Areas
Department of Hydrology and Water Resources, The University of Arizona, Tucson, Arizona

Practical experience with hydrologic model calibration suggests that it is generally impossible to find a single best parameter set whose performance measure differs significantly from other feasible parameters sets. While considerable attention has been given to the development of automatic calibration methods which aim to successfully find a single best set of parameter values, much less attention has been paid to a realistic assessment of parameter uncertainty in hydrologic models. In this paper, we present the Shuffled Complex Evolution Metropolis algorithm (SCEM-UA), which is suited to infer the most likely parameter set and its underlying posterior probability distribution within a single optimization run. The algorithm is related to the successful SCE-UA global optimization algorithm and merges the strengths of the Metropolis Hastings algorithm, controlled random search, competitive evolution and complex shuffling in order to evolve to a stationary posterior target distribution of the parameters. The features and capabilities of the SCEM-UA algorithm are illustrated by means of a hydrologic case study in which the Sacramento Soil Moisture accounting model is calibrated using historical data from the Leaf River watershed near Collins, Mississippi.

1. INTRODUCTION

To calibrate a hydrologic model, the hydrologist must specify values for its parameters in such a way that the behavior of the model closely matches the real system it represents. While some of the parameters can be derived through direct measurements conducted on the real system, others can only be meaningfully inferred by calibration to a historical record of input-output data. Because of the time consuming nature of manual trial-and-error model calibration, there has been a great deal of research into the development of automated (computer based) calibration methods [see e.g., Gupta and Sorooshian, 1994; Yapo et al., 1998; Boyle et al., 2000]. Automatic methods for model calibration seek to take advantage of the speed and power of computers, while being relatively objective and easier to implement than manual methods.

Calibration of Watershed Models
Water Science and Application Volume 6

In the development of suitable automatic calibration approaches, we must consider the fact that the hydrologic model optimization problem suffers from the existence of multiple optima in the parameter space (with both small and large domains of attraction), discontinuous first derivatives and curving multi-dimensional ridges. These considerations inspired Duan et al. [1992] to develop a powerful robust and efficient global optimization procedure, entitled, the Shuffled Complex Evolution (SCE-UA) global optimization algorithm. By merging the strengths of the Downhill Simplex procedure [Nelder and Mead, 1965] with the concepts of controlled random search, systematic evolution of points in the direction of global improvement, competitive evolution [Holland, 1975], and complex shuffling, the SCE-UA algorithm represents a synthesis of the best features of several optimization strategies. Numerous case studies have demonstrated that the SCE-UA global optimization algorithm can reliably find the global minimum in the parameter space for a variety of hydrologic models [e.g., Duan et al., 1992, 1993; Sorooshian et al., 1993; Luce and Cundy, 1994; Gan and Biftu, 1996; Tanakamaru, 1995; Kuczera, 1997; Hogue at al., 2000; Boyle et al., 2000; among many others]. However, it still remains typically difficult, if not impossible, to identify a unique 'best' parameter set, whose performance measure differs significantly from other feasible parameter sets within this region. Estimates of hydrologic model parameters are subject to uncertainty, because the calibration data contain measurement errors, and because the model never perfectly represents the system or exactly fits the data.

Only recently have methods for realistic assessment of hydrologic parameter uncertainty begun to appear in the literature. These include the traditional use of first-order approximations to parameter uncertainty [Kuczera and Mroczkowski, 1998], evaluation of likelihood ratios [Beven and Binley, 1992; Thiemann et al., 2001; see also Misirli et al., this volume], and parametric bootstrapping or Markov Chain Monte Carlo (MCMC) methods [e.g. Tarantola, 1987; Kuczera and Parent, 1998]. In view of the inevitably complicated nature of the hydrologic model an explicit first-order expression for the posterior distribution of the parameters is often not adequate [Kuczera and Parent, 1998; Vrugt and Bouten, 2002]. Therefore, MCMC algorithms have become increasingly popular as a class of general-purpose approximation methods for complex inference, search and optimization problems [Gilks et al., 1996]. Recently, Kuczera and Parent [1998] used the Metropolis-Hastings (MH) algorithm [Metropolis et al, 1953; Hastings, 1970], in a Bayesian inference framework to assess parameter uncertainty for a conceptual watershed model. In fact, the MH algorithm, the earliest and most general class of MCMC

samplers, has proven to be effective in assessing the posterior distribution of the model parameters for a variety of problems. However, the algorithm is fully probabilistic and does not optimally utilize the information gained about the response surface during the evolution process. As a consequence, convergence to the stationary posterior distribution can be slow [Gilks et al., 1996]. An important challenge, therefore, is to design a class of samplers that rapidly converges to the global minimum but resists becoming trapped along the way in a local basin of attraction.

A major weakness of the MH sampler is that it does not share response surface information gained by the individual parallel sequences of points generated during the process of evolving towards a stationary posterior distribution. In examining ways to increase information exchange between the parallel sequences it seems natural to consider the concept of periodic shuffling introduced by Duan et al. [1992] in developing the SCE-UA global optimization strategy (see also the chapter by Duan in this book). Shuffling has been found to have desirable properties, which significantly enhance the efficiency and effectiveness of the SCE-UA search procedure.

In this paper, we describe the Shuffled Complex Evolution Metropolis global optimization algorithm (SCEM-UA). The algorithm is related to the SCE-UA method, but uses the Metropolis Hastings strategy instead of the Downhill Simplex method for population evolution, and is therefore able to infer *both* the most likely parameter set and its underlying posterior probability distribution within a single optimization run. By merging the strengths of the MH algorithm, controlled random search, competitive evolution and complex shuffling, the SCEM-UA is designed to evolve to a stationary posterior target distribution of the parameters. The stochastic nature of the MH annealing scheme avoids the tendency of the SCE-UA algorithm to collapse into a single region of attraction (i.e. the global minimum), while the information exchange (shuffling) between parallel sequences allows the search to be biased in favor of better regions of the solution space. The following sections present a description of the SCEM-UA algorithm, followed by a hydrologic case study in which the features and capabilities of the algorithm are illustrated by calibrating the Sacramento Soil Moisture accounting model (SAC-SMA) using historical data from the Leaf River watershed near Collins, Mississippi.

2. GENERAL BACKGROUND AND OUTLINE OF THE SCEM-UA ALGORITHM

A hydrologic model aims at assessing the relationship between the watershed response or output variable $\hat{y}$, subject to measurement error e, and the input variables X. The hydrologic model η can be cast in a general statistical framework,

$$\hat{y} = \eta(X|\theta) + e \qquad (1)$$

where $\hat{y}(\hat{y}_1, \hat{y}_2, ..., \hat{y}_N)$ denotes a N x 1 vector of model outputs, $X = (X_{11}, ..., X_{Nn})$ is an N x n matrix of input values, $\theta = (\theta_1, \theta_2, ..., \theta_n)$ is a vector of n unknown parameters and e is a vector of statistically independent errors with zero expectation and constant variance σ^2.

We assume that the mathematical structure of the model is fixed and that we can define a uniform prior distribution on the feasible parameter space (e.g. between upper and lower bounds on each of the model parameters). Taking a Bayesian perspective, the aim of model calibration is to infer the posterior probability distribution, $p(\theta|y)$, which describes what is known about the model parameters θ given the observed data y and the prior information.

2.1. Traditional First-Order Approximation

Lets assume that little is known a-priori about the model parameters θ relative to what the experimental data will tell us, i.e. $p_0(\theta|y)$ is uniform on the parameter space. The traditional first-order approximation method is based on a first order Taylor series expansion of the non-linear model equations evaluated at the globally optimal parameter estimates θ_{opt}. The estimated posterior distribution of θ is then expressed as [Box and Tiao, 1973],

$$p(\theta|y) \propto \exp\left[-\frac{1}{2\sigma^2} (\theta - \theta_{opt})^T J^T J (\theta - \theta_{opt}) \right] \qquad (2)$$

where J is the Jacobian or sensitivity matrix evaluated at $\theta = \theta_{opt}$.

If the hydrologic model is linear (or very nearly linear) in its parameters, the posterior probability region estimated by equation 2 will give a good approximation of the actual parameter uncertainty. However, for non-linear models (e.g. conceptual rainfall runoff models such as SAC-SMA), with strong parameter interdependence, this approximation can be quite poor [Kuczera and Parent, 1998; Vrugt and Bouten, 2002]. Besides exhibiting strong and non-linear parameter interdependence, the surface of the posterior parameter distribution $p(\theta|y)$ can deviate significantly from the multi-normal distribution. It may also have multiple local optima and discontinuous derivatives [Duan et al., 1992]. Moreover, the ellipsoid region, defined by equation (2) may represent a very poor approximation of parameter uncertainty, as for example in the case of a strongly hyperbolic banana-shaped curvature in the $p(\theta|y)$ surface.

2.2. Monte Carlo Sampling of Posterior Distribution: The SCEM-UA Algorithm

The Markov Chain Monte Carlo (MCMC) method for assessing parameter confidence intervals in nonlinear models is based on the idea that instead of explicitly computing the probability distribution, $p(\theta|y)$, it is sufficient to approximate the form of the density by drawing a large random sample from $p(\theta|y)$. Diagnostic measures of central tendency and dispersion of the posterior distribution can be estimated by computing the mean and standard deviation of the sample. This directly leads to the question of how to efficiently sample from $p(\theta|y)$. To address this question we have developed a new algorithm, which merges the sampling strategy of the Metropolis Hastings algorithm with the strengths and efficiency of the SCE-UA population evolution method.

The goal of the original SCE-UA algorithm [Duan et al., 1992] is to find a single best parameter set in the feasible space. The SCE-UA begins with a random sample of points distributed throughout the (bounded) feasible parameter space, and uses an adaptation of the Downhill Simplex search strategy to continuously evolve the population toward better solutions in the search space, progressively relinquishing occupation of regions with lower posterior probability. This genetic drift, where the members of the population drift towards a single location in the parameter space (i.e. the mode of $p(\theta|y)$), is typical of many evolutionary search algorithms. By replacing the adapted Downhill Simplex strategy with a Metropolis Hastings strategy, the tendency of the algorithm to collapse into the relatively small region containing the "best" parameter set is avoided. The new algorithm, entitled the Shuffled Complex Evolution Metropolis (SCEM-UA) is developed in collaboration between the University of Arizona and the University of Amsterdam and is presented below.

SETUP

1. To initialize the process, choose the population size s and the number of complexes q. Compute the number of points m in each complex (m = s/q). The algorithm tentatively assumes that the number of sequences is identical to the number of complexes.
2. Generate s samples from the prior distribution $\{\theta_1, \theta_2, ..., \theta_s\}$ and compute the posterior density $\{p^1, p^2, ..., p^s\}$ of each point using a Bayesian inference scheme [Misirli et al., this volume].
3. Rank the points in decreasing posterior probability and store them in array D[1:s,1:n+1] so that the first row of D represents the highest posterior density. The extra column stores the posterior density.

4. Initialize the starting points of the parallel sequences, $S^1, S^2, ..., S^q$, such that S^k is D[k,1:n+1] where k = 1,2,...,q.

5. Partition D into q complexes $C^1, C^2, ..., C^q$, each containing m points, such that the first complex contains every $q(j-1)+1$ ranked point of D, the second complex contains every $q(j-1)+2$ ranked point, and so on, where j = 1,2,...,m.

SEQUENCE EVOLUTION: DO evolve = 1,2,...,q (Loop over all complexes - sequences):
DO β =1,2,...,L
(Loop within complex)

a). Calculate the mean μ^θ and covariance $\sum^k$ of the parameters of C^k.

b). Draw a uniform label Z between 0 and 1.

c). If Z ≥ 0.50, compute a new candidate point according to,

$$\theta^{(t+1)} = N(\theta^{(t)}, c_n^2 \Sigma^k) \qquad (3a)$$

Otherwise if Z < 0.50,

$$\theta^{(t+1)} = N(\mu^\theta, c_n^2 \Sigma^k) \qquad (3b)$$

where $\theta^{(t)}$ is the current draw of S^k associated with posterior density p($\theta^{(t)}$) and c_n is a predefined scaling parameter.

d). If $\theta^{(t+1)}$ is within Θ, compute p($\theta^{(t+1)}$) and go to Metropolis step; otherwise return to step (b)

METROPOLIS STEP: DO

I). Evaluate the ratio α = p($\theta^{(t+1)}$) / p($\theta^{(t)}$).

II). If Z ≤ α then accept the candidate point. If Z > α then remain at the current position, that is $\theta^{(t+1)} = \theta^{(t)}$.

END METROPOLIS STEP

e). Sort the m points in C^k in decreasing posterior density and assign a trapezoidal weight distribution (ρ) to C^k,

$$\rho_i = \frac{2(m+1-i)}{m(m+1)}, i = 1,2,...,m \qquad (4)$$

f). Randomly replace a member of C^k with $\theta^{(t+1)}$ according to the trapezoidal weight distribution defined in Equation (4).

END L DO
END SEQUENCE – COMPLEX EVOLUTION

6. Unpack all complexes C back into D and rank the points in order of decreasing posterior probability.

7. Check Gelman and Rubin (GR) convergence statistic (Appendix A). If convergence criteria are satisfied stop, otherwise return to step 5.

To summarize, the SCEM-UA algorithm begins with an initial population of points (parameter sets) randomly distributed throughout the feasible parameter space. For each parameter set, the posterior density is computed using a Bayesian inference scheme such as the one presented by Misirli, et al., [this volume]. The population is partitioned into q complexes, and in each complex k (k=1,2,...,q) a parallel sequence is launched from the point that exhibits the highest posterior density. A new candidate point in each sequence k is generated using a multivariate normal distribution either centered around the current draw of the sequence (k) or the mean of the points in complex (k) augmented with the covariance structure induced between the points in complex k. The Metropolis-annealing [Metropolis et al., 1953] criterion is used to test whether the candidate point should be added to the current sequence. Subsequently the new candidate point randomly replaces an existing member of the complex using the trapezoidal weight distribution defined in Equation (4). Finally, after a certain number of iterations ($q*L$) new complexes are formed through a process of shuffling. This series of operations results in a robust MCMC sampler that conducts a robust and efficient search of the parameter space. As a basic choice we have adopted the value of the scaling parameter as c_d=2.4/$\sqrt{n}$ [Gelman et al., 1996]. For more information about the SCEM-UA algorithm please refer to Vrugt et al. [2002].

The only variable that needs to be specified by the user is the population size s, which in turn also determines the number of points within each complex (m = s/q). The SCEM-UA algorithm employed for the case study reported in this paper used the values of q = 10 and L = ($m/10$). Preliminary sensitivity of the SCEM-UA algorithm demonstrated that the number of sequences – complexes and adapted relationship between the number of points within each complex (m) and the number of evolutionary steps before reshuffling ($q*L$) works well for a broad range of applications.

3. APPLICATION OF THE SCEM-UA ALGORITHM

We illustrate the application of the SCEM-UA algorithm to hydrologic modeling by using it to calibrate the Sacramento Soil Moisture Accounting model (SAC-SMA) of the National Weather Service River Forecasting System (NWSRFS) using historical data from the Leaf River watershed (1944 km²) located north of Collins, Mississippi.

Approximately 11 years (28 July 1952 to 30 September 1963) of hydrological data from the Leaf River basin were used for model calibration. Because the SAC-SMA model and the Leaf River Basin have been discussed extensively in the literature [e.g. Burnash et al. 1973; Peck, 1976; Kitanidis and Bras, 1980a,b; Thiemann et al., 2001; Boyle et al., 2000; Misirli et al., this volume], the details of these will not be described here. To reduce sensitivity to errors in initialization of the model states, a 365-day warm-up period was used, during which no updating of the posterior density was performed. We used a populations size s of 500 and assumed that the output errors have a heteroscedastic (non-constant) variance that is related to flow level and which can be stabilized using the transformation, $z = [(y+1)\lambda - 1]/\lambda$ with $\lambda = 0.3$. The measurement error standard deviation of the runoff was chosen to be identical to the RMSE value of the most optimal fit (18.00 m^3/s) derived by Boyle et al. [2000].

The algorithm converged to a stationary posterior distribution after 30000 iterations (function evaluations). Figure 1 displays the SCEM-UA estimates of the posterior uncertainty associated with the SAC-SMA parameter estimates given the 11 years of Leaf River calibration data. The parameters ranges in this plot have been normalized according to their prior uncertainty ranges defined as level zero estimates in Boyle et al. (2000). The most likely parameter set is indicated with the dark line. It is interesting to note that almost all of the SAC-SMA parameters are fairly well defined by calibration to the 11-year data set. In particular, the capacity parameters UZTWM, UZFWM, LZTWM, LZFSM and LZFPM are very precisely determined, while parameters ZPERC and REXP (that control percolation), ADIMP (additional impervious area), and the rate parameters LZSK and LZPK are less well determined. Unfortunately, direct comparison of these findings with the result presented in *Boyle et al.* [2000] is difficult, because the results presented here are obtained using the transformed streamflow data. Inspection of the covariance structure induced in the SCEM-UA generated parameter sets revealed that parameter correlations are typically low, confirming that most of the parameters are well determined by calibration to streamflow data. These results seem to contradict the arguments by Jakeman and Hornberger [1993] that only approximately four to five conceptual watershed model parameters can be well defined from rainfall-runoff data. However, it is not clear whether their results are sensitive to the fact that they used only one year of data.

The advantages of the SCEM-UA algorithm over the original SCE-UA algorithm are further demonstrated in Table 1, which compares the "most likely" parameter set found by the SCEM-UA (estimated mode of the posterior distribution) with the optimal parameter values derived

using the SCE-UA global optimization algorithm developed by Duan et al. [1992]. The results show that the SCEM-UA algorithm is able to conveniently derive the posterior distribution of the model parameters, while also successfully identifying the globally optimal parameter values. Comparative testing of the SCEM-UA and SCE-UA algorithms has shown that the SCEM-UA method is less efficient than the SCE-UA (in locating the globally optimal parameter values) for low dimensional problems (n < 6) but provides comparable efficiency when searching higher dimensional parameter spaces.

Finally, the Figures 2a and b present the residuals from the most probable parameter set and the 95% hydrograph prediction uncertainty intervals for the SAC-SMA simulated streamflows associated with the posterior parameter estimates (dark-gray region) and the residual model uncertainty (light gray region), respectively for a portion of the wet calibration year 1953. The solid circles correspond to the observed streamflow data. Note that the 95% streamflow prediction uncertainty ranges (light gray) bracket the observed flows during the entire period, but are quite large. Further, the prediction uncertainty associated with the posterior parameter estimates (dark gray) does not include the observations and displays bias (systematic error) on the long recessions. These indicate that the model structure is in need of further improvements.

Table 1. Most likely parameter sets for the SAC-SMA model derived with the SCE-UA global optimization and SCEM-UA algorithm using 10 years of runoff data (1952-1962) for the Leaf River watershed. Also included are three overall statistics for the calibration period for the selected parameter sets and the 95% Confidence Intervals of the parameters (CI) obtained using the SCEM-UA algorithm.

| Parameter | SAC-SMA | | | |
	SCE-UA	SCEM-UA	CI	
UZTWM	16.40	16.21	12.69	19.74
UZFWM	30.03	30.97	29.27	32.67
LZTWM	263.8	261.6	247.5	275.7
LZFPM	93.58	96.05	79.29	112.8
LZFSM	29.91	25.68	16.24	35.14
ADIMP	0.133	0.118	0.090	0.145
UZK	0.499	0.498	0.479	0.500
LZPK	0.018	0.019	0.017	0.021
LZSK	0.203	0.215	0.181	0.250
ZPERC	249.9	247.8	228.3	267.2
REXP	3.197	3.188	2.515	3.861
PCTIM	0.0002	0.0002	0.0000	0.0004
PFREE	0.148	0.149	0.132	0.165
Bias, %	3.82	3.97		
RMSE, m^3/s	19.38	19.40		
R^2	0.90	0.90		

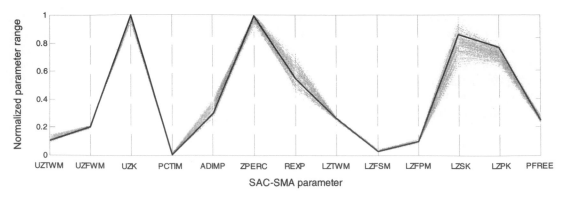

Figure 1. Normalized uncertainty ranges of the SAC-SMA parameters obtained after assimilating and processing 11 years of calibration data of the Leaf River basin with the SCEM-UA algorithm. The most likely parameter set is indicated with the dark line.

4. CONCLUSIONS

We have presented a robust, effective, and efficient Markov Chain Monte Carlo algorithm, which is designed to estimate the most likely parameter set and its underlying posterior probability distribution within a single optimization run. The sampler, entitled the Shuffled Complex Evolution Metropolis algorithm (SCEM-UA), merges the strengths of the Metropolis Hastings algorithm, with controlled random search, competitive evolution and complex shuffling. An illustrative application of the SCEM-UA algorithm to the Sacramento Soil Moisture accounting model (SAC-SMA) has shown that the model parameters can be reasonably well determined by calibration to streamflow data. Our experience with the SCEM-UA algorithm suggests that the method is less efficient than the SCE-UA algorithm in locating the globally optimal parameter values for problems of low dimension, but shows comparable efficiency when searching high dimensional parameter spaces (n>=6). Research aimed at further improvements of the

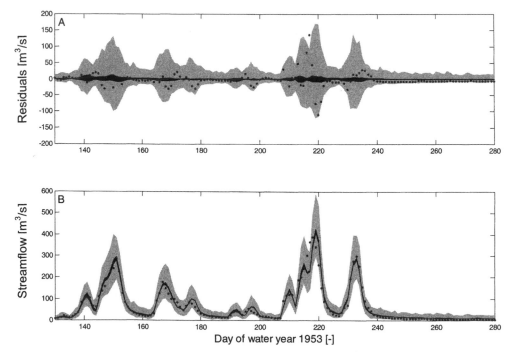

Figure 2. a) Streamflow uncertainties associated with the most probable parameter set derived using the SCEM-UA algorithm. The light gray region denotes model uncertainty, whereas parameter uncertainty is indicated with the dark gray-region. The dots correspond to the observed streamflow data, b) Hydrograph prediction uncertainty associated with the uncertainty in the model (light gray) and parameter estimates (dark-gray region) for the WY 1953.

Shuffled Complex Evolution Metropolis approach, including extensions to multi-criteria problems, is ongoing. These results will be discussed in future papers.

APPENDIX

Many authors have addressed the problem of drawing inferences from MCMC samplers. Gelman and Rubin [1992] demonstrated that it is generally impossible to monitor convergence of an MCMC sampler using a single sequence (one random walk). A strategy recommended by Gelman and Rubin [1992] is therefore to generate several independent parallel sequences, with starting points, $\theta^{(0)}$s sampled from the proposal distribution. Convergence of the MCMC sampler can then be monitored using between sequences as well as within sequence information.

Mathematically we proceed in five steps,

1. Independently simulate q $\geq$ 2 sequences with the SCEM-UA algorithm outlined in section 2.2, each of length 2T, with starting points drawn from the proposal distribution. To diminish the effect of the initial draws, discard the first T draws of each sequence, and focus the attention on the last T.

2. For each parameter of interest, calculate the mean value, $\bar{\theta}_{i.}$, using the $q \cdot T$ drawn values for parameter θ_i,

3. For each parameter θ_i of interest, calculate the variance between the q sequence means, $\bar{\theta}_{i.}$, each based on T samples θ_i

$$B = T\sum_{j=1}^{q} \frac{\left(\bar{\theta}_{i.} - \bar{\theta}_{i..}\right)^2}{q-1} \tag{A.1}$$

and calculate the average of the q within-sequence variances, $s_{\theta_i}^2$, each based on T-1 degrees of freedom,

$$W = \sum_{j=1}^{q} \frac{s_{\theta_i}^2}{q} \tag{A.2}$$

4. Monitor convergence of the MCMC sampler by estimating the factor by which the scale of the current distribution of θ_i might be reduced if T $\rightarrow \infty$. This potential scale reduction is estimated by

$$\sqrt{\hat{R}} \approx \sqrt{\frac{T-1}{T} + \frac{q+1}{qT}\frac{B}{W}} \tag{A.3}$$

and declines to 1 as $T \rightarrow \infty$. $\hat{R}$ is the ratio of the current variance estimate, $\hat{V}$, to the within-sequence variance, W. Due to its minor contribution, the factor to account for the extra variance of the Student's t distribution is omitted from equation (A.3). If the potential scale reduction is

high, then we have reason to believe that proceeding with further draws may improve our inference about the posterior target distribution.

5. Once $\hat{R}$ is near 1 for each of the parameters θ_i, we can conclude that each of the q sequences of T draws is close to the target distribution, p(θ|y). Since a score of 1 is typically difficult to achieve, *Gelman and Rubin* [1992] recommended using a value of 1.2 and less to declare convergence.

Acknowledgements. The Earth Life Sciences and Research Council (ALW) partly supported the investigations of the first author with financial aid from the Netherlands Organization for Scientific Research (NWO). This material is based upon work supported in part by SAHRA (Sustainability of semi-Arid Hydrology and Riparian Areas) under the STC Program of the National Science Foundation, Agreement No. EAR-9876800.

REFERENCES

Beven, K.J., and A.M. Binley, The future of distributed models: Model calibration and uncertainty prediction, *Hydrological Processes*, 6, 279-298, 1992.

Box, G.E.P., and G.C. Tiao, *Bayesian Inference in Statistical Analysis*, Addison-Wesley-Longman, Reading Massachusetts, 1973.

Boyle, D.P., H.V. Gupta, and S. Sorooshian, Toward improved calibration of hydrological models: Combining the strengths of manual and automatic methods, *Water Resources Research*, 36, 12, 3663-3674, 2000.

Burnash, R.J.E., R.L. Ferral, and R.A. McGuire, *A generalized streamflow simulation system*, report, 204 pp., Jt. Fed.-State River Forecast Center, Sacramento, Calif, 1973.

Duan, Q., V.K. Gupta, and S. Sorooshian, Effective and efficient global optimization for conceptual rainfall-runoff models, *Water Resources Research*, 28, 1015-1031, 1992.

Duan, Q., V.K. Gupta, and S. Sorooshian, A shuffled complex evolution approach for effective and efficient global minimization, *Journal of Optimization Theory and Applications*, 76, 3, 501-521, 1993.

Gan, T.Y., and G.F. Biftu, Automatic calibration of conceptual rainfall-runoff models: Optimization algorithms, catchment conditions, and model structure, *Water Resources Research*, 32, 12, 3513-3524, 1996.

Gelman, A., and D.B. Rubin, Inference from iterative simulation using multiple sequences, *Statistical Science*, 7, 457-472, 1992.

Gelman, A., G.O. Roberts, and W.R. Gilks, Efficient Metropolis jumping rules, in: J.M. Bernardo, J.O. Berger, A.F. David and A.F.M. Smith, eds. *Bayesian Statistics V*, Oxford University Press, pp, 599-608, 1996.

Gilks, W.R., S. Richardson, and D. Spiegelhalter (Eds), *Practical Markov Chain Monte Carlo*, Chapmann-Hall, New Yorks, 1996.

Gupta, V.K., and S. Sorooshian, Calibration of conceptual hydrologic models: past, present and future. Invited paper in: Council of Scientific Research Integration (Eds), Trends in Hydrology, Research Trends, Trivandrum, India, pp. 329-346, 1994.

Hastings, W.K., Monte Carlo sampling methods, using Markov Chains and their applications, *Biometrika*, 57, 97-109, 1970.

Hogue, T.S, S. Sorooshian, H.V. Gupta, A. Holz, and D. Braatz, A multistep automatic calibration scheme for river forecasting models, *Journal of Hydrometeorology*, 1, 524-542, 2000.

Holland, J., *Adaptation in natural and artificial systems*, University of Michigan Press, Ann Harbor, 1975.

Jakeman, A., and G. Hornberger, How much complexity is warranted in a rainfall-runoff model? W*ater Resources Research*, 29 (8), 2637-2649, 1993.

Kitanidis, P.K., and R.L. Bras, Adaptive filtering through detection of isolated transient errors in rainfall-runoff models, *Water Resources Research*, 16 (4), 740-748, 1980a.

Kitanidis, P.K., and R.L. Bras, Real-time forecasting with a conceptual hydrological model, 1, Analysis of uncertainty, *Water Resources Research*, 16 (6), 1025-1033, 1980b.

Kuczera, G., Efficient subspace probabilistic parameter optimization for catchment models, *Water Resources Research*, 33, 1, 177-185, 1997.

Kuczera, G., and E. Parent, Monte Carlo assessment of parameter uncertainty in conceptual catchment models: the Metropolis algorithm, *Journal of Hydrology*, 211, 69-85, 1998.

Kuczera, G., and M. Mroczkowski, Assessment of hydrological parameter uncertainty and the worth of multiresponse data, *Water Resources Research*, 34, 6, 1481-1489, 1998.

Luce, C.H., T.W. Cundy, Parameter identification for a runoff model for forest roads, *Water Resources Research*, 30, 4, 1057-1069, 1994.

Metropolis, N., A.W. Rosenbluth, M.N. Rosenbluth, A.H. Teller, and E. Teller, Equations of state calculations by fast computing machines, *J. Chem. Phys.*, 21, 1087-1091, 1953.

Nelder, J.A., and R. Mead, A simplex method for function minimization, Computer Journal, 7, 308-313, 1965.

Peck, E.L., Catchment modeling and initial parameter estimation for the national weather service river forecast system. NOAA Technical Memo NWS Hydro-31, U.S. Department of Commerce, Silver Spring, MD, 60 pp, 1976.

Sorooshian, S., Q. Duan, and V.K. Gupta, Calibration of rainfall-runoff models: Application of global optimization to the Sacramento Soil Moisture accounting model, *Water Resources Research*, 29, 1185-1194, 1993.

Tanakamaru, H., Parameter estimation for the tank model using global optimization, *Trans. Jap. Soc. Irrig. Drainage Reclam. Eng.*, 178, 103-112, 1995.

Tarantola, A., *Inverse problems theory, Methods for data fitting and model parameter estimation*, Elsevier, Netherlands, 1987.

Thiemann, M., M. Trosset, H. Gupta, and S. Sorooshian, Bayesian recursive parameter estimation for hydrological models, *Water Resources Research*, 37 (10), 2521-2535, 2001.

Vrugt, J.A., and W. Bouten, Validity of first order approximations for assessing parameter uncertainty in soil hydrological models, Accepted by *Soil Science Society of America Journal*, 2002

Vrugt, J.A., H.V. Gupta, W. Bouten, and S. Sorooshian, A Shuffled Complex Evolution Metropolis algorithm for optimization and uncertainty assessment of hydrological model parameters, *Submitted to Water Resources Research*, 2002.

Yapo, P.O., H.V. Gupta, and S. Sorooshian, Multi-objective global optimization of hydrologic models, Journal of Hydrology, 204, 83-97, 1998.

Jasper A. Vrugt, and Willem Bouten, Institute for Biodiversity and Ecosystem Dynamics, Nieuwe Achergracht 166, The University of Amsterdam, The Netherlands, Amsterdam, 1018 WV.

Hoshin V. Gupta and Soroosh Sorooshian, SAHRA, NSF STC for the Sustainability of semi-Arid Hydrology and Riparian Areas, Department of Hydrology and Water Resources, Harshbarger, Bldg. 11, University of Arizona, Tucson, AZ 85721, USA.

Bayesian Recursive Estimation of Parameter and Output Uncertainty for Watershed Models

Feyzan Misirli, Hoshin V. Gupta, Soroosh Sorooshian and Michael Thiemann[1]

SAHRA, NSF STC for Sustainability of Hydrology and Semi-Arid Riparian Areas, Department of Hydrology and Water Resources, The University of Arizona, Tucson, Arizona

In any model of a hydrologic system, there is always some uncertainty associated with model structure, parameters, states, and the input and output measurements. Therefore, it is essential to represent this uncertainty in calibration efforts. Bayesian Recursive Estimation (BaRE) is an algorithm being developed towards considering these uncertainties for parameter estimation and prediction within an operational setting. This paper evaluates the current version of the algorithm and provides an application to a watershed for comparison with a conventional deterministic approach. BaRE is tested with different error models and transformation factors. We also introduce a measure called Forecast Range Error Estimate (FREE) to evaluate the model efficiency. Comparison to batch calibration using the Shuffled Complex Evolution (SCE-UA) optimization method indicates that the on-line calibration technique is a powerful tool, especially useful where basins are recently gauged and hydrologic data are not well accumulated. The analyses for this study were done using the HYdrologic MODel (HYMOD) applied to the Leaf River basin in Mississippi.

1. INTRODUCTION

The calibration of conceptual rainfall runoff models is of major interest due to the continual demand for more timely and accurate river forecasts. The goal of model calibration is to adjust the parameter values so that the model is constrained to be consistent with the observed hydrologic data (e.g., streamflow). Because manual calibration is time-consuming and requires unique expertise, automatic calibration techniques have been investigated as an effective alternative. Many studies have focused on finding a unique parameter set which gives the best match of the simulated model output to the observation values (e.g., *Duan et al., 1992, 1993; Sorooshian et al.*, 1993). There are both mathematical

and practical difficulties for finding this "best set", given that there is no perfect model to simulate nature. Therefore, various recent studies have been exploring efficient and reliable ways to summarize the uncertainty in the estimates of parameter values and the subsequent output predictions, while obtaining estimates of the most likely parameter values (e.g., *Beven and Binley*, 1992; *Franks and Beven*, 1997; *Kuczera and Mroczkowski*, 1998; *Bates and Campbell*, 2001; *Thiemann et al.*, 2001; *Kavetski et. al.*, This book). The uncertainty in model simulations is due to several sources: imperfect model structure, incorrect parameter identification, uncertainties in the states, and erroneous input and output measurements. Although batch calibration methods have been shown to provide acceptable calibration results, they do not account for these uncertainties in a satisfactory way.

One other arising issue is the overwhelming number of watersheds without accumulated historical data, which remain to be calibrated for operational flood forecasting. Conventional batch calibration methods assume time-invariant model parameter values and require a considerable amount of data to be used, typically 8-10 years. On-line cal-

---

[1]Now at Riverside Technology Incorporated, Fort Collins, Colorado

Calibration of Watershed Models
Water Science and Application Volume 6

ibration methods can overcome this drawback by allowing forecasts to be generated soon after the first observation becomes available. Although this brings some additional computational cost, advances in computer technology lessen this concern.

In response to these issues, *Thiemann et al.* (2001) developed a Bayesian formulation, which permits the hydrologist to quantify uncertainty about prediction and parameter estimation in an on-line fashion. The method is called Bayesian Recursive Estimation (BaRE). BaRE uses three sources of information for quantifying uncertainty in hydrologic predictions in selecting a suitable parameter set for the model (i.e., model calibration); measured data, physical laws (model), and statistical methods.

In this chapter, we discuss the work of *Thiemann et al.* (2001), explore how to select error model parameters, and conduct a comparison with the batch calibration approach. In the following sections, a brief summary of the Bayesian analysis, description of the BaRE method, and its application to an operational basin will be provided.

2. BAYESIAN ANALYSIS

In Bayesian analysis, uncertainty is quantified probabilistically. *Berger* (1985) gave an excellent review of the Bayesian approach. Assume that we are trying to estimate sample observation y, given the inputs ξ, using a model having an unknown parameter $\theta \in \Theta \subseteq \Re^k (\Re^k$ denotes k-dimensional Euclidean space). Bayesian analysis is performed by combining the prior information ($p(\theta)$) and the sample information y into what is called the posterior distribution of θ given y, from which all inferences are made.

2.1. Prior Information

An important element of Bayesian analysis is prior information concerning θ. The main idea of introducing prior probability is to reflect "before-the-fact" expectations of chance occurrences of an event. It typically does not depend on any currently available inputs or outputs. Characterization of prior probability can be achieved through careful analysis of historical data from another system having similar characteristics. There might be a concern that the prior may dominate and distort the information in data. However, by careful choice of the model structure and appropriate priors, Bayesian analysis can use the information from the data very effectively. When little or no prior information is available, non-informative priors are suggested so as not to favor any possible value of Θ over others. When parameter set Θ has n discrete members, one possible non-informative prior is probability of 1/n assigned to each member.

2.2. Posterior Distribution

Posterior distribution $p(\theta \mid y)$ is the conditional probability distribution of θ given the sample observation y. Noting that θ and y have the joint (subjective) density:

$$h(y,\theta) = p(\theta)p(y \mid \theta) \qquad (1)$$

and y has the marginal (unconditional) density :

$$m(y) = \int_{\Theta} p(y \mid \theta) d\theta \qquad (2)$$

providing that $m(y) \neq 0$:

$$p(\theta \mid y) = \frac{h(y,\theta)}{m(y)} \qquad (3)$$

$p(\theta \mid y)$ reflects the updated beliefs about θ after observing the sample y.

In discrete situations, the formula for $p(\theta \mid y)$ is commonly known as Bayes' theorem, which was introduced by *Bayes* (1763). If there exists a sequence of discrete events $A_1,...,A_n$ with prior probabilities $p(A_i)>0$, and another event B such that $p(B)>0$, then Bayes' theorem states that :

$$p(A_i \mid B) = \frac{p(B \mid A_i).P(A_i)}{\sum_{j=1}^{n} p(B \mid A_j).P(A_j)} \qquad (4)$$

Here $p(A_i \mid B)$ is the conditional distribution of A_i, given that B has occurred. Replacing A_i by θ and B by y, the formula becomes equivalent to the one for posterior distribution.

Although it is simple, this theorem is very useful and is widely used in many statistical applications.

3. BAYESIAN INFERENCE

The idea of Bayesian inference is that the posterior distribution is constructed to summarize all available information about θ (both sample and prior information); therefore, inferences concerning θ could be made solely in terms of the features of this distribution.

3.1. Prediction

To predict the values of y_{T+1}, as yet unobserved outputs, we compute the predictive density of y_{T+1} based on the previous observations (i.e., marginal posterior density of y_{T+1}) as follows:

$$p(y_{T+1} \mid \xi,y) = \int_{\Theta} p(y_{T+1};\theta \mid \xi,y) d\theta \qquad (5)$$

Prediction is done by computing meaningful summary statistics of this density from the region of highest probability density (HPD). A subset of R of the domain of p is called the HPD region of content $1-\alpha$ if $P(R)=1-\alpha$ and $p(y_1) \geq p(y_2)$ for any $y_1 \in R$ and $y_2 \notin R$.

3.2. Estimation

In Bayesian estimation of a real valued parameter $\theta \in \Theta$, we must specify a loss function, $L(\theta, \alpha)$ where $\alpha \in \Theta$ is the true value. The estimate of θ is $\alpha \in \Theta$ that minimizes the posterior expected loss:

$$f(a) = \int_\Theta L(\theta, a) p(\theta \mid \xi, y) d\theta \qquad (6)$$

where, $p(\theta \mid \xi, y)$ is marginal posterior density of θ given as

$$p(\theta \mid \xi, y) = \int p(y_{T+1}; \theta \mid \xi, y) dy_{T+1} \qquad (7)$$

Often, analyses of decision rules are carried out for certain standard losses such as squared-error loss, $(\alpha - \theta)^2$. However, this simple loss function does not typically reflect a useful measure for the calibration of hydrologic models because large errors are penalized too severely.

The robustness of loss functions is questionable. However, because the decisions are functions of uncertain assumptions, this robustness problem is inevitable. Any loss used in the analysis will be uncertain to a degree. It is impossible to obtain a completely accurate specification of the loss function.

3.3. Bayesian Recursive Inference

Thiemann et al. (2001) derived a practical recursive formula for updating information about θ. Supposing that we are at time $t = T$ and that all of the input and output data, y and ξ, are collected up to the current time, the recursive formula was presented as:

$$p(\theta \mid \xi_{T+1}, \xi, y_{T+1}, y) \propto p(y_{T+1} \mid \xi_{T+1}, \xi; \theta) p(\theta \mid \xi, y) \quad (8)$$

4. BAYESIAN RECURSIVE ESTIMATION (BaRE) ALGORITHM

4.1. Basic Formulation

Let η be a mathematical model used to predict an observation y by $\hat{y}$ using *input* ξ and parameter θ as:

$$\hat{y} = \eta(\xi \mid \theta) \qquad (9)$$

and error given by:

$$\varepsilon = y - \hat{y} \qquad (10)$$

Considering time steps, we can write:

$$y_{T+1} = \eta(\xi \mid \theta) + \varepsilon_{T+1} \qquad (11)$$

which is a standard formulation for nonlinear regression.

4.2. Assumptions

There exists a one-to-one and invertible transformation:

$$z = g(y) \qquad (12)$$

such that the measurement errors in the transformed space, given by :

$$v = g(y) - g(\hat{y}) \qquad (13)$$

are mutually independent each having the exponential power density $E(\sigma, \beta)$ described by *Box and Tiao* [1973, Section 3.5];

$$p(v \mid \sigma, \beta) = \omega(\beta) \sigma^{-1} \exp\left[-c(\beta) \left| v / \sigma \right|^{2/(1+\beta)} \right] \qquad (14)$$

where:

$$c(\beta) = \left\{ \frac{\Gamma[3(1+\beta)/2]}{\Gamma[(1+\beta)/2]} \right\}^{1/(1+\beta)} \qquad (15)$$

$$\omega(\beta) = \frac{\{\Gamma[3(1+\beta)/2]\}^{1/2}}{(1+)\beta \ \{\Gamma[(1+\beta)/2]\}^{3/2}} \qquad (16)$$

The shape parameter $\beta \in (-1, 1]$ is fixed and the standard deviation of the measurement errors $\sigma > 0$ is assumed to be unknown but constant with respect to time. As β approaches -1, function approaches uniform distribution. On the other hand, $\beta = 1$ corresponds to double exponential function.

4.3. Recursive Formulation

Following *Box and Tiao* (1973), *Thiemann et al.* (2001) derived the following relationship for the maximum likelihood estimate of the measurement error:

$$\hat{\sigma}_T(\theta)^{2/(1+\beta)} = \frac{T-1}{T} \hat{\sigma}_{T-1}(\theta)^{2/1(1+\beta)} + \frac{c(\beta)}{T(1+\beta)} \left| v_T(\theta) \right|^{2/1(1+\beta)} \qquad (17)$$

A recursive formulation for estimating the posterior density for θ was given as follows:

$$p(\theta \mid \xi, z_{T+1}, z, \beta) \propto N_T(\theta) \mid \xi, z; \beta) \qquad (18)$$

where:

$$N_T = \frac{1}{\hat{\sigma}_T(\theta)} \exp\left[-c(\beta) \left| \frac{v_T(\theta)}{\hat{\sigma}_T(\theta)} \right|^{2/(1+\beta)} \right] \quad (19)$$

4.4. BaRE Algorithm

To approximate the posterior and conditional densities, *Thiemann et al.,* (2001) used a Monte Carlo simulation approach as described in the following algorithm:

Preparation:
Select

- System model $\hat{y} = \eta(\xi \mid \theta)$
- Transformation model $z = g(y)$
- Error model $v \sim E(\sigma, \beta)$
- Kurtosis parameter β (Section 3.5. of *Box and Tiao*, 1973)
- Initial estimate for $\hat{\sigma}_0$ of the error model
- Upper and lower limits for each θ
- Prior probability distribution for parameters $p_0(\theta)$

Sampling

- Sample n different parameter sets $\theta^i, = i=1,..,n$ from a uniform distribution on Θ.

Initialization:

- Set time to zero ($T = 0$)
- Initialize prior $p(\theta^i \mid \xi, z; \beta) = p_0(\theta)$ and error model variance estimate, $\hat{\sigma}_0(\theta^i) = \sigma_0; i=1,....,n$

Prediction of the Output: *(Considers only the model parameter uncertainty at this stage)*:

- Compute transformed model output for each parameter set, $\hat{z}_{T+1}(\theta^i) = g(\eta(\theta^i \mid \xi)); i=1,..,n$
- Sort outputs in ascending magnitude,
- Compute cumulative distribution function of the predicted output in the transformed space.
- Compute appropriate percentiles to define the HPD region for the transformed output and un-transform these back to the original output space.

Prediction of the Output Measurement: *(Includes the additional uncertainty due to structural error and output measurement error as estimated by the error model)*
(No observation available yet)

- Define output region of interest
  - Find minimum and maximum of transformed output, $\hat{z}_{T+1}(\theta^l)$ and $\hat{z}_{T+1}(\theta^u)$ respectively.

- Extend the output range to $\left[a_{T+1}^{\min}, a_{T+1}^{\max} \right]$
  i.e., $a_{T+1}^{\min} = \hat{z}_{T+1}^l - 2\hat{\sigma}(\theta^l)$ and
  $a_{T+1}^{\max} = \hat{z}_{T+1}^u + 2\hat{\sigma}(\theta^u).$

- Discretize new range into n_a (e.g., 100) equally spaced points $b_k, k = 1,..n_a$.
- Compute probability density

$$p(\tilde{z}_{T+1} = b_k \mid \xi, y) = C \sum_{i=1}^n N_{T+1}(\theta^i \mid \xi, \hat{\sigma}_T, \beta) \Big|_{z_{T+1} = b_k} p(\theta^i \mid \xi, y)$$

and the cumulative probability density of the as-yet-unobserved output measurement in the transformed space:

$$p(\tilde{z}_{T+1} \leq b_k \mid \xi, y) = \sum_{i=1}^n p(\tilde{z}_{T+1} = b_k \mid \xi, y)$$

where C is a constant that normalizes the total probability mass to 1.

- Compute appropriate percentiles to define the HPD region for $\tilde{z}_{T+1}$ and un-transform these to the original output space.

Updating: *(When the observation y_{T+1} becomes available)*

- Compute transformed measurement .

$$z_{T+1} = g(y_{T+1})$$

- Update estimates of error model variance $\hat{\sigma}_{T+1}(\theta_i), i=1,...,n$ according to Equation (17).
- Compute posterior density and set it as the prior for the next time step

$$p(\theta^i) \xi, z, z_{T+1}) = C. N_{T+1}(\theta^i \mid \xi, \hat{\sigma}_{T+1}, \beta)\Big|_{z_{T+1}} p(\theta^i \mid \xi, y), \; i = 1,...,n$$

- Set $T = T + 1$ and resume with prediction of output.

MATLAB version of the algorithm is available upon request.

5. APPLICATION

5.1. Case Study: Leaf River Basin

The BaRE method was applied to calibration of a conceptual rainfall-runoff model of the Leaf River basin, a test basin used for many studies in the literature (e.g. *Sorooshian et al.,* 1983; *Brazil,* 1988; *Boyle et al.,* 2000; *Thiemann et al.,* 2001). This humid watershed is 1944 km² and located north of Collins, Mississippi.

A relatively simple, five-parameter conceptual rainfall-runoff model named HYMOD, first introduced by *Boyle et*

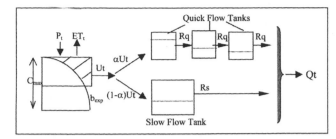

Figure 1. Conceptual diagram of hydrologic model HYMOD.

al. (2000), was used to test the algorithm. Figure 1 shows a diagram of the model. Model parameters are C_{max} and b_{exp} which are the maximum capacity and shape factor of the main soil water storage tank, α, the ratio determining the rate of water flowing through slow and quick flow tanks, and R_s and R_q, referring to the recession constants of the slow flow and quick flow tanks, respectively. Parameter bounds are shown in Table 1.

One year of hydrologic data, Water Year (WY) 1953, was used to test the algorithm and ten years (WY 1954 to 1964) were used for model evaluation. Figure 2 shows the precipitation input for Water Year 1953. One thousand random parameter sets were sampled from the feasible parameter space. A non-informative prior probability of 1/n (n is the sample size, i.e., 1000 in this case) was assigned to each parameter set to initialize the algorithm. Transformation shown in equation (20) was used to deal with heteroscedastic (non-constant) variance of the measurement errors associated with streamflow (*Hogue et al.,* 2000) and the structural errors associated with the model

$$g(y)=[(y+1)^\lambda - 1] / \lambda \qquad (20)$$

λ is referred as the transformation factor. The initial variance of the measurement error was assumed to be 12, twice the variance of the streamflow values (in the transformed space) for the selected water year, for illustrative purposes only (*Thiemann et al.,* 2001).

To provide an objective tool to decide which transformation factor and shape factor to use for the error model, we

focused on two model performance criteria: accuracy and precision. To represent the uncertainty in an efficient way we would ideally like to have the width of prediction bounds as small as possible while containing the streamflow data. For the accuracy measure, we used the simple Daily Root Mean Square (DRMS) error and percent Bias (% Bias) criteria defined as follows:

$$DRMS = \sqrt{\frac{1}{n}\sum_{t=1}^{n}(q_t^{obs} - q_t^{comp})^2} \qquad (21)$$

$$\%Bias = \frac{\sum_{t=1}^{n}(q_t^{comp} - q_t^{obs})}{\sum_{t=1}^{n}(q_t^{obs})} *100 \qquad (22)$$

where q_t^{obs} is the observed streamflow value at time t and q_t^{comp} is the computed streamflow value at time t.

5.2. Measure of Precision: Forecast Range Error Estimate (FREE)

We refer to precision as a characteristic related to the efficiency of the prediction uncertainty bounds in representing the actual distribution of the observed output data. We defined an efficiency criterion called Forecast Range Error Estimate (FREE) to summarize the model performance in terms of both the inclusion of the observed data (desirable as large as possible) and the width of the prediction bounds (to be as small as possible while maximizing inclusion).

Table 1. Parameter bounds and best parameter sets.

	Parameter Bounds	BaRE	SCE-UA
Cmax	1-500	181.91	282.51
Bexp	0.1-2.0	0.150	0.251
Alpha	0.1-0.99	0.667	0.861
Rs	0.00002-0.10	0.0295	0.0100
Rq	0.1-0.99	0.4783	0.465

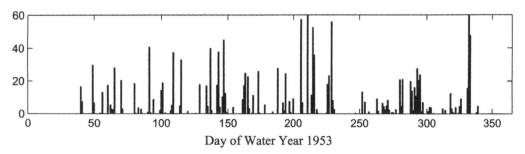

Figure 2. Precipitation input for Water Year 1953.

A positive deviation, FREE_POS, refers to the distance to the prediction boundary if the observation is within the prediction bounds and a negative deviation, FREE_NEG, refers to the same distance if the point is outside the boundary. FREE is the sum, over all time steps, of the absolute values of these two distances.

In mathematical terms,

$$d_t = q_t^{obs} - q_t^{mlh} \tag{23}$$

$$dist_t = \begin{cases} q_t^{max95} - q_t^{obs}, d_t \geq 0 \ (under\ prediction) \\ q_t^{obs} - q_t^{min95}, d_t < 0 \ (over\ prediction) \end{cases} ; t=1,..n \tag{24}$$

If $dist_t$, $\geq 0, q_t^{obs}$ is outside the bounds otherwise, q_t^{obs} is within the bounds

$$FREE\_NEG = \frac{|\ Sum\ of\ negative\ dist_t\ |}{Number\ of\ negative\ dist_t} \tag{25}$$

$$FREE\_POS = \frac{Sum\ of\ positive\ dist_t}{Number\ of\ positive\ dist_t} \tag{26}$$

$$FREE = FREE\_NEG + FREE\_POS \tag{27}$$

where q_t^{mlh} is the maximum likelihood value of streamflow predicted by the BaRE algorithm at time t and $q_t^{max\ 95}$ and q_t^{min95} are upper and lower values of 95 percentile confidence interval at time t, respectively, n is the total number of time periods.

Smaller FREE_NEG means that more points are included within the bounds. Similarly, we desire FREE_POS to be smaller too, such that the bounds are not too wide while containing the data.

5.3. Selection of Output Transformation Factor λ

Deciding which value of the transformation factor λ to use is critical because we do not know the exact nature of the measurement errors except that the higher the streamflow, the larger is the measurement error. To determine the appropriate degree of transformation, we varied λ between 0.1 and 1.0 with an increment of 0.1 (1.0 refers to the un-transformed case) and plotted the changes in FREE, DRMS and %Bias. Figure 3 shows the variation of FREE and its positive and negative components with transformation factor λ. As can be seen from these plots, the results are sensitive to the selection of transformation factor λ. Figure 4 shows that the accuracy measures DRMS and %Bias are relatively insensitive to the selection of λ.

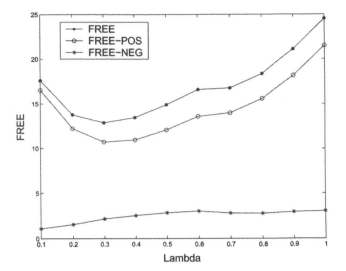

Figure 3. FREE measure and its components for different values of transformation factor λ using the BaRE algorithm for WY 1953 (first 50 days are ignored in calculations).

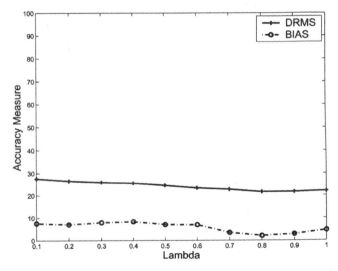

Figure 4. DRMS and % Bias with respect to different values of λ for WY 1953 (first 50 days are ignored in calculations).

Figure 5 provides a visual comparison of the streamflow predictions made using three different transformation factors: 0.1, 0.3, and 1.0. The transformation factor controls both the accuracy of the maximum likelihood prediction of streamflow and the width of the uncertainty bound. Not surprisingly, as λ increases the width of the uncertainty bound decreases for high flows and increases for low flows. However, we know physically that lower flows are associated with smaller measurement uncertainties.

Based on this analysis, we selected a value of 0.3 for the transformation factor λ, which gives the lowest FREE while still being relatively accurate.

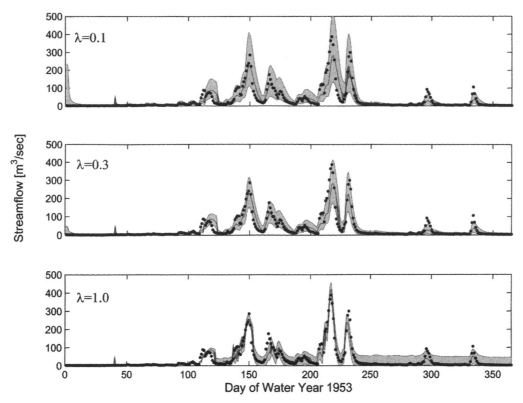

Figure 5. Probabilistic streamflow predictions made using the HYMOD model for the Leaf River basin, Mississippi, (WY 1953). Solid dots denote the measured streamflow, dark regions and light- shaded region indicate the 95% confidence intervals for prediction of "true" streamflow and measured streamflow in the original output space, respectively. λ changes as 0.1, 0.3, 1.0, in order from (a) to (c). (b set to 0 for all cases)

5.4. Selection of Error Model Parameter β

A similar analysis was carried out to determine the appropriate shape parameter, β, of the power density function of the measurement errors in the transformed space. The value of β was varied in the interval (-1 1], λ being fixed at the value of 0.3.

Figure 6 shows the variation of the FREE measure and its components with changing β. The change in accuracy in terms of DRMS and % Bias for several values of β is illustrated in Figure 7. Plots of streamflow prediction bounds for four different β values (-0.95, -0.5, 0, 1.0), in the transformed space, are shown in Figure 8.

When β is very close to -1.0 (corresponding to uniform distribution), the prediction bound of streamflow is extremely wide. However, the prediction bound decreases to a reasonable range very quickly and differs only slightly going towards β equal to 1.0. This is important because it shows that the results are not overly sensitive to a wide range of values for the error model's shape factor. Analyses of these plots and the FREE measure suggest that values of β between 0.0 and 0.5 are reasonable. We chose β equal to 0.0, corresponding to normal distribution for illustration purposes.

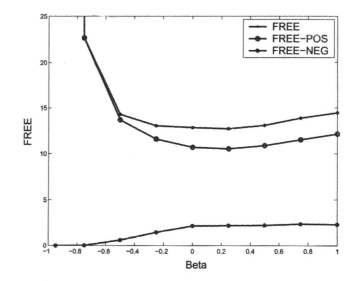

Figure 6. FREE measure and its components for different values of transformation factor β using the BaRE algorithm (WY 1953, first 50 days are ignored in calculations).

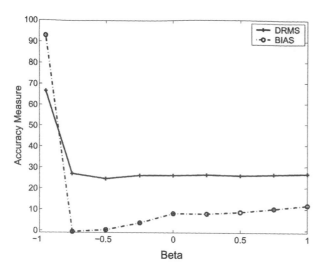

Figure 7. DRMS and % Bias with respect to different values of β for WY 1953 (first 50 days are ignored in calculations).

5.5. BaRE Results

Application of BaRE using HYMOD on the Leaf River basin data results in the 95% confidence interval prediction uncertainty bounds shown in Figure 9. Figure 10 illustrates the uncertainty of the streamflow predictions relative to the maximum likelihood value of streamflow. These plots are shown in the original (un-transformed) output space. It can be seen that the 95% Bayesian confidence intervals for the prediction of the streamflow measurement are relatively narrow while containing most of the observed data. Uncertainty bounds are larger for peak flows and smaller for recessions.

The evolution of the posterior probability distributions for the five model parameters is shown in Figure 11. Note that the probability bounds reduce quickly with the incoming information and, within a short time period (230 days), collapse to a single line. Given the structural simplicity of the HYMOD model, this indicates a major problem of algo-

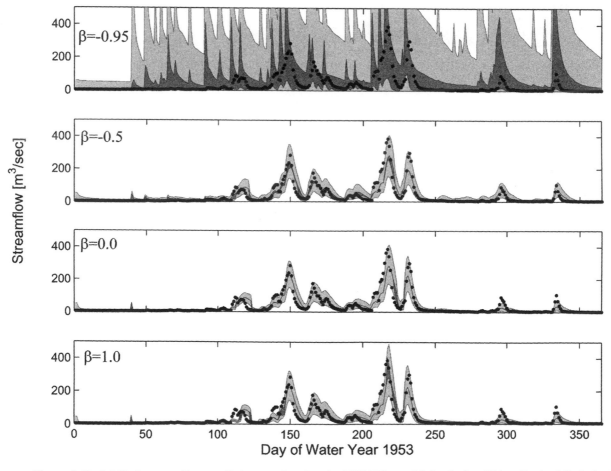

Figure 8. Probabilistic streamflow predictions made using the HYMOD model for the Leaf River Basin, Mississippi, (WY 1953). Solid dots denote the measured streamflow, dark region and light shaded region indicate the 95% confidence intervals for prediction of "true" streamflow and measured streamflow in the original output space, respectively. β changes as -0.95, -0.5, 0., 1.0, in order from (a) to (d). (λ set to 0.3 for all cases).

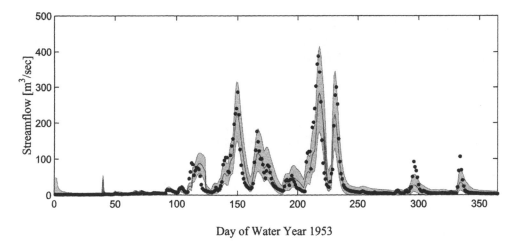

Figure 9. Probabilistic streamflow predictions in the original output space $\beta = 0$ and $\lambda = 0.3$.

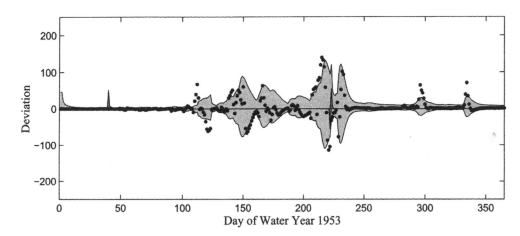

Figure 10. Streamflow uncertainties relative to the most probable forecast. ($\beta = 0; \lambda = 0.3$)

rithm overconfidence, which we believe to be related to not explicitly accounting for model structural error and placing too much confidence in the observed output data. Our on-going research is directed at resolving this problem, with one strategy being to impose an entropy factor on either the prior or the likelihood, thereby attenuating the strength of the information either from the observation itself or from the model simulation up to that time step.

5.6. Comparison With Batch Calibration

To compare the performance of BaRE with conventional calibration using this limited amount (one-year) of data, the SCE-UA method developed by *Duan et al.* (1992, 1993) was applied to the same watershed using HYMOD with DRMS as the objective function. For both cases, the transformation factor λ was set to 0.3. The calibrated model parameters selected by the SCE-UA algorithm are shown by

stars at day 365, in Figure 11. Even though BaRE uses only a discrete sample of 1000 parameters sets in the feasible parameter space, it is promising that its maximum likelihood parameter set is very close to the SCE-UA set at the end of the water year.

Comparative values are shown in Table 1 and the criterion statistics are shown in Table 2. Again, DRMS and %Bias are used as measures of accuracy. The DRMS estimate of model residual standard deviation is also used to construct the 95% confidence interval prediction uncertainty bounds explicit to the SCE-UA algorithm. The streamflow hydrograph corresponding to the optimal parameter set along with the prediction uncertainty bounds is plotted in Figure 12.

5.7. Model Evaluation

For evaluation purposes the best parameter set chosen by SCE-UA method and the most likely parameter set selected by BaRE water year 1953 were used to evaluate the model performance over an independent 10-year period (WY 1953 to WY 1964) for the Leaf River basin. The residual variance estimate (in the transformed space) from the calibration period was used to compute the 95% confidence intervals

for the forecasts. Figure 13 and 14 shows these forecasts for the wettest year (WY 1961) within the 10-year evaluation period. Comparative statistics are shown in Table 2. Note that the model performance is quite similar for both cases. Although BaRE was not developed to provide a single point

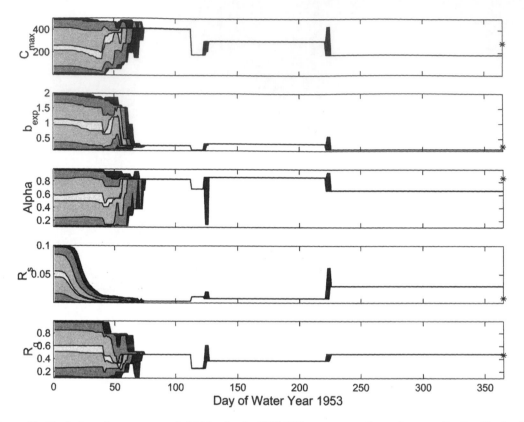

Figure 11. Evolution of parameter probabilities for the HYMOD parameters Cmax, bexp, α, Rs, Rq. Shades from darker to lighter correspond to 99, 95, 68, and 10 percentile confidence intervals, respectively. (* shows the location of SCE-UA optimal parameter set) .

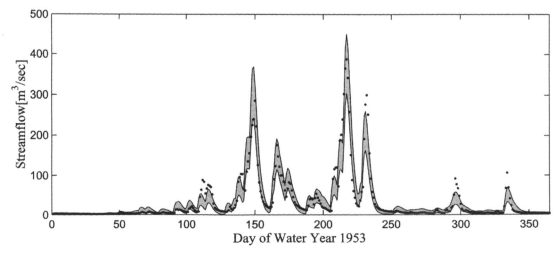

Figure 12. Leaf River basin calibration for Water Year 1953 using the Shuffled Complex Evolution (SCE-UA) algorithm.

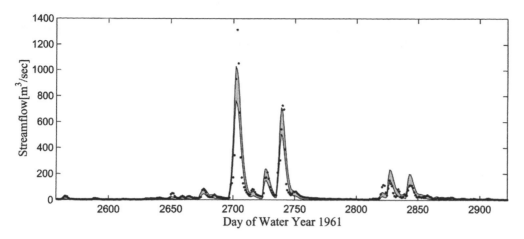

Figure 13. Hydrograph for WY 1961 as a part of 10-year evaluation period (WY 54 to WY 64) generated by BaRE method's most likely parameter set et the end of calibration period.

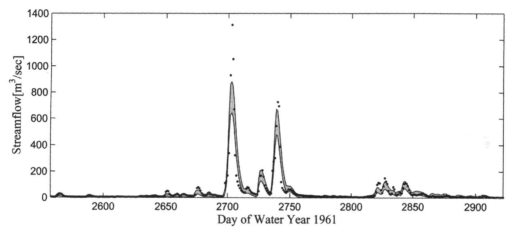

Figure 14. Hydrograph for WY 1961 as a part of 10-year evaluation period (WY 54 to WY 64) generated by SCE-UA method's best parameter set at the end of calibration period.

estimate of the parameters for forecasting, this analysis indicates that the maximum likelihood parameter set provides an acceptable deterministic forecast.

6. SUMMARY AND DISCUSSION

In this paper, we have discussed a Bayesian recursive parameter estimation and output uncertainty prediction approach based on the work of *Thiemann et al.* (2001). To reduce the subjectivity of the algorithm, we used accuracy and precision criteria to select the output transformation factor and the shape factor of the error model. The Daily Root Mean Square (DRMS) and the percent Bias (% Bias) measures were used as measures of accuracy. For precision, we proposed a new measure called the Forecast Range Error Estimate (FREE). The performance of different shape and transformation factors were evaluated with respect to accuracy and precision. The confidence interval estimates of the

forecast prediction were found to be quite sensitive to the choice of transformation factor, λ, whereas it showed little sensitivity to the choice of shape factor, β, within reasonable values (except when approaching a uniform distribution).

The approach was applied to a real watershed streamflow prediction problem, using a relatively simple hydrologic model called HYMOD (*Boyle et al.*, 2000). Despite the use of a simple model, the BaRE algorithm was successful at providing flow predictions close to the observed stream-

Table 2. Summary of statistics.

	Calibration		Evaluation	
	BaRE	SCE-UA	BaRE	SCE-UA
DRMS	25.96	19.77	28.11	27.79
% Bias	7.92	3.67	6.12	15.31
FREE	12.86	7.47	7.35	6.73
FREE_POS	10.7	3.54	2.66	2.01
FREE_NEG	2.16	3.93	4.69	4.72

flow, with reasonable uncertainty estimates. The accuracy of the results compares well with the popular SCE-UA batch calibration method.

The BaRE algorithm can serve as a very useful tool, especially in cases where not enough historical hydrological data have been accumulated or the data have gaps that make conventional batch calibration methods difficult to apply. We are investigating improvements of the algorithm that will provide better estimates of the uncertainty of the hydrologic model parameters and streamflow forecasts by explicitly accounting for the model structural error, and by using progressive re-sampling of the HPD parameter space. The goal is to improve sampling of the HPD parameter space while not collapsing to single point estimates. These improvements will be discussed in future papers.

Acknowledgements. This material is based upon work supported in part by the National Oceanic and Atmospheric Administration (NOAA-GCIP) (grant NA86GP0324), and the Hydrologic Research laboratory of the National Weather Service (grants NA87WHO582 and NA07WH0144, NA77WHO425), by HyDIS program (NASA funded Hydrological Data and Information System grant NAG58503), by NASA (EOS grant NAGW2425), and by SAHRA (Sustainability of Semi-Arid Hydrology and Riparian Areas) under the STC program of the National Science Foundation, agreement EAR-9876800. Special thanks are due to Ms. Corrie Theis for proofreading.

REFERENCES

Bates,C.B. and E.P. Campbell, 2001: A Markov chain Monte Carlo scheme for parameter estimation and inference in conceptual rainfall-runoff modeling, *Water Resources Research, 37,* 4, 937-947.

Bayes, T., 1763: *Phil. Trans. Roy. Soc.,* 53, 370-418.

Berger, J.O., 1985: *Statistical Decision Theory and Bayesian Analysis,* Springer Verlag, New York.

Beven, K.J. and A.M. Binley, 1992: The future of distributed models: model calibration and uncertainty prediction, *Hydrological Processes,* 6, 279-298.

Box, G.E.P. and G.C. Tiao, 1973: *Bayesian Inference in Statistical Analysis,* Addison-Wesley-Longman, Reading, Massachusetts.

Boyle, D.P., H.V. Gupta, and S. Sorooshian, 2000: Toward improved calibration of hydrological models: combining the strengths of manual and automatic methods, *Water Resources Research,* 36, 12, 3663-3674.

Brazil, L.E., 1988: Multilevel calibration strategy for complex hydrologic simulation models, Ph.D. Dissertation, 217p., Colorado State University, Fort Collins.

Duan, Q., V.K. Gupta, and S. Sorooshian, Effective and efficient global optimization for conceptual rainfall-runoff models, *Water Resources Research, 28,* 1015-1031, 1992.

Duan, Q., V.K. Gupta, and S. Sorooshian, A shuffled complex evolution approach for effective and efficient global minimization, *Journal of Optimization Theory and Applications, 76,* 3, 501-521, 1993.

Franks, S.W. and K.J. Beven, 1997: Bayesian estimation of uncertainty in land surface-atmosphere flux predictions, *J. Geophys. Res.* 102(D20), *23,* 991-23, 999.

Hogue, T.S., S. Sorooshian, H.V. Gupta, A. Holz, and D. Braatz, 2000: A multistep automatic calibration scheme for river forecasting models, *Journal of Hydrometeorology, 1,* 524-542.

Kuczera, G., and M. Mroczkowski, 1998: Assessment of hydrological parameter uncertainty and the worth of multiresponse data, *Water Resources Research, 34,* 6, 1481-1489.

Sorooshian, S., Q. Duan, and V.K. Gupta, 1993: Calibration of rainfall-runoff models: application of global optimization to the Sacramento Soil Moisture accounting model, *Water Resources Research, 29,* 1185-1194.

Thiemann, M., M. Trosset, H. Gupta, and S. Sorooshian, 2001: Bayesian recursive parameter estimation for hydrological models, *Water Resources Research, 37,*10,2521-2535.

F. Misirli, Department of Hydrology and Water Resources, College of Engineering and Mines, Harshbarger, Bldg. 11, PO Box 210011, University of Arizona, Tucson, AZ 85721-0011, U.S.A. (e-mail: feyzan@hwr.arizona.edu)

H.V. Gupta, Department of Hydrology and Water Resources, College of Engineering and Mines, Harshbarger, Bldg. 11, PO Box 210011, University of Arizona, Tucson, AZ 85721-0011, U.S.A. (e-mail: hoshin@hwr.arizona.edu)

M. Thiemann, Riverside Technology, Inc., 2290 East Prospect Road, Suite 1, Fort Collins, CO 80525 U.S.A. (e-mail: mt@riverside.com)

S. Sorooshian, Department of Hydrology and Water Resources, College of Engineering and Mines, Harshbarger, Bldg. 11, PO Box 210011, University of Arizona, Tucson, AZ 85721-0011, U.S.A. (e-mail: soroosh@hwr.arizona.edu)

Multiple Criteria Global Optimization For Watershed Model Calibration

Hoshin V. Gupta, Luis A. Bastidas, Jasper A. Vrugt[1], and Soroosh Sorooshian

SAHRA, NSF STC for Sustainability of semi-Arid Hydrology and Riparian Areas
Department of Hydrology and Water Resources, University of Arizona, Tucson, Arizona

The goal of model calibration is to achieve a reduction in model uncertainty by efficiently extracting information contained in the calibration data. *Gupta et al.* [1998] noted that several complementary criteria should be used to extract information about different model components or parameters, thereby enhancing the overall identifiability of the model. The traditional multiple criteria approach has been to select several different criteria and then merge them together into a single function for optimization. However, *Gupta et al.*, [1998] proposed that there is significant advantage to maintaining the independence of the various performance criteria and that a full multi-criteria optimization should be performed to identify the entire set of Pareto optimal solutions. This paper presents a brief overview of the MOCOM-UA algorithm [*Yapo et al.*, 1997] that uses a population evolution strategy (similar to that employed by the SCE-UA algorithm) to converge to the Pareto set via a *single* optimization run. The abilities of the MOCOM algorithm to identify an approximation of the Pareto solution set are illustrated via a simple hydrologic model calibration example.

1. INTRODUCTION

The goal of model calibration is to efficiently extract the information contained in the calibration data, so as to achieve a reduction in the model uncertainty. The process of extracting this information should result in the identification of a smaller parameter region (contained within the feasible parameter space). The greater the information content of the calibration data, and the more efficient the calibration procedure, the smaller the reduced parameter space is expected to be, resulting in a smaller range of possible values on each model forecast. In the limit, however, the size of the reduced parameter space will only approach a unique "point" if there is simultaneously an absence of model structural error (model is perfect) and the measured data are free of systematic biases; in the usual case, the reduced parameter space will remain of finite size.

It has been argued that, in general, many popular conceptual watershed models are over-parameterized and that only a limited subset of their parameters can be identified by means of optimization procedures [*Beck et al.*, 1993; *Beck*, 1994, *Beven*, 1989; *Jakeman and Hornberger*, 1993; among others]. In other words, it has been suggested that several parameter sets can yield very similar results in terms of the objective function value. This phenomenon has been termed "equifinality" by *Beven and Binley* [1992], "equally probable parameter sets" by *van Straten and Keesman* [1991], and "acceptable sets" by *Klepper et al.* [1991]. Such arguments are based on the probabilistic representation of parameter uncertainty.

However, others have argued that significant improvements in the model calibration can be achieved by using additional kinds of information for calibration and/or validation [for example, *De Grosbois*, 1988; *Yan and Haan*, 1991; *Mroczkowski et al.*, 1997] and by exploiting the data in better ways. This view, as stated in *Gupta et al.* [1998],

[1]Institute for Biodiversity and Ecosystem Dynamics, University of Amsterdam, Amsterdam, The Netherlands

Calibration of Watershed Models
Water Science and Application Volume 6
Copyright 2003 by the American Geophysical Union
10/1029/006WS09

raises the issue of complementarity of information; i.e., to improve the identification of the optimal parameter sets it is necessary to identify optimization criteria (objective functions) that measure different (complementary) aspects of system behavior. In principle, different criteria can be selected that are better able to extract information about different model components or parameters, thereby enhancing the overall identifiability of the model.

Of course, the use of multiple objectives within the context of hydrologic modeling and hydrologic model calibration is not new. A common approach has been to establish several different criteria and then merge them together into a single function for optimization [e.g. *Emsellem and de Marsily*, 1971; *Neuman*, 1973; *Yan and Haan*, 1991a,b; etc.]. This is often supplemented by the use of additional observed fluxes and/or state variables to verify the result [*Kuczera*, 1983a,b; *De Grosbois et al.*, 1988; *Hooper et al.* 1988; *Woolhiser et al.*, 1990; *Ambroise et al.*, 1995; *Mroczkowski et al.*, 1997, etc.]. However, *Gupta et al.*, [1998] proposed that there is significant advantage to maintaining the independence of the various performance criteria and that a full multi-criteria optimization should be performed to identify the entire set of Pareto optimal solutions [see also *Gupta et al.*, this volume, "Advances …"]. By analyzing the tradeoffs among the different criteria, the hydrologist is able to better understand the limitations of the current hydrologic model structure, and gain insight into possible model improvements. One way of obtaining an approximation of the Pareto solution set is to construct a weighted sum of the different criteria and to run a number of independent single criteria optimization runs for different values of the weights. This procedure is quite inefficient. *Yapo et al.* [1997] presented an alternative procedure that uses a population evolution strategy (similar to that employed by the Shuffled Complex Evolution (SCE-UA) algorithm) to converge to the Pareto set via a *single* optimization run. The algorithm, entitled Multi Objective COMplex evolution (MOCOM-UA) was developed at the University of Arizona and has been applied successfully in numerous hydrologic and hydrometeorologic model calibration and evaluation studies [see *Gupta et al.*, 1998, 1999; *Yapo et al.*, 1997; *Bastidas et al.* 1999, 2001, 2002; *Boyle et al.*, 2000, 2001; *Wagener et al.*, 2001; *Xia et al.*, 2002; *Meixner et al.*, 2002; *Laplastrier et al.*, 2001; among many others].

This paper presents a brief overview of the MOCOM-UA algorithm and discusses some of its properties. The abilities of the MOCOM algorithm to identify an approximation of the Pareto solution set are illustrated via a simple hydrologic model calibration example. For further examples of its application, please see the references mentioned above and also other chapters in this book [*Boyle et al.*, this volume,

Parada et al., this volume, *Meixner et al.*, this volume, and *Bastidas et al.*, this volume].

2. MULTICRITERIA EVALUATION OF HYDROLOGIC MODELS

Consider a system S for which a hydrologic model H is to be identified. Assume that the mathematical structure of the model is essentially predetermined and fixed and that physically realistic upper and lower bounds on each of the model parameters can be specified a priori (thereby defining the feasible parameter space - i.e., the initial uncertainty in the parameters). Assume also that measurement data on several of the system states and/or output fluxes (say D_1 *through* D_k) may be available which can be used to evaluate the performance of the model. The goal of model calibration now becomes that of finding values for the model parameters θ so that the model-simulated fluxes match all k of these (non-commensurable) measurement data of state variables and/or fluxes as closely as possible.

The following development follows *Gupta et al.* [1998]. Construct the extended data vector $D = \{D_1 …., D_k\}$ and let $y(\theta) = \{y_1(\theta), ……, y_k(\theta) \}$ represent the corresponding vectors of estimated model output fluxes generated using the parameter values θ. The difference between the model-simulated fluxes and the measurement data can be represented by $E(\theta) = G(y(\theta)) - G(D) = \{e_1(\theta), ……, e_k(\theta)\}$, where the function G allows for various user-selected linear or nonlinear transformations (such as log, power, weighting, max, min, median, mean, etc.). The goal, therefore, is to find values for the parameters θ so that E is, in some sense, made as close to "zero" as possible. The standard approach is to define some measure L of the "length" of vector E and to then find the values of the model parameters θ that minimize L. However, given that the individual vector components $e_k(\theta)$ are not directly commensurable (i.e., each represents the model's ability to simultaneously match a different model state variable or output flux), there is no unambiguously "correct" (objective) way in which to minimize the "length" of the error $E(\theta)$. In fact, because the model will, in general, be unable to simultaneously match all aspects of observed system behavior, there will generally be several feasible solutions - each of them reflects a different trade-off in the matching of the various aspects of observed behavior.

Formally, the problem can be posed as a multi-objective optimization problem:

$$\text{minimize } F(\theta) = \text{minimize } \{f^{1,1}(\theta),……, f^{k,m}(\theta)\}$$
$$\text{wrt } \theta \qquad\qquad \text{wrt } \theta$$

where m different norms can be ascribed to each flux simultaneously in an attempt to extract additional information from a single signal. (In the example presented in this paper, m = 1). The solution to this problem consists of $P(\Theta)$, a "Pareto Optimum" set of solutions in the feasible parameter space which defines the minimum parameter uncertainty that can be achieved without stating a subjective relative preference for minimizing one specific component of $F(\Theta)$ at the expense of another. The Pareto set is defined such that any member θ_i of the set has the following properties: (1) For all non-members θ_j, there exists at least one member θ_i such that $F(\theta_i)$ is strictly less than $F(\theta_j)$, and, (2) it is not possible to find θ_j within the Pareto set such that $F(\theta_j)$ is strictly less than $F(\theta_i)$ (by "strictly less" it is meant $f^q(\theta_j) < f^q(\theta_i)$ for all q = 1, . . ., k).

The multi-objective formulation results, therefore, in the partitioning of the feasible parameter space into "good" solutions (Pareto solutions) and "bad" solutions. In the absence of additional information, it is not possible to distinguish any of the "good" (Pareto) solutions as being objectively better than any of the other "good" solutions (i.e., there is no uniquely "best" solution). Further, every member θ_i of the Pareto set will match some characteristics of the system behavior better than every other member of the Pareto set, but the trade-off will be that some other characteristics of the system behavior will not be as well-matched. A powerful advantage of this approach is that it includes the "best" solution for each error component of the vector F, e.g., the classical single objective optimum value for each separate function is an element of the Pareto set.

3. MULTICRITERIA OPTIMIZATION

The multi-criteria optimization problem defined above has been studied extensively in the field of optimization theory [see e.g., Goicoechea et al., 1982; Haimes et al., 1975]. Because the Pareto set typically consists of an infinite number of solutions, most multi-criteria techniques attempt to identify a countable number of distinct solutions distributed within the Pareto region. Classical methods for obtaining such solutions can be categorized as a posteriori methods, a priori methods, and interactive methods. Examples of a posteriori methods (also called generating techniques) include the weighting method [Zadeh, 1963], the g-constraining method [Marglin, 1967], and the goal attainment method [Gembicki, 1974]. Examples of a priori methods include the goal programming and the compromise programming methods [Zeleny, 1974]. Examples of interactive techniques include the surrogate worth trade-off method [Haimes et al., 1975] and the trade-off development method (TRADE) [Goicoechea et al., 1976]. Presentations and discussions of these methods and others can be found in textbooks [Goicoechea et al., 1982; Szidarovsky et al., 1986] and in

review papers [Hipel, 1992; Szidarovsky and Szenteleki, 1987; Yapo et al., 1992; Hendricks et al., 1992].

Although the classical approach is simple to implement, it carries a heavy price: for each discrete Pareto solution, a complete single-criterion optimization problem must be solved. If as in Sorooshian et al. [1993], each single-objective optimization run requires as many as 10,000 function evaluations, a hundred Pareto solutions will require in the neighborhood of a million function evaluations! The MOCOM-UA algorithm provides a much more efficient approach, capable of providing 100 or more Pareto solutions within a single optimization run using only about 10,000 to 20,000 function evaluations.

The MOCOM-UA is a general-purpose global multi-objective optimization algorithm that does not require subjective weighting of the criteria, and provides an efficient estimate of the Pareto solution space with only a single optimization run. The algorithm is related to the SCE-UA population evolution method reported by Duan et al. [1993; see also the chapter by Duan in this book]. For a detailed description and explanation of the method, please see Yapo et al. [1997a,b]. In brief, the MOCOM-UA method involves the initial selection of a "population" of p points distributed randomly throughout the η-dimensional feasible parameter space Θ. In the absence of prior information about the location of the Pareto optimum, a uniform sampling distribution is used. For each point, the multi-objective vector $E(\theta)$ is computed, and the population is ranked and sorted using a Pareto-ranking procedure suggested by Goldberg [1989]. Simplexes of $\eta+1$ points are then selected from the population according to a robust rank-based selection method [Whitley, 1989]. The MOSIM procedure, a multi-objective extension of the Downhill Simplex method [Nelder and Mead, 1965], is used to evolve each simplex in a multi-objective improvement direction. Iterative application of the ranking and evolution procedures causes the entire population to converge towards the Pareto optimum. The procedure terminates automatically when all points in the population become non-dominated. The final population provides a fairly uniform approximation of the Pareto solution space $P(\Theta)$. The MOCOM-UA algorithm is presented below:

1. To initialize the process, choose a population size s >> n where n is the dimension of the problem (experience suggests that s should be at least 100 and can be as large as 500 to get acceptable results).
2. Generate a population D of s points randomly (uniformly) distributed over the feasible parameter space. Compute the function vector F at each point.
3. Sort the s individuals using Pareto ranking (described later) and store the corresponding ranks in R = {r_i, i=1,

... , s}. Set R_{max} to be equal to the maximum rank obtained.

4. If $R_{max} = 1$, then all the points have become mutually non-dominated, so stop. Otherwise evolve the sample population using the multi criteria complex evolution procedure outlined below and return to step 3.

The multi criteria complex evolution algorithm is presented below:

1. Assign a selection probability P_i to each member of the population according to:
   $P_i = (R_{max} - r_i + 1)/([Rmax + 1] \cdot s - sum\{r_i, i=1, \ldots, s\})$
2. Construct A to be the set of points having largest rank, such that $A = \{x_i \in D \mid r_i = R_{max}\}$ and store the relative position of x_i in D in L. Set n_A equal to the number of points in A.
3. Select one point j, from A (without replacement) and n remaining points from D according to the probability distribution P_i, i=1, ... , s with $P_j = 0$ and form simplex S_j. Let $w_j = x_{L(j)} = A_{(j)}$ = point to be evolved. Do this for all j = 1, ... , n_A.
4. Then evolve each simplex $\{S_j\}$, j=1, ... , n_A independently using the MOSIM algorithm presented below. Replace w_j, j=1, ... , n_A into A.
5. Replace A into D using the indices stored in L and return to step 4 of the MOCOM-UA algorithm presented above.

The MOSIM algorithm used by the multi criteria complex evolution procedure is presented below:

1. Sort the simplex $\{S\}$, so that the points are in order of increasing rank. Set S^w to be the member of the simplex having the largest rank and assign $F^w = F(S^w)$. This is the "worst" point and has been target for evolution.
2. Compute the centroid S^g of the simplex after excluding S^w.
3. Attempt a reflection step by computing the reflection point $S^{ref} = \gamma S^g + (1-\gamma)S^w$ using $\gamma=2$ and compute $F^{ref} = F(S^{ref})$. Perform a test for dominance among the points $S^1, \ldots S^{ref}$. If S^{ref} is non-dominated by the other points set $S^{new}=S^{ref}$ and $F^{new}=F^{ref}$ and proceed to step 5. Otherwise proceed to step 4.
4. Compute a contraction step $S^{con} = \gamma S^g + (1-\gamma)S^w$ using $\gamma=0.5$ and compute $F^{con} = F(S^{con})$. Set $S^{new}=S^{con}$ and $F^{new}=F^{con}$ and proceed to step 5.
5. Replace the worst point S^w in $\{S\}$ by S^{new} and store its associated function value F^{new}. Return to step 4 of the complex evolution algorithm presented above.

Like the SCE-UA method, MOCOM-UA treats the global search as a process of natural evolution. The s sampled points constitute a population. Each member of the population is a potential parent with the ability to participate in reproduction. To ensure that the evolution process is competitive, we require that "better" parents have a higher probability of contributing to the generation of offspring, by using a triangular probability distribution function for parent selection. The MOSIM procedure is applied to each simplex to generate the offspring, using the information contained in the simplex to direct the evolution in an improvement direction. Each new offspring replaces the worst point of the current simplex.

Because in a multi-criteria problem several criteria are to be considered simultaneously, ordered ranking of the population by conventional scalar sorting is not possible and the concept of inferiority-superiority is used instead. The special sorting used in MOCOM-UA is called "Pareto ranking" [Goldberg, 1989]. It begins by identifying all non-dominated individuals in the population and assigning them rank "one". These points are then set aside, and the non-dominated points of the remaining set are assigned the rank "two". This procedure is repeated until every point has been assigned a rank. Thus, the smallest ranked points are closest to the Pareto set while the largest ranked points are furthest away.

To illustrate these concepts, Figure 1a shows a set of points sampled from a two-dimensional model parameter space (at the initiation of the MOCOM-UA algorithm), plotted in a two-dimensional function space where the aim is to simultaneously minimize both functions F_A and F_B. The shaded region bounded by the dashed line indicates the actual region in the function space that maps from the entire feasible parameter space. The solid line labeled AB indicates the theoretical Pareto set of solutions. Note that point A minimizes function F_A while point B minimizes function F_B and all other points on the solid line represent different trade-offs in simultaneous minimization of the two functions. All points in the shaded region that do not belong to the Pareto frontier AB are "inferior" or "dominated" points. The best-ranked points in the population are indicated by closed circles and the worst ranked points are indicated by closed squares. Figure 1b illustrates the distribution of the points at the termination of the MOCOM-UA procedure. The points are now all mutually non-dominated and provide an approximation to the location of the Pareto frontier. It should be noted that due to use of a finite number of sample points in the population, it will be typically impossible for the MOCOM-UA to place the points exactly on the Pareto frontier, but the solution can be made to asymptotically approach the theoretical solution with increasing population sizes.

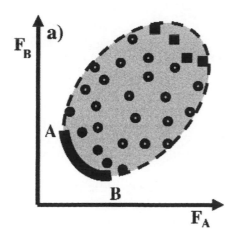
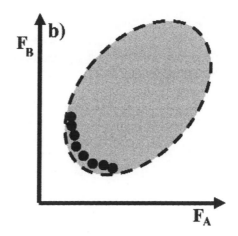

Figure 1. Two-dimensional (F_A, F_B) function space, with shaded region corresponding to the feasible parameter space. a) Solid line AB indicates the theoretical Pareto solution set. Closed circles indicate best-ranked points. Closed squares indicate worst-ranked points. b) Distribution of mutually non-dominated points at the termination of the MOCOM-UA procedure, providing an approximation to the Pareto frontier.

4. A SIMPLE MULTICRITERIA MODEL CALIBRATION CASE STUDY

We illustrate the use of the MOCOM-UA algorithm by means of a simple case study involving calibration of the HyMOD conceptual watershed model using data from the Leaf River watershed near Collins, Mississippi (1950 km2). The illustrative study presented here uses approximately 2 years (28 July 1952 to 30 September 1954) of hydrological data for model calibration. The data, obtained from the Hydrologic Research Laboratory (HRL), consists of mean areal precipitation (mm/day), potential evapotranspiration (mm/day), and streamflow (m3/s). Because the HyMOD model and Leaf River data have been discussed extensively in previous work [*Sorooshian et al.*, 1993; *Duan et al.*, 1993, 1994; *Yapo et al.*, 1996; *Boyle*, 2000; *Vrugt et al.* 2002; *Misirli et al.* in this book], the details will not be described here. To reduce sensitivity to state value initialization, a 65-day warm-up period was used.

Because any conceptual rainfall-runoff model will, in general, be unable to match all the different aspects of the watersheds behavior observed in the measured hydrograph, we follow Boyle et al. [2000] and partition the hydrograph into a driven (D) and non-driven (ND) part, based on information from the measured hyetograph. A pair of Root Mean Squared Error (RMSE) criteria were computed, F_D to measure the ability of the model to simulate the driven portion of the hydrograph response, and F_{ND} to measure the ability of the model to simulate the non-driven portion. The MOCOM-UA optimization algorithm was used to estimate the Pareto set of parameters that simultaneously minimize both F_D and F_{ND} using a population size of 500 points. Figure 2 shows the 500 MOCOM-UA solutions plotted (using dots) in the two-criterion F_{ND} versus F_D criterion space. The trade-off in abil-

ity of the model to simultaneously match the driven and non-driven portions of the hydrograph is clearly illustrated by the Pareto solution set, pointing to structural inadequacies in the model. Figure 2 also shows the individual single-criterion solutions (indicated using the dark circle symbols), obtained by separately calibrating the model to only the driven or the non-driven portions of the hydrograph using the SCE-UA global optimization algorithm [see chapter by Duan in this book]. These represent the theoretical end points of the Pareto solution set. Note that the final population of parameter sets obtained by MOCOM-UA provides a fairly uniform estimate of the middle

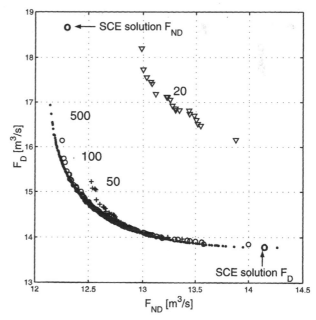

Figure 2. F_D versus F_{ND} Pareto solutions estimated by the MOCOM-UA algorithm. For explanation see the text.

region of the Pareto frontier, but does not represent the two ends well. We have found this inability to uniformly cover the extreme parts of the Pareto region to be a characteristic failing of the current version of MOCOM-UA algorithm. Methods to improve the algorithm are being investigated.

The MOCOM-UA algorithm has one parameter (the population size *s*) that must be specified by the user. Figure 2 illustrates the sensitivity of the MOCOM-UA solution to this parameter. The figure shows the Pareto frontier estimates obtained using the four different population sizes of 20, 50, 100 and 500. The population size of s=20 (triangle symbols) provides a very poor approximation to the Pareto frontier, and the Pareto estimates move closer to the origin with increasing population size. However, the improvement in the estimate of the Pareto set when increasing *s* from 100 to 500 members (open circle symbols and dot symbols respectively) is marginal, while resulting in a considerable increase in the number of function evaluations (from 1907 to 14994) required for algorithm convergence. In other studies we have generally found that a population size of 250 works well for most multi-criteria optimization problems having between two and four optimization criteria.

For completeness, Figure 3 shows a plot of the trade-off uncertainty in the simulated hydrographs associated with the 500 Pareto solutions (light-gray region) and the observed data (circles) for a portion of the Water Year 1953. Although the model generally simulates the variations in the observed hydrograph very well, the inability of the model to properly simulate portions of the data, particularly the long slow recessions (even for parameter sets on the Pareto frontier that give the smallest values for F_{ND}) indicates that attention may need to be given to improving the model components that control this portion of the simulated response.

5. SUMMARY

Significant improvements in model calibration can be achieved by using multiple sources of information and by exploiting the data in better ways. We, among others, have suggested (see also Gupta et al., the first chapter of this book) that the identification of model parameters can be improved by using multiple optimization criteria that measure different (complementary) aspects of system behavior. A common approach has been to establish several different criteria and then merge them together into a single function for optimization. However, there is significant advantage to maintaining the independence of the various performance criteria, since a full multi-criterion optimization will allow an analysis of the tradeoffs among the different criteria and enable the hydrologist to better understand the limitations of the current hydrologic model structure. This chapter presents a brief overview of the MOCOM-UA algorithm [Yapo et al., 1998] and discusses some of its properties. The algorithm uses a population evolution strategy, similar to that employed by the SCE-UA algorithm, to converge to the Pareto set via a *single* optimization run. The abilities of the MOCOM algorithm to identify an approximation of the Pareto solution set were illustrated via a simple hydrologic model calibration example. In general the algorithm performs well, but fails to properly approximate the extreme ends of the Pareto frontier, suggesting that further improvement of the methodology is warranted. For problems having between two and four optimization criteria, we recommend using a population size of approximately 250. The MOCOM-UA code can be obtained by contacting the authors.

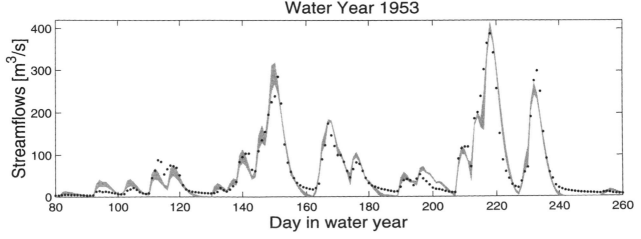

Figure 3. Hydrograph uncertainty ranges (shaded region) associated with the Pareto solution set for part of Water Year 1953. Circles correspond to the observed streamflows.

Acknowledgements. We gratefully acknowledge the hard work, dedication and innovative contribution of the numerous graduate students and colleagues who have worked with us on research related to model calibration. Partial funding for this research was provided, over the years, by several grants, including recent awards by the Hydrologic Research Laboratory of the U.S. National Weather Service [grants NA85AA-H-HY088, NA90AA-H-HY505, NA47WH0408, NA57WH0575, NA77WHO425, NA87WH0581, NA87WHO582, NA86GPO324, NA07WH0144], NOAA [grant NA86GP0324], NSF [grants ECE-86105487, BCS-8920851, EAR-9415347, EAR-9418147, EAR-9876800], and SAHRA [NSF-STC grant EAR-9876800].

REFERENCES

Ambrosie, B., J.L. Perrin, and D. Reutenauer, Multicriterion validation of semidistributed conceptual model of the water cycle in the Fecht Catchment, *Water Resources Research*, 31(6), pp. 1467-1481, 1995.

Bastidas, L., H.V. Gupta, S. Sorooshian, W.J. Shuttleworth and Z.L. Yang, Sensitivity Analysis of a Land Surface Scheme using Multi-Criteria Methods, *Journal of Geophysical Research-Atmospheres*, 104(D16), 19481-19490, 1999.

Bastidas, L.A., H.V. Gupta, O.L. Sen, Y. Liu, S. Sorooshian, and W.J. Shuttleworth, Comparative evaluation of land surface models using multi-criteria methods, submitted to *Journal of Geophysical Research-Atmospheres*, 2001.

Bastidas, L. A., S. Sorooshian, and H. V. Gupta, Emerging Paradigms in the Calibration of Hydrologic Models, In *Mathematical Models of Large Watershed Hydrology – Volume I*, V. P. Singh, D. Frevert, Editors, pp. 25-56, Water Resources Publications, LLC, Englewood, Colorado, 2002

Beck, M.B., Understanding uncertain environmental systems, in *Predictability and Nonlinear Modeling in Natural Sciences and Economics*, J. Grasman and G. van Straten, Eds., Kluwer Academic Publishers, Dordrecht, Netherlands, pp.294-311, 1994.

Beck, M.B., A.J. Jakeman, and M.J. McAleer, Construction and evaluation of models of environmental systems, *In Modeling Change in Environmental Systems*, A.J. Jakeman, M.B. Beck, and M.J. McAleer, Eds., pp. 3-36, J. Wiley & Sons, New York, 1993.

Beven, K., Changing ideas in hydrology-the case of physically based models, *Journal of Hydrology*, 105, 157-172, 1989.

Beven, K.J., and A.M. Binley, The Future of Distributed Models: Model Calibration and Predictive Uncertainty, *Hydrological Processes*, 6, 279-298, 1992.

Boyle, D.P., Multicriteria calibration of hydrologic Models, Ph.D. Dissertation, Department of Hydrology and Water Resources, University of Arizona, Tucson, 2001

Boyle, D.P., H.V. Gupta, and S. Sorooshian, Toward Improved Calibration of Hydrologic Models: Combining the strengths of Manual and Automatic Methods, *Water Resources Research*, 36(12) 3663-3674, 2000.

De Grosbois, E., R.P. Hooper, and N. Christophersen, A Multisignal Automatic Calibration Methodology for Hydrochemical Models-A Case-Study of the Birkenes Model, *Water Resources Research*, 24(8), 1299-1307, 1988.

Emsellem, Y., G. de Marsily, An Automatic Solution For The Inverse Problem, *Water Resources Research*, 7(5), p. 1264-1271, 1971.

Gembicki, F.W., Vector optimization for control with performance and parameter sensitivity indices, Ph.D. Thesis, Case Western Reserve University, Cleveland, Ohio, 1974.

Goicoechea, A., L. Duckstein, and M.M. Fogel, Multiobjective Programming In Watershed Management: A Study Of The Charleston Watershed, *Water Resources Research*, 12(6), p. 1085-1092, 1976.

Goicoechea, A., D.R. Hasen, and L. Duckstein, *Multi-Objective Decision Analysis with Engineering and Business Applications*, John Wiley, NY, 1982.

Goldberg, D.E., Genetic Algorithms in Search, Optimization, and Machine Learning, Addison- Wesley Publishing Co., Reading, MA, 1989.

Gupta, V.K., S. Sorooshian, and P.O. Yapo Towards Improved Calibration of Hydrologic Models: Multiple and Non-commensurable Measures of Information, *Water Resources Research*, 34, 751-763, 1998.

Gupta, H.V., L. Bastidas, L., S. Sorooshian, W.J. Shuttleworth and Z.L. Yang, Parameter Estimation of a Land Surface Scheme using Multi-Criteria Methods, *Journal of Geophysical Research*, Atmospheres, 104 (D16), 19491-19503, 1999.

Haimes, Y.Y., W.A. Halls, and H.T. Freedman, *Multiobjective Optimization in Water Resources Systems: The Surrogate Worth Trade-off Method*, Elsevier Scientific Publishing Company, NY, NY, 1975.

Hendriks, M.W.B., J.H. de Boer, A.K. Smilde, and D.A. Doornbos, Multicriteria Decision Making, *Chemometrics and Intelligent Laboratory Systems*, 16, pp. 175-191, 1992.

Hipel, K.W., 1992: Multiple Objective Decision Making In Water Resources, *Water Resources Bulletin*, 28(1), 3-12.

Hooper R.P., A. Stone, N. Christophersen, E. De Grosbois, and H.M. Seip, Assessing the Birkenes Model of Stream Acidification Using a Multisignal Calibration Methodology, *Water Resources Research*, 24(8), 1308-1316, 1988.

Jakeman, A., and G. Hornberger, How Much Complexity Is Warranted In A Rainfall-Runoff Model?, *Water Resources Research*, 29(8), 2637-2649, 1993.

Klepper, O., H. Scholten, and J.P.G. van de Kamer, "Prediction Uncertainty in an Ecological Model of the Oosterschelde Estuary," *Journal of forecasting*, Vol 10, 191-209, 1991.

Kuczera, G., Improved Parameter Inference In Catchment Models. 1. Evaluating Parameter Uncertainty, *Water Resources Research*, 19(5), p. 1151-1162, 1983a.

Kuczera, G., Improved parameter inference in catchment models. 2. combining different kinds of hydrologic data and testing their compatibility, *Water Resources Research*, 19(5), p. 1163-1172, 1983b.

Leplastrier, M. A.J. Pitman, H. Gupta, and Y. Xia, 2001: Exploring the Relationship Between Complexity and Performance in a Land Surface Model using the Multi-Criteria Method, submitted to *Journal of Geophysical Research*.

Marglin, S.A., Public Investment Criteria, Massachusetts Institute of Technology Press, Cambridge, Massachusetts, 1967.

Mroczkowski, M., G.P. Raper, and G. Kuczera, The quest for more powerful validation of conceptual catchment models, *Water Resources Research*, 33, pp. 2325-2335, 1997.

Nelder, J.A., and R. Mead, A Simplex Method for Function Minimization, *Computer Journal*, 7(4), 308-313, 1965.

Neuman, S.P., Calibration Of Distributed Parameter Groundwater Flow Models Viewed As A Multiple-Objective Decision Process Under Uncertainty, *Water Resources Research* 9(4), pp. 1006-1021, 1973.

Sorooshian, S., Q. Duan, and V.K. Gupta, Calibration of Rainfall-Runoff Models: Application of Global Optimization to the Sacramento Soil Moisture Accounting Model, *Water Resources Research*, 29, 1185-1194, 1993.

Szidarovsky, F., M.E. Gershon, and L. Duckstein, Techniques for Multiobjective Decision Making in Systems Management, Advances in Industrial Engineering, V 2, Elsevier, NY, NY, 1986.

Szidarovsky, F.M., and K. Szenteleki, A Multiobjective Optimization Model For Wine Production, *Appl. Math. Comput.*, 22, 255-275, 1987.

van Straten, G., and K.J. Keesman, Uncertainty Propogation and Speculation in projective Forecasts of Environmental Change: A Lake-Eutrophication Example, Journal of Forecasting, Vol 10, 163-190, 1991.

Wagener, T., D.P. Boyle, M.J. Lees, H.S. Wheater, H.V. Gupta, and S. Sorooshian, A Framework for Development and Application of Hydrological Models, *Hydrology and Earth Systems Science*, European Geophysical Society, 2001.

Whitley, D., The Genitor Algorithm And Selection Pressure: Why Rank-Based Allocation Of Reproductive Trials Is Best, Proceedings of the Third International Conference on Genetic Algorithms, 116-121, 1989.

Woolhiser, D.A., R.E. Smith, and D.C. Goodrich, A kinematic runoff and erosion manual: documentation and user manual, ARS 77, U.S. Department of Agriculture, 1990.

Xia, Y., A.J. Pitman, H.V. Gupta, M. Laplastrier, A. Henderson-Sellers, and L.A. Bastidas, Calibrating a Land Surface Model of Varying Complexity Using Multi-Criteria Methods and the Cabauw Data Set, *Journal of Hydrometeorology*, Vol. 3, 181-194, 2002.

Yan, J. and C.T. Haan, Multiobjective Parameter Estimation for Hydrologic Models–Weighting of Errors, *Transactions of the ASAE*, 34(1), 135-141, 1991a.

Yan, J. and C.T. Haan, Multiobjective Parameter Estimation for Hydrologic Models–Multiobjective Programming, *Transactions of the ASAE*, 34(3), 848-856, 1991b.

Yapo, P.O., N. Buras, and F. Szidarovszky, Applications Of Multiobjective Decision Making In Water Resources Management, *Pure Mathematics and Applications*, Ser. C, 3(1-4), 77-112, 1992.

Yapo, P.O., H.V. Gupta, and S. Sorooshian, Multi-Objective Global Optimization for Hydrologic Models, *Journal of Hydrology*, 204, 83-97, 1997.

Yapo, P.O., H.V. Gupta, and S. Sorroshian, A Multiobjective Global Optimization Algorithm with Application to Calibration of Hydrologic Models, HWR Technical Report No. 97-050, Department of Hydrology and Water Resources, The University of Arizona, Tucson, AZ 85721, 1997.

Zadeh, L.A., Optimality and non-scalar valued performance criteria, *IEEE Transactions*, C8, 1, 1963.

Zeleny, M., A Concept of Compromise Solutions and the Method of the Displaced Ideal, *Computers and Operations Research*, 1(4), 479-496, 1974.

Hoshin V. Gupta, Luis A. Bastidas, Soroosh Sorooshian, SAHRA, NSF STC for the Sustainability of semi-Arid Hydrology and Riparian Areas, Department of Hydrology and Water Resources, Harshbarger, Bldg. 11, University of Arizona, Tucson, AZ 85721, USA.

Jasper A. Vrugt, Institute for Biodiversity and Ecosystem Dynamics, University of Amsterdam, The Netherlands, Nieuwe Achtergracht 166, Amsyterdam, 1018 WV

Hydrologic Model Calibration in the National Weather Service

Michael B. Smith[1], Donald P. Laurine[2], Victor I. Koren[1], Seann M. Reed[1],
and Ziya Zhang[1]

Comprehensive procedures have been developed by the NWS for calibration of the conceptual hydrologic models used in river forecasting. These procedures are designed to achieve model parameters that are consistent between calibration and operational forecasting. Using these procedures, model parameters are derived using calibration data sets in a way that minimizes biases and errors when used in operational forecasting using real time estimates of precipitation, temperature, and evaporation. An overview of the data analysis techniques and manual calibration steps for rainfall-runoff models is presented. Future enhancements to the calibration process will also be discussed.

1. INTRODUCTION

The National Weather Service (NWS) has a mandate to provide forecasts for the Nation's rivers. To fulfill this mission, the NWS uses its River Forecast System (NWSRFS) at 13 River Forecast Centers (RFCs) to provide daily stage forecasts at over 4,000 points. Research and development to support the NWSRFS is conducted within the Hydrology Lab (HL) of the NWS Office of Hydrologic Development (OHD). Within the NWSRFS are algorithms for hydrologic and hydraulic models as well as procedures for data ingest, display and analysis of results, and other functions. Interested readers are referred to Stallings and Wenzel [1995], Larson et al., [1995], Fread et al., [1995], and Monroe and Anderson [1974] for more information regarding the structure and mission of the NWS river forecasting program.

While calibration of hydrologic models is widely considered a standard step in any application, the mandate assigned to the NWS to forecast the Nation's rivers has immense implications regarding the calibration and implementation of hydrologic and hydraulic models on a national scale. Since the introduction of calibration procedures over two decades ago, [*Brazil and Hudlow*, 1981], a great deal of effort has been directed toward improving and streamlining the calibration procedures. The purpose of this paper is to present an overall view of the current NWS hydrologic model calibration process, from deriving the input data sets

Calibration of Watershed Models
Water Science and Application Volume 6
Copyright 2003 by the American Geophysical Union
10/1029/006WS10

to incorporating the calibrated parameters and other information into the operational forecasting system. Contained within the NWSRFS are also hydrologic and hydraulic channel routing algorithms. However, the calibration of these models is beyond the scope of this paper.

Figure 1 presents the major components of the NWSRFS and shows that the Calibration System (CS) is a significant component of the entire functional structure. In the CS, time series of historical forcings are prepared and model parameters are calibrated. In the Operational Forecast System (OFS), real time data are used with the calibrated hydrologic and hydraulic models to produce forecast river stages several days into the future. The Interactive Forecast Program (IFP) allows the hydrologist to make run-time adjustments to account for non-standard conditions. The historical time series of precipitation, temperature, and potential evaporation are used to generate a suite of long term probabilistic forecasts weeks or months into the future in the Ensemble Streamflow Prediction system (ESP). Statistical procedures are used to quantify the uncertainty of these forecasts within a designated window.

The primary rainfall-runoff model used for operational forecasting in the NWS is the Sacramento Soil Moisture Accounting (SAC-SMA) model. Methods described in this paper will address the calibration of the parameters of the SAC-SMA. Interested readers are referred to Koren et al., [this volume], Burnash et al., [1973], Burnash [1995] and Finnerty et al., [1997] for more complete descriptions and applications of the SAC-SMA model.

Basically, the SAC-SMA is a two layer conceptual model of a soil column, with several modifications to account for the spatial variability of certain processes. Six types of

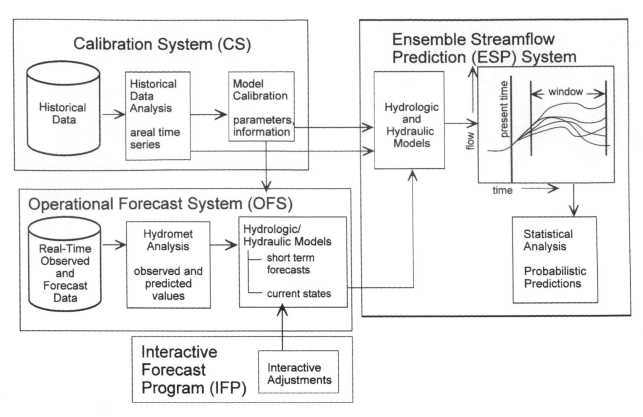

Figure 1. Main Components and Data Flow within the NWSRFS (adapted from Johnson et al., [1999])

runoff can be generated to form a complete runoff hydrograph. Each of the two layers in the SAC-SMA contains a tension water and free water component. Rain falling on the soil column first encounters the upper zone. Here, rain falling on any impervious areas generates impervious area runoff, while rain falling on the non-impervious areas of the basin first encounters the upper tension water storage. After filling this reservoir, excess soil water enters the upper zone free water. Water in this free water storage can percolate into the lower zone storages or flow out as interflow. If the upper zone free water fills completely, then excess soil water flows out as surface runoff. Most percolated water flows into the lower zone tension water storage, although some can go directly to free water storages in the lower zone. Upon filling the lower zone tension water storage, all soil water moves into the two lower zone free water storages. These two free water storages generate fast and slow responding base flow. The combination of these two base flows is designed to model a variety of hydrograph recessions. The SAC-SMA also has a variably–saturated area component from which saturation excess runoff is generated.

The snow model within NWSRFS is the snow accumulation and ablation model (Snow-17) developed by Anderson [1973]. Snow-17 is a conceptual model of a point snow column with an areal depletion curve used to determine the portion of the area being modeled that has snow cover. Snow-17 uses temperature as an index for the amount of energy exchange at the snow-air interface with separate equations for rain-on-snow and non-rain melt and for heat transfer during periods when melt is not occurring. The major Snow-17 parameters that are calibrated include those that control the seasonal variation of non-rain melt events, and areal depletion curve parameters. Overall results from Snow-17 compare favorably to those obtained with a complete energy balance model except during unusual meteorological situations such as periods with high dew-points and wind speeds [*Anderson*, 1976]. Further research is underway in HL to compare Snow-17 with energy balance models.

2. THE VALUE OF CALIBRATION IN THE NWS

Model calibration leads to obvious major benefits for operational forecasting. With a well calibrated model, real time simulations should more closely follow observed streamflow, resulting in more accurate forecasts with a longer lead-time into the future. Such well-calibrated models should require fewer run-time modifications to keep the models on track. Also, models that can simulate historical conditions with a known degree of accuracy allow for reliable probabilistic forecasts to be made for predictions of streamflow and other variables weeks or months into the future.

In addition, the manual calibration process also allows the user to develop a much deeper understanding of the data and the models and their limitations. This process is a chance for the calibrator to develop an understanding of the sensitivity of model parameters, interactions between parameters, effects of hydrologic inputs, and the knowledge of system mechanics. Calibration is an important evolutionary step in the development of an effective hydrologic forecaster who will be better able to apply the models for operational use.

Operationally, the forecaster is often required to make many adjustments to the hydrologic model to account for model and data errors. These adjustments are critical to the ability of the system to properly forecast future events. There are many options available for the forecaster to accomplish the same result in simulating the forecast hydrograph. Through guided interactive trial and error calibration, the hydrologist gains knowledge and experience to be used in selecting the proper adjustment.

Another important value gained through calibration is an understanding of the physical process occurring in the watershed and how well the calibrated model simulates those processes. Most river basins are very complex. For example, a typical basin in the western U.S. can be affected by reservoir operations, diversions, agricultural consumptive use, return flows, and basin geographical variability. Many of the current hydrologic models can only approximate these physical processes. Through calibration, the forecaster can build an awareness of model limitations and basin processes. The information gained provides the level of confidence the forecaster places on the forecast procedure.

Two basic approaches exist for the calibration of model parameters. The first is a manual trial-and-error method [*Anderson*, 2002], and the second involves the use of automated optimization programs. For the conceptual models currently in use in the NWS, interactive manual calibration that follows a logical strategy is recommended in order to preserve the physical basis of the model parameters and an appropriate variability of the parameters among watersheds in a basin. Automatic optimization can be used in conjunction with the manual steps. At various points in the calibration process, many tools including geographic information system (GIS) based programs are available to assist the hydrologist.

3. REGIONAL APPROACH FOR CALIBRATING A RIVER BASIN

It is usually recommended that data analysis and calibrations be performed on a large area or river basin basis rather than on an individual watershed for several reasons. Details of this procedure can be found in publications by Anderson [2002] and a comprehensive calibration training video developed by the NWS in conjunction with the Hydrologic Research Center [*Hydrologic Research Center*, 1999]. First, the meteorologic processes that control the development of precipitation, temperature, and evaporation variations occur on a scale much larger than a typical watershed, especially in mountainous areas. In order to properly understand these processes, the data analysis should be performed on a regional and not watershed scale. Such an approach also facilitates more efficient historical data retrieval and analysis since many precipitation, temperature, and other stations are common to several watersheds in a basin. If historical data analysis is performed on a watershed by watershed basis, then redundant downloading and processing of station data occurs.

Second, it is much more likely to achieve a realistic and consistent set of parameters using a strategy that examines the spatial variability of physiographic features and hydrograph response to guide the variation of parameters among watersheds within a basin. Physiographic features which affect model parameters such as topography, type of soils, and vegetation can be viewed to note areas of significant similarity or difference. Such qualitative information can be used to subjectively determine how model parameters can be expected to vary across a basin. If such physiographic features appear to be spatially invariant, the analyst can expect that the calibrated parameters from one watershed can be used as reasonable starting points for calibration of a neighboring watershed, resulting in a more efficient calibration effort.

Observed streamflow data show the integrated effects of all basin features and can also be used to qualitatively assess the spatial variability of hydrologic model parameters across watersheds in a basin. Observed discharges can be scaled to the drainage area of one of the watersheds and then plotted on semi-log scale. Hydrographs that show similarities in base flows and storm runoff indicate that the hydrologic model parameters could be quite similar to one another.

4. SOURCES OF DATA FOR MODEL CALIBRATION

The primary source of historical data for model calibration is the National Climatic Data Center (NCDC), which collects and maintains an archive of measurements of precipitation, temperature, evaporation, and other meteorologic variables. Data in digital form are readily available starting in 1948, with recent efforts underway to convert the entire period to digital format. In HL, recent work has begun to develop direct Internet links to the NCDC archive through web servers [G. Bonnin, NWS HL, personal communication, 2001]. Such efforts should alleviate the need for the NWS to maintain its own archive of the NCDC data sets.

Streamflow data have traditionally been available from the United States Geological Survey (USGS) in the form of mean daily flow. For many years, these were the only observed streamflow data available for hydrologic model calibration in the NWS. Mean daily flow data are derived from hourly or sub-hourly streamflow measurements (unit values) that have been quality controlled. Recently, the USGS has been making available unit value streamflow data from its local field offices for research and calibration needs. These data are provisional in that no quality control procedures have been performed on them. Calibration at sub-daily time steps is critical as the NWS moves to hydrologic modeling at finer spatial and temporal time scales as part of its distributed modeling efforts for river and flash flood forecasting [Zhang et al., 2001; Smith et al., 1999]. Other data available from the USGS are limited peak flow data, as well as reservoir pool elevation data.

An emerging source of data for model calibration is the archive of operational real time data collected each day at the 13 RFCs. As part of their operations, RFCs can receive thousands of observations of temperature, precipitation, and streamflow each day. These data are stored in custom data bases within NWSRFS and are processed to generate daily river forecasts. On a regular basis, a copy is made of these data and stored in the NOAA Hydrologic Data System (NHDS) to become available for future calibration efforts and research studies [Pan et al., 1998; Bonnin, 1996]. Data from the NWS series of WSR-88D Next Generation Radar platforms (NEXRAD) are also included in the NHDS archive. Individual RFCs often maintain their own archive of their operational data files. A limited set of utilities are available for converting these data from an operational format to a standard format used in the calibration system. Currently, efforts are underway to develop a consistent RFC archive data base design [D. Page, NWS HL, personal communication, 2002].

Other sources of data used for calibration include the Natural Resources Conservation Service (NRCS) and its Snowpack Telemetry (SNOTEL) system. SNOTEL provides year round temperature and precipitation data in remote, mountainous areas primarily in the western United States.

5. MAJOR STEPS IN THE CALIBRATION PROCESS

The calibration process is comprised of the following three general steps:

1. Analysis of historical data and derivation of time series of observed precipitation, temperature, and potential evapotranspiration.

2. Calibration of hydrologic model parameters so that simulated streamflow agrees with observed data.

3. Implementing the calibrated parameters and data analysis information into the operational forecast system. This step will not be explicitly discussed here. The interested reader is referred to Anderson [2002] for details on this important issue.

5.1. Analysis of Historical Data

5.1.1. Overview. Analysis of historical data to derive multi-year time series of mean areal precipitation, temperature, and potential evaporation proceeds according to the steps shown in Figure 2. Time series of precipitation are derived using the Mean Areal Precipitation preprocessor (MAP) while corresponding mean areal time series of temperature and potential evaporation are produced using the MAT and MAPE preprocessors, respectively. These time series are then used as forcings in the calibration of hydrologic model parameters. Henceforth, the acronyms MAP, MAT, and MAPE will denote both the times series of data as well as the preprocessor that computes them.

As shown in Figure 2, different analysis procedures are available for each variable depending on whether the area is non-mountainous or mountainous. For precipitation, an area is non-mountainous if the long term annual or seasonal station means are within a range of ±5%. If the range is greater than this, the mountainous area analysis should be used. Similarly, this criteria applies to the analysis of temperature and evaporation as well. In non-mountainous areas, it is assumed that any station can be used to estimate missing data at another station without making any adjustments for differences in magnitude. Moreover, spatial averages of the variables can be computed using station weights that are based solely on their location in the x,y plane. In non-mountainous areas, the station weights always sum to a value of 1.0.

Terrain differences are usually the main factors requiring the use of mountainous area procedures for analyzing precipitation, temperature, and evaporation. In these procedures, long term station means are accounted for in the estimation of missing data and information other than simple station location is used to derive station weights for the computation of areal averages. Station weights in mountainous areas usually sum to a value greater than 1.0. In mountainous areas, watersheds are frequently sub-divided in order to properly model the accumulation and ablation of the snow cover.

The HL-developed Calibration Assistance Program (CAP) contains data sets and tools that are primarily used for the analysis of mountainous areas. CAP is a national ArcView GIS- based suite of tools that facilitates the derivation of basin sub-divisions, model parameters, potential

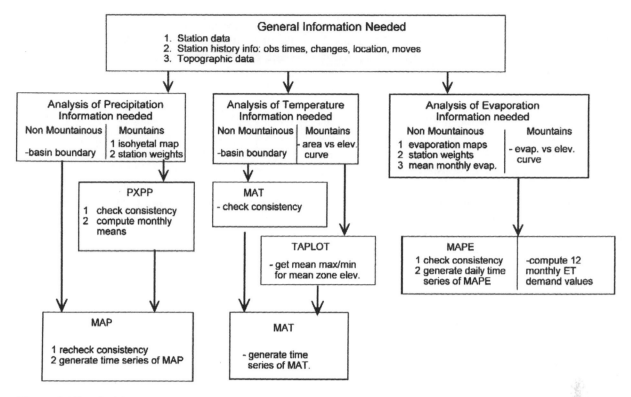

Figure 2. Historical Data Analysis Steps and NWSRFS Programs.

evaporation estimates, and mean areal estimates of precipitation and temperature. Moreover, CAP also contains satellite snow cover maps that can be used in conjunction with observed streamflow hydrographs and area elevation curves to identify different regimes that occur over the watershed. CAP also contains vegetation maps to help identify such regimes. The Appendix provides more details about the functions available in CAP.

It is very important that the resultant time series of precipitation, temperature, and evaporation are properly scaled to accurately represent what actually occurred in nature. Tests have shown that 10% variations in MAP time series can result in variations in simulated streamflow volumes of nearly 25% [*Anderson*, 2002]. Similar results have been reported for biases in the MAT time series. Even a bias of a few degrees can cause a significant shift in the timing of snowmelt. Parameters calibrated using such biased forcings will be distorted and lead to sub-optimal forecasts. Also, the MAP, MAT, and MAPE time series derived in calibration are used for ensemble forecasting, so biases in these time series can lead to degraded ensemble forecasts.

5.1.2. Station selection. Analysis begins with the selection of stations to be used to derive the historical time series of mean areal precipitation, temperature, and in some cases evaporation. For precipitation, it is advisable to look for hourly and daily stations with at least 5 and preferably 10 or more years of complete data. Since precipitation can be quite spatially variable, all stations located in the basin having good quality data are selected, as well as some outside the basin. For each daily station, it is important to note the location of hourly stations so as to have information needed to disaggregate the daily values. In mountainous areas, stations further out from the basin may need to be used to adequately represent higher elevations. In the current NWSRFS, only daily max/min temperature observations are used to generate historical MAT time series. Given that temperature is generally less spatially variable than precipitation, fewer temperature stations are required. Generally, all stations within and near the basin with long records are selected for analysis. Additional stations outside the basin with shorter periods of record are used if needed to properly represent high elevations. In contrast to precipitation stations, temperature stations with a long period of record are needed due to limitation in the current processing programs. Pan evaporation sites and stations with meteorological data to compute potential evaporation are selected to properly represent the variation in evaporation over the basin.

5.1.3. Quality control. Quality control procedures are applied to the station data prior to the derivation of time series of mean areal values of the variables. These procedures are necessary to avoid having a bias between one period of the calibration record and another. If an individual station receives considerable weight in the computation of the mean areal value, then inconsistencies in the station data will be translated to the mean areal value time series. Parameters calibrated from a period before the inconsistency will not be the same as parameters calibrated using the period after the inconsistency. If the period used for calibration does not reflect the current status of the station, then operational results will be biased.

The main quality control procedure for precipitation, temperature and potential evaporation data is to check the consistency of a station using double mass analysis. Double mass analysis can be performed in the MAP, MAT, and MAPE programs. Inconsistences appear as long term shifts in the general slope of the station accumulation curve. Such inconsistencies can result from man-made changes in station location, changes in equipment (e.g., addition of a shield to a rain gage), or changes in station exposure due to surrounding vegetation. Shifts due to such changes should generally be corrected. Thus, station history information is critical to identifying man-made effects that must be corrected. Double mass plots are displayed of stations grouped geographically, so that any shifts in one station can be compared to nearby stations. Such grouping helps identify when natural climatic changes occur in a region. Shifts such as these should not be corrected. Sharp discontinuities in double mass plots often indicate bad raw data values. In general, if there is any doubt as to whether a consistency correction is necessary, it is advisable not to make the correction. Pan et al., [1998] have developed the Interactive Double Mass Analysis (IDMA) tool, which is a graphical user interface to automate the generation of double mass plots and correction factors.

5.1.4. Generation of mean areal precipitation time series. As seen in Figure 2, non-mountainous and mountainous area procedures can be used to derive time series of mean areal values of precipitation. In non-mountainous areas, any station can be used to estimate missing data at other stations. Also, simple station weighting schemes such as Thiessen polygons or inverse distance squared weighting can be used to compute time series of mean areal precipitation values using the MAP program.

For analysis of precipitation in mountainous areas, additional steps are needed that require the use of the Preliminary Precipitation Processing Program (PXPP). The primary function of PXPP is to relate precipitation stations having different periods of record by generating

monthly means of all stations. These monthly means are then used to condition the estimates of missing data. The basic equation for missing data estimation in both PXPP and MAP is:

$$P_x = \frac{\sum\limits_{i=1}^{i=n} \frac{\overline{P_x}}{\overline{P_i}} \cdot P_i \cdot w_{i,x}}{\sum\limits_{i=1}^{i=n} w_{i,x}} \qquad (1)$$

where P_x is the precipitation at the station being estimated, P_i is the precipitation at the estimator station, n is the number of estimating stations, i is the station being used as an estimator, $\overline{P_x}$ is the mean monthly precipitation for station x, $\overline{P_i}$ is the mean monthly precipitation for station i, and $w_{i,x}$ is the station weight, computed as:

$$w_{i,x} = \frac{1}{d_{i,x}^{\ 2}} \qquad (2)$$

where $d_{i,x}$ is the distance from station x to estimator i.

The use of monthly means in Equation 1 attempts to account for orographic effects in areas having significant elevation range. In non-mountainous areas it is assumed that there is little spatial variability in long term station means. Hence, the ratio of station monthly means drops out of Equation 1 in such cases. Other functions in PXPP include double mass analysis and the generation of consistency corrections. In addition, PXPP also contains correlation analyses to aid in the determination of station weights in mountainous areas.

In mountainous areas, an isohyetal analysis is needed to compute a long term mean areal precipitation value over a watershed in order to understand the relationship between the precipitation measured at stations and that which usually occurs over an area. The long term mean areal value is used to derive weights for stations that are used to compute not only the historical MAP time series but also weights for stations that are used operationally but were not in the historical data set. This procedure helps ensure that biases are not introduced between the calibration and operational systems when different stations are used.

Isohyetal maps are available from several sources. In some rare cases, such maps may have been developed as part of a previous study. The method outlined by Peck and Brown [1962] to create isohyetal maps can also be used using some of the output from PXPP. Recently, monthly digital isohyetal maps for large parts of the country have been developed by Oregon State University for the Natural Resources Conservation Service (NRCS) as part of the

Precipitation- elevation Regressions on Independent Slopes Model (PRISM) project [*Daly et al.*, 1994]. Data from PRISM are contained in CAP.

Once an isohyetal map is obtained or derived, the map is analyzed and modified to ensure that it is appropriate for the period of analysis and computation of the MAP time series. Water balance analyses for the watershed in question and surrounding watersheds can also be used to validate the isohyetal map.

After any elevation zones or sub-divisions are derived, annual or seasonal mean areal precipitation values need to be computed for each watershed or zone using the procedures described above. These means are then used in the computation of station weights. For this computation, Equation 3 is used:

$$W_{i,s} = R_{i,s} \cdot \frac{\overline{A}_s}{\sum\limits_{i=1}^{i=N} \left[\overline{S}_{i,s} \cdot R_{i,s} \right]} \quad (3)$$

where W is the station weight, i is the station whose weight is being computed, s is the season of the year, N is the total number of stations with weight, R is the relative station weight, $\overline{A}$ is the long term areal mean precipitation, and $\overline{S}$ is the station long term mean precipitation.

The determination of the relative weights R is a subjective process and is based on the user's knowledge of the basin as well as types, coverage, and directions of storms, and station intercorrelation. Typically, station weights W sum to a value greater than one to reflect the fact that the gages tend to catch less than the basin as a whole. Both seasonal and annual station weights can computed.

Equation 3 provides the user with a method of calibrating a basin with data from one set of precipitation stations and then operationally forecasting with a potentially different set of stations. Because of the use of the term $\overline{A}$ derived from the isohyetal analysis, stations that were not part of the historical network can be added to the operational network without introducing a bias into the computations. New stations can be added to the operational network after an estimate of the long term station mean $\overline{S}$ is derived.

After using PXPP to perform consistency checks and compute monthly means, program MAP is used with the station weights $W_{i,s}$ to compute a time series of mean areal precipitation values. Usually, a 6 hour time step is selected, although other time steps can be specified.

The use of gridded precipitation estimates from the NEXRAD platforms presents similar challenges. Research in the NWS and elsewhere has shown that hydrologic model parameters calibrated using rain gage data are most likely not directly applicable for use with radar data [*Johnson et al.*, 1999; *Smith et al.*, 1999; *Bradley and Kruger*, 1998, *Finnerty et al.*, 1997]. As a result, re-calibration of raingage based model parameters should be considered for use with radar precipitation estimates. However, the period of NEXRAD data available for calibration is not long enough for effective calibration in most areas of the country. Also, changes in processing algorithms may have contributed to time-dependent inconsistencies in the data, making calibration of continuous hydrologic models a difficult task. Consequently, use of NEXRAD data for calibration and forecasting continues to be an active area of research.

5.1.5. Generation of mean areal temperature time series. Time series of temperature are primarily needed for use with Snow-17 as well as frozen ground computations in the SAC- SMA. The main program for computing time series of mean areal temperature is the MAT program. MAT contains procedures for computing missing data and for computing consistency corrections. Using observed daily maximum and minimum temperatures and station weighting schemes, MAT computes a 6-hour time series of mean areal temperatures for a watershed. An assumed diurnal variation is used to convert the daily maximum-minimum temperatures into a 6 hour time series. Equation 4 is the general equation for estimating missing maximum and minimum temperatures within the MAT program for both non-mountainous and mountainous areas:

$$T_x = \frac{\sum\limits_{i=1}^{i=n} \left[\overline{T}_x - \overline{T}_i + T_i \right] \cdot w_{x,i}}{\sum\limits_{i=1}^{i=n} w_{x,i}} \quad (4)$$

where T is the computed maximum or minimum temperature value, $\overline{T}$ is the mean value, x is the station being estimated, i is the estimator station, n is the number of estimator stations, and w is the weight applied to each estimator, computed as:

$$w_{x,i} = \frac{1.0}{d_{x,i} + F_e \cdot \Delta E_{x,i}} \quad (5)$$

where $d_{x,i}$ is distance, $\Delta E_{x,i}$ is elevation difference and F_e is the elevation weighting factor. In non-mountainous areas, distance is the dominant factor in determining which stations are used as estimators of missing data, so a value of zero is used for F_e in Equation 5. For the computation of time series of mean areal temperature, station weights in non- mountainous areas are automatically computed using an inverse distance scheme.

In order to compute mean areal values of temperature over a mountainous watershed or sub-area of a watershed, seasonal variations of maximum and minimum temperature with elevation are developed on a regional basis. These temperature-elevation relationships are developed using the program TAPLOT program as shown in Figure 2. Straight line relationships are generally fitted to the data for each month which should represent physically realistic lapse rates, generally in the range of 0.3 to 0.8 °C/100m. Temperature-elevation relationships are primarily used to extrapolate temperatures from lower to higher elevations due to a general lack of high elevation data in most basins.

The computation of time series of MAT in mountainous areas centers around the use of a synthetic station established at the mean elevation of each watershed or zone. A synthetic station is one with no observed data. All data for the station is estimated from surrounding real stations. and the synthetic station is assigned a predetermined weight of 1.0. Mean monthly max and min temperatures for the synthetic station are derived from the temperature-elevation plots and are used in Equations 4 and 5 to estimate missing data for the synthetic station. Using this method, an MAT time series is derived that reflects the average conditions within each elevation zone. As with precipitation, procedures are used to minimize biases between the calibration and operational station networks. The same synthetic stations and corresponding mean monthly temperatures must be defined in the operational system and given full weight. New stations can be added to the operational network after computing a best estimate of the long term mean monthly maximum and minimum temperatures.

5.1.6. Generation of evapotranspiration data. The SAC-SMA requires evapotranspiration demand (ET Demand) as input. ET Demand is the evaporation that occurs given that moisture is not limiting and considering both the type and activity of vegetation. Thus, while PE is defined for an actively growing grass surface, ET Demand is based on the actual vegetation in the area and how active that vegetation is given the time of the year and other factors. As shown in Figure 2, two methods for generating estimates of ET Demand for calibration and operational forecasting are available. In the first method, the SAC-SMA will accept a daily PE value in conjunction with a seasonal adjustment curve. The second method uses mean monthly values of PE and a seasonal adjustment curve.

In the first approach, daily estimates of potential evaporation demand are computed using meteorological data from synoptic stations and the Penman [1948] equation. In this case, the net radiation is estimated from sky cover data according to the method of Thompson [1976]. In these cases, time series of mean areal estimates of potential evaporation are computed using the MAPE program using a simple distance weighting scheme or user defined station weights. However, skycover measurements at some stations have recently been discontinued so that reliable values of daily PE can no longer be computed. In light of this, research is underway in HL to investigate new methods of computing PE. One requirement for a new method is that any data used for calibration must have the same statistical properties as the data used for operational forecasting.

In the second method, monthly estimates of PE are developed from published tables of evaporation pan measurements and other information [*Farnsworth and Thompson*, 1982]. Average monthly pan evaporation values from stations in and around the basin are used to derive an average monthly curve. The pan coefficient is applied to create an evaporation demand curve. Finally, the curve is adjusted to show the average effects of transpiration, resulting in an ET Demand curve. Traditionally, vegetative effects were estimated based on a users knowledge of the type of vegetation in the basin. Recently, a methodology to derive these monthly adjustment factors based on Normalized Difference Vegetative Index (NDVI) greenness fraction data has been developed in HL. This procedure has been incorporated into CAP.

In non-mountainous areas, estimates of PE are adjusted to the evaporation maps derived by Farnsworth and Peck [1982]. In mountainous areas, these maps aren't of sufficient detail to determine PE for individual watersheds or sub-watersheds. In these areas, the recommended approach is to derive a basin-wide relationship between PE and elevation. Water balance computations are then performed and the MAP time series is adjusted to achieve a correct water balance.

With either method of computing evapotranspiration, procedures are designed to ensure that the long term mean areal value of potential evapotranspiration used in calibration is the same used in operational forecasting. This is accomplished by using the free water surface evaporation maps published in Farnsworth and Peck [1982] as a standard.

5.2. Hydrologic Model Calibration

The next major step is to use the MAP and MAT time series and PE estimates as observed forcings to calibrate the hydrologic model parameters. Primarily, manual techniques are used and a systematic and proven strategy is followed for calibrating each of the parameters within the SAC-SMA and SNOW-17 models [*Anderson*, 2002; *Hydrologic Research Center*, 1999]. While at times intensive, manual calibration provides the user with an opportunity to learn the inner workings of the hydrologic models. Consequently, the user will be better equipped to use the models in an operational setting.

The primary program used for model calibration is the Manual Calibration Program (MCP). MCP is basically the same as the OFS with the main exception being that the hydrologic models are executed over multi-year calibration periods rather than multi-day or multi-week forecast periods. MCP allows for the computation of a number of goodness-of-fit statistics. A significant enhancement to the manual calibration process has been the recent development of the Interactive Calibration Program (ICP). This tool is a powerful graphical user interface for executing MCP. ICP displays the simulated and observed hydrographs for the run as well as the SAC-SMA and Snow-17 model states and runoff components for the entire run period. Plate 1 shows the main displays within ICP. In this display, the hydrograph from February 1, 1979 to April 28, 1979 is presented, along with the corresponding SAC-SMA runoff components and states in the various SAC-SMA storages. In the extreme upper pane is displayed the rainfall hyetograph with the computed runoff. Below that, the 6 runoff components from the SAC-SMA are displayed as a percentage, allowing the user to clearly see which components comprise a streamflow response at any time. Below that is a pane that shows the states of the tension and free water storages in the SAC-SMA. Lastly, the bottom pane shows the computed and observed hydrographs. A similar display is available for the Snow-17 model.

ICP has the capability to display previous simulations so that the effects of an individual parameter change can be easily identified. With ICP, parameter changes and subsequent model runs and output displays can be performed in seconds. In spite of this high turnaround speed, it is advised that the user pause before displaying the new simulation and ask himself: "What effect should I see with this change?" If the expected result is not achieved, the user is encouraged to investigate potential causes rather than quickly making another parameter change. In this way, the process of manual calibration produces a set of optimum parameters and gives the user more expertise with the inner workings of the models.

Along with visual inspection fo the hydrographs, goodness-of-fit statistics are computed to guide the process as well as to determine when the calibration phase is completed. While a large number of statistics are computed, the dominant statistics are overall, seasonal, and flow interval biases. Also, the accumulation of the differences between simulated and observed flows over time should be examined. A check of the statistics as well as visual evaluation of the simulated hydrographs are performed after parameter changes to help guide the user through the process. A variety of statistical measures are available for evaluating the final results, but these are not usually helpful when making individual parameter changes.

Before parameter calibration begins, calibration and verification periods need to be selected. Experience has shown that the SAC-SMA needs at least an 8 year period for parameter calibration in wet areas. In drier regions, a longer period may be necessary in order to obtain enough events to consistently force all the model components. If possible, a calibration period is identified that contains a number of large precipitation events as well as several extended periods where base flow is dominant. Such a period is necessary to ensure that all the SAC-SMA components are activated a number of times. For verification, an independent period containing flows outside of the range in the calibration period is selected. Such a period allows the user to understand how the model might behave in an operational setting with extreme events.

A suggested strategy for calibrating watersheds within a basin has been developed by Anderson [2002]. Generally, the watershed with the best data and fewest complications is calibrated first. Next, other headwater areas with minimal complications, as well as downstream local areas where a good local hydrograph can be generated, are calibrated. These calibrations use the spatial assessment information to determine which previously calibrated parameters should be used as initial values. Here, only parameters that need to be changed are adjusted. Lastly, parameters are assigned to remaining watersheds from a calibrated area with similar hydrologic conditions. Minor adjustments to parameters are sometimes possible to remove biases, but a full calibration is not possible for these basins.

Anderson's [2002] proposed strategy should not only result in realistic and spatially consistent parameters, but should also greatly reduce the amount of time required for calibration. After the initial headwater calibrations, subsequent calibrations should require less effort because the process is generally one of making adjustments to only a few parameters and not performing a full calibration.

Initial values for some of the SAC-SMA model parameters can be derived through analysis of the observed streamflow data [*Anderson*, 2002; *Burnash*, 1995; *Peck*, 1976]. Typically, good initial values of the base flow withdrawal coefficients can be reliably obtained through hydrograph analysis, as can the size of the upper zone tension water storage. In some cases, initial values of the sizes of the lower zone baseflow storages can be obtained. Alternatively, initial SAC-SMA parameters derived using the method discussed in Koren et al., [this volume] can be used. (The interested reader is referred to Koren et al., [2000] for more details, while the work of Duan et al., [2001] discusses an application of these initial parameters). The NWS also provides guidelines for selecting initial parameter values for the snow model based on forest cover, typical amount of snow experienced, and other information.

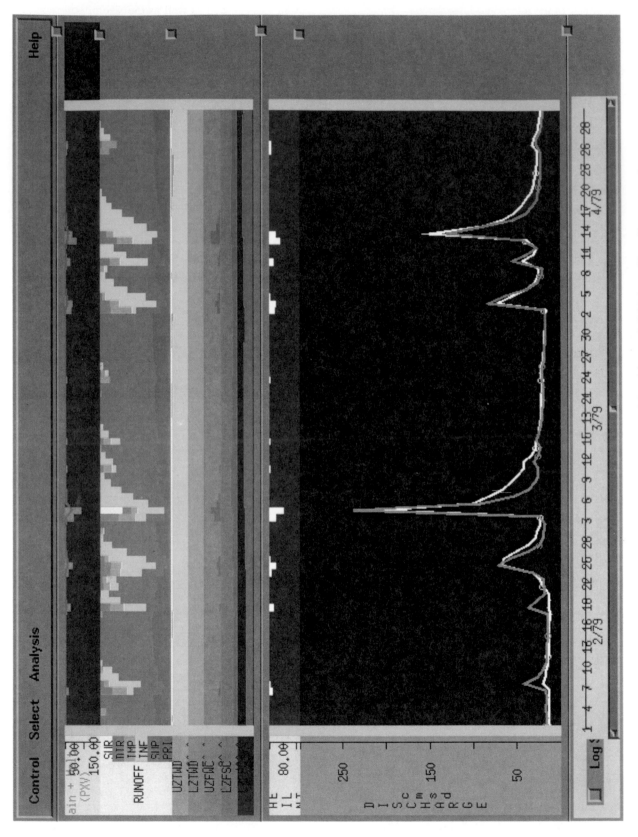

Plate 1. The ICP Display of the SAC-SMA Model Showing Runoff Components, Soil Moisture Zone Contents, and Corresponding Simulated and Observed Hydrographs.

Once initial values of the SAC-SMA and SNOW-17 models are obtained for the first headwater area, manual calibration with ICP proceeds according to a systematic strategy developed by Anderson [2002]. Without a systematic procedure and knowledge of the model, parameters could be derived that are out of a reasonable range yet still provide acceptable statistics [Burnash, 1995]. Examples of the effects of changing each of the SAC-SMA and Snow-17 model parameters have been developed to help the user understand how the hydrographs change [Hydrologic Research Center, 1999].

The main steps in this strategy are summarized as follows:

1. Remove gross overall errors, usually volume errors. The annual percent bias statistic should be within ±10 percent before continuing (± 20 percent in areas with small amounts of annual runoff). Large volume errors are usually caused by initial parameter values way out of range, or large errors in the MAP, MAT or MAPE time series. Errors such as these should be corrected because they will mask the effect of any subsequent parameter changes and will hinder the calibration process.

2. Obtain reasonable simulation of observed baseflow. Even though the models will be used primarily for flood forecasting, the best calibration results are obtained when the entire model is calibrated properly, starting with baseflow. Since the volume of storm runoff is linked to the percolation of water to the lower zone in the SAC-SMA, the proper simulation of baseflow will result in an improved simulation of fast response storm runoff. The size of the two baseflow storages are adjusted as well as the withdrawal coefficients. At this point it is advisable to check for the existence of riparian vegetation effects. Important considerations here are the proper division between fast and slower baseflow responses.

3. Adjust major snow model parameters. Here, parameters governing the melt rates and volume of snowmelt runoff are examined. Also, if a significant number of events where the form of precipitation is not typed correctly (i.e., rain vs. snow), it may be necessary to change the MAT time series so that the major events have the correct form of precipitation. These changes are sometimes necessary due to deficiencies caused by the use of a constant diurnal temperature variation and only daily maximum and minimum temperatures. Model parameters cannot be adjusted to account for mistyping of events.

4. Examine SAC-SMA tension water capacities. These two storages act as thresholds to govern the timing of runoff generation as well as the size of soil moisture deficits that can occur in response to evapotranspiration. To determine the proper values of these parameters, one looks at the time when the deficits are greatest, and then how the model responds when the deficits are filled.

5. Adjust parameters governing the generation of fast response or storm runoff. The proper separation between baseflow and fast response runoff is examined under a wide range of soil moisture conditions. In addition, parameters affecting the separation of fast response runoff into interflow and surface runoff are adjusted.

6. Final adjustments. These involve changes to the unit hydrograph, adding in riparian vegetation effects, and adjustments to the evapotranspiration demand function.

Often the user will need to return to a previous step and refine a parameter value that was previously adjusted. Due to the interactions of many of the SAC-SMA parameters, adjustments to one parameter to achieve a primary effect will also result in a secondary effect that must be corrected through subsequent interactive parameter adjustment. Following the calibration of the initial headwater area, the strategy described earlier for calibrating the remaining watersheds in the basin is followed.

A common question regarding model calibration is 'when is the calibration finished?' The NWS recommends several criteria for evaluating when a calibration is finished:

1. All biases between the simulated and observed hydrographs have been removed such that only random error remains. There should be no seasonal trends. Also, there should be no trends in peak flow estimation, as well as trends at low and intermediate levels.

2. Each parameter properly models the intended portion of the streamflow hydrograph. In such cases, the models should have a better chance of operationally simulating events outside the calibration period.

3. Parameter sets should have a logical spatial pattern among watersheds. Calibrated parameter sets for the watersheds in a basin should logically follow the assessments derived by examining the spatial variation of physiographic features and hydrologic response across a basin.

RFCs are encouraged to derive statistical criteria of their own in order to assess the quality of their calibrations to meet program goals. The following statistical criteria are also suggested as targets:

1. Overall percent bias within ±5%
2. Random variation in monthly biases (ie, no positive or

negative seasonal trends greater than ±5%, especially during periods of high flows.)

3. Flow interval biases within ± 5%.

Certainly, there are many cases in which the calibration cannot achieve the suggested criteria. To a large degree, this is governed by the availability and quality of the historical data [Burnash, 1995] and the variability of the meteorological conditions. In such cases, it is recommended that potential causes for the final statistics be documented. These data are then retained and made available during operational forecasting to assist the forecaster. Moreover, it should be expected that some trends exist simply due to the limitations of lumped modeling and the use of an index to compute snowmelt.

HL is heavily engaged in the development of distributed models to take advantage of spatial variability of precipitation and physiographic features. Approaches including semi- lumped (i.e., sub-basins) [Boyle et al., 2001; Smith et al., 1999] have been developed that show improvement in hydrograph simulation is some cases. Recently, a gridded distributed model has been developed and tested [Zhang et al., 2001]. Accompanying the continuing development of distributed approaches and increasing availability of spatial data sets comes the corresponding problem of parameter estimation and calibration. To a large extent, the calibration problem centers on the need to adjust the parameters in each computational element when observed streamflow is available only at the basin outlet.

Within HL, several approaches for distributed model calibration are being evaluated. One approach scales the a-priori estimates for each element using a ratio of the spatially averaged a-priori estimates to the calibrated lumped parameters.

6. AUTOMATIC CALIBRATION

Long standing collaboration with university research has led to a successful integration of automatic optimization procedures into the NWS calibration system. These efforts have tried to remove the subjectivity and reduce overall time required by manual calibration. Research by Duan et al., [1992] and Sorooshian et al., [1993] has led to the incorporation of the shuffled complex evolution (SCE) scheme into the suite of available optimization procedures. Also, an adaptive random search algorithm developed by Brazil [1989] has been made available. These methods optimize a single objective function, such as the daily root mean square error (DRMS). Gupta et al., [1999] concluded that automatic calibration methods have progressed to the point where they may be expected to perform with a level of skill approaching that of a well-trained hydrologist. This does not mean that the skill of the hydrologist is no longer necessary, but rather that more confidence may be placed in the use of these automatic tools to assist in the calibration process.

Automatic optimization has been used in NWS field offices in several ways. In some cases, automatic calibration is used to fine tune a parameter set after manual calibration is complete. Other field personnel use automatic methods to evaluate the parameters at an existing stage in the manual calibration process. Cooperative research between the NWS and the University of Arizona has led to the development of a step-wise procedure that mimics the steps recommended for manual calibration [Hogue et al., this volume; Hogue et al., 2000]. In the Multi-Step Automatic Calibration Scheme (MACS) procedure, base flow parameters are first optimized by minimizing the log objective function. In step two, the optimized base flow parameters are fixed and the parameters governing the generation of fast response runoff are optimized using the Root Mean Square Error (RMSE) criteria. Lastly, the fast response parameters optimized in step two are fixed and the base flow parameters are adjusted. Results with the MACs procedure have shown to be comparable or slightly better than results from manual calibrations in certain cases [Hogue et al., this volume]. The MACS approach also somewhat addresses the limitation of using one objective function for all parts of the hydrograph.

In spite of the advances in more efficient and powerful search algorithms, several drawbacks have limited the use of automated methods. One of the main limitations of automatic approaches is that no one objective function to be minimized works well with all parts of a streamflow hydrograph [Boyle et al., 2000]. Also, Burnash [1995] and Boyle et al., [2000] stated that automatic calibration may tend to result in parameters that have conceptually invalid values. Yet another concern is that unlike the approach developed by Anderson [2002], current automatic calibration techniques cannot be guided to produce spatially consistent parameter sets among watersheds in a basin. To address this concern, Koren et al. [this volume] propose a method of conditioning the parameter search space using a-priori estimates of model parameters.

7. MODEL CALIBRATION AT THE RIVER FORECAST CENTERS

While the development of strategies and tools is critical to the efficiency of model calibration, the success of any effort also hinges on the organizational structure within an RFC. Through the use of guidelines, teams, and peer review, calibrations are better and more consistent across basins. As an example, the Northwest RFC (NWRFC) will be used to illustrate how the calibration effort can be organized to meet objectives in an efficient manner and generate consistent results.

Prior to any work, the calibration team first determines how to split the region into calibration areas. Hydrologic similarities, basin soil characteristics, common geological attributes, and meteorological considerations are used to define regions. Actual river locations to calibrate are determined by availability of data, user requests, and complexity of hydrologic processes.

Specialists in data analysis then use the techniques described earlier to create data files that are used by all calibrators. Template files for MAP and MAT are produced for the calibrators. A TAPLOT is run to provide common temperature-elevation plots for deriving means for the MAT synthetic station. Evaporation data are summarized.

The calibration group then reviews data, basin hydrology, geological factors, and lists of perspective calibration points in each region. A set of guidelines is prepared and distributed to each calibrator. These guidelines provide limits for model parameters, basin splits, model selection, and criteria for determining when a calibration is complete. These guidelines are reviewed and possibly modified during the course of the calibration process. All forecasters are then assigned calibrations from the list of prospected sites.

The calibrator follows the procedures outlined in this paper. Upon completion of a calibration, it is submitted to a peer review group for acceptance. The calibrator can sit in these reviews and often benefits from the experience and dialog presented during these meetings. The review group is made up of 2 to 3 of the most experienced calibrators.

Experience has shown that the review process has ensured consistency and hydrologically sound calibrations.

The final step in the process is to transfer the calibration into operations. Considerable knowledge has been gained during the one to two week period the calibrator has spent calibrating. It is during this phase, parameter sensitivities and specific or unique hydrologic processes are discussed. The calibrator and the review team provide new forecast point training before a point is placed into operations.

8. ILLUSTRATION OF THE CALIBRATION PROCESS

As an illustration, the procedures outlined in this paper were applied to the Oostanaula River above the USGS gage in Rome, Georgia, and used to develop an NWS calibration training video [*Hydrologic Research Center*, 1999]. Figure 3 shows that the watershed contains several headwater and local basins. The letters indicate the sequence in which the basins are calibrated. The watersed boundaries define the areas draining to RFC forecast points. While the USGS maintains streamflow gages at points 1 and 2 in Figure 3, these points are not currently forecasted by the RFC and thus are not explicitly modeled.

Observed streamflow data from the basins were normalized and plotted to note similarities in response amongst the watersheds. Historical data analysis was performed for the basin as a whole, rather than as separate steps for each watershed. A network of 31 daily and 11 hourly raingages was used in the analysis of precipitation. An MAT time series was not developed as snow is not hydrologically important in this region. Given the spatial variability of long term precipitation station means, a mountainous area analy-

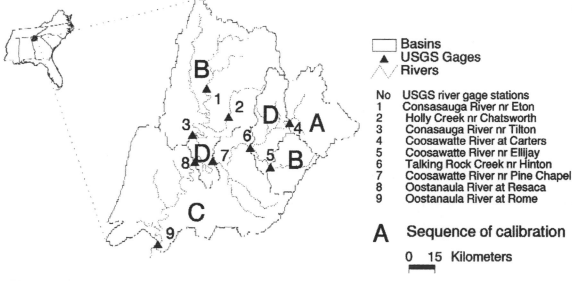

Figure 3. The Oostanaula River Basin above the USGS Gage at Rome, Georgia.

sis was performed. An isohyetal map was derived using the method of Peck [1962] and subsequently integrated to derive the long term mean areal precipitation value $\bar{A}$ in Equation 3 for each watershed. Using Equation 3, precipitation station weights were derived for each of the watersheds and 6-hour MAP time series were developed. USGS mean daily flow data were retrieved for the calibration process.

Following the strategy proposed by Anderson [2002], the Coosawatte River above the USGS gage near Ellijay, Ga. was selected as the watershed to calibrate first, as it had the best data and fewest complications. Initial values of the SAC-SMA were derived from analysis of the observed hydrographs and appear in Table 1.

As discussed earlier, Step 1 in the calibration process is to remove large errors in the simulation which might mask needed parameter changes. Examining the hydrograph plots in ICP revealed that the model greatly over-simulated the high flows and underestimated the low flows. This indicated that too much surface runoff was being generated and too frequently. Thus, the UZFWM parameter was changed from 40mm to 80mm. After running the model with this parameter change, the ICP display revealed that the amount of surface runoff was indeed reduced. Seasonal bias statistics for this simulation appear in column 1 of Table 2. These results

Table 1. Initial SAC-SMA Parameter Values for the Coosawattee River above the USGS Gage in Elijay, Georgia

Parameter	Initial Value
Size of Upper Zone Tension Water Storage, mm (UZTWM)	100.0
Size of Upper Zone Free Water Storage, mm (UZFWM)	40.0
Upper Zone Free Water Withdrawal Coefficient (UZK)	0.2
Percent of Basin that is Impervious (PCTIM)	0.04
Percent of Basin that is Variably Impervious (ADIMP)	0.0
Size of Lower Zone Tension Water Storage, mm (LZTWM)	150.0
Size of Primary Baseflow Storage, mm (LZFPM)	600.0
Size of Supplemental Baseflow Storage, mm (LZFSM)	125.0
Primary Baseflow Withdrawal Coefficient (LZPK)	0.002
Supplemental Baseflow Withdrawal Coefficient (LZSK)	0.04
Percolation Function Parameter (ZPERC)	25.0
Percolation Function Parameter (REXP)	2.0

indicate a rather large bias in November, December and January. To reduce this bias, changes were made in the ET Demand curve. The resulting statistics shown in column 2 of Table 2 indicate that the bias was reduced by this change.

Step 2 of the calibration process is to adjust the base flow parameters. Visual examination of the hydrographs at this point in the process revealed that too much primary baseflow and not enough supplemental baseflow was being generated. To correct these tendencies, the primary baseflow storage was decreased from 600mm to 480mm, and the supplemental baseflow storage was increased from 125mm to 250mm. The PFREE parameter was changed from 0.1 to 0.4 to increase the amount of baseflow recharge, and the value of LZSK was reduced from 0.04 to 0.02 to slow down the withdrawal of supplemental baseflow. After these parameter changes were made, the model was re-run and the bias statistics in column 3 of Table 2 were generated. These results and a visual examination of the simulation in ICP revealed a closer agreement of the simulated and observed hydrographs.

Step 3 in the process is to examine the size of the upper and lower tension water storages. As before, visual examination of the hydrograph plots and storage components in the ICP displays revealed the needed changes. The size of the upper zone tension water storage, UZTWM, seemed appropriate, but the size of the lower zone storage, LZTWM, needed to be increased from 150 to 250mm. After making this change, the ICP plots revealed that the lower zone was filling up at the appropriate time in the fall and that base flow recharge during the winter was being

Table 2. Monthly Percent Bias Statisctis for the Calibration of the Coosawattee River at Ellijay, Georgia.

Mon.	Percent Bias					
	Trial					
	1	2	3	4	5	6
Oct.	20.5	15.5	7.9	2.6	-1.1	-3.2
Nov.	38.5	27.5	11.4	8.6	6.2	-0.1
Dec.	51.4	36.2	21.1	17.7	17.0	10.1
Jan.	39.5	26.8	15.6	11.9	12.9	7.2
Feb.	27.0	13.5	12.8	9.4	11.0	7.4
Mar.	12.0	-0.6	-3.3	-5.0	-2.8	-3.2
April	-4.8	-11.4	-7.2	-9.3	-8.9	-6.5
May	-15.7	-18.5	-6.7	-9.1	-9.5	-4.9
June	-26.6	-28.9	-9.2	-13.0	-13.6	-8.3
July	-8.4	-10.8	1.3	-1.5	-3.0	1.5
Aug.	-5.0	-7.3	2.8	-1.3	-4.0	0.1
Sept.	9.0	6.4	10.3	6.15	2.4	4.6
Overall	11.4	2.9	3.3	0.4	0.2	0.0

modeled appropriately. Column 4 of Table 2 shows the seasonal bias statistics at this point in the calibration process.

The fourth step in the calibration strategy is to examine the division of fast and slow response portions of the hydrographs. Primarily, this involves an analysis of the parameters controlling the percolation function, or the potential rate at which water moves from the upper zones to the lower zones. The flow interval bias figures in column 1 of Table 3 indicate that at this point in the process, the model was overpredicting low flows and underpredicting the larger events. To mitigate these effects, the percolation parameters were changed to reduce percolation during the wetter periods and increase percolation during the drier periods. Also during this step, the size of the variable impervious area was changed from 0 to 5% of the total basin area to increase peak flows during intermediate rainfall events. After making these changes, the flow interval bias statistics in column 2 of Table 3 indicate that while the model still over predicts the low flows, the simulation of intermediate and large events was more appropriate. Visual analysis of the simulation also showed a better division of fast and slow runoff. Column 5 in Table 2 indicates an improvement in the overall bias statistic.

To finish the calibration of the Coosawattee River headwater basin, final adjustments are made to reduce seasonal and overall biases. Column 5 of Table 2 shows that at this point in the calibration process, large positive biases existed in the winter with large negative biases occurring in the early summer. The ET-Demand curve was adjusted to increase evapotranspiration demand in the fall, winter, and spring while reducing this demand in April, May, and June. The final seasonal bias statistics in column 6 reveal reasonable monthly biases with a very good overall bias.

Table 4 presents final values of the bias and correlation coefficient for the calibration and verification periods for the Coosawattee River. These statistics quantify the good-

ness of fit that was also apparent in a visual examination of the simulated and observed hydrographs.

Following the sequence in Figure 3, the calibrated Ellijay parameters were used as starting values for the calibration of Talking Rock Creek at Hinton, Ga.. In this calibration, only the parameters that needed adjustment were changed to produce an acceptable fit. Column 1 of Table 5 shows the flow interval bias statistics that resulted by using the Ellijay parameters as initial values for the Hinton basin. As expected from the analysis of the normalized observed flows, these results indicate that the Ellijay parameters oversimulated the lower flows and under-estimate the high flow events for the Hinton basin. As a first step to correcting these biases, the percolation curve was lowered to generate more fast response runoff. The LZFSM parameter was changed from 400mm to 285mm while the LZFPM parameter was adjusted from 480mm to 340mm. Both a visual examination of the computed and observed hydrographs and the results in Column 2 of Table 5 indicate these changes to the percolation curve produced a better fit. After these percolation changes, visual examination of the computed and observed hydrographs for the Hinton basin showed that more surface runoff needed to be generated. To achieve this, the size of the upper zone free water storage (UZFWM) was reduced from 80mm to 55mm. The result-

Table 4. Statistics for the Cazlibration and Verification Periods for the Coosawattee River at Ellijay, Georgia.

Statistic	Period		
	Calibration WY[a] 75-84	Verification WY 64-74	Verification WY 95-93
Bias (%)	-0.6	3.5	-3.5
Correlation (daily flows)	0.940	0.935	0.914

[a]WY denotes Water Year, i.e., October 1 to September 30.

Table 3. Monthly Percent Bias Statisctis for the Calibration of the Coosawattee River at Ellijay, Georgia.

Flow Interval, cms.	Percent Bias	
	Trial	
	1	2
0.0 - 3.5	8.9	9.5
3.5 - 5.5	1.0	0.3
5.5 - 10.0	-1.5	-3.3
10.0 - 20.0	2.1	0.4
20.0 - 50.0	2.7	2.8
50.0 - 140.0	-7.0	-2.3
> 140.0	-17.3	-8.3

Table 5. Flow Interval Bias Statistics for the Calibration of Talking Rock Creek at Hinton, Georgia.

Flow Interval, cms	Percent Bias			
	Trial			
	1	2	3	4
0.0 - 1.8	64.2	39.1	45.6	4.4
1.8 - 2.8	21.9	3.55	8.2	-8.6
2.8 - 5.0	16.0	4.05	7.1	-4.9
5.0 - 10.0	7.2	5.91	5.0	-2.35
10.0 - 25.00	-4.6	11.0	4.4	6.7
25.00 - 75.00	-23.7	-10.5	-15.2	-7.2
>75.00	-46.1	-30.8	-4.0	-5.9

1 - Initial run using Ellijay Parameters

ing flow interval bias statistics are shown in column 3 of Table 5 and show improvement in the bias figures for large events with a corresponding degradation in the bias statistics for medium flow levels. A visual examination of the hydrographs also confirmed the use of a smaller value of UZFWM for large events.

To complete the calibration of the Hinton basin, the value of upper zone free water withdrawal coefficient, UZK, was increased from 0.2 to 0.3 to generate a faster interflow response. Also, the RIVA parameter was increased to model the effects of riparian vegetation evaptranspiration in the dry summer months. In addition, the amount of constant impervious area was reduced so that the model did not respond as much to every rainfall event. Column 4 of Table 5 presents the final flow interval bias statistics after these last parameter changes were made.

Calibration of the entire watershed continues according to the sequence shown in Figure 3. In this step, the final Hinton parameters are used as the initial parameters for the Conasauga River at the USGS gage in Tilton, Ga.. As before, only the parameters which need to be changed are adjusted. For this watershed, the analysis of the simulated and observed hydrographs revealed that more fast response runoff was needed. The statistics for the initial run in Table 6 also show that the base flows were greatly overpredicted and the larger events were underpredicted.

To generate more fast response runoff, the percolation function was reduced to 40% of its initial value by reducing the LZFSM parameter from 285mm to 115mm and the LZFPM parameter from 340mm to 135mm. Also during this step, the value of UZK was increased from 0.3 to 0.4 to speed up the generation of interflow. After these changes, the subsequent simulation improved both visually and statistically. However, the hydrographs showed that subsequent changes to the supplemental base flow needed to be made. These included changing the LZSK parameter from 0.02 to 0.03 and increasing the size of the supplemental base flow storage, LZFSM, from 115mm back to 135mm. As seen in column 2 of Table 6, the flow interval bias statistics for the second trial have been reduced across all flow ranges. The corresponding seasonal bias statistics for this simulation are shown in column 1 of Table 7.

Visual examination of the hydrographs after these changes revealed that the summer events were being overpredicted, indicating that the size of the upper zone tension water storage parameter UZTWM need to be increased from 100mm to 150mm. In addition, the ET-Demand curve was adjusted to reduce the seasonal bias shown in column 1 of Table 7. After these changes were made, improved seasonal bias statistics resulted and are shown in column 2 of Table 7. Visual inspection of the simulations also showed

improvement from these changes. However, as seen in column 2 of Table 7, a large bias still remained for the summer months. The shape of the simulated and observed hydrographs during this period indicated that riparian vegetation evapotranspiration effects were present. To model these effects, the value of the RIVA parameter was increased from 0.0 to 0.05, indicating that 5% of the basin area was affected by riparian vegetation.

After making the change to the RIVA parameter, the model was re-run and the flow interval statistics in column 3 of Table 6 were generated. The corresponding seasonal bias statistics for this simulation are shown in column 3 of Table 7. These results show acceptable levels of bias. Along with a visual examination of the simulation, these statistics showed that the calibration of the Tilton basin was complete.

Table 6. Flow Interval Bias Statistics for the Calibration of the Conasauga River at Tilton, Georgia.

Flow Interval, cms.	Percent Bias		
	Trial		
	1	2	3
0.0 - 5.0	180.8	23.7	3.1
5.0 - 15.0	88.5	3.2	-4.1
15.0 - 50.0	19.5	1.5	4.0
50.0 - 100.0	-8.9	9.0	8.3
100.0 - 200.0	-27.4	-2.2	-2.2
200.0 - 400.0	-33.7	-1.9	-2.7
>400.0	-34.2	-5.5	-5.7

1 - Initial run using Hinton Parameters

Table 7. Seasonal Bias Statistics for the Conasauga River at Tilton, Georgia.

Month	Percent Bias		
	Trial		
	1	2	3
Oct.	29.5	10.8	7.3
Nov.	7.25	0.5	-0.3
Dec.	1.0	1.5	1.4
Jan.	-0.6	4.9	4.9
Feb.	-7.4	-0.5	-0.6
Mar.	-6.5	-0.2	-0.3
April	-3.4	1.9	1.8
May	4.5	3.6	2.6
June	8.8	4.3	1.0
July	11.6	-2.8	-7.6
Aug.	21.2	-0.2	-7.9
Sept.	43.6	14.3	-7.8
Overall	1.7	2.2	1.9

Following the calibration sequence shown in Figure 3, the local area above the USGS gage in Rome, Ga. was calibrated. In Figure 3, this is the area between points 8 and 9. To begin this calibration, the calibrated parameters from the most hydrologically similar basin were used as initial values. The Tilton parameters were used as this basin was most similar to the Rome local area. To define the local area hydrograph, the unit hydrograph from Tilton was used but scaled to reflect the drainage area of the Rome local area. The channel routing parameters were obtained from the RFC operational files. The steps involved in calibrating this local area were also applied to the calibration of the local area between points 6 and 7 in Figure 3.

After minor modification to the channel routing parameters, it was clear that the Tilton SAC-SMA parameters worked quite well in the simulation of the Rome local area. Only a decrease in the RIVA parameter from 0.05 to 0.0 was required as the Rome local area showed no evidence of evapotranspiration from riparian vegetation. The final flow

interval bias statistics in Table 8 and the final seasonal bias statistics in Table 9 show the adjustment to the RIVA parameter resulted in good statistics. The large bias in the lowest flow interval of Table 8 reflects the noise in the derived local area observed hydrograph. The local area at Rome, Ga. represents only about 25% of the total drainage area above this point.

As stated earlier, the final step in the calibration of a watershed is to assign calibrated parameters to remaining areas. These final areas are usually local areas that are too small compared to the total drainage area to generate a reasonable local area observed hydrograph against which to calibrate. The regions labeled D in Figure 3 were treated in this manner.

Table 10 presents the parameters that resulted from the calibration of the Oostanaula watershed. It can be seen that after the initial calibration of the Ellijay basin, only a few parameters were adjusted in the subsequent calibration of neighboring basins.

This example shows the efficiency of the regional approach to calibration in that only the Ellijay watershed required a significant amount of manual calibration effort. Far less time was required to calibrate the other watersheds due to the use of calibrated parameters as starting points for hydrologically similar watersheds. Equally important is the fact that the final calibrated parameters exhibit a spatial pattern that is quite reasonable considering the spatial variation in physiographic features as well as the comparison of observed streamflow hydrographs.

Table 11 presents two of the summary statistics for the entire period of record. It can be seen that in all basins, the

Table 8. Final Flow Interval Bias Statistics for the Calibration of the Local Area above Rome, Georgia.

Flow Interval, cms.	Final Percent Bias
0.0 - 1.0	4712.1
1.0 - 10.0	20.3
10.0 - 20.0	-3.2
20.0 - 75.0	-5.14
75.0 - 150.0	-10.0
150.0 - 300.0	-13.9
>300.0	-2.0

Table 9. Seasonal Bias Statistics for the Calibration of the Local Area above Rome, Georgia.

Month	Percent Bias
Oct.	2.3
Nov.	2.6
Dec.	-1.0
Jan.	-4.4
Feb.	-7.6
Mar.	-5.6
April	-2.2
May	8.14
June	13.4
July	2.6
Aug.	-4.6
Sept.	4.2
Overall	-2.0

Table 10. Parameter Summary for the Calibration of the Oostanaula River Basin.

Parameter	Ellijay Initial Value	Ellijay Final Value	Hinton Final	Tilton Final	Rome Local Area
UZTWM, mm	100.0	100.0	100.0	150.0	150.0
UZFWM, mm	40.0	80.0	55.0	55.0	55.0
UZK	0.2	0.2	0.3	0.4	0.4
PCTIM	0.04	0.04	0.02	0.02	0.02
ADIMP	0.0	0.05	0.05	0.05	0.05
LZTW, mm	150.0	250.0	250.0	150.0	150.0
LZFPM, mm	600.0	480.0	340.0	135.0	135.0
LZFSM, mm	125.0	400.0	285.0	135.0	135.0
LZPK	0.002	0.002	0.002	0.002	0.002
LZSK	0.04	0.02	0.02	0.03	0.03
ZPERC	25.0	80.0	80.0	80.0	80.0
REXP	2.0	3.0	3.0	3.0	3.0
PFREE	0.1	0.4	0.4	0.4	0.4

bias statistic is well within the recommended target criteria. Good values of the correlation coefficient were also obtained.

9. SUMMARY

The National Weather Service has continued to develop a comprehensive set of procedures and tools to facilitate the calibration of hydrologic models. A logical and systematic strategy has been developed for manual calibration of the SAC-SMA and Snow-17 models for an individual watershed. Moreover, an efficient strategy for calibrating the additional watersheds within a river basin has been developed. This strategy leads to sets of parameters that make sense considering the spatial variability of physiographic features amongst the watersheds in a basin. In addition, the use of guidelines, teams, and peer reviews at RFCs has proven to generate quality calibrations in an efficient manner.

Through manual calibration, the hydrologist is able to learn the inner workings of the model and understand the sensitivities to different forcings and model parameters. In this way, the hydrologist is better prepared for the forecasting environment in which unusual conditions may be encountered. A thorough understanding of the model function is important when a user is making short term and long term forecasts.

10. RECOMMENDATIONS

While a number of tools and recommended procedures have been developed and fielded in the NWS, numerous enhancements to the existing programs should be made. Past efforts such as the development of ICP and IDMA have proven that significant reductions can be realized in the amount of time required for calibration. Additional enhancements are critical considering the national scope of the model calibration and implementation efforts in the NWS.

Perhaps what is most needed is a re-design of the Calibration System so that the functions of the MAP, MAT,

Table 11. Statistical Summary of the Calibration of the Oostanaula River Basin: Percent Bias and Correlation Coefficient R.

Statistic	Watershed and Period			
	Ellijay WY 64-93	Hinton WY 75-93	Tilton WY 49-93	Rome WY 49-93
Bias (%)	-0.2	-1.5	-1.1	-0.1
R (daily flows)	0.928	0.941	0.951	0.988

MAPE, PXPP, TAPLOT, MCP/ICP, CAP and other programs are retained but more efficiently linked, perhaps under the umbrella of one parent tool. Such a re-design would incorporate the latest data handling and display techniques. Currently, the programs mentioned exist as standalone entities, requiring a good bit of data manipulation on the part of the user. Also, some of the functional redundancy could be eliminated. For example, each of the MAP, MAT, MAPE, and PXPP programs contains a double mass analysis capability. A re-designed Calibration System might need to have only one such capability.

In addition to software enhancements, continued research in several areas is also needed. Research related to automatic optimization is necessary, especially in the development of multi-objective calibration strategies. More objective methods could be developed for deriving station weights in mountainous areas, as well as determining the need to make consistency corrections to station data. In addition, refined methods of computing estimates of evapotranspiration are needed. The use of instantaneous temperatures to compute an MAT time series for calibration would greatly reduce the errors resulting from the use of a fixed diurnal variation in conjunction with daily maximum and minimum temperature values.

Acknowledgements. The authors are indebted to Dr. Eric Anderson, formerly of the Hydrology Lab, for developing much of the information regarding calibration of the NWS snow and rainfall-runoff models during his long and productive career with the NWS. Dr. Anderson is currently a consultant to the NWS Hydrology Lab.

APPENDIX: CURRENT CAPABILITIES AND DATA SETS AVAILABLE IN THE CALIBRATION ASSISTANCE PROGRAM (CAP)

1. CAPABILITIES

The CAP is an Arc/View application. Its capabilities include the following.

1. Derive area-elevation curves
2. Sub-divide basins based on elevation zones
3. Derive elevation-precipitation plots
4. Display defined zones on top of other data layers (e.g. precipitation, elevation)
5. Compute basin mean, max, and min values of: (may also compute for each elevation zone defined within a basin)
   5.1 precipitation (monthly, annual, and seasonal)

5.2 potential evaporation (monthly, annual, and seasonal)
5.3 potential evaporation adjustment factors
5.4 percent forest
5.5 percent of each forest type
5.6 soil-based estimates for 11 SAC-SMA parameters
6. Display selected NOHRSC historical snow images from (1990-1995)

2. DATA SETS

1. Digital Elevation Model (DEM) 4km resolution
2. Percent of forest cover on 1km grid
3. Forest type on 1km grid
4. PRISM annual and monthly precipitation grids.
5. Annual and monthly potential evaporation grids
6. Soil type, depth, and texture for 11 layers
7. Snow cover grids for 1990-1995
8. State boundaries
9. EPA River Reach Files (RF1)

REFERENCES

Anderson, E.A., 'Calibration of Conceptual Hydrologic Models for Use in River Forecasting', *NOAA Technical Report*, NWS 45, Hydrology Laboratory, 2002, copies available upon request.

Anderson, E.A.,'A Point Energy and Mass Balance Model of a Snow Cover', *NOAA Technical Report NWS 19*, February, 1976, copies available upon request.

Anderson, E.A., 'National Weather Service River Forecast System - Snow Accumulation and Ablation Model', *NOAA Technical Memorandum NWS HYDRO-17*, November, 1973, copies available upon request.

Armstrong, B.B., 'Derivation of Initial Soil Moisture Accounting Parameters from Soil Properties for the National Weather Service River Forecast System', NOAA Technical Memorandum NWS Hydro 37, 1978, copies available upon request.

Bonnin, G., 'The NOAA Hydrologic Data System', 12th International Conference on Interactive Information and Processing System (IIPS) for Meteorology, Oceanography, and Hydrology, Atlanta, Georgia , January 28-February 2, 1996.

Boyle, D.P., H.V. Gupta, S. Sorooshian, V. Koren, Z. Zhang, and M. Smith,'Towards improved streamflow forecasts: The value of semi-distributed modeling,' *Water Resour. Res.* Vol. 37, No. 11, 2749-2759, 2001.

Boyle, D.P., H.V. Gupta, and S. Sorooshian,'Toward Improved Calibration of Hydrologic Models: Combining the Strengths of Manual and Automatic Methods', *Water Resour. Res.*, Vol. 36, No. 12, 3663-3674, 2000.

Bradley, A.A., and A. Kruger, 'Recalibration of Hydrologic Models for use with WSR-88D Precipitation Estimates', preprints the special Symposium on Hydrology, AMS, 302-305, 1998.

Brazil, L.E.,'Multilevel Calibration Strategy for Complex Hydrologic Simulation Models', *NOAA Technical Report, NWS 42*, Hydrology Lab, Silver Spring, Md, 1989, copies available upon request.

Brazil, L.E., and M.D. Hudlow, 'Calibration Procedures Used with the National Weather Service River Forecast System', in *Water and Related Land Resource Systems*, edited by Y.Y. Haimes and J. Kindler,pp 457-466, Pergamon, Tarrytown, N.Y., 1981.

Burnash, R.J.C., 'The NWS River Forecast System - Catchment Model', Chapter 10, in *Computer Models of Watershed Hydrology*, Vijay P. Singh, editor, Water Resources Publications, 1995.

Burnash, R. J. C., R. L. Ferral, and R. A. McGuire, 'A generalized streamflow simulation system - Conceptual modeling for digital computers', *Technical Report, Joint Federal and State River Forecast Center, U.S. National Weather Service and California Department of Water Resources, Sacramento, California,* 204pp. 1973.

Daly, C., R.P. Neilson, and D.L. Phillips, 'A Statistical-Topographic Model for Mapping Climatological Precipitation over Mountainous Terrain', *J. Applied Meteorology*, Vol. 33, No. 2, February, pp. 140-158, 1994.

Duan, Q., J. Schaake, and V. Koren, '*A Priori* Estimation of Land Surface Model Parameters', *In Land Surface Hydrology, Meteorology, and Climate: Observation and Modeling, V. Lakshmi, et al. (Ed.), Water Science and Application, Vol. 3, AGU, Washington, DC,* 77-94, 2001.

Duan, Q., S. Sorooshian, and V.K. Gupta,'Effective and Efficient Global Optimization for Conceptual Rainfall-Runoff Models', *Water Resour. Res.*, 28(4), 1015-1031, 1992.

Farnsworth, R.K., and E.L. Peck, 'Evaporation Atlas for the Contiguous 48 United States', *NOAA Technical Report NWS 33*, U.S. Department of Commerce, 1982, copies available upon request.

Farnsworth, R.K., and E.S. Thompson,'Mean Monthly, Seasonal, and Annual Pan Evaporation for the United States', *NOAA Technical Report NWS 34*, U.S. Department of Commerce, 1982, copies available upon request.

Finnerty, B.D., M.B. Smith, D.J. Seo, D.J., V.I. Koren, and G.E. Moglen, 'Space-Time Scale Sensitivity of the Sacramento Model to Radar-Gage Precipitation Inputs', *J. Hydrology*, 203, 21-38, 1997.

Fread, D. L., R.C. Shedd, G.F. Smith, R. Farnsworth, C.N. Hoffeditz, L.A. Wenzel, S.M. Wiele, J.A. Smith, and G.N. Day,'Modernization in the National Weather Service River and Flood Program', *Weather and Forecasting*, Vol. 10, No. 3, September, 477-484, 1995.

Gupta, H.V., S. Soorooshian, and P.O. Yapo, 'Status of Automatic Calibration for Hydrologic Models: Comparison with Multilevel Expert Calibration', *J. of Hydrologic Engineering*, Vol. 4, No. 2, pp. 135-143, April, 1999.

Hogue, T., S. Sorooshian, S., H. Gupta, A. Holz, and D. Braatz, 'A Multistep Automatic Calibration Scheme for River Forecasting Models', *J. Hydrometeorology*, Vol. 1, No. 6. December, pp. 524-542, 2000.

Hydrologic Research Center (HRC), Producer: "Calibration of the Sacramento Soil Moisture Accounting Model, Demonstration of an Interactive Calibration Approach." NOAA Video Series: VHS Video (5hrs and 15min) and Companion Notebook (200 pp.). National Oceanic and Atmospheric Administration, National Weather Service, Office of Hydrologic Development, Silver Spring, MD, May, 1999, copies available upon request.

Johnson, D., M. Smith, V.I. Koren, and B. Finnerty, 'Comparing Mean Areal Precipitation Estimates from NEXRAD and Rain Gauge Networks', *J. Hydrologic Engineering*, Vol. 4, No. 2, April, 117-124, 1999.

Koren, V.I., B.D. Finnerty, J.C. Schaake, M.B. Smith, D.J. Seo, and Q.Y. Duan, 'Scale Dependencies of Hydrologic Models to Spatial Variability of Precipitation', *J. Hydrology*, 217, 285-302, 1999.

Koren, V.I., M.B. Smith, D. Wang, and Z. Zhang, 'Use of Soil Property Data in the Derivation of Conceptual Rainfall-Runoff Model Parameters', 15th Conference on Hydrology, American Meteorological Society, Long Beach California, 103-106, 2000.

Larson, L.W., R.L. Ferral, E.T. Strem, A.J. Morin, B. Armstrong, T.R. Carroll, M.D. Hudlow, L.A. Wenzel, G.L. Schaefer, and D.E. Johnson, 'Operational Responsibilities of the National Weather Service River and Flood Program', *Weather and Forecasting*, Vol. 10, No. 3, September, 465-476, 1995.

Monroe, J.C., and E.A. Anderson, 'National Weather Service River Forecasting System', *ASCE J. Hydraulics Division*, Vol. 100, No. HY5, pp. 621-630, 1974

Pan, J., G.M. Bonnin, R.M. Motl, and H.Friedeman, 'Recent Developments in Data Analysis Quality Control and Data Browsing at the National Weather Service Office of Hydrology', 78th Annual AMS Meeting, Phoenix, Arizona, January, 1998.

Peck, E.L, 'Catchment Modeling and Initial Parameter Estimation for the National Weather Service River Forecast System', NOAA Technical Memorandum NWS HYDRO-31, 1976.

Peck, E.L, and M.J. Brown, 'An Approach to the Development of Isohyetal Maps for Mountainous Areas', *J. Geophys. Res.*, Volume 67, No. 2, February, 681-694, 1962.

Penman, H.L., 'Natural Evaporation from Open Water, Bare Soil, and Grass', Proceedings, Royal Society, Series A, Vol. 193, 1948.

Smith, M.B., V.I. Koren, B.D. Finnerty, and D. Johnson, '*Distributed Modeling: Phase 1 Results*', *NOAA Technical Report NWS 44*, February, 1999. Copies available upon request.

Sorooshian, S., Q. Duan, and V.K. Gupta, 'Calibration of Rainfall-Runoff Models: Application of Global Optimization to the Sacramento Soil Moisture Accounting Model', *Water Resour. Res.* 29(4), pp. 1185-1194, 1993.

Stallings, E.A., and L.A. Wenzel, 'Organization of the River and Flood Program in the National Weather Service', *Weather and Forecasting*, Vol. 10, No. 3, September, 457-464, 1995.

Thompson, E.S., 'Computation of Solar Radiation from Sky Cover', *Water Resour. Res.*, Vol. 12, No. 5, pp. 859-865, 1976.

Zhang, Z., V. Koren, M. Smith, and S. Reed, 'Application of a Distributed Modeling System Using Gridded NEXRAD Data', Proceedings, Fifth Int'l. Symposium on Hydrological Applications of Weather Radar -Radar Hydrology-, pp. 427-432, Nov. 19-22, Heian-Kaikan, Kyoto, Japan, 2001.

[1]Michael Smith, [1]Victor Koren, [1]Seann Reed, and [1]Ziya Zhang, Hydrology Laboratory, Office of Hydrologic Development, NOAA/National Weather Service, 1325 East-West Highway, Silver Spring, Maryland, 20910, USA.

[2]Donald Laurine, Northwest River Forecast Center, NOAA/National Weather Service, 5241 NE 122 Avenue, Portland, Oregon, 92730-1089,USA.

[1]Hydrology Laboratory, Office of Hydrologic Development, NOAA/National Weather Service

[2]Northwest River Forecast Center, NOAA/National Weather Service

A Process-Oriented, Multiple-Objective Calibration Strategy Accounting for Model Structure

Richard Turcotte

Centre d'expertise hydrique du Québec, Québec, Canada

Alain N. Rousseau, Jean-Pierre Fortin, and Jean-Pierre Villeneuve

Institut National de la Recherche Scientifique, INRS-ETE (formerly INRS-Eau), Québec, Canada

The Quebec Hydrological Expertise Center and Hydro-Quebec are developing an operational forecasting system for 3-hour stream flow predictions using a distributed hydrological model. The system will be implemented on several southern Quebec basins characterized by quick response times and requiring timely forecasts for dam management. This paper introduces the first steps toward the general development of a calibration strategy using a processes-oriented, multiple-objective, approach accounting for model structure. The calibration objectives are used to sequentially minimize errors between observed and modeled: (i) prolonged summer drought recessions, (ii) annual and monthly flow volumes, (iii) summer and fall high flows, (iv) high flow synchronization, and (v) spring runoff resulting from snowmelt. Specific groups of parameters are assigned to each one of these objectives. Parameters affecting objectives characterized by long time scales are calibrated first while those characterized by short time scales are calibrated last. Any loss in model performance is compensated by readjusting previously calibrated parameters. Repeating the process until a satisfactory model performance is reached. A preliminary, manual, application with the distributed hydrological model HYDROTEL clearly illustrates the need to pursue our work as all underlying concepts and theories withstood this first test.

1. INTRODUCTION

There exists an increasing demand for stream flow forecast systems based on distributed and, as much as possible, physically-based hydrological modeling. In Quebec, Canada, two of the major organizations involved in real-time hydrological forecasting, the *Centre d'expertise hydrique du Québec* (CEHQ: the Quebec Hydrological Expertise Center) and Hydro-Québec (HQ) have undertaken the development of an operational forecasting system for

Calibration of Watershed Models
Water Science and Application Volume 6

several basins located in the southern part of the province. These basins have quick hydrological responses (less than 2 days) and dam systems which need timely stream flow forecasts for management purposes. To meet this requirement, the forecasting system relies on a distributed, and physically-based (with some conceptual approaches remaining), hydrological model – HYDROTEL [Fortin et al., 2001a]. To facilitate the implementation of the forecast system on these basins, there is a need to develop and implement a model calibration procedure that requires minimal labor involvement while accounting for hydrological processes, multiple objectives, and model structure.

A calibration strategy accounting formally for model structure and, hence, modeled processes and their mathematical representations, is likely to improve our understanding of the origins of the embedded parameter interactions and ultimately reduce computational requirements. This latter goal is highly important since the computational requirements associated with the use of HYDROTEL are somewhat larger than those of other models used operationally for hydrological forecasting–particularly if a calibration procedure requiring several simulation runs were used (e.g., the multiple–objective formulation of Madsen [2000] or Gupta et al. [this book]). Moreover, like other researchers [e.g., Boyle et al., 2000; Madsen, 2000], we believe that an automatic calibration strategy based on a single objective does not necessarily lead to satisfactory hydrological calibrations. Indeed, during a manual calibration, hydrologists frequently and spontaneously adjust their calibration strategy with respect to more than one objective without formally documenting their approach.

This paper is organized in three sections (sections 2–4). Section two puts into perspective our work with respect to recent developments on calibration strategies. Section three lays down the concepts and theoretical foundations behind the proposed calibration strategy. It is noteworthy to point out that this paper solely focuses on the development of the proposed calibration strategy within the context of a manual calibration and that the development of a more automatic yet interactive calibration, including an optimization procedure, will be reported in a future publication. Finally, section four focuses on a preliminary application of the proposed calibration strategy with the hydrological model HYDROTEL.

2. BACKGROUND

Over the course of the last decade, a great deal of research and development has been done to answer the challenges of model calibration. Mathematical optimization strategies based on genetic algorithms [e.g., Wang, 1991] or a combination of these with a direct search method such as the shuffled complex evolution (SCE–UA) algorithm of Duan et al. [1992], stochastic approaches such as the simulated annealing approach of Sumner et al. [1997], and Bayesian approaches such as those of Bates and Campbell [2001] and Thiemann et al. [2001] were developed. Intercomparison studies of some of these strategies highlighted the strengths and advantages of SCE–UA, when applicable [Sorooshian et al., 1993; Gan and Biftu, 1996; Kuczera, 1997; Thyer et al., 1999]. Other studies focused on assessment criteria or goodness–of–fit measures [Legates and McCabe, 1999] and convergence criteria [Isabel and Villeneuve, 1986]. General calibration methods based on maximum likelihood functions [Sorooshian et al., 1983; Beven and Binley, 1992], multiple objective strategies [Yapo et al., 1998; Madsen, 2000; Boyle

et al., 2000], process oriented strategies [Harlin, 1991; Zhang and Lindström, 1997], and a priori estimation of model parameters [Eckhardt and Arnold, 2001] were developed. Moreover, specific problems and issues related to distributed models were discussed by Refsgaard [1997] and Ambroise et al. [1995]. Regionalization techniques based on regression analyses between watershed characteristics and model parameters were also studied by Fernandez et al. [2000] and Yokoo et al. [2001].

In our view, there exist at least two promising strategies in model calibration that need to be exploited and based on: (i) formal multiple objectives, and (ii) hydrological processes Regarding multiple objectives the works of Yapo et al.[1998], Madsen [2000], and Boyle et al. [2000], among others discussed in this book (e.g. Gupta et al. [this volume]; Boyle et al.[this volume]; Meixner and Bastidas [this volume]) are worth examining at this point. These authors proposed calibration strategies based on optimization of multiple objectives. These strategies provide a set of Pareto optimal solutions that are all equally good solutions in the sense they either provide an optimal solution for one of the objectives or a particular combination of objectives. This means that the ensuing forecasting of stream flows must be done according to a potential ensemble (or scenario) of parameter values included in the Pareto set. This may not meet the requirements of several operational forecasting systems which normally produce a single forecast not an ensemble of potential forecasts. The other problem we can point out is that these methods are computationally intensive.

On the other hand, Harlin [1991], Zhang and Lindström [1997] and Khu [1998] developed process–oriented calibration strategies for the HBV and NAM hydrological models, respectively. Harlin [1991] directly linked the period of the year where specific processes, and their corresponding parameters, dominate, and evaluated for these periods specific statistics. Meanwhile, Zhang and Lindström [1997] modified Harlin's strategy by further accounting for relationships between parameters (and the equations that include them) and by considering impacts on simulated hydrographs. We think that these strategies represent a good starting point and something to build on. Nevertheless, neither Harlin [1999], Zhang and Lindström [1997] or Khu [1998] provided specific instructions on how to navigate between objectives. As mentioned by Gan and Biftu [1996] and briefly tested by Khu [1998], the subdivision of the calibration exercise into several sub–problems, allows for the use of simple optimization methods. The underlying hypothesis of the proposed calibration strategy assumes that it is possible to meet multiple objectives using a process–oriented calibration approach.

3. PROPOSED CALIBRATION STRATEGY

3.1. Basic Concepts

The goal of the proposed strategy is to simultaneously, and as closely as possible, satisfy a certain number of objectives while accounting in an optimal way for model structure. A hydrological model simplifies to a certain extent real–world processes and, thus, it would be wise to exploit them during the calibration exercise as long as these simplifications are integral parts of the model.

For this presentation, we shall identify calibration objectives under two groups: (i) those related to model responses with respect to characteristic time scales of modeled processes and (ii) those related to periods of the year where specific processes dominate. It is likely that some parameters may affect more than one objective while others may have a limited impact. For example, increasing soil water storage capacity will affect total water balance, while modifying surface roughness will solely affect hydrograph shape. Following these observations, it would make sense to calibrate first those parameters affecting the largest numbers of objectives while keeping for last the least impacting parameters. This strategy assumes that earlier adjustments will not be disrupted by the last series of adjustments and if so they will not significantly compromise the earlier objectives that were met. Therefore, we believe that a sequential adjustment of model parameters is promising; starting with those affecting objectives characterized by long time scales and terminating with those characterized by short time scales. For those parameters affecting intermediate time–scale objectives, we propose to end with those related to the snowmelt process.

3.2. Hydrologic Processes and Model Parameters

Numerous calibration objectives have been reported in the literature. For example, calibration of the well known HSPF model is usually obtained after the following objectives are met [see Jacomino and Fields, 1997]: (i) annual water balances, (ii) monthly water balances, and (iii) other short time scale objectives. This approach includes the main calibration objectives. Having mentioned that, it might be interesting to further detail these objectives to cash in, in an optimal way, on the possibilities of the model.

It is noteworthy that, in general, the number of objectives is less than the number of model parameters. This means that there is a group of parameters for each sought after objective.

3.2.1. Prolonged summer drought recessions. Under the hydrometeorological conditions of southern Quebec, the characteristic time scales of prolonged summer droughts are of the order of a few weeks. Prolonged summer droughts undeniably affect large–scale water balances that are mostly controlled by the evapotranspiration process (see section 3.1.2). Hence, the intensity of the corresponding base flows will have an impact on available soil moisture. The more intense summer base flows are the larger the large–scale water balances will be and the smaller the soil moisture level and evapotranspiration will be. The necessity of matching observed and simulated base flows to close large–scale water balances has been systematically addressed by Szilagyi and Parlange [1999]. Madsen [2000] also showed using numerical simulations that small departures from observed large–scale water balances resulted in small differences between observed and simulated low flows.

3.2.2. Annual and summer flow volumes. The evapotranspiration process is linked to short time scales but its overall influence is generally felt on flows characterized by large time scales. Evapotranspiration depends, among others, on variables related to available soil moisture and meteorological conditions – where the later conditions can change very quickly. This is why we prefer to calibrate first those parameters related to prolonged summer drought recessions since this directly affects available soil moisture. However, as far as summer high flows (i.e., summer runoffs) are concerned, the indirect effect of evapotranspiration is too important to contemplate the idea of calibrating the parameters related to these flow conditions before adjusting the parameters controlling the evapotranspiration process. Thus, calibration of the evapotranspiration process is the focus of this second calibration objective.

3.2.3. Summer and fall high flows. During the summer, the bulk of high flows is mostly controlled by the infiltration capacity and the total soil water content–where the later depends on large time–scale water volumes already adjusted with prolonged summer drought recession and evapotranspiration. This means we can separately calibrate those parameters controlling individual high flows (e.g., soil water profile and infiltration capacity) by analyzing summer and fall high flow events. It is important to underline that a large variety of high flow intensities be part of this analysis.

3.2.4. High flow synchronization. The differences between the shape of observed and simulated hydrographs can be reduced by adjusting the parameters controlling the transfer or routing of water within a watershed. Short time scales usually characterize this transfer which is strongly influenced by surface roughness. This adjustment tends to be performed once large time–scale objectives are met and, thus, the related parameters may be quasi optimal at the beginning of this calibration step.

3.2.5. Winter recessions. In general, it is possible to build on the calibration of other parameters affecting the snow–free periods of the year (and related processes) to adjust independently the parameters controlling snowmelt. Soil water contribution and melting at the snow–soil surface interface represent the two processes governing winter recessions. In theory, it should be possible to reduce the errors between observed and simulated winter low flows by solely adjusting the parameters controlling melting at the snow–soil surface interface while keeping in mind that measured winter flows are usually inaccurate. This is primarily due to the phenomena of ice formation and movement which strongly disrupt flow measurements. For this reason, we have decided to forego this calibration step, although a winter recession objective remains a sound objective from a theoretical point of view.

3.2.6. Spring runoffs resulting from snowmelt. Once the calibration of the parameters controlling the shape of the hydrographs and the melting at the snow–soil interface is completed successfully, we can start the calibration of those parameters controlling the melting of the overall snowpack. The degree of soil saturation resulting from the melt is a function of how much snow has melt up to that point in time (i.e., the so called memory of the watershed). It does not depend on the areal distribution of the melt itself. Therefore, it is easy to separate the evolution of the snowmelt, on a computational time–step basis, from other modeled processes.

3.3. Objective Functions. The proposed calibration strategy requires selection of an objective function to compute a numerical measure of the error between simulated output and observed watershed output for each objective. This approach allows for a continuous assessment of model performance throughout the calibration process.

For prolonged summer drought recessions, we propose the use of an objective function that assesses the relative differences between simulated and observed flows over the course of the summer period. Furthermore, within the summer period, we only consider time intervals where daily stream flows are continuously decreasing (e.g., seven days) and less than a specified watershed threshold value. An interactive choice rather than a fully automatic selection of recession intervals is preferable. The objective function which we refer to as RV–R (recession volume residuals) must converge towards

zero. Similarly, we propose the use of such an objective function to evaluate the relative differences between simulated and observed annual and summer flows. We refer to these functions as AV–R and SUV–R, respectively.

For the assessment of high flow objectives, we select the square root of the mean square error (RMSE). For summer and fall high flows, we only consider time intervals of the snow–free period where daily stream flows are larger than a specified watershed threshold value but again this automatic choice must be validated interactively by the user. These objective functions are referred to as SUF–RMSE (summer/fall) and WSP–RMSE (winter/spring).

Calibration of high flow synchronization represents the final step in the calibration of summer and fall objectives. At this point, calibration of all other summer and fall objectives should be quasi optimal. The RMSE could then be used for the overall calibration period. As pointed out by Legates and McCabe [1999], this kind of coefficient is more sensitive to high–flow errors than low–flow errors, hence, well suited to highlight synchronization problems.

4. PRELIMINARY APPLICATION OF THE PROPOSED STRATEGY WITH HYDROTEL

4.1. HYDROTEL

HYDROTEL [Fortin et al., 2001a] is a physically–based, distributed, hydrological model which was designed to use available remote sensing and GIS data. The model consists of six computational modules which are run in a cascade (i.e., in a decoupled manner) at each time step. These modules are: estimation of meteorological variables, snow accumulation and melt, potential evapotranspiration, vertical water budget, flow on relatively homogeneous hydrological units (RHHUs), and channel routing. The RHHU, which is the computational unit used to calculate the vertical water budget, corresponds to a very small drainage unit, delineated using a digital elevation model (DEM) and a digital river and lake network [Turcotte et al., 2001]. Although Fortin et al. [2001a] provide a detailed description of the model, we herewith take a few paragraphs to describe key modules and their calibration parameters.

Two options are available for interpolation of meteorological variables on each RHHU: Thiessen polygons and weighted mean of three nearest meteorological stations with due care for vertical gradients. Where a threshold air temperature which is used for separation between solid and liquid precipitations is considered as a calibration parameter (see Table 1).

A mixed, degree–day–energy–budget, approach is used to simulate daily variations of mean snowpack characteristics (thickness, water equivalent, mean density, thermal deficit, liquid water content, and temperature). This approach

Table 1. Selected algorithms and model parameters for operational stream flow forecasting in southern Quebec.

Modules	Algorithms	Calibration parameters
Interpolation of meteorological variables	Weighted mean of three nearest meteorological stations	- Threshold air temperature for separation between solid and liquid precipitations (TPPN)
Snowpack evolution	Mixed degree-day-energy-budget approach	- Melt factors for open areas, deciduous forests and coniferous forests (FF-O, FF-F, FF-R) - Threshold air temperatures for melt for open areas, deciduous forests and coniferous forests (SF-O, SF-O, SF-R) - Melt rate at the snow-soil interface (TFSN) - Compaction coefficient (CC)
Potential evapotranspiration	Hydro-Québec (Fortin, 2000)	- PET multiplication factor (FETP)
Vertical water budget	(BV3C)	- Depth of the lower boundaries of the three soil layers (Z1, Z2, Z3) - Recession coefficient (CR)
Flow on RHHUs	Kinematic wave equation	- Manning's roughness coefficient for forest areas, open areas, and water (n-F, n-O, n-E)
Channel routing	Diffusive wave equation	- Manning's roughness coefficient for rivers (n-R)

requires melt factors, compaction coefficients, and threshold air temperatures for melt. They are considered calibration parameters.

An empirical equation developed by Hydro–Quebec which solely requires air temperatures [Fortin, 2000] may be used to estimate potential evapotranspiration (PET). This equation is particularly useful in cases where only air temperatures are available. Despite this crude approximation, this equation has withstood remarkably well the test of time under Quebec conditions. A multiplication factor is currently used as calibration parameter to adjust PET values.

The vertical water budget allows for partitioning between surface runoff, soil water redistribution, and actual evapotranspiration (AET). The computational algorithm requires division of the soil column of a RHHU into three layers (see Figure 1) where each soil layer may be considered as a reservoir with physical proprieties such as saturated hydraulic conductivity; wilting point and drainage capacity in terms of soil water content; porosity; and matrix potential at saturation. Note that each RHHU is also characterized with physiographic properties such as slope and land use percentages. All these characteristics are not generally adjusted. Soil water redistribution and surface runoff strongly depends on soil layer depths. These depths along with the base flow recession coefficient are, thus, considered as calibration parameters.

The downstream transfer of available water at each time step within a RHHU, as computed by the vertical water budget, is simulated using a geomorphological unit hydrograph (GUH) accounting for the internal drainage structure of each RHHU. The shape of this unit hydrograph is determined by routing a reference depth of water over all DEM cells of a RHHU according to a kinematic wave model. The model accounts for the topographic and land use characteristics of the RHHU. Channel routing is performed using a diffusive wave model. When the river segment associated with a RHHU is a lake or reservoir, the continuity equation is used along with stage–discharge relationships for water routing. All these flows are strongly dependent on values of the Manning's roughness coefficient and these values are calibrated.

4.2. Customization of the Proposed Calibration for HYDROTEL

As mentioned earlier, there is no coupling in HYDRO-TEL between the downstream surface transfer of available water and the vertical budget on a RHHU. Moreover, the effect of snow and the calibration of snow characteristics can be easily circumscribed to a short period of the year. However, the calibration of prolonged summer drought recessions, annual and summer flow volumes, and summer and fall high flows strongly depends on those parameters controlling the evapotranspiration process and the vertical water budget (Z1, Z2, Z3 and CR). Despite this inherent coupling, we think it is possible to assign specific groups of parameters to each one of the above calibration objectives.

Without loss of continuity – detailed explanations will follow – the recession coefficient (CR) is associated with the prolonged summer drought recession objective; the

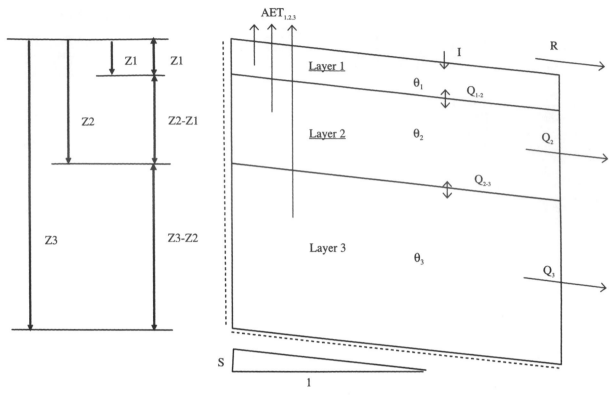

Z1, Z2 and Z3 : depths of the lower boundary of soil layers 1, 2 and 3

Q_2 and Q_3 : outflows of soil layers 2 and 3

Q_{1-2} and Q_{2-3} : water fluxes between soil layers 1-2 and 2-3

θ_1, θ_2 and θ_3 : water content of soil layers 1, 2 and 3

R : surface runoff

S: slope

I: infiltration

$AET_{1,2,3}$: actual ET of soil layers 1, 2, and 3

Figure 1. Schematic diagram of the vertical water budget algorithm.

depth of the third soil layer and the multiplication factor for PET (Z3 and FETP) are associated with the annual and summer flow volume objective; the depths of the first and second soil layers (Z1 and Z2) are associated with the summer and fall high flow objective. Note that any gain in model performance obtained through adjustments of CR must be conserved. Hence, any loss in model performance due to adjustments to Z3 and FETP should be compensated by readjusting CR. Similarly, any loss in model performance due to adjustments to Z1 and Z2 should be compensated by readjusting Z3, FETP and CR. Repeating this calibration process until a satisfactory model performance is reached. The computational requirements associated with this iterative process which aims at a simultaneous attainment of multiple objectives are somewhat minimized through the fact that it is not necessary to globally iterate over all objectives. Figure 2 introduces a schematic representation of the proposed approach.

4.2.1. Prolonged summer drought recessions. A theoretical analysis and several ad–hoc simulation trials have shown that the summer recession is mostly controlled by flows out of the second and third soil layers. These flows are governed by the following equations:

$$Q_3 = CR \; (Z3 - Z2) \; \theta_3 \qquad (1)$$

where

Q_3: flow out of the third soil layer [$L^2 \cdot t^{-1}$]

CR: recession coefficient [$L \cdot t^{-1}$]

Z2: depth of the lower boundary of the second soil: layer [L]

θ_3: water content of the third soil layer at saturation [$L \cdot L^{-1}$]

Z1: depth of the lower boundary of the first soil layer (i.e., soil layer thickness) [L]

$$Q_2 = K \; (\theta_2) \; \sin(\tan^{-1}) \; (S) \; (Z2 - Z1) \qquad (2)$$

where

Q_2: flow out of the second soil layer [$L^2 \bullet t^{-1}$]

K: hydraulic conductivity as a function of soil water content [$L \cdot t^{-1}$]

Z3: depth of the lower boundary of the third soil layer [L]

θ_2: water content of the second soil layer at saturation

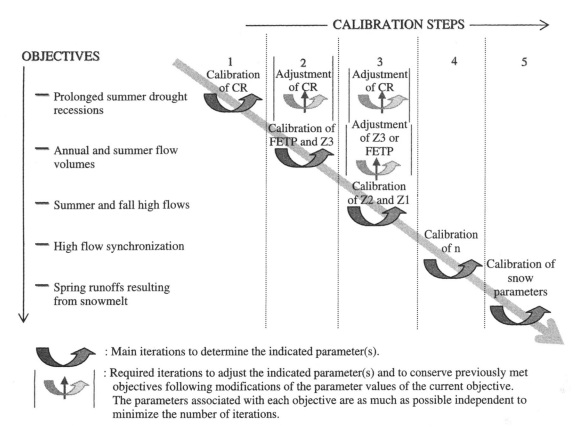

Figure 2. Schematic diagram of the proposed calibration approach.

[$L \cdot L^{-1}$]
S: slope of the RHHU [$L \cdot L^{-1}$]

Calibration of the recession period should be obtained by simultaneous adjustments of the above parameters in equations (1) and (2), that is CR, Z1, Z2 and Z3. The flow intensity also depends, albeit to a smaller extent, on the soil water contents of these two soil layers which depend themselves on evapotranspiration and, thus, on FETP. In principle, the calibration could be performed by determining one or several combinations of parameter values all equally good. However, this is nothing but an over parameterization of the recession. A close look at equation (1) reveals that perhaps the recession could be calibrated by solely adjusting CR. This should always be the case except for cases where the contribution of the second layer is too important, that is, when the soil thickness of the second layer is too large so the flow can in no way be compensated by reducing CR.

To circumvent this problematic situation, we suggest to start the calibration by choosing representatives values of Z1, Z2, Z3 and FETP with respect to other model applications [Fortin et al., 1998, 1999, 2001b]. In other words, this means that Z3 should correspond to the depth actually contributing to summer base flow. To avoid the aforementioned

problem, the thickness of the second layer should be bounded as well.

4.2.2. Annual and summer flow volumes. As previously mentioned, the depth of the third soil layer and the multiplication factor for PET (Z3 and FETP) are associated with the annual and summer volume objective. Since the proposed calibration strategy requires iteration by definition, we must also add that this objective also depends on CR, Z2 and Z1 values. These affect the base flow rate and, indirectly, the available soil moisture needed for the evapotranspiration process. Nevertheless, FETP and Z3 represent the primary parameters for this second calibration objective with the former parameter having a more significant effect than the latter parameter. Now as far as FETP is concerned, it raises a peculiar problem since it is impossible to increase summer PET while reducing winter PET. It is so because the actual algorithm does not offer this possibility yet. It only allows for global and systematic, time independent, adjustments. This is why we recommend to find combinations of optimal values of FETP and Z3 that will minimize the residuals during summer months only. This usually produces reasonable and satisfactory intra–annual modulations of the evapotranspiration process. In passing, to preserve earlier gains in model performance for the recession objective, it is noteworthy any adjustments of FETP and Z3 values will require a posteriori adjustments on CR.

4.2.3. Summer and fall high flows. In addition to the depths of the first and second soil layers (Z1 and Z2), the hydraulic conductivity of the second soil layer at saturation represents the other parameter affecting summer and fall high flows. This soil property is used as an upper bound for water infiltration in the second layer as long as the soil water content of the soil surface is not too restricting. Following this conceptualization, a reduction or increase of the first soil layer thickness will create the bottleneck effect that arises when saturation takes place or near saturation conditions occur at the soil surface. On the other hand, the thickness of the second layer affects directly interflow intensity. An amplified influence of the second layer will have a tendency to produce an increased drying rate and, consequently, favor increased water infiltration in the first soil layer. Note that in HYDROTEL, the hydraulic conductivity at saturation is a soil property obtained from soil texture surveys (i.e., database). It is not currently considered as a calibration parameter. At this point we can not isolate or assign these parameters to specific sub–objectives and, thus, we must consider they are tightly coupled. This means the summer and fall high flow objective will be met in a two–dimensional parameter space.

4.2.4. High flow synchronization. The downstream transfer of available water through RHHUs and river segments depends on the value of Manning's roughness coefficient for various land uses and river beds. We propose a multidimensional search within the lower and upper bounds values of the Manning's roughness coefficient. For example, the value of the Manning's roughness coefficient for forest should always be larger than that for open areas. Calibrated values must also be closed to those found in the literature.

4.2.5. Spring runoff resulting from snowmelt. All parameters related to the evolution of the snowpack, including the threshold air temperature for separation between solid and liquid precipitation (TPPN), are calibrated within this objective, although the melt rate at the snow–soil interface may be calibrated under another objective (see section 3.2.5). It is true that a deeper analysis could lead to the identification of sub–objectives related to snow accumulation and melt. Nevertheless, at this point, we favor a simultaneous calibration approach under a single objective. The search for a set of parameters must be done within the lower and upper bounds of physically meaningful values. For example, the melt factor for open areas must always be larger than that for deciduous and coniferous forests.

4.2. Synthesis and Preliminary Calibration Results. Table 2 summarizes the conclusions reached in sections 3 and 4. These conclusions and the following calibration results represent a first step toward a complete definition and validation of the calibration strategy proposed in this paper. These results are preliminary as they illustrate the strengths of the underlying hypotheses and theory.

Table 3 presents preliminary calibration results of an HYDROTEL application on the Chaudiere river basin, southern Quebec, north of the border between Maine and Quebec [Fortin et al., 2001b]. More specifically, this first test was performed for the sub–basin having for outlet the Sartigan dam (see Figure 3). This test is not indicative of the overall model calibration on the Chaudiere river basin. Nevertheless, it was used as a means of exploring the potential of the proposed calibration strategy. A daily computational time step was used for this application. The calibration was performed using three–year long meteorological series (1997–1999) recorded at three stations (Hilaire, Beauceville, and Lac–Mégantic). Without loss of continuity, we present the calibration of the first three groups of parameters. The relative independence between the two other groups of parameters is such that it was not deemed necessary to include their numerical calibration in a first crack at the proposed strategy.

For the first step, we manually calibrated the recession coefficient with a relative residual criteria on summer flow volumes of 1%. This objective was met after five trials. The second step consisted in the manual calibration of Z3 and FETP. True to our theoretical analysis, adjustments of FETP allowed for significant gains in model performance. Minor adjustments to Z3 improved to a lesser extent the intra–annual modulations which resulted in a marginal overestimation of annual volumes (1%). Meanwhile, these adjustments could not lower below 5% an overestimation of summer volumes. Note that each new combination of FETP and Z3 values required posteriori adjustments of CR which, in passing, required two or three trials. Finally, at the third step, only one iteration was done to improve model performance with respect to summer high flows. This iteration led to FETP and CR values which preserved gains in model performance obtained in steps one and two.

These preliminary numerical results are quite acceptable and illustrate a need to pursue our developmental work as all foundation hypotheses passed this first application test with HYDROTEL.

5. CONCLUSION AND FUTURE WORK

This paper introduced the first steps toward a process–oriented, multiple objectives, hydrological calibra-

Table 2. Calibration objectives and parameters for HYDROTEL and corresponding objective functions

Steps	Objectives	Parameters	Objective functions
1	Minimize the errors in prolonged summer drought recessions	Recession parameter (CR)	RV-R : residuals between observed and simulated flow volumes for the summer time steps for stream flows less than Q_e and included in N-day period of continuous decreasing flows
2	Minimize the errors in annual and monthly flow volumes	Parameters controlling flow volumes (FETP, Z3)	AV-R : residuals between observed and simulated flow volumes throughout all periods of the year SUV-R : residuals between observed and simulated flow volumes during the summer
3	Minimize the errors in summer and fall high flows	Parameters partitioning of surface runoff, water redistribution and fluxes within the soil (Z1, Z2)	SUF-RMSE : square root of the mean square error of summer and fall stream flows greater than Q_c
4	Minimize the high flow synchronization errors	Parameters governing transfer rates (n)	RMSE : square root of the mean square error of the overall calibration period
5	Minimize the errors in spring runoff resulting from snow melt	Parameters governing snow melt (FF, SF, CC) and winter recession (TFSN)	WSP-RMSE : square root of the mean square error of winter and spring stream flows

Q_e :   Threshold stream flow for recession conditions
Q_c :   Threshold stream flow for high flow conditions
N   :   Number of continuous days where stream flows meet strong recession conditions

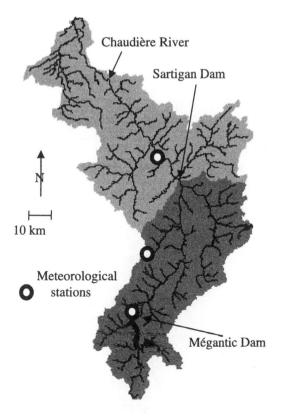

Figure 3. Geographic location of the Sartigan Dam, Chaudiere river basin.

tion strategy accounting for model structure. A preliminary application of the proposed strategy with the hydrological model HYDROTEL [Fortin et al., 2001a] was presented. Test results clearly illustrated the need to pursue our developmental work as all foundation hypotheses withstood this first application with HYDROTEL. Indeed, calibration of the first three groups of parameters associated with the first three objectives (i.e., prolonged summer drought recessions; annual and monthly flow volumes; and summer and fall high flows) was successfully achieved.

In future work, we will further test the proposed groups of calibration parameters. Because the proposed calibration strategy is linked to the characteristic time scales of the modeled hydrological processes, we will conduct a sensitivity analysis in the frequency domain of each parameter. Short time scale parameters should solely affect the high frequencies of the signal. Moreover, for each calibration parameter we will validate the sensitive periods of the year. Following these tests, we will conduct a complete manual calibration of HYDROTEL This exercise will help identify all hidden pitfalls associated with the proposed strategy and understand the general behavior of the chosen objective functions as well as the number of iterations actually required to preserved earlier model performances. As a first guess, we pretend that this number will remain relatively

Table 3. Preliminary test results of a first application of the proposed calibration strategy

Steps	Calibration objectives and objective functions								Iteration
	RV-R (%)*	Prolonged summer recession	AV-R (%)	SUV-R (%)*	Annual and monthly volumes	SUF-RMSE (m³/s)	Summer and fall high flows	Other objectives	
	41.50	CR= 5 e-6			Z3= 2.25; FETP= 1		Z1= 0.05; Z2= 0.25		Initial
1		Iterations for calibration of CR	↰						
	0.95	CR= 1.5 e-6	-3.7		Z3= 2.25; FETP= 1		Z1= 0.05; Z2= 0.25		Final/ Initial
2		Adjustments of CR for each combination of Z3 and FETP ←			Iterations for the calibration of Z3 and FETP ↰				
	0.98	CR= 2.15 e-6	-0.5	4.6	Z3= 2.0; FETP= 1.05	4.4	Z1= 0.05; Z2= 0.25		Final/ Initial
3		Adjustments of CR for each combination of Z3 and FETP ←			Adjustments of FETP for each new combinations of Z1 and Z2 ←		Iteration for the calibration of Z1 and Z2 ↰		
	-0.95	CR= 2.75 e-6	-0.7	4.9	Z3= 2.0; FETP= 1.05	3.8	Z1= 0.05; Z2= 0.20		After one iteration **
4								Calibration of the other parameters ***	

* A 1% tolerance level on residuals was arbitrarily chosen.

** For testing purposes only one iteration was done.

*** In a first approximation, calibration of the other parameters can be done independently.

small. Finally, based on the experience gained through these tests, we will implement an automatic version of the proposed calibration strategy. It is clear that this automatic procedure will need some interaction with the user in order to choose, for example, the right time periods to evaluate a specific objective. To facilitate this interaction, a graphical user interface will be designed.

Acknowledgements. The authors wish to thank Alain Royer of INRS–ETE and Pierre Lacombe of CEHQ for their computer expertise and assistance during the realization of this work.

REFERENCES

Ambroise, B., Perrin, J.L., Reutenauer, D., Multicriterion validation of a semidistributed conceptual model of the water cycle in the Fecht Catchment (Vosges Massif, France), *Water Resources Research*, 31 (6) 1467–1481, 1995.

Bates, B.C., Campbell, E.P., A Markov chain Monte Carlo scheme for parameter estimation and inferences in conceptual rainfall–runoff modeling, *Water Resources Research*, 37 (4) 937–947, 2001.

Beven, K.J., Binley, A., The future of distributed models : Model calibration and uncertainty prediction, *Hydrological process*, 6, 279–298, 1992.

Boyle, D.P, Gupta, H.V., Sorooshian, S., Toward improved calibration of hydrologic models : Combining the strenghs of manual and automatic methods. *Water Resources Research*, 36 (12), 3663–3674, 2000.

Duan, Q., Sorooshian, S., Gupta, V., Effective and efficient global optimization for conceptual rainfall–runoff models, *Water Resources Research*, 28 (4) 1015–1031, 1992.

Eckhardt, K., Arnold, J.G., Automatic calibration of a distributed catchment model, *Journal of hydrology*, 251, 103–109, 2001.

Fernandez, W., Vogel, R.M., Sankasubramanian, A., Regional calibration of a watershed model. *Hydrological Sciences–Journal–des Sciences Hydrologiques*, 45(5), 2000.

Fortin J.P., Turcotte R., Bouffard F., Beaudoin S., Gauthier Y., Bernier M. Perrier, R., Slivitzky, M., Simulation diverses d'apports naturels au réservoir Kénogami, Rapport confidentiel No. R–502. INRS–Eau, Sainte–Foy, Québec. 144 p. + annexes, 1998.

Fortin, J.P., Turcotte, R., Gauthier, Y., Royer, A., Bernier, M., Simulations de crues maximales probables et prévisions des apports sur le bassin de la rivière Mitis par le modèle HYDRO-

TEL, Rapport d'étape, Rapport No R–551a. INRS–Eau, Sainte–Foy, 103 p., 1998.

Fortin, J.P., Turcotte, R., Royer, A., Ajustement du modèle HYDROTEL pour la prévision d'apports sur le bassin de la rivière Mitis, Rapport final, Rapport No R–554. INRS–Eau, Sainte–Foy, 37p., 1999.

Fortin, J.P., Turcotte, R., Massicotte, S., Moussa, R., Fitzback, J., Villeneuve, J.P., A Distributed watershed model compatible with remote sensing and GIS data. Part 1 : Description of the model. *Journal of Hydrologic Engineering, American Society of Civil Engineering.* 6(2), pages 91–99, 2001a.

Fortin, J.P., Turcotte, R., Massicotte, S., Moussa, R., Fitzback, J., Villeneuve, J.P., A Distributed watershed model compatible with remote sensing and GIS data. Part 2 : Application to the Chaudière watershed. *Journal of Hydrologic Engineering, American Society of Civil Engineering.* 6(2), pages 100–108, 2001b.

Fortin, V., Le modèle météo–apport HSAMI: historique, théorie et application, Rapport de recherche, révision 1.5, Institut de recherche d'Hydro–Québec (IREQ), 68 p., 2001.

Gan, T.Y, Biftu, G.F., Automatic calibration of conceptual rainfall–runoff models: Optimization algorithms, catchment conditions, and model structure, *Water Resources Research,* 32 (12) 3513–3524, 1996.

Harlin, J., Development of a process oriented calibration scheme for the HBV hydrological model. *Nordic Hydrology,* 22, 15–36, 1991.

Isabel, D., Villeneuve, J.P., Importance of the convergence criterion in the automatic calibration of hydrological models. *Water Resources Research,* 22 (10), 1367–1370, 1986.

Jacomino, V.M.F, Fields, D.E., A Critical approach to the calibration of a watershed model, *Journal of the American Water Resources Association,* 33(1), 143–154, 1997.

Khu, S.T, Automatic calibration of NAM model with multi–objectives consideration, D2K Technical report 1298–1, National University of Singapore/ Danish Hydraulic Institue, 41 p.

Kuczera, G., Efficient subspace probabilistic parameter optimization for catchment models. *Water Resources Research,* 33(1), 177–185, 1997.

Legates, D.R., McCabe Jr., G.J., Evaluating the use of "goodness–of–fit" measures in hydrologic and hydroclimatic model validation. *Water Resources Research,* 35(1), 233–241, 1999.

Madsen, H., Automatic calibration of a conceptual rainfall–runoff model using multiple objectives. *Journal of hydrology,* 235, 276–288, 2000.

Refsgaard, J.C., Parameterization, calibration and validation of distributed hydrological models. *Journal of hydrology,* 198, 69–97, 1997.

Sorooshian, S., Gupta, V.K., Fulton, J.L. Evaluation of maximum likelihood parameter estimation techniques for conceptual rainfall–runoff models: influence of calibration data variability and length on model credibility. *Water Resources Research,* 19(1), 251–259, 1983.

Sorooshian, S., Duan, Q., Gupta, V.K., Calibration of rainfall–runoff models: application of global optimization to the Sacramento Soil Moisture Accounting model. *Water Resources Research,* 29(4), 1185–1194, 1993.

Sumner, N.R., Fleming, P.M., Bates, B.C., Calibration of a modified SFB model for twenty–five Australian catchments using simulated annealing. *Journal of Hydrology,* 197, 166–188, 1997.

Szilagyi, J., Parlange, M., A geomorphology–based semi–distributed watershed model. *Advances in Water Ressources,* 23, 177–187, 1999.

Thiemann, M., Trosset, M., Guota, H., Sorooshian, S., Bayesian recursive parameter estimation for hydrologic models. *Water Resources Research,* 37(10), 2521–2535, 2001.

Thyer, M., Kuczera, G. Bates, B. Probabilistic optimizations for conceptual rainfall–runoff models : A comparison of the suffled complex evolution and simulated annealing algorithms, *Water Resources Research,* 35(3), 767–773, 1999.

Turcotte, R., J.–P. Fortin, A. N. Rousseau, S. Massicotte, et J.–P. Villeneuve. (2001). Determination of the drainage structure of a watershed using a digital elevation model and a digital river and lake network. *Journal of Hydrology,* 240: 225–242.

Wang, Q.J. The Genetic Algorithm and its Application to Calibrating Conceptual Rainfall–Runoff Models. *Water Resources Research,* 27(9), 2467–247, 1991.

Yapo, P.O., Gupta, H.V., Sorooshian, S., Multi–objective global optimization for hydrologic models. *Journal of hydrology,* 204, 83–97, 1998.

Yokoo, Y., Kazama, S. Sawamoto, M., Nishimura, H., Regionalization of lumped water balance model parameters based on multiple regression, *Journal of Hydrology,* 246, 209–222, 2001.

Zhang, X., Lindström, G., Development of an automatic calibration scheme for the HBV hydrological model, *Hydrological processes,* 11, 1671–1682, 1997.

R. Turcotte, CEHQ, Centre d'expertise hydrique du Québec, 675, boul. René–Lévesque Est, box 20,Québec (Québec), Canada, G1R 5V7

(Richard.Turcotte2@menv.gouv.qc.ca)

A.N. Rousseau, J.P. Fortin, J.P. Villeneuve, INRS–ETE; Institut National de la Recherche Scientifique, Eau, Terre et Environnement, 2800, rue Einstein CP 7500, Québec (Québec), Canada, G1V 4C7

A Multi-Step Automatic Calibration Scheme for Watershed Models

Terri S. Hogue, Hoshin V. Gupta, Soroosh Sorooshian, and Claire D. Tomkins[1]

Department of Hydrology and Water Resources, The University of Arizona, Tucson, Arizona

As evidenced by the papers presented within this monograph, optimization methods have advanced significantly over the last few decades. Although used extensively by the research community, operational hydrologists have been hesitant to implement improved automatic calibration techniques due to previously reported problems with single-step, single-objective optimization. A Multi-step Automatic Calibration Scheme (MACS) is presented which utilizes varying objective functions in a step-by-step approach to optimize parameters for NWS rainfall-runoff models, specifically the Sacramento Soil Moisture Soil Moisture Accounting (SAC-SMA) and SNOW-17 models. Results are presented for operational basins within three National Weather Service (NWS) River Forecast Center (RFC) regions. The Leaf River in Mississippi, the South River in Iowa, and the Flint River in Georgia are calibrated with the MACS procedure and compared against RFC manual calibration. Additionally, the MACS procedure is compared against previously reported calibration methodologies on the Leaf River basin. Parameters obtained with the MACS procedure demonstrate improved, quality calibrations, comparable to RFC simulations and other existing optimization methods. The RFCs are currently in the process of calibrating numerous watersheds to the SAC-SMA and SNOW-17 models. The MACS procedure offers a time-saving, reliable approach for obtaining quality calibrations for forecast points within their area of responsibility.

1. INTRODUCTION

The National Weather Service (NWS), under the direction of the National Oceanic and Atmospheric Administration, is charged with "providing accurate and timely hydrologic information and forecasts for watersheds and rivers throughout the United States" [*Brazil and Hudlow*, 1981]. The NWS River Forecast Centers (RFCs) are the responsible parties for this federal mandate. Thirteen RFCs issue river forecasts for approximately 4,000 locations located throughout the United States [*Ingram*, 1996]. The NWS RFCs are currently in the

midst of a national modernization effort, with the goal of improving their hydrologic forecasts and mitigating the loss of life and property caused by flooding. Congress has allocated funding for this modernization through the Advanced Hydrologic Prediction System (AHPS). As part of AHPS, the RFCs are implementing the NWS River Forecast System (NWSRFS), which includes the Sacramento Soil Moisture Accounting Model (SAC-SMA) and the SNOW-17 model [*Anderson*, 1973] as the main routines for the rainfall-runoff modeling of river systems. As part of the implementation of AHPS, the hydrologic models within NWSRFS (SAC-SMA and SNOW-17) must be calibrated to the numerous river forecast points within each of the RFCs.

Traditional calibration within the NWS has included a sophisticated, highly interactive manual procedure to estimate parameter values. The NWSRFS includes an Interactive Calibration Program (ICP) [*NWS*, 1999] for modelers to evaluate these calibrations both visually and statistically. The hydrologist endeavors to match hydrograph

---

[1] Now at the Department of Management Science and Engineering, Stanford University, Stanford, California.

Calibration of Watershed Models
Water Science and Application Volume 6
Copyright 2003 by the American Geophysical Union
10/1029/006WS12

characteristics such as peak flow, flood volumes, recessions, and base flow. This highly interactive process is time-consuming and labor-intensive, with a typical calibration taking from a few days to a few weeks for an experienced calibrator with thorough knowledge of the watershed system and the rainfall-runoff model. While manual calibration has been the norm in most operational settings, automatic optimization routines have seen extensive use by the research community over the last two to three decades. Several reasons exist for the hesitation by modelers to implement automatic calibration procedures within operational hydrology, including conceptually unrealistic parameter values, poor model performance upon evaluation of the parameters (vs. the calibration period), and the inability of the algorithms to find a "single" optimum parameter set [*Gupta and Sorooshian*, 1994; *Duan et al.*, 1993; *Gupta et al.*, 1998]. Research within the last few years has resulted in the development of global search procedures and multi-objective optimization routines that have resulted in more reliable tools for hydrologists to estimate model parameters via automatic routines [*Brazil*, 1988; *Duan et al.*, 1992, 1993; *Sorooshian et al.*, 1993; *Gupta et al.*, 1998; *Yapo et al.*, 1998]. The Multi-step Automatic Calibration Scheme (MACS) presented here uses these tools. The procedure incorporates the global search algorithm, Shuffled Complex Evolution-University of Arizona (SCE-UA) developed by *Duan et al.* [1992, 1993], and a step-by-step process, all within the NWSRFS, to obtain a "best" parameter set for use in NWS rainfall-runoff models. The goal of the development of the MACS procedure is to provide a time-saving, reliable, automatic calibration technique that is comparable in quality to current RFC practices, and which is available to operational hydrologists as an alternative to the time-consuming manual calibration procedure. With a typically MACS calibration taking 3-4 man hours, the savings in time are significant, allowing the hydrologists to perform other necessary RFC responsibilities. The remaining topics of this chapter include development of the MACS procedure, application of MACS to several operational basins within the NWS RFCs, and a discussion of results and conclusions.

2. DEVELOPMENT OF MACS

2.1 Models

The models calibrated as part of this study are within the NWSRFS and include the SAC-SMA model and the SNOW-17 model (where relevant). The SAC-SMA is a conceptual rainfall-runoff model utilizing 16 parameters (13 of which are typically calibrated) to describe the flow of water through the soil zone (Table 1). The model has been

Table 1. SAC-SMA and SNOW-17 parameter descriptions.

SAC-SMA	Description
UZTWM	Upper zone tension water max. storage (mm)
UZFWM	Upper zone free water max. storage (mm)
LZTWM	Lower zone tension water max. storage (mm)
LZFPM	Lower zone free water primary max. storage (mm)
LZFSM	Lower zone free water suppl. max. storage (mm)
UZK	Upper zone free water lateral depletion rate (day^{-1})
LZPK	Lower zone prim. free water depletion rate (day^{-1})
LZSK	Lower zone suppl. free water depletion rate (day^{-1})
ADIMP	Additional impervious area (decimal fraction)
PCTIM	Impervious fraction of the watershed (fraction)
ZPERC	Maximum percolation rate (dimensionless)
REXP	Exponent of the perco. equation (dimensionless)
PFREE	Fraction of water percolating directly to lower zone free water storage (%)
RIVA	Riparian vegetation (decimal fraction)
SIDE	Ratio of deep recharge to channel baseflow (fraction)
RESERV	Fraction of lower zone free water not transferable to lower zone tension water (%)

SNOW-17	Description
SCF	Snow correction factor (dimensionless)
MFMAX	Maximum melt factor (mm/ C/6 hr)
MFMIN	Minimum melt factor (mm/ C/6 hr)
UADJ	Wind function factor (mm/mb/6 hr)
SI	Water equivalent maximum (mm)
Areal Depletion Curve	
MBASE	Melt base temperature (C)
NMF	Maximum negative melt factor (mm/mb/6 hr)
TELEV	Elevation of temperature series (m)
DAYGM	Average daily ground melt (mm)
PLWHC	Percent liquid water-holding capacity (%)
PXTEMP	Rain/Snow temperature index (C)

Additional Parameters (usually not optimized)	
EFC	Effective forest cover (decimal fraction)
PXADJ	Precipitation adjustment factor (dimensionless)

described extensively in the literature [*Burnash*, 1995; *Hogue et al.*, 2000; *Sorooshian et al.*, 1993] and is also illustrated in other studies throughout this volume [*Smith et al.*, this volume; *Boyle et al.*, this volume]. The SNOW-17 model was originally developed by *Anderson* [1973, 1998] and is used to model snow accumulation and ablation. The model is conceptual, using temperature as an index to the energy exchange occurring in a snowpack and, subsequently, the amount of snowmelt that will occur within a basin. The SNOW-17 model contains 12 parameters, six of which are considered major, having the most impact on snow processes, and six of which are considered minor, having less of an effect on snowmelt (Table 1). The calibration algorithm used for the MACS procedure is the SCE-UA, a search algorithm that has been demonstrated to be effective and efficient in finding the global optimum within the parameter space [*Duan et al.*, 1992, 1993; *Sorooshian et al.*, 1993; *Gan and Biftu*, 1996; *Kuczera*, 1997; *Cooper et al.*, 1997; *Franchini et al.*, 1998; *Freedman et al.*, 1998; *Thyer et al.*, 1999]. The algorithm typically searches for the minimum of the response surface for a single criterion, resulting in a single "best" set solution. Because the SCE-UA has been discussed extensively in previous publications [*Duan et al.*, 1992, 1993], specifics of the search algorithm will not be presented here [*Duan, this volume*]. The SCE-UA is one of six search algorithms available in the automatic OPTimization program [OPT3] within the NWSRFS. OPT3 also contains several choices of objective functions, including Daily Root Mean Square Error (DRMS), sum of the squares of the LOGarithms (LOG), and Heteroscedastic Maximum Likelihood Estimator (HMLE), among others. In development of the MACS procedure, the limitations (specifically, a 16-parameter single calibration maximum) of the existing OPT3 code defined the process that was developed for calibration within the NWSRFS system. Revisions are underway within the OPT3 code to allow for simultaneous calibration of more than the current 16 parameters.

2.2 Methodology

Because the OPT3 system is a single-objective optimization system, a step-by-step process was developed using various objective functions for different parameters (or hydrograph characteristics) to emulate a multi-criteria/multi-objective approach. There are 28 parameters that need to be estimated when calibrating both the SAC-SMA and SNOW-17 models. Given the constraints of the current NWSRFS, only 16 parameters can be optimized in one calibration run. Of the 16 SAC-SMA parameters, three of the parameters, RIVA, SIDE, and RSERV, can typically be set to established values [*Burnash*, 1995]. The PCTIM parameter also can be established from regional maps and local hydrologic information. For the SNOW-17 model, the minor parameters, along with the areal depletion curve, were set at values obtained from the RFC and were not optimized. These parameters can also be estimated from model documentation [Anderson, 1973, 1978] or obtained from historical snow data for the basin. This left the four major parameters (SCF, MFMAX, MFMIN, SI) for calibration. Three additional miscellaneous parameters: EFC, PXADJ and UADJ, were not optimized and set to pre-established literature values. In summary, a total of 16 parameters: 12 from the SAC-SMA model and four from the SNOW-17 model were considered for optimization in this study. The parameters used at each of the MACS steps, along with the objective function chosen for optimization, are detailed in Table 2.

where LOG (Eq. 1) and DRMS (Eq. 2) are defined as:

$$LOG = \Sigma (LOG_{Q_{sim,t}} - LOG_{Q_{obs,t}})^2 \qquad (1)$$

$$DRMS = \sqrt{\frac{1}{n} \left(\sum_{t=1}^{n} (Q_{sim,t} - Q_{obs,t})^2 \right)} \qquad (2)$$

where Qsim,t = simulated flows, and Qobs,t = observed flows at time step t.

The multi-step approach of MACS was designed to follow the NWS manual calibration approach and is described as follows:

Step 1

In the initial calibration phase of a basin, the NWS modeler typically attempts to estimate lower zone (primarily baseflow) parameters of the SAC-SMA. MACS imitates this process by running the initial optimization with the LOG objective function to model recessions and lower flow values. All 16 parameters, 12 of the SAC-SMA and four of the SNOW-17 model (Table 2), are calibrated in this first run. The use of the LOG criterion places strong weighting on the low-flow portions of the hydrograph to provide good estimates of the lower zone parameters. However, by computing the criterion over the entire hydrograph and optimizing all of the parameters, this step also helps to loosely constrain the remaining (upper zone) model parameters into the region that provides coarse fitting of the peaks.

Step 2

The second step of the MACS process emphasizes the estimation of parameters that influence higher flow events.

Table 2. Parameters optimized during MACS.

Model	Step1	Step 2	Step 3
SAC-SMA	UZTWM	UZTWM	--
	UZFWM	UZFWM	--
	UZK	UZK	--
	ADIMP	ADIMP	--
	ZPERC	ZPERC	--
	REXP	REXP	--
	LZTWM	--	LZTWM
	LZFSM	--	LZFSM
	LZFPM	--	LZFPM
	LZSK	--	LZSK
	LZPK	--	LZPK
	PFREE	--	PFREE
SNOW-17	SCF	SCF	--
	MFMAX	MFMAX	--
	MFMIN	MFMIN	--
	SI	SI	--
OBJ. FX.	LOG	DRMS	LOG

Lower zone parameters estimated in the first step are held constant, and a second optimization is run using the DRMS function with ten of the model parameters (Table 2). The DRMS objective function is used to provide stronger emphasis on reproduction of the peak flows. Once these upper zone and snow parameters are estimated, they may be fine-tuned manually or held as estimated, but they are not optimized further. This second step significantly decreases overall percent bias on the study basins.

Step 3

Once parameters are obtained in Step 2, a third calibration is run to fine-tune baseflow parameters with the new upper zone and snow parameters. Only the six SAC-SMA lower zone parameters are optimized again using the LOG objective function (holding the ten parameters from step 2 constant). Once the optimized values are obtained for the parameters, the modeler may fine-tune the estimates manually using local expertise and knowledge of the system.

Step 4

As a final but optional step, a check of monthly biases may reveal trends that call for an adjustment of previously estimated ET parameters. The current version of OPT3 does not allow for automatic optimization of the ET demand curve. A manual fine-tuning or adjustment of these parameters, using monthly errors as a guide, may produce more accurate streamflow during all seasons.

3. APPLICATION OF MACS

3.1 Study Basins

The MACS procedure was originally developed and tested on basins within the North Central River Forecast Center (NCRFC). The watersheds in this region are modeled using both the SAC-SMA along with the SNOW-17 model for snowmelt and are represented as a lumped system (1-elevation band) with a single set of parameters used for the entire basin. The MACS procedure has since been tested by this research group on several operational RFC forecast points within the U.S. representing various hydrologic regimes, including the Southeast River Forecast Center (SERFC), Lower Mississippi River Forecast Center (LMRFC), Colorado Basin River Forecast Center (CBRFC), and Alaska River Forecast Center (AKRFC). The CBRFC and AKRFC watersheds tested with MACS represent a distributed-type modeling system with multi-tiered watersheds (2-3 elevation bands or zones) for a single forecast point. Each elevation band in these systems is represented with its own SAC-SMA and SNOW-17 mod-

els, increasing the dimensionality of the calibration problem three-fold. These results, along with results for all the RFC basins tested to date, are discussed further in *Hogue et al.*, [2002]. Results from the application of MACS to several "lumped" RFC forecast points are presented here, including the Leaf River near Collins, Mississippi (LMRFC), the South River at Ackworth, Iowa (NCRFC), and the Flint River at Culloden, Georgia (SERFC). The Flint and Leaf rivers involve calibration of the SAC-SMA model only, while the South River (headwater of the Des Moines River) involves calibration of both the SAC-SMA and SNOW-17 models. Basin area, mean daily flow, and the time periods used for calibration and evaluation are shown in Table 3 for the study watersheds.

3.2 Results

All three watersheds in this analysis were modeled using a split-sampling technique. A selected period (based on previous analyses of basin climatology and consultation with the RFC) of approximately 11 years was used for optimization, and a final "best" parameter set was obtained. This parameter set was then tested over a longer period of data to evaluate the performance of the calibration methodology. When using the MACS procedure, ranges for parameter bounds are obtained from the RFCs to ensure the calibration procedure obtains values that are "physically realistic", appropriate to the regional hydrology. Along with visual inspection of the hydrograph, several statistics were evaluated, including overall DRMS (Eq. 2), Percent Bias (%Bias) (Eq. 3), and correlation coefficient (Rcoeff) (Eq. 4). Similar to the NWS RFC calibration procedures, monthly %Bias and flow group %Bias were also assessed.

$$\% \ Bias = \left(\sum_{t=1}^{n} Q_{sim,t} - Q_{obs,t} \Big/ \sum_{t=1}^{n} Q_{obs,t} \right) * 100 \qquad (3)$$

$$Rcoeff = \frac{\sigma_{Q_{Obs},Q_{sim}}}{\sigma_{Q_{obs}} \ \sigma_{Q_{sim}}} \qquad (4)$$

The Leaf River basin in Mississippi has been used extensively by this research group (and others) in testing and

Table 3. Basin statistics and data periods used for study.

Basin	Calib. Period	Eval. Period	Area (km^2)	Mean Daily Flow (cms)
Leaf River	1953-63	1956-93	1944	32.41
South River	1971-81	1948-93	1192	7.28
Flint River	1977-88	1950-92	4880	62.80

evaluating calibration techniques with the SAC-SMA model [*Brazil*, 1988; *Yapo et al.*, 1996; *Boyle et al.*, 2000; *Thiemann* et al., 2001]. The LMRFC also has a forecast point at this location and has calibrated the basin using RFC manual calibration techniques. The performances of parameter values obtained from previous calibration methods were also analyzed to compare with MACS and the current RFC parameters, including Brazil's three-stage interactive multi-level calibration procedure [*Brazil*, 1988], the BaRE (Bayesian Recursive Estimation) "maximum likelihood" parameters [*Thiemann et al.*, 2001; *Misrili et al.*, this volume], and the SCE-UA parameter values [*Thiemann et al.*, 2001]. Three of the methods (MACS, SCE-UA, and Brazil-IMC) were all calibrated using the same data period (WY 1953-63), while the BaRE procedure typically uses less data and was calibrated using approximately 1.5 years of this same data period (WY 1953). Table 4 displays parameter values obtained from the various methods, along with the DRMS value for the calibration period. As illustrated, different combinations of parameters have resulted from the various methods. While all of the schemes present acceptable solutions and show similar DRMS values during the calibration period, the SCE-UA and MACS obtain slightly lower errors (%Bias and DRMS). The parameters obtained from the various calibration methods were then tested over a longer timeseries to analyze performance of the parameters.

Evaluating these parameters over a longer historical time frame allows a better indication of overall calibration performance and detection of model divergence [good model performance during calibration period and poor performance during evaluation]. Table 5 displays the statistics for the evaluation period (WY 1956-93) for all calibration methods. The MACS and BaRE methodologies exhibit better performances over this given time period. The MACS has the lowest %Bias (and DRMS) and the highest correlation between simulated and observed streamflows (Rcoeff). BaRE is similar in performance, which is actually quite notable, given that the on-line recursive method used only ~1.5 years of data to find a reliable parameter set [*Thiemann et al.*, 2001]. The SCE-UA, Brazil-IMC, and RFC are all similar in performance, with slightly higher DRMS and %Bias, but still fairly good correlation of modeled and observed flows.

Along with overall %Bias, performance of the calibrated parameters was also evaluated by analyzing %Bias for selected streamflow ranges or intervals (NWSRFS STAT-QME program). As illustrated in Figure 1, all of the calibration methods select parameters that are not as precise at the lower flow interval (1.02-3.25 cms), although the BaRE algorithm performs fairly well in this range. As the flows increase, all of the methods perform fairly consistently and lower %Bias to typically less than ±10%. Figure 2 also shows monthly %Bias for all calibration methods for the Leaf River basin. Again, all of the methodologies show consistency during the late winter and into the early summer months, with lower %Biases (also higher flow season). As the year progresses, it is observed that all of the models have trouble predicting flow (over-simulating) during the late summer and fall months. This is consistent with the flow interval biases seen in Figure 1. This is typically the drier season for this region, and the model [and selected parameters] has difficulty simulating these low flows.

Hydrographs for a portion of one water year (days 180-280) of the Leaf River basin are displayed in Figures 3 and 4. Figure 3 displays the RFC and MACS calibrations against observed flows, while Figure 4 displays SCE, BaRE, and Brazil parameters against observed flows. All of the methods show similar visual performance for most of the evaluation period, including this water year (WY 1962). The RFC and MACS show very similar performance. The RFC over-predicts a flow event around day 210, and both methods under-simulate the event around days 185 and 190. However, both methods catch the general trend of the

Table 4. Comparison of parameters for the Leaf River Basin.

Parameter	RFC	MACS	SCE-UA	Brazil	BaRE
UZTWM	45	52.9	14.089	9.00	33.61
UZFWM	20	55.1	63.825	39.8	76.12
UZK	0.310	0.345	0.100	0.20	0.332
ADIMP	0.05	0.108	0.363	0.250	0.266
LZTWM	120	179	238	240	236
LZFSM	40	71.5	3.19	40	132
LZFPM	100	142	99.8	120	124
LZSK	0.06	0.042	0.019	0.200	0.089
LZPK	0.0065	0.005	0.021	0.006	0.015
ZPERC	55	250	250	250	117
REXP	2.50	4.44	2.46	4.270	4.95
PCTIM	0.005	0.007	0.00	0.003	0.016
PFREE	0.30	0.196	0.021	0.024	0.146
RIVA *	0.01	0.01	0.01	0.01	0.01
SIDE *	0.30	0.30	0.30	0.30	0.30
RESERV *	0.00	0.00	0.00	0.00	0.00
DRMS	**	19.6[a]	18.2[a]	20.3[a]	21.8[b]

* Set to fixed value
** Calibration period not applicable for RFC calibration

Table 5. Statistics for the Leaf River (WY 1956-93).

	RFC	MACS	SCE	Brazil	BaRE
DRMS	36.72	23.75	28.05	28.88	24.93
%Bias	10.42	3.20	4.08	6.78	3.43
Rcoeff	0.879	0.936	0.911	0.911	0.928

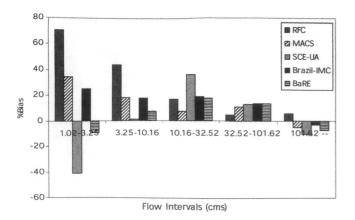

Figure 1. Flow Group %Biases for the Leaf River (WY 1956-93).

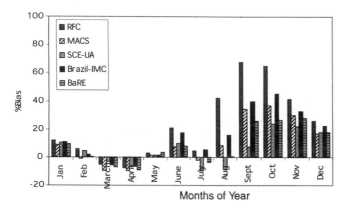

Figure 2. Monthly %Biases for the Leaf River (WY 1956-93).

rising limbs and recessions and match other flow events throughout the year fairly well. In Figure 4, all presented methods (SCE, BaRE, and Brazil) under-simulate the event at day 185. The Brazil and BaRE methods slightly over-predict on day 210. The Brazil parameters tend to catch the

long recession from day 220 to 240 slightly better than the BaRE or SCE. The SCE generally captures peak flows well, but as noted in other automatic single-step batch calibration methods, tends to over-simulate on nearly all recessions. All of these issues become relevant to the hydrologist who uses these parameters in forecasting future flow events with the model.

The South River in Iowa and the Flint River in Georgia were calibrated with only the MACS as part of an overall evaluation of this automated procedure within operational basins (CBRFC, AKRFC, SERFC, and NCRFC). Table 6 displays overall DRMS and %Bias for the evaluation periods for RFC and MACS. Statistical comparisons were not made between RFC and MACS for the calibration period because different time periods were used for calibration. On both the South River and the Flint River, the MACS calibration performs well as compared to the RFC calibrations (similar DRMS values) with MACS having a significantly lower %Bias on the South River basin.

Figures 5 and 6 display flow interval %Bias and monthly %Bias on the South River (headwaters of the Des Moines River in Iowa). Generally, both the RFC and MACS over-simulate on low flows and perform better on higher flow groups (13.8 cms and higher). Both calibrations also tend to over-simulate flow in the wet, spring months. This is proba-bly due to inadequate snowmelt representation within the basin. Other than the months of January and September, the MACS calibration has slightly better performance through-out the year. Figure 7 displays days 100-300 for one year of runoff simulation for the South River (WY 1973). Both the RFC and MACS tend to under-simulate on some peak flow events (days 130 200), but catch the general trend of the flows fairly well. The MACS procedure tends to do a better job of simulating the falling limb and recessions of the most flow events.

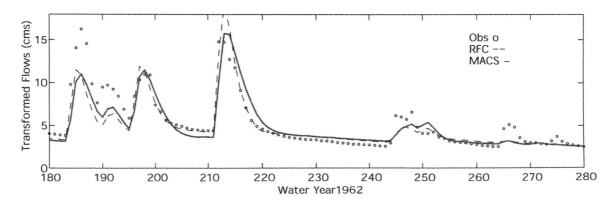

Figure 3. Leaf River basin RFC and MACS calibrations for days 180-280 (WY 1962), where:

$$Transformed\ flow = \frac{(flow + 1)^\lambda - 1}{\lambda}$$

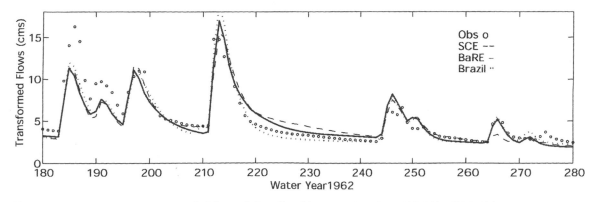

Figure 4. Leaf River basin SCE, BaRE, and Brazil calibrations for days 180-280 (WY 1962), where transformed flows are as described in Figure 3.

Table 6. Statistics for the South and Flint rivers (eval. period).

Basin	DRMS RFC	DRMS MACS	%Bias RFC	%Bias MACS
South River	18.72	15.94	17.84	4.55
Flint River	33.61	29.03	-1.65	-0.63

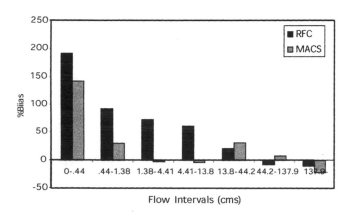

Figure 5. Flow Group %Biases for the South River (WY 1948-93).

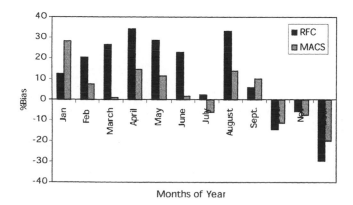

Figure 6. Monthly %Biases for the South River (WY 1948-93).

The MACS procedure was also run on the Flint River within the SERFC (near Culloden, Georgia). Only the SAC-SMA model was calibrated in this region, as snowfall is not a normal part of the region climatology. Figures 8 and 9 show flow interval %Bias and monthly %Biases, respectively, for both the MACS and RFC calibrations. Again, both the MACS and RFC over-simulate on very low flows (1.81-5.65 cms), but perform much better on higher flow intervals (Figure 8). Looking at the performance of the two calibrations throughout the year, except for the fall months of September and October, the MACS calibration performs slightly better and produces lower %Bias throughout the seasons. The hydrograph for days 100-300 from WY 1983 is depicted in Figure 10, and it is evident that both the RFC and MACS perform very well in this basin. Both sets of calibrated parameters catch nearly all of the peak flow events and also simulate recessions better than on the South River in the NCRFC.

4. DISCUSSION AND CONCLUSIONS

The analysis presented in this paper demonstrates the success and applicability of an automated step-wise approach for use in the calibration of watershed systems. The MACS procedure established parameters for the Leaf River, South River and Flint River that provided comparable, and sometimes improved, calibrations to the RFC manual parameters. Of the five calibration methods tested for the Leaf River, the MACS actually provided the lowest overall %Bias and DRMS and the highest correlation between observed and simulated flows. MACS also showed improved overall %Bias and DRMS on the South and Flint rivers. Hydrographs for the South, Flint, and Leaf River calibrations illustrate similar quality simulations with RFC and MACS parameters. The MACS procedure has been tested on several basins within various NWS RFC regions and demonstrates consistency in finding parameter sets that provide

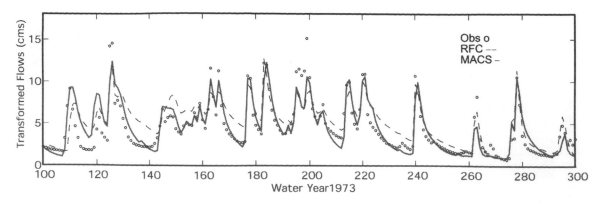

Figure 7. South River basin RFC and MACS calibrations for days 100-300 (WY 1973), where transformed flows are as described in Figure 3.

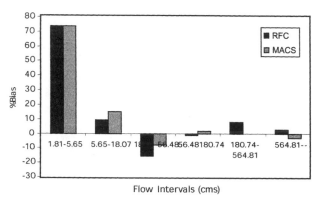

Figure 8. Flow Group %Biases for the Flint River (WY 1950-92).

Figure 9. Monthly %Biases for the Flint River (WY 1950-92)

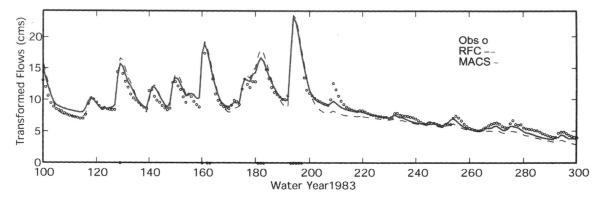

Figure 10. Flint River basin RFC and MACS calibrations for days 100-300 (WY 1983), where transformed flows are as described in Figure 3.

quality, comparable calibrations to RFC manual calibrations, in both lumped and multi-tiered watershed systems.

Under the NWS AHPS modernization effort, the RFCs are under increased pressure to provide timely, quality calibrations for implementation of the SAC-SMA model into the NWSRFS. The obvious advantage of MACS, and other automated technologies, is the savings in time for the operational hydrologist. A typical MACS calibration takes 3-4

hours of personnel time, while a manual calibration may take as long as 2-3 days (or more) for an experienced hydrologist [Holz, 1999, Personal Communication]. MACS can be used to help fine-tune an existing set of parameters or can be used to find an initial set of parameters that can than be fine-tuned using RFC expertise to arrive at a final set of parameters. The MACS procedure was developed within the existing NWSRFS, with current limitations of the coding incorporated into the methodology (specifically the 16-parameter optimization limit). Changes in the coding of OPT3 are underway within the Office of Hydrologic Development, and adaptations in MACS will be tested and reported in due course

The MACS procedure provides reliable, quality calibrations, comparable to the RFC manual calibrations. With over 4,000 river forecast points within the United States to calibrate, MACS is an available, straightforward procedure that can be used as a tool in this task. With the advancement and improvement of automatic calibration techniques and the nearly exponential growth in available computing power, there is little reason for operational hydrologists not to take advantage of existing technology to aid in their calibration efforts. Several studies [Boyle et al., 2001; Hogue et al., 2000] have now shown that automatic procedures can produce comparable, sometimes improved, calibrations to traditional manual calibration techniques. Implementation of these methods into operational hydrology has been slow. It is time for automated calibration technology to become a part of operational hydrologic forecasting routines. Feedback and dialogue on the ongoing development and application of automated calibration procedures to operational watershed systems is encouraged.

Acknowledgements. The ideas for this research rose out of discussions and cooperation with many branches of the NWS, including the Office of Hydrologic Development's Hydrology Laboratory, the NCRFC, SERFC, AKRFC and CBRFC. We especially want to thank Dean Braatz, Andrea Holz, Reggina Garza, Ben Balk, Dave Brandon, Edwin Welles, and Mike Smith for their helpful input and comments. The material in this work was partially supported by grants from the Hydrologic Laboratory of the National Weather Service [Grants NA87WHO582 and NA07WH0144], the National Science Foundation Graduate Research Trainee Program [Grant DGE-935029#3], the Salt River Project Doctoral Fellowship in Surface Water Hydrology, the NASA EOS Graduate College Fellowship, and SAHRA [Sustainability of semi-Arid Hydrology and Riparian Areas] under the STC program of the National Science Foundation, agreement EAR-9876800.

REFERENCES

Anderson, E.A., 1973, National Weather Service River Forecast System – Snow Accumulation and Ablation Model, NOAA Technical Memorandum: NWS Hydro-17, U.S. National Weather Service.

Anderson, E.A., 1978: Snow Cover Energy Exchange–NWSRFS Manual and Documentation II.2.1-1, NOAA Office of Hydrology, U.S. National Weather Service.

Boyle, D.P., H.V. Gupta, and S. Sorooshian, 2000: Towards Improved Calibration of Hydrologic Models: Combining the Strengths of Manual and Automatic Methods, *Water Resources Research*, 36(12), 3663-3674.

Brazil, L.D., 1988: Multilevel Calibration Strategy for Complex Hydrologic Simulation Models, Ph.D. Dissertation, Department of Civil Engineering, Colorado State University, Fort Collins, CO.

Brazil, L.E. and M.D. Hudlow, 1981: Calibration Procedures Used with the National Weather Service River Forecast System. In: *Water and Related Land Resource Systems*, Y. Y. Haimes and J. Kindler [Editors], Pergamon Press, New York, pp. 457-566.

Burnash, R.J.C., 1995: The NWS River Forecast System–Catchment Modeling, V. P. Singh (Editor), *Computer Models of Watershed Hydrology*, Water Resources Publications, Highlands Ranch, CO, pp. 311-366.

Cooper, V.A., V.T.V. Nguyen, and J.A. Nicell, 1997: Evaluation of Global Optimization Methods for Conceptual Rainfall-Runoff Model Calibration, *Water Science and Technology*, 36(5), 53-60.

Duan, Q., V.K. Gupta, and S. Sorooshian, 1992: Effective and Efficient Global Optimization for Conceptual Rainfall-Runoff Models, *Water Resources Research*, 28, 1015-1031.

Duan, Q., V.K. Gupta, and S. Sorooshian, 1993: A Shuffled Complex Evolution Approach for Effective and Efficient Global Minimization, *Journal of Optimization Theory Application*, 76, 501-521.

Franchini, M., G. Galeati, and S. Berra., 1998: Global Optimization Techniques for the Calibration of Conceptual Rainfall-Runoff Models, *Hydrological Sciences Journal*, 43(3), 443-458.

Freedman, V.L., V.L. Lopes, and M. Hernandez, 1998: Parameter Identifiability for Catchment-Scale Erosion Modeling: a Comparison of Optimization Algorithms, *Journal of Hydrology*, 207(1-2), 83-97.

Gan, T.Y. and G.F. Biftu, 1996: Automatic Calibration of Conceptual Rainfall-Runoff models: Optimization Algorithms, Catchment Conditions, and Model Structure, *Water Resources Research*, 32(12), 3513-3524.

Gupta, V.K. and S. Sorooshian, 1994: A New Optimization Strategy for Global Inverse Solution of Hydrologic Models, In: *Numerical Methods in Water Resources*, A. Peters et al. (Editors), Kluwer Academy Press, Norwell, MA.

Gupta, V.K., S. Sorooshian, and P.O. Yapo, 1998: Towards Improved Calibration of Hydrologic Models: Multiple and Non-commensurable Measures of Information, *Water Resources Research*, 34, 751-763.

Hogue, T.S., S. Sorooshian, H. Gupta, A. Holz, and D. Braatz, 2000: A Multistep Automatic Calibration Scheme for River Forecasting Models, *AMS Journal of Hydrometeorology*, 1, 524-542.

Hogue, T.S., C.D. Tomkins, H.V. Gupta, and S. Sorooshian, 2002: Application of a Multi-step Automatic Calibration Scheme to NWS River Forecasting Models, for submission to *Journal of Hydrometeorology*.

Holz, A., 1999: NWS NCRFC Hydrologist, Personal Communication.

Ingram J., 1996: Lesson Taught by Floods in the United States of America, Presented at ICSU SS/IDMDR Workshop on River Flood Disasters, Koblenz, Germany, November, 26-28, 1996.

Kuczera, G., 1997: Efficient Subspace Probabilistic Parameter Optimization for Catchment Models, *Water Resources Research*, 33(1), 177-185.

NWS, 1999: National Weather Service River Forecast System (NWSRFS) User's Manual, NOAA–National Weather Service, Office of Hydrology, Silver Spring, MD.

Sorooshian, S., Q. Duan, and V.K. Gupta, 1993: Calibration of Rainfall-Runoff Models: Application of Global Optimization to the Sacramento Soil Moisture Accounting Model, *Water Resources Research,* 29, 1185-1194.

Thiemann, M., M. Trosset, H. Gupta, and S. Sorooshian, 2001: Bayesian Recursive Parameter Estimation for Hydrologic Models, *Water Resources Research,* 37, 2521-2535.

Thyer, M., G. Kuczera, and B.C. Bates, 1999: Probabilistic Optimization for Conceptual Rainfall-Runoff Models: a Comparison of the Shuffled Complex Evolution and Simulated Annealing Algorithms, *Water Resources Research,* 35(3), 767-773.

Yapo, P.O., H.V. H.HGupta, and S. Sorooshian, 1996: Calibration of Conceptual Rainfall-Runoff Models: Sensitivity to Calibration Data, *Journal of Hydrology,* 181, 23-48.

Terri S. Hogue, Department of Hydrology and Water Resources College of Engineering and Mines, Harshbarger, Bldg. 11, PO Box 210011, University of Arizona, Tucson, AZ 85721-0011

Hoshin V. Gupta, Dept. of Hydrology and Water Resources, College of Engineering and Mines, Harshbarger, Bldg. 11, PO Box 210011, University of Arizona, Tucson, AZ 85721-0011

Soroosh Sorooshian, Dept. of Hydrology and Water Resources, College of Engineering and Mines, Harshbarger, Bldg. 11, PO Box 210011, University of Arizona, Tucson, AZ 85721-0011

Claire D. Tomkins, Dept. of Management and Science Engineering, Terman Engineering Center, 3rd floor, Stanford University, Stanford, CA 94305-402

Hydrologic-Hydraulic Calibration and Testing in an Impacted Flood Plain: Forensic Hydrology

Hugo A. Loaiciga

Department of Geography, University of California, Santa Barbara, California

Hydrologic models contain parameters that are critical to their predictive accuracy. Calibration is a procedure aimed at determining model parameters that reproduce measured variables over a wide range of hydrologic conditions: average, dry, and wet. Four decades of experience in hydrologic model calibration has produced a bounty of experience, and frustration, about the models' limited ability to predict accurately in the presence of highly variable and/or extreme hydrologic inputs and spatially heterogeneous watersheds. Special challenges to effective model calibration and testing arise when the hydrologic system under consideration, be it a watershed or flood plain, undergoes changes so that its input-response characteristics become transient. This situation raises interesting theoretical and practical challenges to the calibration and implementation of a hydrologic model across non-steady hydrologic regimes. This chapter reviews the problem of hydrologic-hydraulic calibration in impacted flood plains and provides an example of the possibilities available and the obstacles posed to hydrologists in this unique setting. The relevance of effective hydrologic model calibration and the testing of its predictive skill are demonstrated within the context of forensic hydrology, a branch of hydrology that supports legal investigations and that deals with the study of flood events with the objective of determining probable causes and sources of human-induced contributions to flood damages.

CHANGING WATERSHEDS

Consider a watershed where stream flow is measured at a specific location. Suppose that land use changes during a time span of twenty years caused by vegetation removal and by conversion of a portion of the watershed from cropland to other functions that render its surface less permeable. The effect of land-use change on the stream flow hydrograph at the gauging location may manifest itself as depicted in Figure 1. For the same rainfall intensity, duration, and spatial coverage, it is seen in Figure 1 that land-use change produces (1) an increasing flow peak, (2) steeper rising and falling hydrograph limbs, and (3) shorter times to peak flows. Although not evident from Figure 1, the type of land-use change being entertained usually produces a larger total volume of stream flow passing through the gauging location. From a water-resources management point of view, the cited land-use raises challenges. For example, the flood stage (h) at any location may rise over time for the same level of stream flow. This is illustrated in Figure 2 by an upward shift of the rating curve (i.e., the flood stage vs. flow function). Likewise, the flood-frequency function is also shifted upwards, as it is shown in Figure 3, wherein the 100-yr flood peak increases from Q^*_{100} to Q_{100}. Upwards shifts of the rating and flood frequency curves of the types shown in Figures 2 and 3, respectively, generally heighten the flood risk [*Loaiciga, 2001*].

TRANSIENT HYDROLOGIC RESPONSE

From a hydrologic modeling perspective, a transient intermediate rainfall-runoff response—such as that shown in Figure 1—poses potentially serious difficulties for model

Calibration of Watershed Models
Water Science and Application Volume 6

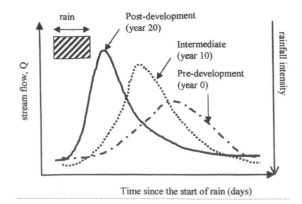

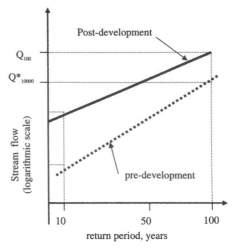

Figure 1. Assumed stream-flow hydrographs at a gauging station at various stages of flood-plain development.

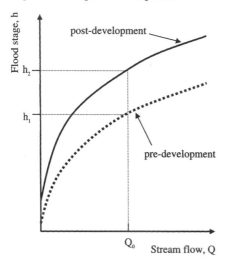

Figure 2. Upward displacement of the rating curve caused by flood-plain development.

Figure 3. The flood-frequency function is displaced upwards by flood-plain changes.

in which | | denotes a norm (for example, the root mean square error defined by the hydrographs x_1 and x_2, see *Amorocho and Espildora* [1973] for a review of closure norms in hydrology).

MODEL CALIBRATION

Hydrologists work with models (H) of the system prototype (J). The former produces an approximate hydrologic response $x^\wedge = H(v^\wedge)$ from the error-corrupted measurement $v^\wedge$ (= v + e, where e is measurement noise) of the input v (see the lower branch of Figure 4). In addition, the system response x is measured with some error. The measurement operator M converts the system response x into the measured value x*, i.e., x* = x + e = M [x] in the upper branch of Figure 4. A hydrologic model H is considered calibrated whenever the following condition holds:

$$\lim_{e \to 0} P\big[\,|x^* - H(v + e)| \leq a\,\big] \to 1 \qquad (2)$$

in which **a** is a positive closure criterion.

Calibration is carried out based on measured inputs and outputs that contain a certain amount of error in them. Rainfall fields are particularly difficult to measure accurately over space and time [*Larson and Peck*, 1974; *Groisman et al.*, 1994]. The same is true of high flows that exceed calibrated rating curves and overflow into adjacent flood plains. Therefore, the hydrologist may be faced with the task of calibrating a hydrologic model when the input data (e.g., rainfall) are biased. A probable outcome is that the model parameters must be assigned unrealistic values in order to match model predictions to measured watershed response. Models that are calibrated in this fashion perform poorly when used

calibration and testing (see *Lapointe et al.* [1998]).Consider the upper branch of the schematic of Figure 4. A (vector, in bold face) input (v, say, rainfall) induces a system response (the vector x, say, the stream flow hydrograph), in which the transformation of v into x is effected by the system response function J, or prototype. Under transient conditions caused by land-use changes, the system response J is time-dependent. Thus, for two times (or time periods, in which case time would be a vector-valued variable) t_1 and t_2, t_2 unequal to t_1, the same input v would produce unequal responses; that is, $x_1 = J_1(v)$ which differs from $x_2 = J_2(v)$ in some suitably defined sense. Taking into account the stochastic uncertainties in inputs and in hydrologic response, the existence of a transient hydrologic response may be stated in probabilistic terms (P[ ] denotes the probability of an argument; a is a positive closure criterion):

$$\lim_{a \to 0} P\big[\,|J_1(v) - J_2(v)| \leq a\,\big] \to 0 \qquad (1)$$

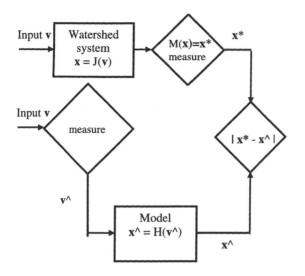

Figure 4. A distributed input **v** and the random outputs **x*** and **x^** produced by the watershed (the prototype) and the watershed model, respectively.

to predict watershed response with rainfall or other stress or state conditions that depart from those used during calibration. In practice, it is common to see the representation of hydrologic processes, the parameterizations of constitutive and empirical relationships, as well as the parameter values themselves, be subject to change during calibration. Although the refinement of a model's structure is necessary during model building, it is not acceptable once the model has been identified. To do so during calibration introduces a faulty circular logic [*Demeritt*, 2001] and raises questions about the model's intrinsic validity. In some instances a hydrologic model may be properly structured, yet, the quality of the input (and/or output) data to the model may be so compromised that calibration becomes an exercise in futility.

MODEL VALIDATION (OR TESTING)

A hydrologic model that is calibrated at time t_1 (= H_1) is validated if it approximates the prototype's response (= **x***) at a time t_2 unequal to t_1. Mathematically:

$$\lim_{e_2 \to 0} P\big[|x^* - H_1(v_2 + e_2)| \le a\big] \to 1 \qquad (3)$$

in which a is a positive closure criterion and e_2 is measurement error.

Under transient watershed conditions, the prototype (or watershed) changes over time. Although the model can undergo successive calibrations with data measured at specific times, model validation as defined in equation (3) is not possible because of the changing nature of the watershed response system. Depending on the degree of change in the

system response, the predictive skill at time t_2 of a model calibrated at time t_1 may be seriously degraded. For example, if the following condition holds, the model's successful predictions would become a type of Bernoulli process:

$$\lim_{e_2 \to 0} P\big[|x^* - H_1(v_2 + e_2)| \le a\big] \to 0.5 \qquad (4)$$

Equation (4) implies that there is at best a 50 % chance of meeting the desired predictive skill. The reader may find a more in-depth analysis of model calibration and validation in *Loaiciga* et al. [1996], as well as a critique of model testing in the earth sciences in *Oreskes et al.* [1994]. *Demeritt* [2001] provides a lucid analysis of model calibration under climate-change forcing.

FORENSIC HYDROLOGY

Forensic hydrology was defined in *Loaiciga* [2001] as " a branch of hydrology that supports legal investigations and that deals with the study of flood events with the objective of determining probable causes and sources of human-induced contributions to flood damages". Forensic flood studies are becoming commonplace in the United States as its flood plains continue to be modified in various ways, with more frequent and severe damages to property being the typical outcome. The following sections summarize a forensic hydrology study that deals with increasing flood damages in an impacted flood plain. Model calibration and testing under transient conditions in an impacted flood plain are illustrated via the case study.

THE STUDY AREA

On March 10, 1995, intense rain fell on the San Luis Obispo Creek watershed of California causing property damages at several sites within the Creek's flood plains and margins. We focus our attention on the lower reach of the San Luis Obispo Creek, where several buildings were damaged by water levels that exceeded the 100-yr flood stage. The San Luis Obispo Creek watershed has a mean annual precipitation of 559 mm and a mild west-coast Mediterranean climate with warm and dry summers (June through September) and wet, cool, season (November through March). There is high inter-annual precipitation variability characterized by unusually intense storms during strong El Niño phenomena and by protracted droughts that may last a decade [*Loaiciga et al.*, 1993]. The drainage area is 217 km² and ground elevations range from sea level to about 800 m. The lower San Luis Obispo Creek underwent numerous physical changes between 1969 –the year of the historical flood on record- and 1995. Those changes included: (1) flood-plain filling (in Avila Farms and RV Park), (2) orchard planting (Avila Farms), (3) levee construction (Avila Farms), (4) greater vegetation density in the creek channel, and (5) bridge construction (Ontario Bridge). Other possible flood-impact factors were considered but ruled out as implausible. One such factor was the possible contribution of tidal backwater effects to high water levels. It turned out that the highest flood level on March 10, 1995, occurred during low tide. A 1994 brush fire was entertained as a likely contributor to increased runoff in March of 1995. It was established, however, that that fire burned only 3% of the watershed, in a remote region in the San Luis Obispo Creek's headwaters. The next section presents a forensic analysis of the contributions of the various flood-plain changes to high water levels on March 10, 1995, and of the procedure to calibrate and test a hydraulic model of the impacted flood plain.

HYPOTHESES

Two hypotheses were put forward to explain the damage-causing flood levels in the lower San Luis Obispo Creek watershed. According to the first one, unusually intense rainfall and wet antecedent conditions caused the high water levels on March 10, 1995. All the buildings that were damaged were located outside the 100-yr flood zone for the lower San Luis Obispo Creek. Thus, according to this hypothesis, the flood of March 10, 1995, exceeded the 100-yr event. The second hypothesis was that the flood of March 10, 1995, was not nearly as large as the 100-yr event, but, rather, that flood-plain changes modified the hydraulic properties of the channel and the right and left overbanks (or terraces, in geomorphologic jargon), thus causing high water levels for a moderate flow peak.

Compounding matters in this case was the fact that there were no paired stream gage records (i.e., time-discharge records) at any location in the study area. The estimated 100-yr flow at the point of interest in the lower San Luis Obispo Creek was estimated to be between 580 and 700 m^3 s^{-1} in a study by the *U.S. Army Corps of Engineers* [1974] following the floods of March 1969. The March 1969 flood was rated as a 100-yr event by the *U.S. Corps of Engineers* [1974]. *Waananen and Crippen* [1977] compiled regionalized statistical equations that predict the 100-yr flood peaks (as well as flood peaks with various other return periods) in terms of drainage area, mean annual precipitation, and elevation index. These equations estimated a 100-yr flood peak equal to 544 m^3 s^{-1} for the lower San Luis Obispo Creek.

RAINFALL ANALYSIS

The maximum 3-hr and 24-hr rainfalls on March 10, 1995, were 28.4 mm and 124.8 mm, respectively. The 3-hr rainfall depth is relevant because the time of concentration in the lower San Luis Obispo Creek is about 2.5 hours. It is not surprising then that the flood peak on March 10, 1995, occurred about 17:30, the center point of the maximum 3-hr rainfall depth. The 24-hr rainfall depth was calculated to provide another reference about the statistical nature of the rainfall events that affected the San Luis Obispo Creek on March 10, 1995. The maximum historical 3-hr and 24-hr rainfalls are 45.7 mm and 155.0 mm, respectively, which occurred in March 1969. It is evident that the damage-causing storms of March 1969 were more severe than those of March 1995. Antecedent conditions were wet in both instances. The National Oceanic and Atmospheric Administration –NOAA- [1972] estimated the 5-yr, 3-hr, rainfall depth in the San Luis Obispo Creek to be 41.8 mm, while the 10-yr, 24-hr, rainfall was estimated at 124.3 mm. These *NOAA* [1972] data indicate that the March 10, 1995, 3-hr maximum rainfall was less than the 5-yr return event, while the 24-hr maximum rainfall on that same date had a return interval of about 10 years. Although no impossible, it is unlikely that rainfall with a return interval of less than five years could have caused a flood peak in excess of the 100-yr event. This is important evidence against the (first) hypothesis that the peak flood of March 10, 1995, was an extremely rare event, an "act of God" in legal parlance. Let us pursue the testing of this hypothesis with a hydrologic-hydraulic model.

FLOOD SIMULATION: A HYDROLOGIC MODEL

The Hydrologic Modeling System (HMS, Hydrologic Engineer Center of the *U.S. Army Corps of Engineers*, [2001]) was used to simulate floods in the lower San Luis Obispo Creek. Rainfall was estimated from radar-reflectivity data and adjusted with rain-gage data. The radar-estimated rainfall was averaged over 2 km x 2 km cells in a grid that extended throughout the entire study area with a 15-minute temporal resolution and input to HMS. The rain gages were sparsely scattered throughout and on the vicinity of the watershed and several were not recording, thus reporting total depths only. Therefore, there was uncertainty in the accuracy of the radar estimated, spatially-distributed, rainfall input to the model (see *Legates* [2000] for further analysis of the accuracy of radar-estimated rainfall). Another shortcoming in carrying out the HMS simulations was the fact that there were no stream-flow measurements with which to calibrate the model parameters. Instead, soil, hydrograph, and routing HMS parameters were assigned (using professional judgment) from observed or inferred hydrologic/hydraulic basin characteristics for March 1995. The San Luis Obispo Creek was subdivided into 59 sub-basins, each of which was assigned soil, hydrograph, and routing HMS parameters. This produced a large parameter set that had to be inferred to carry out the HMS stream flow simulations.

The HMS-simulated flow for March 10, 1995 had a peak flow equal to 595 m^3 s^{-1} and was timed at 18:00, only minutes after the actual peak-flow time (17:30). The HMS-simulated flow peak would put the March 1995 on the order of a 100-yr event, if the U.S. Army Corps of Engineers [1974] estimates quoted earlier were accepted.

A second HMS simulation was carried out, whereby the inferred 1969 watershed conditions were used to assign a new set of parameters (soil, hydrograph, flood routing) and the March 10, 1995, radar-estimated rainfall was used to drive the model. Similar wet antecedent conditions prevailed in March 1969 and March 1995. Recall also that the flood of March 1969 was considered a 100-yr event. The simulated hydrograph for March 1969 watershed conditions was essentially equal to the March 10, 1995, simulated hydrograph. It was stated in the previous section that the maximum 3-hr rainfall was 28.4 mm in March 1995, while it was 45.7 mm in March 1969. Is it possible that the lighter rain of March 10, 1995, could have produced a 100-yr flood peak just as the heavier rain of March 1969 did with 1969 watershed conditions? This would be plausible only if the rainfall losses were large enough in 1969 to exact the same effective rainfall from more intense storms events than those of March 10, 1995. This is unlikely given the wet antecedent conditions that prevailed in all of March 1969, part of a wet El Niño year. Instead, the previous results obtained with HMS suggest that the assigned HMS parameters for 1969 and 1995 conditions were poorly chosen in the absence of complete rainfall-stream flow data with which to carry out thorough model calibration.

The first hypothesis could not be substantiated via modeling in view of the inconsistencies that arose from model simulations with uncertain parameters and rainfall estimates. Therefore, we pursued the second hypothesis, which relies on hydraulic changes (and heavy rain) in the flood plain to explain the high water levels on March 10, 1995.

CALIBRATION OF A HYDRAULIC MODEL

The only reliable and accurate data measured in March 1995 were water levels recorded at various points along the lower San Luis Obispo Creek. Channel and flood plain hydraulic conditions have been estimated for 1969 and 1995 conditions from surveys of vegetative cover, degree of mean-

dering, and other observable features that determine hydraulic roughness. It was decided then to implement a hydraulic model—The U.S. Army Corps of Engineers' HEC-2 model [U.S. Army Corps of Engineers, 1990]—to simulate water levels with the roughness conditions of March 1969 and those of March 1995. The flood peak on March 10, 1995, was unknown. Therefore, a series of HEC-2 simulations had to be run until a flood rate was found that reproduced the measured water levels. This is tantamount to model calibration as described above. March 1995 hydraulic roughness was used for that purpose. The identified flood rate was considered the best estimate of the March 10, 1995, flood peak. Subsequently, that same flood rate was simulated in HEC-2 with the March 1969 hydraulic roughness conditions. The difference in water levels between March 1969 and March 1995 were then attributed to increasing hydraulic roughness between those two periods. The line of reasoning followed was that had the 1969 flood-plain (pre-development) hydraulic roughness persisted through 1995, then, the flood levels would have been much lower in the latter year. Once the March 10, 1995, flood peak was estimated (see next section for actual flood-peak values), the change in hydraulic roughness was apportioned among the various impact factors identified earlier (i.e., vegetation, flood-plain filling, etc.). Finally, the contributions of impact factors to changes in flood levels between March 1969 and March 1995 were determined by HEC-2 simulations.

The approach outlined in the previous paragraph has an appealing simplicity. It involves a low number of parameters to be estimated. Those are limited to flood-plain hydraulic roughness. It bypasses the need to estimate the 1969 flood peak. In addition, it does not involve rainfall estimates, always a major obstacle in watersheds with low-density rain gage networks. By its simplicity, this approach complies with the principle of parsimony, or Occam's razor (after William of Ockham, c. 1285 – c. 1349, a philosopher): "given several possible alternative explanations to an event, the best explanation is the simplest one". The following section presents the results of this parsimonious approach, which, by the way, strengthened our second hypothesis—stated earlier—as the most reasonable explanation to the high flood levels that occurred on March 10, 1995. Recall that such hypothesis proposed that a combination of heavy rain and flood-plain changes in the lower San Luis Obispo Creek between 1969 and 1995 induced the damage-causing water levels on March 10, 1995.

RESULTS OF THE HYDRAULIC SIMULATIONS

Table 1 shows the hydraulic roughness coefficients (Manning's N) used in the HEC-2 hydraulic simulations of water levels in the lower San Luis Obispo Creek. Pre-development (March 1969) and post-development (March 1995) coefficients are reported in Table 1. Table 2 contains the HEC-2-calculated water levels (in m, above mean sea level) at all cross-sections. It also shows that the estimated flow rate downstream of cross section 27+54 was 405 m^3 s^{-1} while it was 377 m^3 s^{-1} upstream of that location, as seen in Table 2. A tributary to the San Luis Obispo at that cross section accounted for the change in stream flow there. These estimated stream flows are much lower than the *U.S. Army Corps of Engineers* [1974] and the *Waananen and Crippen* [1977] estimates of the 100-yr flood peak. *Loaiciga* [2001] argued that they were more consistent with a 50-yr flood magnitude. The post-development (March 1995) water levels reported in Table 2 within cross sections 30+96 and 36+38 were within ± 2 cm of the recorded flood levels on March 10, 1995. The resemblance between calculated and measured water levels confers a reasonable degree of confidence that the estimated flood peak for March 10, 1995, is fairly close to the actual—yet unknown—one. Thus, our second hypothesis concerning the cause of flood damages on March 10, 1995, appears well substantiated by the available evidence and our hydraulic calculations.

Most of the damage that took place on March 10, 1995, along the lower San Luis Obispo Creek was concentrated within cross sections 30+96 and 36+38. Of particular interest is cross-section 36+38, which was chosen as the reference cross section with the purpose of calculating the individual contributions to flood hazard (posed by heightened flood levels) by several impact factors (or flood-plain changes) cited earlier. Valuable property and significant flood damages occurred at and in the vicinity of cross section 36+38.

The pre-development (March 1969) water levels reported in Table 2 provide a baseline from which to ascertain the human-induced rises in flood level between March 1969 and March 1995. If flood-plain conditions had not changed from their pre-development status, the water levels that would have occurred on March 10, 1995 would have not caused damages. For example, the non-damaging water level at cross section 36+38 is 7.68 m. This elevation is slightly below the lower floor level of buildings at that location. Table 2 indicates that if pre-development flood-plain conditions had persisted through March 1969, the water level at cross section 36+38 would have been 6.91 m on March 10, 1995, well below the damage threshold elevation. Instead,

Table 1. Values of N (Manning's N) hydraulic roughness for pre- and post-development conditions in the lower San Luis Obispo Creek, California.

X-section	N values, pre-development, 1969			N values, post-development, 1995		
	LOB[a]	Channel	ROB[b]	LOB	Channel	ROB
22+86	0.025	0.025	0.025	0.11	0.11	0.11
24+78	0.025	0.025	0.025	0.11	0.11	0.11
25+95	0.025	0.025	0.025	0.11	0.11	0.11
26+69	0.025	0.025	0.025	0.11	0.11	0.11
27+29	0.025	0.025	0.025	0.11	0.11	0.11
27+30	0.025	0.025	0.025	0.11	0.11	0.11
27+39	0.025	0.025	0.025	0.11	0.11	0.11
27+40	0.025	0.025	0.025	0.11	0.11	0.11
27+54	0.025	0.025	0.025	0.11	0.11	0.11
28+65	0.035	0.035	0.035	0.13	0.13	0.13
30+96	0.035	0.035	0.035	0.14	0.14	0.14
33+68	0.035	0.035	0.035	0.15	0.15	0.15
35+40	0.035	0.035	0.035	0.15	0.15	0.15
36+38	**0.035**	**0.035**	**0.035**	**0.15**	**0.15**	**0.15**
39+11	0.035	0.035	0.035	0.15	0.15	0.15
40+91	0.035	0.035	0.035	0.15	0.15	0.15
41+51	0.035	0.035	0.035	0.15	0.15	0.15
42+71	0.035	0.035	0.035	0.15	0.15	0.15
43+41	0.035	0.035	0.035	0.15	0.15	0.15
43+42	0.035	0.035	0.035	0.15	0.15	0.15
43+51	0.035	0.035	0.035	0.15	0.15	0.15
43+52	0.035	0.035	0.035	0.15	0.15	0.15
44+84	0.035	0.035	0.035	0.15	0.15	0.15
46+76	0.035	0.035	0.035	0.12	0.12	0.12

[a]LOB: left overbank; [b] ROB: right overbank; cross-section location is measured in m.

Table 2. Calculated water levels (above mean sea level) in the lower San Luis Obispo Creek, California, pre- and post-development conditions.

X-section	Pre-development, 1969[a]		Post-development, 1995	
	Flow Q ($m^3 s^{-1}$)	Water level (m)	Flow Q ($m^3 s^{-1}$)	Water level (m)
22+86	405	4.96	405	7.47
24+78	405	4.98	405	7.73
25+95	405	5.48	405	7.88
26+69	405	5.62	405	7.98
27+29	405	5.61	405	8.05
27+30	405	5.57	405	8.03
27+39	405	5.61	405	8.14
27+40	405	5.75	405	8.22
27+54	405	5.89	405	8.30
28+65	377	6.02	377	8.45
30+96	377	6.51	377	8.75
33+68	377	6.83	377	8.95
35+40	377	6.88	377	9.02
36+38	377	6.91	377	9.06
39+11	377	6.97	377	9.18
40+91	377	7.05	377	9.28
41+51	377	6.99	377	9.35
42+71	377	7.90	377	9.66
43+41	377	8.07	377	9.79
43+42	377	8.07	377	9.79
43+51	377	8.08	377	9.88
43+52	377	8.08	377	9.89
44+84	377	8.19	377	10.16
46+76	377	8.64	377	10.47

[a]Pre-development and post-development conditions were defined by channel roughness in Table 1.

with the hydraulic conditions that prevailed on March 10, 1995, the actual water level was 9.06 m, or 1.38 m above the damage threshold elevation of 7.68 m.

CONTRIBUTIONS TO FLOOD HAZARD: "BEFORE OR AFTER" VS. "WITH OR WITHOUT"

It was stated above that flood-plain changes between 1969 and 1995 caused a water-level rise from 6.91 m to 9.06 m, or 2.15 m. Of this level rise, 1.38 m was above the damage threshold. These elevations correspond to cross section 36+38, the reference location. The hydraulic roughness coefficients were subjected to an incremental analysis based on the flood-plain changes that modified the lower San Luis Obispo Creek between 1969 and 1995. Those changes or impact factors included (1) farming operations (levee, filling in of farm land, orchard plantation, fencing, induced greater vegetation density in the creek channel), and (2) non-farming impacts (bridge construction, filling in for a recreational vehicle park). Farming impact factors were located within the same geo-graphical area. They also preceded non-farming factors in time, which allowed their separate treatment in the apportionment of their contributions to flood hazard as outlined below. As the roughness coefficients were varied, the predicted HEC-2 water levels increased accordingly. The flood-level variation associated with each increment in hydraulic roughness (and hence, with the identified impact factors) was noted and used to calculate the proportional contributions of the impact factors to the flood-level rise.

Table 3 summarizes the results obtained by means of the incremental analysis. It is seen there that 53 % and 47 % of the total water-level change were apportioned to farming and non-farming impact factors, respectively. These contributions were produced by a hydraulic analysis of flood-plain changes that took place between 1969 and 1995.

This type of analysis relied on the "before or after" approach, whereby flood impacts are ascertained starting with a baseline condition (pre-development) and then with post-development conditions following a chronological pathway of flood-plain changes. One could calculate the contributions to flood-level changes using the "with or without approach", whereby the post-development condition is considered a baseline. Impact factors are then dropped one at a time, and new water levels are calculated after a factor is dropped. The percentage contributions of the various impact factors can then be calculated. The "before or after" and the "with or without" approaches yield the same percentage contributions when the sequence in which impacts factors are added in the former equals that in which factor are dropped in the latter. Otherwise, the approaches produce different percentage contributions. Therefore, the chronology of flood-plain changes as well as the geographical locations of those changes, that is, the ability to cluster individual impact factors or the need to separate them in the hydraulic analysis are paramount to the outcome of the flood-hazard analysis.

MODEL TESTING

The testing (or "validation") of the implemented HEC-2 model is beset by the transient nature of the flood plain under consideration, as discussed in earlier sections of this article. In March of 2001, however, there was heavy rainfall in the San Luis Obispo Creek watershed and water levels were measured accurately at two locations in the study area. Furthermore, a newly installed stream gauging station measured the stream flow at a cross section located a few meters above the discharge point of See Creek into the San Luis Obispo Creek. This allowed, for the first time, to test the validity of the HEC-2 model calibrated with the 1995 flood event. To this end, the 1995 hydraulic roughness was adjusted to year 2001 conditions by conducting field observations of the creek's channel and overbanks as of March of that same year. Hydraulic roughness had changed from

March 1995 conditions by several improvements that took place between 1995 and 2001 (e.g., vegetation and debris removal, tree cutting in the farm orchard). The independently measured stream flow rate and the adjusted hydraulic roughness were input into the HEC-2 model and water levels were simulated and compared with the measured water levels. Table 4 shows results. It is seen there that at the two cross sections where water levels were measured (33+68 and 36+38), the HEC-2 calculated and the measured levels are equal. This suggests that with the hydraulic roughness adjustments made in March 2001, the hydraulic model has excellent predictive skill. If the March 2001 flood-plain conditions are maintained, the calibrated and tested model can be reliably used to predict flood levels associated with large stream flows.

SUMMARY AND OTHER IMPORTANT ISSUES

The previous considerations on the type of approach adopted to sort out the contributions to flood hazard by various impact factors cannot be divorced from the view that the Courts have on issues bearing on these type of cases. For example, case law may establish precedents that dictate which approach is likely to prevail in a legal context. Statutory law plays an important role also on what types of hydraulic-hydrologic analyses may be viable in Court. The quality of data, key to determine the weight of evidence, takes a leading role in legal proceedings.

Notice, in addition, that the analysis of flood-hazard contributions presented above does not translate necessarily into liabilities to the various impact factors (or their agents) associated with flood damages caused by the specific flood event considered in this case. That is, the proposed contributions to flood-level rise do not necessarily translate into

Table 3. Flood factors and their contributions to flood level rise, March 10, 1995 (cross section 36+38).

Impact factor	Hydraulic roughness	Water level, h (m)	Change in h (m)	% contribution to change in h
Pre-development condition (1969)	0.035	6.91	---	---
Farming factors	0.085	8.04	1.13 (= 8.04-6.91)	53
Non-farming factors	0.15	9.06	1.02 (= 9.06-8.04)	47
Total			2.15 (= 9.06-6.91)	100

Table 4. Calculated water levels for March 2001 flood, with new N values.

| Cross-section[a] | N, hydraulic roughness values, March 2001 | | | Flow and water level, March 2001 | |
	LOB[b]	Channel	ROB[c]	Flow (m^3s^{-1})	Water level calculated (measured) (m)
22+86	0.09	0.09	0.09	269	4.78
24+78	0.09	0.09	0.09	269	5.78
25+95	0.09	0.09	0.09	269	6.17
26+69	0.09	0.09	0.09	269	6.37
27+29	0.09	0.09	0.09	269	6.50
27+30	0.09	0.09	0.09	269	6.50
27+39	0.09	0.09	0.09	269	6.56
27+40	0.09	0.09	0.09	269	6.59
27+54	0.09	0.09	0.09	269	6.68
28+65	0.11	0.11	0.11	269	6.95
30+96	0.10	0.10	0.10	252	7.46
33+68	0.10	0.10	0.10	252	7.65 (7.65)
35+40	0.10	0.10	0.10	252	7.70
36+38	0.10	0.10	0.10	252	7.73 (7.73)
39+11	0.10	0.10	0.10	252	7.82
40+91	0.10	0.10	0.10	252	7.90
41+51	0.10	0.10	0.10	252	7.94
42+71	0.10	0.10	0.10	252	8.32
43+41	0.13	0.13	0.13	252	8.51
43+42	0.13	0.13	0.13	252	8.51
43+51	0.13	0.13	0.13	252	8.53
43+52	0.13	0.13	0.13	252	8.53
44+84	0.14	0.14	0.14	252	9.01
46+76	0.12	0.12	0.12	252	9.57

[a]Cross-section location is measured in m; [b]LOB: left overbank; [c]ROB: right overbank;.

Court-accepted contributions to flood-damages. Not only can the proposed contributions be challenged on technical grounds, but on legal grounds as well. Besides legal constraints –such as statues of limitation, the intentional or accidental nature of an impact factor, waivers of liability to certain types of agricultural activities, etc.- the ultimate outcome of a civil (or criminal) case related to flood damages rests with juries, and how they perceive the totality of the evidence in any particular set of circumstances.

REFERENCES

Amorocho, J., and B. Espildora, Entropy in the assessment of uncertainty in hydrologic systems and models, *Water Resour. Res., 9*, 1511-1522, 1973.

Demeritt, D., The construction of global warming and the politics of science, *Annals Am. Assoc.Geographers, 91(2)*, 307-337, 2001.

Groisman, P.Y. and D. R. Legates, The accuracy of United States precipitation data, *Bull. Am. Meteor. Soc., 75(3)*, 215-227, 1994.

Lapointe, M.F., Y. Secretan, S.N. Driscoll, and M. Leclerc, Response of the Ha! Ha! River to the flood of July 1996 in the Saguenay Region of Quebec: large scale avulsion in a glaciated valley, *Water Resour. Res., 34(9)*, 2383-2392, 1998.

Larson, L. and E. L. Peck, Accuracy of precipitation measurements for hydrologic modeling, *Water Resour. Res., 10(4)*, 857-863, 1974.

Legates, D.R., Real-time calibration of radar precipitation estimates, *The Professional Geographer, 52(2)*, 235-246, 2000.

Loaiciga, H.A., Flood damages in changing flood plains: a forensic-hydrology case study, *J. Am. Water Resour. Assoc., 37(2)*, 467-478, 2001.

Loaiciga, H.A., L. Haston, and J. Michaelsen, Dendrohydrology and long-term hydrologic phenomena, *Rev. Geophys., 31(2)*, 151-171, 1993.

Loaiciga, H.A., J.B Valdes, R. Vogel, J. Garvey, and H. Schwarz, Global warming and the hydrologic cycle, *J. Hydrol., 174(1-2)*, 83-128, 1996.

National Oceanic and Atmospheric Administration (NOAA), *Precipitation-depth frequency maps for California*, National Weather Service, US Department of Commerce, Washington, D.C., 1972.

Oreskes, N., K. Shrader-Frechete, and K. Belitz, Verification, validation, and confirmation of numerical models in the earth sciences, *Science, 263*, 641-646, 1994.

US Army Corps of Engineers, *HEC-2 Water surface profiles*, Hydrologic Engineering Center, Davis, California, 1990.

US Army Corps of Engineers, *Hydrologic modeling system HEC-HMS Version 2.1*, Hydrologic Engineering Center, Davis, California, 2001.

Waananen, A.O. and J.R. Crippen, Magnitude and frequency of Floods in California. *Water Resources Investigations Report 77-21,* U.S. Geological Survey, Menlo Park, California, 96 pp., 1977.

Loaiciga, Hugo A., Department of Geography, Room 3611 Ellison Hall, University of California, Santa Barbara, California 93106-4060 USA.

Multicriteria Calibration of Hydrologic Models

Douglas P. Boyle

Department of Hydrologic Sciences, Desert Research Institute, University and Community College System of Nevada, Reno, Nevada

Hoshin V. Gupta and Soroosh Sorooshian

Department of Hydrology and Water Resources University of Arizona, Tucson, Arizona

The level of spatial and vertical detail of important hydrologic processes within a watershed that needs to be represented by a conceptual rainfall-runoff (CRR) model in order to accurately simulate the streamflow is not well understood. The paucity of high-resolution hydrologic information in the past guided the direction of CRR model development to more accurately represent processes directly related to the vertical movement of moisture within the watershed rather than the spatial variability of these processes. As a result, many of the CRR models currently available are so complex (vertically), that expert knowledge of the model and watershed system is required to successfully estimate values for model parameters using manual methods. Newly available, high-resolution hydrologic information may provide insight to the spatial variability of important rainfall-runoff processes. However, effective and efficient methods to incorporate the data into the current modeling strategies need to be developed. In this work, we use a new hybrid multicriteria calibration approach to investigate the benefits of different levels of spatial and vertical representation of important watershed hydrologic variables with CRR models.

1. INTRODUCTION AND BACKGROUND

Conceptual rainfall-runoff (CRR) models have become widely used for streamflow forecasting as the demand for timely and accurate forecasts has increased. CRR models provide an approximate, lumped description of the dominant sub-watershed scale processes that contribute to the overall watershed scale hydrologic response of the watershed system. In their most basic form, CRR models transform rainfall into runoff with two main components, precipitation excess generation and flow routing. Precipitation excess is generated as a function of the vertical movement of moisture (precipitation, evaporation, transpiration, and losses to the system) into and out of the watershed. The flow routing component involves the movement of the excess precipitation over the land surface and along stream and channel networks to the outlet of the watershed.

The variability of the excess generation process within the watershed is related to the level of spatial variability of the soil properties, vegetation type, and precipitation rates throughout the watershed. In the past, high-resolution information describing these characteristics was not readily available. As a result, performance improvements of CRR models were primarily focused on improving the representation of processes directly related to the vertical movement of moisture within the watershed rather than the spatial variability of these processes. Now that remotely sensed, high-resolution, hydrologic data are now becoming available in the United States through a variety of different sources, The incorporation of these high-resolution data sets, in particular, the NEXRAD stage III data, into current modeling procedures is considered highly desirable by hydrologists. The

Calibration of Watershed Models
Water Science and Application Volume 6
Copyright 2003 by the American Geophysical Union
10/1029/006WS14

development of efficient and effective methods for doing so is an active area of research.

A simple method to incorporate the high-resolution data into the modeling process is to average the information over the entire area of the watershed and proceed with the current lumped model application. The main advantage of this approach is that the existing modeling structure does not need to be modified to use the new data. The main disadvantage is clearly the loss of the spatial distribution of information as well as the potential for further modeling improvement and understanding. Another, strategy is the "semi-distributed" approach in which the watershed is partitioned into a network of hydrologic units based on the spatial variability of the precipitation. The main disadvantage of this strategy is the increase in model complexity and parameters parallel to the increase in partitioning. For complex, highly parameterized models, as the number of hydrologic units is increased, the calibration procedure quickly becomes intractable. Further, many of the parameters may not be supported (identifiable) by the information contained within the observed data, remotely sensed or otherwise.

The hydrologic modeling problem can be partitioned into three main components; hydrologic model structure, hydrologic data, and parameter estimation procedures. Successful development and application of any hydrologic model requires careful consideration of each component and its relevance to the overall modeling problem. In the following sections, a new hybrid multicriteria calibration approach that combines the strength of automatic and manual calibration methods is presented and used to investigate the benefits of representing different levels of spatial and vertical representation of important watershed hydrologic variables within CRR models.

2. MUTICRITERIA PARAMETER ESTIMATION METHODOLOGY

The multi-criteria approach to calibration presented in detail by *Boyle et al.* [2000] combines the strengths of both automated and manual calibration methods. The approach involves the identification of several characteristic features of the observed streamflow hydrograph, each representing a distinct (preferably unique) aspect of the behavior of the watershed. In brief, the hydrograph is partitioned into three components based on the reasonable assumption that the behavior of the watershed is inherently different during periods "driven" by rainfall and periods without rain. Further, the periods immediately following the cessation of rainfall and dominated by interflow can be distinguished from the later periods that are dominated by baseflow. The streamflow hydrograph can, therefore, be partitioned into

three components (Figure 1), which we call "driven" (Q_D), "non-driven-quick" (Q_Q), and "non-driven-slow" (Q_S). The time steps corresponding to each of these components are identified through an analysis of the precipitation data and the time of concentration for the watershed. The time steps with non-zero rainfalls, lagged by the time of concentration for the watershed, are classified as driven. Of the remaining (non-driven) time steps, those with streamflows lower than a certain threshold value (e.g., mean of the logarithms of the flows) are classified as "non-driven-slow", and the rest are classified as "non-driven-quick". For each of the components, the closeness between the model outputs and the corresponding observed values is estimated separately using the RMSE statistic, resulting in three evaluation criteria, designated as FD (driven), FQ (non-driven quick), and FS (non-driven slow), respectively.

An important characteristic of the multi-objective problem is that it does not, in general, have a unique solution. Because of errors in the model structure (and other possible sources), it is not usually possible to find a single unique solution that simultaneously minimizes all of the criteria. Instead, it is common to have a "Pareto set" of solutions with the property that moving from one solution to another results in the improvement of one criterion while causing a deterioration in one or more others. The Pareto set represents the minimum uncertainty that can be achieved for the parameters via calibration, without subjectively assigning relative weights to the individual model responses. The size

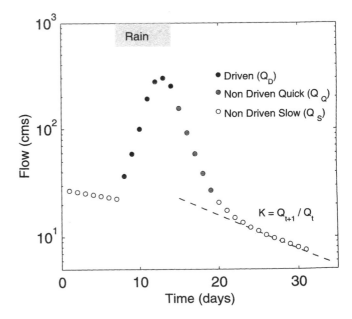

Figure 1. Partitioning of the observed hydrograph into three components: Q_D, Q_Q, and Q_S. The dashed line shows how the observed hydrograph can be used to estimate the recession constant, K.

and properties of this set are related to errors in the model structure and data. In this work, we used the Multi-Objective COMplex evolution (MOCOM-UA) algorithm [*Yapo et al.*, 1998; *Bastidas et al.*, 1999] to solve the multi-criteria optimization problem. MOCOM is a general-purpose multi-objective global optimization algorithm that provides, in a single optimization run, a set of points that approximate the Pareto set. For details, the reader is referred to *Gupta et al.* [1998] and *Yapo et al.* [1997, 1998].

3. INCORPORATING SPATIAL INFORMATION

3.1. Overview

The focus of this section is to provide an assessment of the potential improvements in rainfall-runoff model performance that can be achieved by semi-distributed modeling of a watershed using radar-based (NEXRAD) remotely sensed precipitation data. *Boyle et al.* [2001] examined the relative benefits of spatially distributing the model input (precipitation), structural components (soil moisture and streamflow routing), and surface characteristics (parameters). The CRR model used was the NWS Sacramento Soil Moisture Accounting (SAC-SMA) model [*Burnash et al.*, 1973] applied to the Blue River watershed near Blue, Oklahoma. The study was designed to complement the NWS semi-distributed studies on the Blue River by expanding our understanding of the specific benefits associated with different levels of spatial representation of the model. The multi-criteria framework presented in Section 2 for application to lumped hydrologic models was used to calibrate the semi-distributed model in terms of three objective measures designed to reflect the different observable characteristics of watershed behavior (peak flow and timing, quick recession, and baseflow). Multi-criteria performance comparisons among the different model applications were used to evaluate the benefits of various types and degrees of spatial complexity. Results from an independent manual calibration conducted by the NWS were used in the study as the basis for an evaluation of the strengths and weaknesses of the automatic multi-criteria calibration approach [*Gupta et al.*, 1998; *Boyle et al.*, 2000].

3.2. Methods

The NWS created a digital elevation model (DEM) of the Blue River watershed from 100 meter (cell size) elevation data. The watershed was partitioned into eight subwatersheds based on an analysis of DEM stream connectivity data (stream channel structure), and the variability of the high-resolution soil property information available from the

USDA State Soil Geographic Database (STATSGO) for the resulting subwatersheds. Mean Areal Precipitation (MAP) values for each of the eight subwatersheds were estimated from the 4 x 4 km NEXRAD Stage III hourly precipitation data. Unit hydrographs for each subwatershed were developed in conjunction with the DEM, using the methodology described by Smith et al. [1999], to route the simulated channel inflow to the outlet of the watershed. For the lumped conceptualization, the unit hydrograph was derived from the subwatershed unit hydrographs.

The NWS applied the SAC-SMA model in both lumped and semi-distributed (eight subwatersheds) forms to the Blue River watershed. In the lumped case, the channel inflow was computed at each time step for the entire watershed and then routed to the outlet with a single unit hydrograph. In the semi-distributed case, the soil-moisture computations were made separately for each subwatershed, and the resulting simulated channel inflows were then routed independently to the outlet of the watershed and combined to compute the total simulated streamflow for the entire watershed. The NWS used a sophisticated, highly interactive manual procedure to estimate values for 13 of the SAC-SMA parameters (four were set to default values) [*Anderson*, 1997] for the lumped watershed case and values for 104 parameters (13 for each of the eight subwatersheds) for the semi-distributed case. The reader is referred to *Boyle et al.* [2000] for a detailed description of the NWS parameter estimation procedure.

A series of lumped and semi-distributed applications of the SAC-SMA model to the Blue River watershed was made to investigate the improvements in model performance associated with various levels of spatial representation of model input (precipitation), structural components (soil moisture and streamflow routing computations), and surface characteristics (parameters). Each model application was designed to isolate the effects of the different levels of spatial representation in terms of specific desirable watershed behaviors (driven flow–"peaks and timing", non-driven quick flow–"quick recession" responses, and non-driven slow–"baseflow" responses). The calibration data set (precipitation, PET, and streamflow) used was the same as that used in the NWS manual calibration approach for the lumped and 8 subwatershed cases. Model calibration and evaluation of the performance improvements for each application were performed using the multi-criteria approach described above. For each case, the Pareto optimal solution space for the three criteria (FD, FQ, and FS) was estimated by 500 solutions generated using the MOCOM algorithm.

In Case 1, (LUMP-ALL) the SAC-SMA model was applied in a lumped configuration (precipitation P, soil moisture computations S, and streamflow routing computations R, were all lumped) to the Blue River watershed. This case served as a benchmark for performance comparisons with the other cases, in which the SAC-SMA model was applied in varying levels of spatial distribution to the 8-subwatershed configuration used by the NWS.

In Case 2 (DIST-PS), the precipitation and soil-moisture computations were spatially distributed among the subwatersheds, but the routing was treated as lumped. In this application, soil-moisture computations were performed separately to compute separate channel inflow sequences for each subwatershed, but these were combined into a total channel inflow for the entire watershed before routing to the outlet of the watershed using a single unit hydrograph. In Case 3 (DIST-PSR), the precipitation, soil-moisture computations, and streamflow routing computations were spatially distributed among the subwatersheds to assess the additional benefit of distributed routing. In this application, the channel inflow computed for each subwatershed was independently routed to the outlet of the watershed with separate unit hydrographs and then combined to estimate the total runoff from the watershed. Note that, in Cases 2 and 3, the model parameters were treated as lumped (all the subwatersheds were assigned the same values of the 13 calibration parameters) and only the spatial distribution of the model input and structural components was investigated. Other cases, not covered here, were also tested to investigate the value of spatially distributed precipitation and model parameters. The reader is referred to *Boyle et al.* [2001] for details.

Finally, to further investigate the effects of spatial representation, all cases were repeated using a smaller number of subwatersheds (i.e., the entire watershed was partitioned into a 3-subwatershed configuration). In this new configuration

(also provided to us by the NWS), the original subwatersheds 1, 2, and part of 3 were combined to form the new subwatershed 1 of the 3-subwatershed configuration. Similarly, 4, 5, and parts of 3 and 6 were combined to form the new subwatershed 2, while 7, 8, and part of 6 were combined to form the new subwatershed 3. The mean areal precipitation and PET for each of the three new subwatersheds were estimated by the NWS using the same methods mentioned previously.

3.3. Results

Main text is 10 point type, single column width at 8.5 cm, with full justification on both left and right margins. Use hyphenation. The NWS manual calibration studies were used as benchmarks for evaluation of the automatic calibration studies described above. The manual calibration results are shown in the multicriteria format in Figure 2a-c. Figures 2a-c present the results for each case using two-dimensional projections of the three-criteria solution space (NWS lumped case = large open square and NWS semi-distributed case = large open circle). Clearly, the semi-distributed application results in an improvement in the model's ability to simulate the observed flow in terms of FQ and FS, as compared with the lumped application. There is a slight decrease, however, in the model's ability to simulate the driven flows measured by FD.

The results of the multi-criteria automatic calibration of Case 1 (LUMP-ALL) are also shown in Figures 2a-c, as a three-criteria trade-off surface represented by the set of 500 Pareto optimal solutions (indicated by the light-gray dots). The inability of the model to simultaneously match all three aspects of the hydrograph is clearly illustrated. For exam-

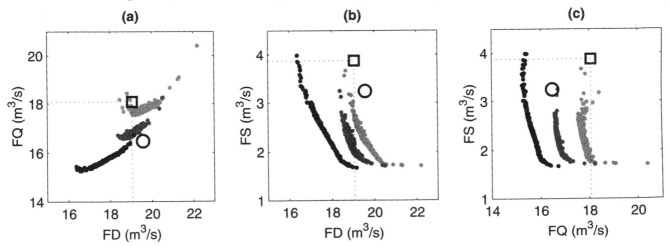

Figure 2. Pareto solutions obtained with the automatic multi-criteria approach to calibrate the SAC-SMA model: (a-c) two-dimensional projections of objective space. Marked points correspond to, respectively, 500 Pareto solutions for Case 1 (light-gray points), Case 2 (dark-gray points), and Case 3 (black points). NWS manual calibration results for lumped (square) and semi-distributed (open circle).

ple, Figure 2b illustrates the smoothly varying trade-off between the model's ability to match the driven (QD) and the non-driven-slow (QS) portions of the hydrograph (similarly see Figure 2c and, to a lesser extent, Figure 2a).

A visual comparison of the 500 Pareto solutions with the NWS lumped solution (open square) in Figures 2a-c shows that the automatic approach provides a closer fit of the baseflow responses (FS) and, to a lesser extent, the quick recession responses (FQ). In terms of the peaks and timing (FD), however, most of the 500 Pareto solutions are inferior to the NWS lumped solution.

The results of the multi-criteria automatic calibration of Case 2 (DIST-PS) are shown in Figures 2a-c. Note that, in this case, the channel inflows for all the sub-watersheds are lumped together and routed to the outlet using a single unit hydrograph. The results for the 8-subwatershed configuration did not give better results than the 3-subwatershed configuration. Therefore, the results presented here will draw primarily from the results of the 3-subwatershed study. Comparison of the solutions for this case (dark-gray dots) with the lumped case (Case 1, LUMP-ALL, light-gray dots) indicates a significant benefit to allowing the precipitation input and the soil-moisture computations to be distributed. In particular, the ability of the model to simulate the quick recession responses (FQ) and, to a lesser extent, the peaks/timing (FD) has been improved. However, there appears to be no additional impact on the model's ability to simulate the baseflow responses (FS).

A visual comparison of the DIST-PS results with the NWS lumped solution in Figures 2a-c clearly shows that the automatic approach provides a closer fit to the observed data in terms of all three criteria FD, FQ, and FS. Further, comparison of the DIST-PS results with the NWS semi-distributed solution shows that most of the 500 Pareto solutions provide a better fit to the baseflow (FS) and peaks/timing (FD), while providing a comparable fit to the quick recession (FQ).

The results of the multi-criteria automatic calibration of Case 3 (DIST-PSR) are also shown in Figures 2a-c. In this case, the precipitation, soil-moisture computations, and channel routing are all treated separately for each subwatershed. Again, the results for the 8-subwatershed configuration did not give better results than the 3-subwatershed configuration, and results are therefore only presented for the latter configuration. The 500 Pareto optimal parameter sets (black dots) show that routing the channel inflow independently from each subwatershed to the outlet of the watershed improves the model's ability to simulate both the quick recession responses (FQ) and the peaks/timing (FD). Once again, there is no additional improvement in the model's ability to simulate the baseflow responses (FS). A visual comparison of the 500 Pareto solutions for this case with the NWS lumped and semi-distributed solutions (Figures 2a-c) clearly shows that the automatically calibrated semi-distributed model DIST-PSR provides a much better reproduction of the watershed response, in terms of all three criteria FD, FQ, and FS.

4. INCORPORATING VERTICAL INFORMATION

4.1. Overview

The focus of this section is to provide an assessment of the potential improvements in streamflow simulation that can be achieved through various levels of representation of the vertical movement of moisture through the watershed using CRR models in lumped applications. The relative benefits of different levels of vertical model structure (direct runoff, upper soil moisture storage, and the percolation process) are examined with a simple hydrologic model, HYMOD [*Boyle*, 2001; *Wagener et al.*, 2001]. HYMOD consists of a variety of different excess generation (interception storage, tension storage, free storage, etc.), percolation, and streamflow routing functions that can be put together in different combinations to describe the different hydrologic behaviors of the watershed system. The multicriteria approach described in *Boyle et al.* [2000] (see Section 2 above) for application to hydrologic models was used to calibrate each CRR model in terms of three objective measures designed to reflect the different observable characteristics of watershed behavior (peak flow and timing, quick recession, and baseflow). Multicriteria performance comparisons among the different model applications were used to evaluate the benefits of various types and degrees of vertical model complexity. Results obtained from a lumped application of the SAC-SMA model were used as a benchmark for comparison with results from this study.

4.2. Methods

The automatic multicriteria approach outlined in Section 2 was used to estimate values for the parameters of the SAC-SMA flood forecast model using an 11-year period (WY 1952–1962 inclusive) of historical data from the Leaf River watershed (1950 km^2) located north of Collins, Mississippi. Forty consecutive years of data (WY 1948-88) are available for this watershed, representing a wide variety of hydrologic conditions. The details of the Leaf River data have been discussed previously in the literature [e.g., *Burnash et al.*, 1973; *Peck*, 1976; *Brazil and Hudlow*, 1981; *Sorooshian and Gupta*, 1983; etc.].

The general configuration of HYMOD for the purposes of this study is shown in Figure 3. The watershed is partitioned into two areas, pervious and impervious, by means of a single parameter, percent impervious area (PCTIM). Precipitation (rainfall) falling on the impervious portion of the watershed becomes direct runoff available for routing along with the surface runoff component, to the outlet of the watershed. Precipitation falling on the pervious portion of the watershed enters the upper soil moisture zone (UZ). The UZ consists of two components, tension water storage and free water storage. Tension water storage must be completely satisfied before moisture can move to the free water portion of the UZ. Soil moisture within the UZ tension and free water storages is available to satisfy the potential evapotranspiration (PET) demand. Saturated excess is generated from the free water storage and then combined with the direct runoff to estimate the quick (or surface) runoff. The surface runoff is then routed through a series of NUMQ linear reservoirs, each with the same recession coefficient, KQ, to the outlet of the watershed. Soil moisture percolates from the UZ free water storage to the lower soil moisture zone (LZ) free water storage. The moisture in the LZ free water is routed through a single linear reservoir, with recession coefficient KS, to estimate the slow (or baseflow) runoff at the outlet of the watershed. The quick and slow flows are

then combined to estimate the total streamflow at the outlet of the watershed.

Boyle [2001] examined sixty different applications (or cases) of HYMOD to the Leaf River watershed to investigate the improvements in model performance associated with various levels of vertical detail describing the movement of moisture through the soil (UZ tension and free water storages, percolation process, and pervious area). Each model application was designed to isolate the effects of different levels of vertical model complexity in terms of specific desirable watershed behaviors (driven flow–"peaks and timing", non-driven quick flow–"quick recession" responses, and non-driven slow–"baseflow" responses). The multi-criteria approach described in Section 2 was used to calibrate each of the different applications of HYMOD in terms of the three objective measures driven flow (FD), non-driven quick flow (FQ), and non-driven slow flow (FS). For each modeling case, the Pareto optimal solution space for the three criteria (FD, FQ, FS) was estimated by 500 solutions generated using the MOCOM algorithm.

In Cases 1-30, presented here, a simple bucket loss (BL) model (see Figure 4) was used to describe the functional

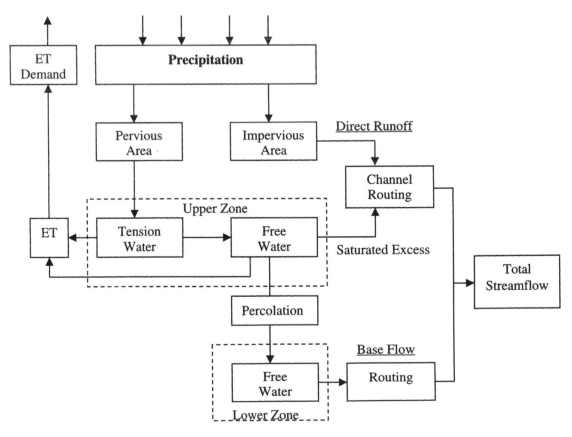

Figure 3. HYMOD watershed model.

relationship of the UZ free water storage. In Cases 31-60 (not presented here) the Moore loss model, described in detail by Moore [1985], was substituted for the simple bucket. The BL model generates saturated excess when the soil moisture level in the tank is greater than parameter FMAX. Ten different functional relationships describing the percolation between the UZ and LZ free water storages were also tested in this study. The ten functions are described by the following five equations:

$$PERC(t)_{1-2} = min(contents), PMAX)$$

$$PERC(t)_{3-4} = PFRAC*contents(t)$$

$$PERC(t)_{5-6} = min(PFRAC*contents(t), PMAX)$$

$$PERC(t)_{7-8} = min(BETA(1 + ZETA(LZDR(t))^{REXP}, \\ 1.0))*contents(t)$$

$$PERC(t)_{9-10} = min(BETA(X\_FREE(t)/FMAX) \\ (1 + ZETA(LZDR(t))^{REXP},$$

$$1.0))*contents(t)$$

where contents(t) is the UZ saturated excess at time t ($PERC(t)_{1,3,5,7,9}$) or the free water storage at time t ($PERC(t)_{2,4,6,8,10}$), LZDR(t) is the lower zone deficiency ratio (1-contents of lower zone free water storage/maximum contents of LZ free water storage) at time t, X_FREE(t) is the UZ free water contents, FMAX is the maximum contents of UZ free water storage, and PMAX, PFRAC, BETA, ZETA, and REXP are calibration parameters.

Each of these ten functions provides a unique conceptualization of the relationship between the UZ and LZ free water contents and the percolation process. In general, the complexity of the conceptual relationships ranges from low in $PERC(t)_{1-2}$ and $PERC(t)_{3-4}$ to high in $PERC(t)_{7-8}$ and $PERC(t)_{9-10}$. The functional relationships in $PERC(t)_{1-2}$ and $PERC(t)_{3-4}$ describe the amount of UZ free water storage that can be percolated to LZ free water storage as a maximum amount and fraction, respectively, of the contents. $PERC(t)_{5-6}$ describes the percolation process as a fraction of the UZ free water storage that can be percolated to LZ free water storage with a maximum value for a given time step. $PERC(t)_{7-8}$ and $PERC(t)_{9-10}$ approximate the complex percolation process used in the SAC-SMA model. The primary difference between the latter two being that the function $PERC(t)_{9-10}$ allows the UZ free water storage contents (X_FREE(t)/FMAX) and the lower zone (LZDR(t)) to influence the percolation rate while $PERC(t)_{7-8}$ is influenced by the lower zone (LZDR(t)).

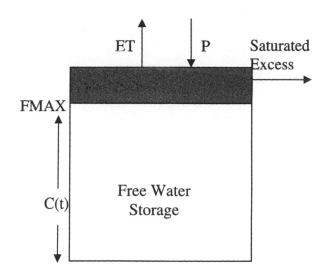

Figure 4. Representation of Upper Zone free water with simple bucket model.

The influence of the UZ tension water storage and the percent impervious area of the watershed were also investigated. In Cases 1-10, the bucket loss representation of the UZ free water storage was combined, separately, with each of the ten percolation functions (see Table 1). In each of these cases, the size of the tension water storage tank and the percent impervious area of the watershed were set to zero (no UZ tension water storage and no impervious area in the watershed). In Cases 11-20, Cases 1-10 were repeated with the UZ tension water storage component but no impervious area. Finally, in Cases 21-30, the UZ tension water storage and the impervious area components of HYMOD were considered.

4.3. Results

The results obtained with the SAC-SMA model (Case 0) were used as a benchmark for comparison with the studies described above. The ranges of the 500 Pareto solutions for the SAC-SMA model, in terms of the three criteria FD, FQ, and FS, are shown as black dots with the corresponding gray shaded area in the multi-criteria format in Figure 5a-c.

Figure 5a-c presents the results for Cases 1-30 (application of HYMOD using the simple bucket representation of the UZ free water storage). Notice that in terms of the criteria FD, the results for all the Cases (except Case 10) are inferior to all of the solutions obtained with the SAC-SMA model. In addition, the results for many of the cases are inferior to all of the solutions obtained with the SAC-SMA model in terms of FQ. These results indicate that the model structures in Cases 1-30 are not representing hydrologic processes important to the simulation of the larger flows in the same way that the SAC-SMA model does. On the other hand, all of the thirty of the cases contain solutions that are

Table 1: Modeling Cases 1-30 using the simple bucket loss (BL)

Case #	Percolation (1-10)	UZ Tension (T)	Impervious Area. (I)
1	1	N	N
2	2	N	N
3	3	N	N
4	4	N	N
5	5	N	N
6	6	N	N
7	7	N	N
8	8	N	N
9	9	N	N
10	10	N	N
11	1	Y	N
12	2	Y	N
13	3	Y	N
14	4	Y	N
15	5	Y	N
16	6	Y	N
17	7	Y	N
18	8	Y	N
19	9	Y	N
20	10	Y	N
21	1	Y	Y
22	2	Y	Y
23	3	Y	Y
24	4	Y	Y
25	5	Y	Y
26	6	Y	Y
27	7	Y	Y
28	8	Y	Y
29	9	Y	Y
30	10	Y	Y

Y = Yes
N = No

superior to those obtained with the SAC-SMA model in terms of the FS criterion. This result indicates that the LZ representation used by each of the Cases 1-30 may be adequate to represent the important hydrologic processes required to simulate the lower flows.

The results for Cases 1-10 (simple bucket loss representation of UZ free water storage with no UZ tension water storage or impervious area) are also shown in Figures 5a-c. A visual comparison of the results shows that Cases 1,3, 5, and 7 are superior to Cases 2, 4, 6, and 8 in terms of both FD and FQ indicating that the model performs better (at least in this configuration) with the percolation source as the saturated excess rather than the UZ free water storage contents. Notice that the results for Cases 7 and 9 (as well as 17 and 19, and 27 and 29) appear to be identical for all three criteria, FD, FQ, and FS. The fact that these results are not unique is a consequence of combining the simple bucket model with percolation relation PERC(t)9-10. When the contents(t) variable is set to be the saturated excess (as it is

in Cases 7, 9, 17, 19, 27, and 29) the only time there can be percolation is at times when there is saturated excess (when X_FREE(t)/FMAX = 1.0). This effectively makes $PERC(t)_{7-8} = PERC(t)_{9-10}$ for all thirty of the simple bucket applications. This was not a relevant issue in Cases 31-60 (not shown) since saturated excess can occur in the Moore loss representation without X_FREE(t)/FMAX = 1.0.

From visual inspection of Figure 5a-c it can be clearly seen that the results for Case 10 are superior in terms of fitting the FD and FQ criteria than the results for any other of the Cases. Figure 6a-c presents the results for Cases 9 and 10 and the SAC-SMA model in two-dimensional projections of the three-criteria solution space. The SAC-SMA solutions (black dots) are clearly superior in terms of FD to most of the solutions in Case 10 (dark gray dots) and all of the solutions in Case 9 (light gray dots). Many of the solutions for Case 10, however, are at least as good as the SAC-SMA solutions in terms of the FQ and FS criteria (the lower values of FS criteria for Case 10 are out of the plot-

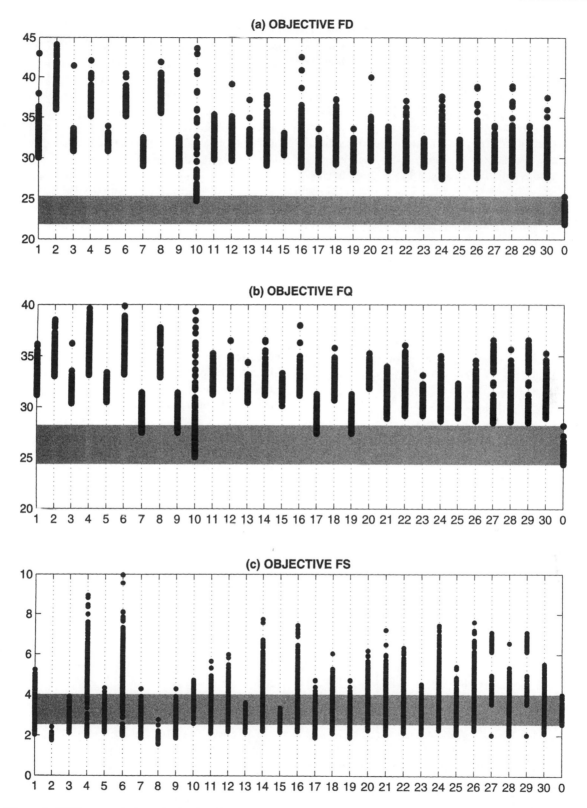

Figure 5. 500 Pareto solutions obtained with the automatic multi-criteria approach (black dots) to calibrate Cases 1-30 using the simple bucket model: (a) objective function FD, (b) objective function FQ, (c) objective function FS. The gray shading represents the SAC-SMA results.

ting range indicating a very large trade-off between the fitting of the FD and FS criteria). Many of the Case 9 solutions are superior to both the SAC-SMA and Case 10 solutions in terms of the FS criteria. Also note that the solutions for Case 9 have much less trade-off in fitting the FD and FS, and FQ and FS criteria, although the fit to both FD and FQ is inferior compared with Case 10 and the SAC-SMA model.

A visual inspection of Figures 5a-c indicates that the addition of the tension water storage in Cases 11-20 has generally improved the results slightly in terms of FD and FQ. In addition, the differences between Cases 1,3, 5, and 7 and Cases 2, 4, 6, and 8, in terms of both FD and FQ, have been significantly reduced. Further improvement is made in Cases 21-30, in terms of FQ criteria, when the pervious area component is added to the model. In none of these cases, however, did the addition of the UZ tension water storage or the impervious area component substantially improve the fitting of the FS criterion.

The three cases that appear to have the "best" solutions for each of the three main model configurations are Case 10 (Cases 1-10), Case 19 (Cases 11-20), and Case 30 (Cases 21-30). Figures 7a-c present the results for Case 10 (light gray dots), Case 19 (dark gray dots) and Case 30 (black dots) in two-dimensional projections of the three-criteria solution space. Notice that the solutions for these cases have very little trade-off in terms of the FD and FQ criteria. From the figure it can clearly be seen that many of the solutions for Case 10 are superior to those in Cases 19 and 30. The trade-off range, however, is dramatically different for Cases 19 and 30 compared to that of Case 10 (improvement is FS is possible with very little cost in terms of FD and FQ).

5. SUMMARY AND CONCLUSIONS

Many of the CRR models used to simulate streamflow consist of highly complex functional relationships to describe the movement of moisture vertically through the soil. These models are often very difficult to calibrate due to the large number of parameters and complex relationships within the model. Further, the large number of parameters may limit the manner in which newly available high-resolution (spatial) hydrologic information can be incorporated into the hydrologic modeling process, thereby limiting the actual benefit(s) of the potential added information contained within the new data. As a result, there is a real need to understand the specific benefits associated with increased representation of the movement of moisture vertically through the soil. With this new understanding, new models can be developed from new and existing modeling concepts, with parsimonious model structures that represent only those response modes that are identifiable within the available data.

The primary objective of this chapter was to present a new hybrid multi-criteria calibration approach that combines the strength of automatic and manual calibration methods and use the new approach to investigate the benefits of different levels of spatial and vertical representation of important watershed hydrologic variables with conceptual rainfall runoff models. This chapter explores the specific improvements in streamflow simulation that can be achieved through various levels of vertical model structure (direct runoff, upper soil moisture storage, and the percolation process). This was accomplished through application and calibration of two CRR models, SAC-SMA and HYMOD, with a variety of different combinations of excess generation (interception storage, tension storage, free storage, etc.), percolation, and streamflow routing functions.

In this work, it has been demonstrated how multi-criteria methods provide a useful framework for the systematic investigation of appropriate model complexity. In addition, the applicability of the multi-criteria automatic calibration methods to the calibration of CRR models with increased model complexity has been demonstrated in this study.

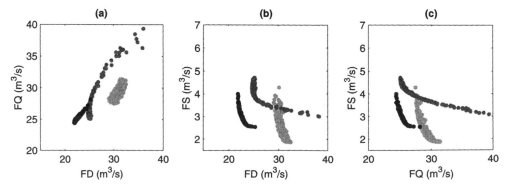

Figure 6. Pareto solutions obtained with the automatic multi-criteria approach to calibrate the HYMOD model: (a-c) two-dimensional projections of objective space. Marked points correspond to, respectively, 500 Pareto solutions for Case 9 (light-gray points), Case 10 (dark-gray points), and SAC-SMA model (black points).

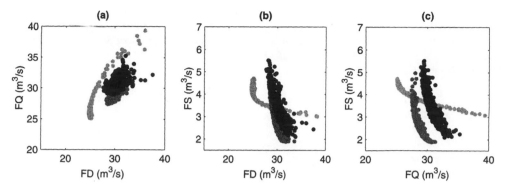

Figure 7. Pareto solutions obtained with the automatic multi-criteria approach to calibrate the HYMOD model: (a-c) two-dimensional projections of objective space. Marked points correspond to, respectively, 500 Pareto solutions for Case 10 (light-gray points), Case 19 (dark-gray points), and Case 30 (black points).

Manual calibration of the different model structures would have required a significant effort since little was known (no "expert" knowledge) about the behavior and performance of many of the model structures prior to testing. Instead, the automatic approach allowed efficient and "consistent" estimation of parameters (and hence model performance) with a minimal amount of effort (5-10 minutes for each case on a Sun workstation).

The effectiveness and efficiency of the automatic approach allowed rapid investigation of the specific benefits associated with different levels of vertical and spatial model structural complexity, including impervious area contribution, UZ tension and free water storage, and percolation computations. Based on the results from this study, the next logical research direction should include an investigation in which the spatial resolution and vertical structural complexity of the CRR model application is investigated simultaneously.

Acknowledgments. Partial financial support for this research was provided by the National Science Foundation (EAR-9418147), the Hydrologic Research Laboratory of the National Weather Service (grants NA47WG0408 and NA77WH0425), the Sustainability of Water Resources in Semi-Arid Regions (SAHRA), the National Aeronautics and Space Administration (NASA-EOS grant NAGW2425), and by the Desert Research Institute (New Faculty Support 6224-640-4833).

REFERENCES

Anderson, E.A., "Hydrologic model calibration using the Interactive Calibration Program (ICP)." Hydrologic Research Laboratory, U.S. National Weather Service, Silver Spring, Maryland, 1997.

Bastidas, L.A., H.V. Gupta, and S. Sorooshian, "The Multi-Objective COMplex evolution algorithm, MOCOM-UA, User's Guide." Department of Hydrology and Water Resources, The University of Arizona, Tucson Arizona, 1999.

Boyle, D.P., "Multicriteria calibration of hydrologic models." Ph.D. Dissertation., Department of Hydrology and Water Resources, University of Arizona, Tucson, Arizona, USA, 2001.

Boyle, D.P., H.V. Gupta, and S. Sorooshian, (2000). "Toward improved calibration of hydrologic models: Combining the strengths of manual and automatic methods." *Water Resources Research*, 36(12), 3663-3674.

Boyle, D.P., H.V. Gupta, S. Sorooshian, V. Koren, Z. Zhang, and M. Smith, "Toward improved streamflow forecasts: Value of semidistributed modeling," *Water Resources Research*, AGU, 36(11), 2749-2759, 2001.

Brazil, L.E. and M.D. Hudlow, "Calibration procedures used with the National Weather Service Forecast System." in *Water and Related Land Resource Systems*, edited by Y.Y. Haimes and J. Kindler, pp. 457-466, Pergamon Press, New York, 1981.

Burnash, R. J.C., R.L. Ferral, and R.A. McGuire, "A generalized streamflow simulation system: conceptual modeling for digital computers." U.S. Department of Commerce, National Weather Service and State of California, Department of Water Resources. Sacramento, California, 1973.

Gupta, H.V., S. Sorooshian, and P.O. Yapo, "Towards improved calibration of hydrologic models: multiple and non?commensurable measures of information." *Water Resources Research*, 34(4), 751-763, 1998.

Moore, R.J., "The probability-distributed principle and runoff production at point and basin scales." *Hydrol. Sci. J., 30*, 273-297, 1985.

Peck, E. L., "Catchment Modeling and Initial Parameter Estimation for the National Weather Service River Forecast System." NOAA Technical Memorandum NWS HYDRO-31. Office of Hydrology. Washington, D.C. National Weather Service. Silver Spring, Maryland, 1976.

Smith, M.B., V. Koren, D. Johnson, B.D. Finnerty, and D.J. Seo, "Distributed modeling: Phase 1 results." *NOAA Tech. Rep. NWS 44*, National Hydrologic Research Lab, 210p, 1999.

Sorooshian, S. and V.K. Gupta, "Automatic calibration of conceptual rainfall-runoff models: the question of parameter observability and uniqueness." *Water Resources Research*, 19(1), 251-259, 1983.

Yapo, P.O., H.V. Gupta, and S. Sorooshian, A multiobjective global optimization algorithm with application to calibration of hydrologic models, *HWR Tech. Rep.* 97-050, Department of Hydrology and Water Resources, The University of Arizona, Tucson, 1997.

Yapo, P.O., H.V. Gupta, and S. Sorooshian, "Multi-objective global optimization for hydrologic model." *Journal of Hydrology, 204,* 83-97, 1998.

Wagener, T., D.P. Boyle, M.J. Lees, H.S. Wheater, H.V. Gupta, and S. Sorooshian, "A framework for development and application of hydrological models," *Hydrology and Earth System Science,* European Geophysical Society, 5(1), 13-26, 2001.

D.P. Boyle, Department of Hydrologic Sciences, Desert Research Institute, University and Community College System of Nevada, 2215 Raggio Parkway, Reno, Nevada, 89512-1095, USA. (dboyle@dri.edu)

H.V. Gupta and S. Sorooshian, Department of Hydrology and Water Resources University of Arizona, Tucson, Arizona, 85721, USA. (hoshin@hwr.arizona.edu)

Multi-Resolution Calibration Methodology for Hydrologic Models: Application to a Sub-Humid Catchment

Laura M. Parada, Jonathan P. Fram, and Xu Liang

Department of Civil and Environmental Engineering, University of California, Berkeley, California

Wavelet analysis allows for calibration of hydrologic models at multiple temporal scales, thus accounting for the time-variant fluctuations imbedded in streamflow data. This investigation further evaluates the incorporation of a multi-resolution framework for construction of objective functions to the shuffled complex evolution (SCE-UA) optimization algorithm. One routing and six soil parameters of the Variable Infiltration Capacity (VIC-3L) hydrologic model were optimized to fit daily streamflow observations for a mid-sized sub-humid catchment in Northern California. Calibration was performed with Root Mean Square Error (RMSE) and Heteroscedastic Mean Likelihood Error (HMLE), and over two periods consisting of 1024 and 2048 days, respectively. Our evaluation suggests that multi-resolution optimization is likely to yield better results during validation than its single-scale counter-part, and it may, at least, perform equivalently. Moreover, the improvements obtained by the multi-resolution approach during validation are observed in terms of RMSE and HMLE regardless of which of these was selected for calibration. In this regard, the multi-resolution paradigm constitutes a more robust alternative for calibration than its traditional single-scale counter-part since it may render better or equivalent results during validation regardless of the choice of cost function or calibration period.

1. INTRODUCTION

Reliable predictions of water and energy budgets at the land surface are central to water resource planning, climate simulation, and numerical weather forecasting. Macroscale hydrologic models provide these predictions by simulating surface water and energy fluxes at scales ranging from small watersheds to large continental river basins. However, many of these models are based on abstract conceptual representations of watershed characteristics or the physical processes inherent to the water and energy budgets. Therefore, their performance tends to depend on parameter optimization. This task is not a trivial one since the parameters of interest have a tendency to interact in a highly non-linear and complex manner, which results in feasible spaces that are usually non-convex, rough, and exhibit multiple local as well as global optima [*Duan et al.*, 1992, 1994].

Effective and efficient techniques have been developed in the past decade that make automatic calibration a viable option. These include the shuffled complex evolution (SCE-UA) [*Duan et al.*, 1992, 1994; *Sorooshian et al.*, 1993] and multi-objective complex evolution (MOCOM-UA) [*Gupta et al.*, 1998; *Yapo et al.*, 1998] global optimization algorithms, as well as the Bayesian generalized likelihood uncertainty estimation technique [*Beven and Binley*, 1992; *Cameron et al.*, 1999; *Freer et al.*, 1996] among others. Generally, the success of optimization methodologies depends critically on the length and quality of the observed time series, and on the choice of the objective function used to evaluate the proximity of the simulated and observed time series. In particular, the Root Mean Square Error (RMSE) and Heteroscedastic Mean Likelihood Error (HMLE) objective measures have been extensively used and inter-compared in the literature [e.g. *Gupta et al.*, 1999; *Sorooshian et al.*, 1993; *Yapo et al.*, 1998].

Calibration of Watershed Models
Water Science and Application Volume 6
Copyright 2003 by the American Geophysical Union
10/1029/006WS15

Wavelet analysis is a powerful tool for processing non-periodic multi-scaled signals. *Smith et al.* [1998] and *Saco and Kumar* [2000] applied it successfully to the categorization of streamflow response modes over several time-scales in the United States. They determined that large, wet regions with high snowfall respond to climatological variables and physiological characteristics in distinctly identifiable ways over time spans ranging from one and a half months to a year. On the other hand, catchments with opposite qualification do so over shorter time scales on the order of one week to a month.

Liang et al. [manuscript 2002] developed a new scheme to construct objective functions for model calibration by applying the multi-resolution framework of wavelet theory. They incorporated wavelet analysis into the objective function formulation scheme of the SCE-UA optimization algorithm, and tested it using the Three-Layer Variable Infiltration Capacity (VIC-3L) hydrologic model.

In this paper, we further explore and evaluate the use of multi-resolution optimization [*Liang et al.,* manuscript 2002], as incorporated to the SCE-UA optimization algorithm, with the VIC-3L model for a mid-sized (510-km²) sub-humid catchment with annual precipitation on the order of 900 mm. The multi-resolution optimization approach is compared to the single-scale one with RMSE and HMLE used as objective measures. Calibration is performed over two distinct periods consisting of 1024 and 2048 observations, respectively.

2. MULTIRESOLUTION OPTIMIZATION METHODOLOGY

The multi-resolution analysis of wavelet theory allows for a signal, such as streamflow time-series, to be decomposed into various resolutions (e.g. time scales) so that the scale-variant fluctuations imbedded in it may be captured and analyzed. In particular, orthogonal wavelet transforms have been evaluated and successfully applied in the literature to describe the variability present in geophysical time series at various temporal scales [e.g. *Kumar and Foufoula-Georgiou,* 1994; *Liang et al.,* manuscript 2002; *Saco and Kumar,* 2000]. *Daubechies* [1988] and *Mallat* [1989a,b] provide a mathematically rigorous description of these techniques. To keep this paper self-contained, the general framework of the multi-resolution optimization approach presented by *Liang et al.* [manuscript 2002] is briefly described and summarized here.

2.1. Concepts of Multiresolution Analysis with Wavelets

Orthogonal wavelets are the building blocks of a series-decomposition similar to the more familiar Fourier trans-

formation. They permit the behavior of a signal at the original, finest scale, with index m = 0, to be represented by its behavior at a coarser scale (m = M, M > 0) plus some details arranged hierarchically from scale M to zero. The approximation at scale 2^m contains all the information needed to represent the signal at the next coarser scale, 2^{m+1}. If f_m is used to approximate the original signal f(t) (e.g. daily streamflow) at scale 2^m, this representation can be expressed as:

$$f_m = \sum_{n=-\infty}^{+\infty} <f,\phi_{m,n}> \phi_{m,n} ; \qquad (1)$$

where m and n are scale and location parameters, respectively. The inner products $<f,\phi_{m,n}>$ give the approximations at the scale with index m and are called smooth coefficients. $\phi_{m,n}$ denote basis functions, which are dilations and translations of the scaling function $\phi(t)$, and can be expressed as:

$$\phi_{m,n}(t) = 2^{-\frac{m}{2}} \phi\left(2^{-m}t - n\right). \qquad (2)$$

The scaling function satisfies:

$$\int \phi(t)\, dt = 1 . \qquad (3)$$

In going from any given scale to the next (coarser) aggregation level, some information about the signal is lost. This information is called details and can be expressed as:

$$f_m' = \sum_{n=-\infty}^{+\infty} <f,\psi_{m,n}> \psi_{m,n} ; \qquad (4)$$

where the inner products $<f,\psi_{m,n}>$ give the "details" that appear at scale 2^{m-1} (i.e., a finer scale than 2^m) and are called wavelet coefficients. $\psi_{m,n}$ denote basis functions, which are translations and dilations of the wavelet function , and can be expressed as:

$$\psi_{m,n}(t) = 2^{-\frac{m}{2}} \psi\left(2^{-m}t - n\right). \qquad (5)$$

The wavelet function satisfies:

$$\int \psi(t)\, dt = 0 . \qquad (6)$$

Thus, the signal approximated at scale 2^{m-1} can be expressed as:

$$f_{m-1} = f_m + f_m' . \qquad (7)$$

Details from all scales are necessary for complete reconstruction of the original signal. The Filter Banks Algorithm [*Mallat* 1989a, 1989b] is an efficient technique, which allows for hierarchical decomposition or reconstruction of a given signal into or from its wavelet representation at various time scales.

2.2. The Haar Scaling Function

Wavelet transforms have the capability to accurately locate irregular features in both the time and frequency domains. However, a trade-off exists such that improved localization in the time domain leads to a poorer localization in the frequency domain. The choice of wavelet function determines which of the two domains is given more emphasis. Therefore, its selection is crucial and should be dictated by the application at hand.

Smith et al. [1998] evaluated several wavelet functions. They determined that single-peaked wavelets, such as Haar, provide a good approximation to streamflow records at the daily scale since these are usually single-peaked as well. The Haar wavelet is employed in this study. The Haar scaling and wavelet functions are given by:

$$\phi(t) = \begin{cases} 1 & 0 \leq t < 1 \\ 0 & \text{otherwise} \end{cases} \quad (8a)$$

$$\psi(t) = \begin{cases} 1 & 0 \leq t < 1/2 \\ -1 & 1/2 \leq t < 1 \\ 0 & \text{otherwise} \end{cases} \quad (8b)$$

2.3. Description of the Multi-Resolution Optimization Methodology

The hydrologic processes driving the water and energy fluxes of interest occur within distinct time frames. Previous work by *Smith et al.* [1998] and *Saco and Kumar* [2000] evinces that streamflow responds in distinctly identifiable ways at different time scales due to the influence of climatological (e.g. snow fall, and precipitation type) and physiographic (e.g. topography and soil characteristics) variables.

Wavelet analysis allows for calibration to be performed at multiple temporal scales, thus accounting for the time-variant fluctuations imbedded in streamflow data. The framework for multi-resolution optimization presented by *Liang et al.* [manuscript 2002], which is based on wavelet theory, is briefly described and summarized here, with further modifications incorporated into their original work.

For the case in which L time series of observations (e.g. evaporation, runoff, and net radiation), indexed by j, are to be used simultaneously for optimization, each with its corresponding set of N_j objective functions, $\{f_j(\theta)\}$, with respect to a set of model parameters, θ, the general form of the cost function, F, is given by:

$$\min \text{ (with respect to } \theta) \; F(\theta) = \{f_1(\theta), ..., f_L(\theta)\}; \quad (9)$$

Each objective function, G_i^j, in the set $f_j(\theta)$ is optimized by considering k_i^j wavelet scales:

$$\min \text{ (with respect to } \theta) \; f_j(\theta) = \left\{ G_1^j(\theta), ..., G_{N_j}^j(\theta) \right\}; \quad (10)$$

$$G_i^j(\theta) = \sum_{m=0}^{k_i^j} \alpha_{i,m}^j g_{i,m}^j(\theta); \quad (11)$$

where the summation in (11) is over temporal scales, such that $g_{i,m}^j(\theta)$ is the objective function value corresponding to $G_i^j(\theta)$ evaluated at scale 2^m. is the weighing coefficient for $g_{i,m}^j(\theta)$ at that same scale. The weighing coefficients are such that:

$$\sum_{m=0}^{k_i^j} \alpha_{i,m}^j = 1. \quad (12)$$

The multi-objective complex evolution (MOCOM-UA) optimization technique allows for (9) and (10) to be evaluated for several time-series and objective functions [*Gupta et al.*, 1998; *Yapo et al.*, 1998]. The shuffled complex evolution (SCE-UA) methodology permits this evaluation to be performed with a single time-series of observations and one objective function [*Duan et al.*, 1992, 1994; *Gupta et al.*, 1999; *Sorooshian et al.*, 1993]. The use of wavelet transforms introduces multiple time frames into the problem, as outlined in (11) and (12).

3. CASE STUDY

The multi-resolution paradigm described in section 2 was implemented to the objective function formulation of the SCE-UA optimization technique. It was then applied for the simultaneous calibration of one routing and six soil parameters of the Three-Layer Variable Infiltration Capacity (VIC-3L) hydrologic model [*Cherkauer and Lettenmaier*, 1999; *Liang and Xie*, 2001; *Liang et al.*, 1994, 1996a, 1996b, 1999; *Wood et al.*, 1997] to fit streamflow observations for a sub-humid catchment at multiple time scales. This section describes the application of the multi-resolution approach and provides an analysis of the results.

3.1. Three-Layer Variable Infiltration Capacity (VIC-3L) Model

VIC-3L may be operated in full energy and water balance mode or in water balance mode only. In both cases, the model is at least driven by precipitation, maximum and minimum daily temperature (daily time-step) or temperature for every sub-daily time step. The full energy and water balance mode, which is employed in this study, additionally requires wind speed as an input. This mode results in the prediction of soil moisture states at different soil layers, as well as land surface water and energy fluxes.

The VIC-3L model characterizes the soil column as consisting of three soil layers denoted layer 0 or thin surface layer, layer 1 or upper layer, and layer 2 or lower layer. Layer 0 is a thin surface layer (~10 cm) that captures the rapid dynamics of the upper soil layers by allowing for quick bare soil evaporation following small rainfall events [*Liang et al.*, 1996a]. The upper soil zone (layer 1) is designed to represent the dynamic behavior of the soil column responding to rainfall events. The lower layer (layer 2) characterizes the slower dynamics of inter-storm deep soil moisture and baseflow processes. Baseflow from the lower soil layer is determined by the nonlinear Arno model formulation (Figure 1). Vegetation exerts important controls on the exchange of water and energy at the land surface and is therefore explicitly incorporated into the model in a simple yet reasonable manner. VIC-3L also has the ability to simulate frozen soils [*Cherkauer and Lettenmaier*, 1999].

A distinguishing characteristic of this model is that it can represent sub-grid scale heterogeneity in soil properties and precipitation, and hence evaporation, soil moisture, and runoff due to its application of variable moisture and infiltration capacities [*Liang and Xie*, 2001; *Liang et al.*, 1994, 1996b]. Spatial probability distributions of the beta form are used to characterize the available soil moisture capacity and infiltration capacity rate as functions of the relative saturated and unsaturated areas of a grid cell respectively as shown in Figure 2 [*Liang et al.*, 1994]. When precipitation plus the amount of soil moisture is in excess of the soil moisture capacity, saturation-excess runoff occurs. The version of VIC-3L used for this study also includes a new feature to dynamically simulate the generation of infiltration-excess runoff such that the effects of soil heterogeneity are considered and the parameterization is consistent with the model's representation of saturation-excess runoff [*Liang and Xie*, 2001].

The VIC-3L model has been extensively tested and successfully applied to basins of various sizes [e.g. *Liang and Xie*, 2001; *Liang et al.*, 1994, 1996a; *Lohmann et al.*, 1998a; *Nijssen et al.*, 1997, 2001; O'Donnell et al., 2000; *Wood et al.*, 1997]. It has been shown to perform quite well in humid climate environments, and consistently well in the Project for Intercomparison of Land-surface Parameterization Schemes (PILPS) [e.g. *Chen et al.*, 1997; *Liang et al.*, 1998; *Lohmann et al.*, 1998b; *Wood et al.*, 1998].

3.2. Description of Optimized Parameters

One routing and six soil parameters of VIC-3L were optimized. These are listed in Table 1, together with their corresponding physical meanings and feasible ranges. Illustrations for the six soil parameters are provided in Figures 1 and 2. Smaller values of b imply decreased heterogeneity. Typical values for this parameter are usually less than 2.

3.3. Implementation of Multi-Resolution Optimization Approach and Formulation of Objective Functions

Since calibration was performed with a single time series, i.e. streamflow, (L = 1), and only one objective function was minimized at a time ($N_1 = 1$), the general form of the multi-resolution optimization problem, given by (9), (10) and (11) reduces to:

$$\min \text{ (with respect to } \theta) \ F(\theta) = \sum_{m=0}^{k} \alpha_m g_m(\theta) ; \qquad (13)$$

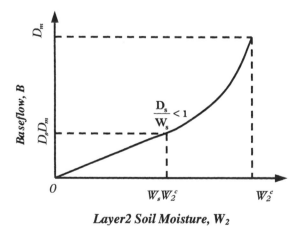

Layer2 Soil Moisture, W_2

Figure 1. Baseflow curve: D_s, D_m, and W_s are as defined in Table 1. W_2 is the total soil moisture capacity over a grid cell for layer 2. It is given by the product of the depth of this layer and the soil porosity.

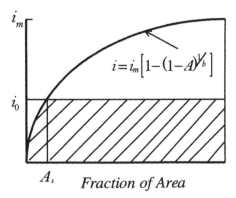

Figure 2. Variable moisture capacity curve: A_s represents the fractional grid cell area that is saturated, i and i_m are the point moisture and maximum point moisture capacity for a grid cell respectively, and i_o is the point moisture capacity corresponding to grid areas that are just saturated. Thus, the saturated area of a grid cell has a moisture capacity $i \le i_o$ while the remaining non-saturated area has $i \ge i_o$.

Table 1. Description of optimized parameters.

Parameter	Units	Physical meaning	Feasible range
b	N/A	Variable moisture capacity curve parameter	0 to 5
D_s	Fraction	Fraction of D_m where non-linear baseflow begins (baseflow curve)	0 to 1
D_m	mm/day	Maximum velocity of baseflow (baseflow curve)	1 to 40
W_s	Fraction	Fraction of maximum soil moisture where non-linear baseflow occurs (baseflow curve)	0.001 to 1
D_1	Meters	Thickness of soil moisture layer 1	0.1 to 1
D_2	Meters	Thickness of soil moisture layer 2	0.5 to 4
K_{stor}	N/A	Routing parameter	0.5 to 20

where the summation in (13) is over temporal scales, such that $g_m(\theta)$ is once again the objective function evaluated at scale 2^m, and k is the number of wavelet scales considered.

Two objective functions were chosen for optimization, Root Mean Square Error (RMSE) and Heteroscedastic Mean Likelihood Error (HMLE). These are defined as follows for the single time-scale approach:

$$RMSE(\theta) = \sqrt{\frac{1}{N}\sum_{n=1}^{N}[q_n^{sim}(\theta) - q_n^{obs}(\theta)]^2} \; ; \qquad (14)$$

$$HMLE(\theta,\lambda) = \frac{\frac{1}{N}\sum_{n=1}^{N}w_n(\lambda)[q_n^{sim}(\theta) - q_n^{obs}(\theta)]^2}{\left[\prod_{n=1}^{N}w_n(\lambda)\right]^{\frac{1}{N}}} \; ; \qquad (15)$$

where N denotes the number of observations available, n indexes the time step, and q_n^{sim} and q_n^{obs} are the model simulated and observed streamflow, respectively. In (15), $w_n(\lambda)$ is the weight assigned to the data value at time n, and is given by the observed streamflow at that time raised to the $2(\lambda-1)$ power, where λ is a parameter associated with HMLE. In this study, λ was fixed to 0.5. *Sorooshian and Dracup* [1980] provide a thorough definition of the HMLE cost function. The RMSE criterion tends to focus on minimization of peak flow errors, while HMLE is more consistent across all flow regimes [*Yapo et al.*, 1998].

A normalized form of (11) is introduced for the multiscale approach as in *Liang et al.* [manuscript 2002]:

$$g_{RMSE,m}(\theta) = RMSE_m(\theta) = \frac{N_m^{-\frac{1}{2}}\|q_m^{sim} - q_m^{obs}\|}{N_m^{-\frac{1}{2}}\|q_m^{obs}\|}$$
$$= \sqrt{\frac{\sum_{n=1}^{N_m}(q_{m,n}^{sim} - q_{m,n}^{obs})^2}{\sum_{n=1}^{N_m}(q_{m,n}^{obs})^2}} \; ; \qquad (16)$$

where N_m provides the number of observations available at scale 2^m. If the objective function value computed at each temporal scale is weighed by its fractional contribution to the total objective function value by ascribing the weighing coefficients in (13) as fractional errors, the total cost func-

tions for the multi-scale optimization problem corresponding to (15) and (16) become:

$$F_{RMSE}(\theta) = \sum_{m=0}^{k}\left[\frac{g_{RMSE,m}}{\sum_{m=0}^{k}g_{RMSE,m}}g_{RMSE,m}\right] \; ; \qquad (17)$$

$$F_{HMLE}(\theta,\lambda) = \sum_{m=0}^{k}\left[\frac{g_{HMLE,m}}{\sum_{m=0}^{k}g_{HMLE,m}}g_{HMLE,m}\right] \; ; \qquad (18)$$

where $g_{HMLE,m}$ denotes the HMLE cost function given in (15) evaluated at scale 2^m.

3.4. Site Description & Data Sources

The VIC-3L model in full energy and water balance mode was applied to the 510-km^2 drainage area upstream of U.S. Geological Survey (USGS) gauging station near Lower Lake, California (site number 11451500). The streamflow records for this station were used for calibration and validation of the VIC-3L model. Calibration was performed on two periods of different duration. The first period (denoted period 1 from here on) extends from October 21, 1952 through August 10, 1955 (1024 observations). The mean annual precipitation for the study area in this period is 780 mm, and the corresponding observed runoff ratio is 0.447. The second period (denoted period 2 from here on) extends from January 1, 1950 through August 10, 1955 (2048 observations). The mean annual precipitation and observed runoff ratio for the study area during this period are 880 mm and 0.442, respectively. Figure 3 displays the mean monthly precipitation and observed mean monthly streamflow corresponding to period 2. Validation was conducted from August 11, 1955 through March 19, 1961 (2048 observations). During this time interval, the mean annual precipitation and runoff ratio for the study area were 940 mm and 0.445, respectively. Figure 4 displays the mean monthly precipitation and observed mean monthly streamflow for the validation period. Most of the precipitation in the catchment of interest occurs during winter cyclonic storms.

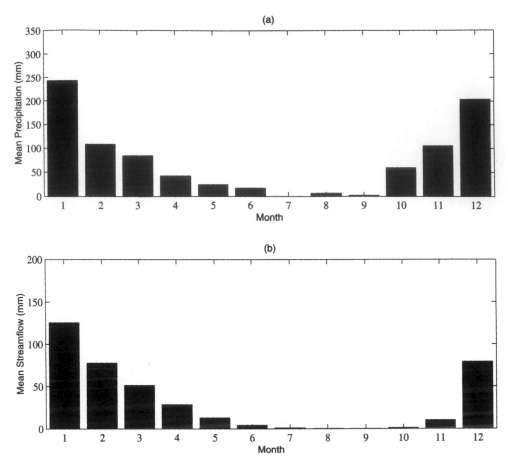

Figure 3. Mean monthly precipitation (a) and streamflow (b) for calibration period 2 (1/1/1950 through 08/10/1955, 2048 days).

Rainfall events usually begin in November and occur frequently through mid-April. Surface runoff is observed during this period. Runoff becomes negligible as the rainfall season ends. During the rest of the year, dry conditions prevail and streamflow is very low.

Daily gridded meteorological data (precipitation, wind speed, and minimum and maximum daily temperature) to one-eighth degree were obtained from the web site (http://www.hydro.washington.edu/Lettenmaier/gridded_data/) of the Surface Water Modeling group at the University of Washington. *Maurer et al.* [2001] describe the methodology used to compile and process this data set. Soil and vegetation parameters to the same resolution were also acquired through the same source.

Ten grid cells were partially included within the study area. The runoff from each of these cells was obtained and weighted by the area of each cell located within the catchment divided by the total catchment area. The runoff from all cells was aggregated and routed to the outlet point by within-catchment routing. Two to four vegetation classes were assigned to each cell within the catchment, the pre-

dominant ones being woodland, wooded grassland, and evergreen needle-leaf forest. The range of mean–cell elevation for the area of interest is 477 to 1127 meters.

3.5. Evaluation of Multi-Resolution Optimization Approach

To evaluate the benefits and robustness of the multi-resolution optimization procedure, calibration of the VIC-3L model was performed on two periods of different duration, as described in section 3.4, with both RMSE and HMLE as given by single scale formulations (14) and (15) and the corresponding multi-resolution ones (17) and (18). For the multi-resolution approach, the eighth wavelet scale (i.e. m = 8) was the coarsest resolution considered for optimization. Tables 2 and 3 list the optimal parameter sets obtained through single and multi-scale calibration with RMSE and HMLE during periods 1 and 2 respectively. Validation for all cases was conducted during a single period consisting of 2048 observations.

The percent improvement (I) measure shown below was utilized to establish several comparisons:

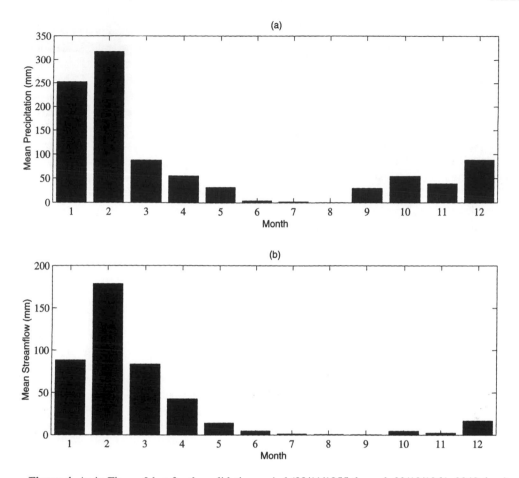

Figure 4. As in Figure 3 but for the validation period (08/11/1955 through 03/19/1961, 2048 days).

$$I = \frac{F\big(\theta_b \mid \{q_m^{sim,b}\}, \{q_m^{obs}\}\big) - F\big(\theta_a \mid \{q_m^{sim,a}\}, \{q_m^{obs}\}\big)}{\overline{q_m^{obs}}} \times 100 ; \quad (19)$$

where F is given by (14) or (15) evaluated for the scale with index m, given the time series of observed streamflow values, $\{q_m^{obs}\}$, and those of simulated ones, $\{q_m^{sim,a}\}$ and $\{q_m^{sim,b}\}$, at that resolution. $\overline{q_m^{obs}}$ is the mean observed streamflow at the

same scale. This equation yields the improvement, if positive, obtained by simulating the observed process with $\{q_m^{sim,a}\}$, and its corresponding parameter set θ_a, over simulating it with $\{q_m^{sim,b}\}$, and its parameter set θ_b.

3.5.1. Evaluation of Multi-Resolution Paradigm

Tables 4 and 5 list improvements for the entire validation period obtained by the use of the multi-resolution approach

Table 2. Optimal parameter values for calibration during period 1 (10/21/1952 through 08/10/1955, 1024 days). SS and MS denote single-scale and multi-scale optimization, respectively.

Parameter	RMSE (SS)	RMSE (MS)	HMLE (SS)	HMLE (MS)
b	0.196	0.260	0.106	0.104
D_s	0.022	0.029	0.998	0.991
D_m	7.903	10.455	7.221	7.012
W_s	1.000	0.980	0.001	0.001
D_1	0.819	0.733	1.000	1.000
D_2	0.591	0.846	0.527	0.524
K_{stor}	1.291	1.363	1.001	1.000

Table 3. As in Table 2 but for calibration during period 2 (1/1/1950 through 08/10/1955, 2048 days).

Parameter	RMSE (SS)	RMSE (MS)	HMLE (SS)	HMLE (MS)
b	0.202	0.226	0.091	0.213
D_s	0.003	0.014	0.950	0.009
D_m	8.599	9.870	31.544	9.876
W_s	1.000	1.000	0.001	1.000
D_1	0.727	0.620	1.000	0.692
D_2	0.661	0.789	2.973	0.795
K_{stor}	1.328	1.400	1.002	2.525

for calibration during periods 1 and 2 respectively. That is, $\{q_m^{sim,a}\}$ in (19) is the model-simulated discharge for the validation period obtained using the parameters derived by means of multi-resolution optimization and $\{q^{sim,b}\}$ is the equivalent for the parameters derived using single-scale calibration (i.e. the conventional approach). Improvements were computed with both RMSE and HMLE regardless of which of these cost functions was minimized during calibration.

Multi-resolution optimization yields improvements during validation in three of the four study cases, namely when calibration is performed with RMSE and 1024 observations, and when it is conducted with RMSE and HMLE with 2048 observations. When calibration is performed with HMLE and 1024 observations, the implementation of the multi-resolution paradigm performs equivalently to its single scale counter-part. These statements hold regardless of whether the objective functions used to compute the improvements are the same as those minimized during optimization. Figure 5 illustrates the performance of VIC-3L during three different rainfall seasons in the validation period when the parameters obtained by single and multi-scale calibration with RMSE and 1024 observations are used. This study case is the one for which improvements can be more easily noticed by visual inspection since they correspond to better representations of peaks as well as recessions. For all other cases, improvements were seen to result mainly from better characterizations of recession periods.

There are two main conclusions to draw from these observations. First, the use of multi-resolution optimization appears likely to yield better results during validation than its single scale counter-part, and it may, at least, perform equivalently. Second, when improvements are obtained in validation with the multi-resolution approach, these are observed in terms of RMSE and HMLE regardless of which of these objective measures was selected for calibration. In terms of the multi-objective optimization terminology commonly used in the literature [*Gupta et al.*, 1998; *Yapo et al.*, 1998], this implies that the multi-resolution methodology may define an optimal Pareto front that is better, in terms of both objective functions, than the optimal Pareto front obtained by traditional single-scale optimization.

3.5.2. Evaluation of Objective Functions

Tables 6 and 7 list improvements for the entire validation period obtained by the minimization of HMLE rather than RMSE in calibration during periods 1 and 2 respectively. That is, $\{q_m^{sim,a}\}$ in (19) is the model-simulated discharge for the validation period obtained using the parameters derived by optimization with HMLE, and $\{q^{sim,b}\}$ is the equivalent for the parameters derived by calibration with RMSE. For this comparison, both and correspond to parameter sets derived by using either single-scale or multi-scale optimization.

The validation improvements obtained by single-scale optimization with HMLE as an objective measure during periods 1 (Table 6) and 2 (Table 7) are negative if computed with RMSE and positive for the daily (m = 0) and two-day (m = 1) scales if determined with HMLE. This indicates that with the single-scale approach a trade-off exists at the daily and two-day scales between the use of parameters derived by optimization with RMSE and HMLE. Conversely, the validation improvements obtained by multi-resolution optimization with HMLE as a cost function during period 1 are negative for all scales regardless of whether RMSE or HMLE are used to compute them. The validation improvements obtained by multi-resolution optimization with HMLE during period 2 are, in turn, predominantly positive regardless of whether they are computed with RMSE or HMLE. This implies that no tradeoff exists between the

Table 4. Percent improvements, if positive, during validation obtained by the use of multi-resolution optimization for calibration during period 1 (10/21/1952 through 08/10/1955, 1024 days). Column headings indicate the objective measures used for calibration, as well as those employed to evaluate improvements (shown with Italics and in parenthesis).

Scale index (m)	RMSE (*RMSE*)	RMSE (*HMLE*)	HMLE (*RMSE*)	HMLE (*HMLE*)
0	67.4	18.8	-0.18	0.01
1	57.2	15.7	-0.20	0.01
2	32.8	10.3	-0.15	0.01
3	10.1	4.23	-0.10	0.01
4	2.82	-2.14	-0.09	0.01
5	-3.58	-7.12	-0.05	0.01
6	1.20	-6.32	-0.01	0.01
7	-1.57	-7.38	0.00	0.01
8	-3.67	-6.78	0.00	0.01

Table 5. As in Table 4 but for calibration during period 2 (1/1/1950 through 08/10/1955, 2048 days).

Scale index (m)	RMSE (*RMSE*)	RMSE (*HMLE*)	HMLE (*RMSE*)	HMLE (*HMLE*)
0	13.6	5.22	73.4	10.9
1	13.3	4.62	80.7	11.5
2	8.50	3.15	73.5	14.1
3	5.99	1.28	73.6	17.4
4	3.70	-0.48	79.2	23.7
5	-0.45	-3.29	62.3	29.9
6	-1.28	-3.79	54.1	30.8
7	-2.07	-3.44	46.5	34.3
8	-1.61	-2.93	38.5	33.4

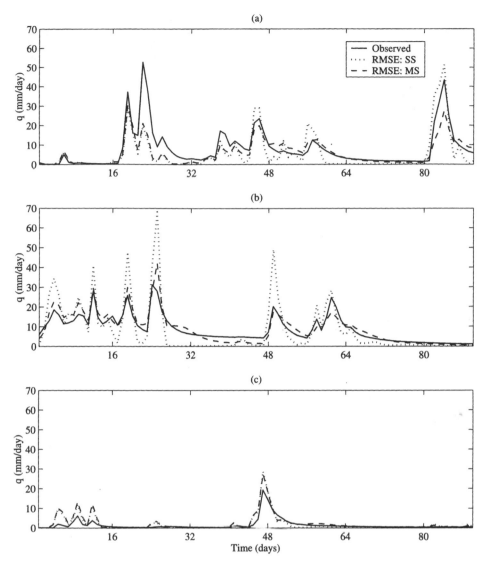

Figure 5. Validation results for the case in which calibration was performed with RMSE and during period 1 (10/21/1952 through 08/10/1955, 1024 days). Subplots display 90-day periods starting December 1, 1955 (a), February 1, 1958 (b), and January 1, 1959 (c). SS and MS denote single-scale and multi-scale optimization respectively.

use of parameters derived by multi-scale calibration with RMSE and HMLE. The parameters obtained by multi-resolution optimization with RMSE yield the best results in validation if calibration is performed during period 1. On the other hand, if multi-scale calibration is conducted during period 2, the parameter set derived by minimization of HMLE yield the best results during validation. Figure 6 illustrates the performance of VIC-3L for three rainfall seasons in the validation period when the parameters obtained by multi-scale calibration with RMSE during period 1 and HMLE during period 2 are used. The model performance is equally good in both cases.

Tables 8 and 9 list improvements for the entire validation period obtained by conducting calibration with 2048 obser-

vations (i.e. during period 2) instead of 1024 observations (i.e. during period 1) with RMSE and HMLE used as objective measures respectively. That is, $\{q_m^{sim,a}\}$ in (19) is the model-simulated discharge for the validation period obtained using the parameters derived through optimization during period 2, and $\{q^{sim,b}\}$ is the equivalent for the parameters derived by calibration during period 1. For this comparison, both $\{q_m^{sim,a}\}$ and $\{q^{sim,b}\}$ correspond to parameter sets derived by using the same cost function (i.e. RMSE or HMLE) and either single-scale or multi-scale optimization. As shown in Table 8, the validation results obtained by using the parameter sets derived through optimization with RMSE and 2048 observations are worse than the corresponding results for calibration with the same objective

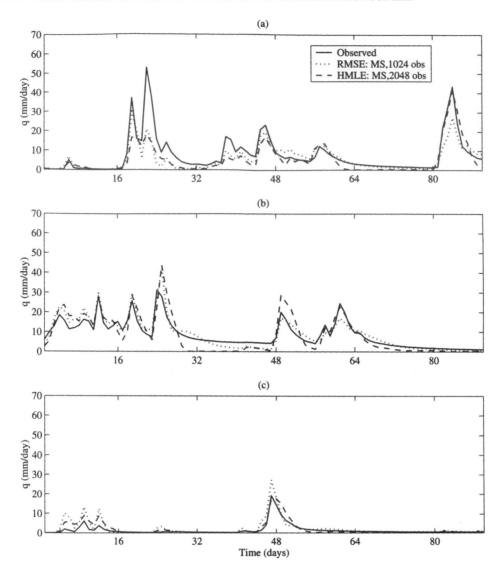

Figure 6. Validation results for the case in which multi-resolution calibration was performed during period 1 (10/21/1952 through 08/10/1955, 1024 days) with RMSE and during period 2 (1/1/1950 through 08/10/1955, 2048 days) with HMLE. Subplots display 90-day periods starting December 1, 1955 (a), February 1, 1958 (b), and January 1, 1959 (c). MS denotes multi-scale optimization.

measure and 1024 observations regardless of whether the single-resolution or multi-resolution approach are used. On the contrary, Table 9 shows that the results obtained by the use of parameters derived with HMLE improve if calibration is conducted with 2048 instead of 1024 observations. Figures 7 and 8 depict the performance of VIC-3L for three rainfall seasons in the validation period with the parameters obtained by multi-scale calibration during periods 1 and 2 with RMSE and HMLE respectively. These confirm the results presented in Tables 8 and 9.

The results exposed thus far indicate that it is possible for multi-scale calibration with either RMSE or HMLE to yield optimal parameter sets, as determined during validation,

even if a trade-off between the two was associated with the corresponding parameter sets obtained by single-scale optimization. Whether optimal parameter sets are derived by multi-scale minimization of RMSE or HMLE is seen to depend on the period over which calibration is conducted. In practice, validation results are not available prior to the choice of a period and a cost function for use in optimal parameter estimation. In spite of this, the multi-resolution paradigm constitutes a more robust alternative for calibration of hydrologic models than its traditional single-scale counter-part since it seems likely to yield better results during validation regardless of the chosen cost function or calibration period.

4.0. Summary and Conclusions

Wavelet analysis was incorporated into the SCE-UA optimization algorithm and applied to the calibration of one routing and six soil parameters of the VIC-3L hydrologic model for a sub-humid catchment in Northern California. The full energy and water balance mode of VIC-3L was utilized at a daily time step. Calibration was preformed based on two objective functions, Root Mean Square Error (RMSE) and Heteroscedastic Mean Likelihood Error (HMLE), and over two periods consisting of 1024 and 2048 days respectively. Validation for all cases was conducted over a period of 2048 days.

The results obtained indicate that it is possible for multi-scale calibration with either RMSE or HMLE to yield optimal parameter sets for validation in terms of both objective measures even if a trade-off between the two was associat-

ed with the corresponding parameter sets obtained by single-scale optimization. Whether optimal parameter sets are derived by multi-scale minimization of RMSE or HMLE is seen to depend on the period over which calibration is conducted. In practice, validation results are not available prior to the choice of a period and a cost function for use in optimal parameter estimation. In spite of this, the multi-resolution paradigm constitutes a more robust alternative for calibration of hydrologic models than its traditional single-scale counter-part since it appears likely to yield better results during validation, and it may, at least, perform equivalently. Moreover, when improvements are obtained in validation with the multi-resolution approach, these are observed in terms of RMSE and HMLE regardless of which of these objective measures was selected for calibration.

Table 6. Percent improvements, if positive, during validation obtained by using HMLE versus RMSE as an objective measure for calibration during period 1 (10/21/1952 through 08/10/1955, 1024 days). Column headings indicate whether single scale or multi-scale results are being compared, as well as the objective functions employed to evaluate improvements (shown with Italics and in parenthesis).

Scale index (m)	Single scale (RMSE)	Single scale (HMLE)	Multi-scale (RMSE)	Multi-scale (HMLE)
0	-9.40	3.91	-77.0	-14.8
1	-22.1	0.60	-79.6	-15.1
2	-42.6	-5.80	-75.5	-16.1
3	-67.7	-13.8	-78.0	-18.0
4	-77.5	-21.1	-80.4	-19.0
5	-63.9	-27.4	-60.4	-20.3
6	-55.7	-28.2	-57.0	-21.8
7	-47.6	-31.9	-46.0	-24.5
8	-38.3	-31.3	-34.6	-24.5

Table 7. As in Table 6 but for calibration during period 2 (1/1/1950 through 08/10/1955, 2048 days).

Scale index (m)	Single scale (RMSE)	Single scale (HMLE)	Multi-scale (RMSE)	Multi-scale (HMLE)
0	-3.86	9.28	55.9	14.9
1	-17.0	5.31	50.4	12.2
2	-37.7	-2.12	27.3	8.85
3	-63.7	-11.9	3.92	4.19
4	-75.2	-21.4	0.27	2.80
5	-64.2	-30.6	-1.49	2.66
6	-56.1	-32.1	-0.72	2.46
7	-48.7	-35.5	-0.18	2.28
8	-39.7	-34.3	0.41	1.95

Table 8. Percent improvements, if positive, during validation obtained by conducting calibration during period 2 (1/1/1950 through 08/10/1955, 2048 days) versus period 1 (10/21/1952 through 08/10/1955, 1024 days) with RMSE as an objective measure. Column headings indicate whether single scale or multi-scale results are being compared, as well as the objective functions employed to evaluate improvements (shown with Italics and in parenthesis).

Scale index (m)	Single scale (RMSE)	Single scale (HMLE)	Multi-scale (RMSE)	Multi-scale (HMLE)
0	-5.60	-5.29	-59.4	-18.8
1	-5.23	-4.63	-49.1	-15.7
2	-5.23	-3.60	-29.6	-10.7
3	-4.26	-1.73	-8.40	-4.68
4	-2.56	0.36	-1.67	2.02
5	0.16	3.27	3.30	7.10
6	0.39	4.05	-2.09	6.58
7	1.16	3.66	0.66	7.60
8	1.47	3.13	3.54	6.98

Table 9. As in Table 8 but for calibration with HMLE.

Scale index (m)	Single scale (RMSE)	Single scale (HMLE)	Multi-scale (RMSE)	Multi-scale (HMLE)
0	-0.06	0.08	73.5	10.9
1	-0.15	0.09	80.8	11.6
2	-0.33	0.09	73.3	14.2
3	-0.25	0.11	73.5	17.5
4	-0.29	0.12	79.0	23.8
5	-0.16	0.08	62.2	30.0
6	-0.01	0.09	54.1	30.9
7	0.06	0.07	46.5	34.4
8	0.08	0.07	38.5	33.4

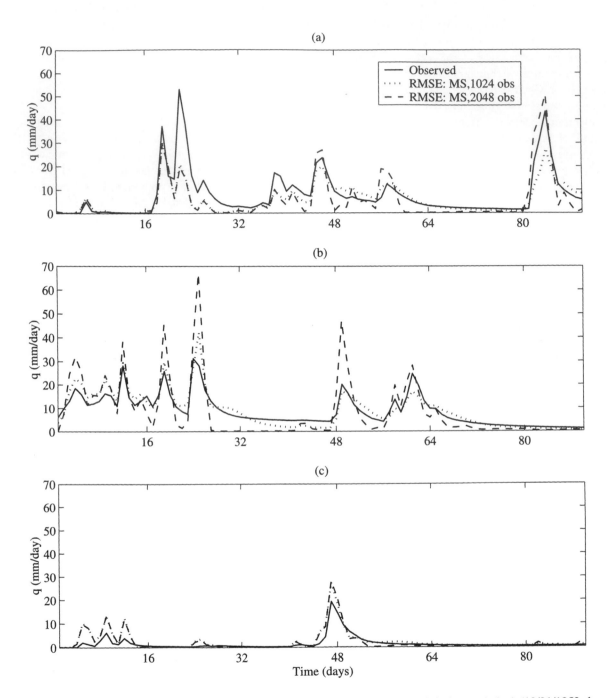

Figure 7. Validation results for the case in which calibration was performed during periods 1 (10/21/1952 through 08/10/1955, 1024 obs) and 2 (1/1/1950 through 08/10/1955, 2048 obs) with the multi-resolution methodology and RMSE as an objective measure. Subplots display 90-day periods starting December 1, 1955 (a), February 1, 1958 (b), and January 1, 1959 (c). MS denotes multi-scale optimization.

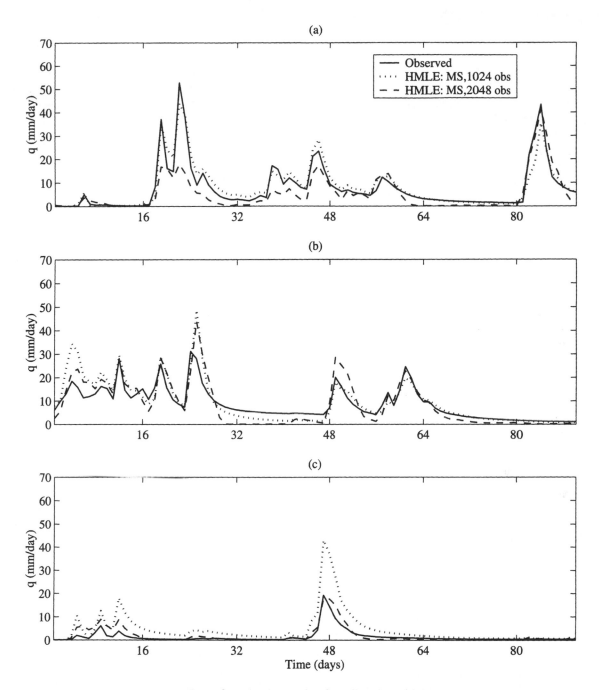

Figure 8. As in Figure 7 but for calibration with HMLE.

Acknowledgements. The authors thank Dr. Dennis P. Lettenmaier and his group for providing the vegetation and soil information for conducting this study. The research reported herein is partially supported by the National Oceanic and Atmospheric Administration under grant NA86GP0595 to the University of California, Berkeley.

REFERENCES

Beven K. J., Changing ideas in hydrology: the case of physically based models. *J. Hydrol.*, 105, pp 157-172, 1989.

Beven, K. J., and A. M. Binley, The future of distributed models: Model calibration and uncertainty prediction, *Hydrol. Process.*, (6), pp. 279-298, 1992.

Cameron, D. S., K. J. Beven, J. Tawn, S. Blazcova, and P. Naden, Flood frequency estimation by continuous simulation for a gauged upland catchment (with uncertainty), *J. Hydrol.*, pp. 219, 169-187, 1999.

Chen, T. H. et al., Cabauw experimental results from the project for intercomparison of land-surface parameterization schemes, *J. Clim.*, 10(6), pp. 1194-1215, 1997.

Cherkauer, K. A., and D. P. Lettenmaier, Hydrologic effects of frozen soils in the upper Mississippi River basin, *J. Geophys. Res.*, 104(D16), 19, pp. 599-619, 610, 1999.

I. Daubechies, Orthonormal bases of compactly supported wavelets, *Commun. Pure Appl. Math.*, vol. 41, pp. 909-996, Nov. 1988.

Duan, Q., S. Sorooshian, and V. K. Gupta, Effective and efficient global optimization for conceptual rainfall-runoff models, *Water Resour. Res.*, 28(4), pp. 1015-1031, 1992.

Duan, Q., V. K. Gupta, and S. Sorooshian, Optimal use of the SCE-UA global optimization method for calibrating watershed models, *J. Hydrol.*, 158, 265-284, 1994.

Freer, J., K. Beven, and B. Ambroise, Bayesian estimation of uncertainty in runoff prediction and the value of data: An application of the GLUE approach, *Water Resour. Res.*, 32(7), pp. 2161-2173, 1996.

Gupta, H. V., S. Sorooshian, and P. O. Yapo, Toward improved calibration of hydrologic models: Multiple and noncommensurable measures of information, *Water Resour. Res.*, 34(4), pp. 751-763, 1998.

Gupta, H. V., S. Sorooshian, and P. O. Yapo, Status of automatic calibration for hydrologic models: Comparison with multilevel expert calibration, *J. Hydrol. Eng.*, 2(4), pp. 135-143, 1999.

Kumar, P., and E. Foufoula-Georgiou, Wavelet analysis in Geophysics: An introduction, in *Wavelets in Geophysics*, edited by E. Foufoula-Georgiou and P. Kumar, pp. 1-43, Academic, San-Diego, Calif., 1994.

Liang, X, D.P. Lettenmaier, E.F. Wood, and S.J. Burges, A simple hydrologically based model of land surface water and energy fluxes for general circulation models, *J. Geophys. Res. 99(D7)*, 14, pp. 415-428, 1994.

Liang, X, D.P. Lettenmaier, and E.F. Wood, Surface soil moisture parameterization of the VIC-2L model: evaluation and modification, *Global Planet. Change 13*, pp. 195-206, 1996a.

Liang X., D. P. Lettenmaier, and E. F. Wood, A one-dimensional statistical dynamic representation of subgrid spatial variability of precipitation in the two-layer variable infiltration capacity model, *J. Geophys. Res.*, 101(D16), 21, pp. 403-421, 1996b.

Liang, X., et al., The project for intercomparison of land-surface parameterization schemes (PILPS) phase 2(c) Red-Arkansas river basin experiment: 2. Spatial and temporal analysis of energy fluxes, *Global Planet. Change*, 19, pp. 137-159, 1998.

Liang, X., E.F. Wood, and D.P. Lettenmaier, Modeling ground heat flux in land surface parameterization schemes, *J. Geophys. Res.*, 104(D8), pp. 9581-9600, 1999.

Liang, X. and Z. Xie, A new surface runoff parameterization with subgrid-scale soil heterogeneity for land surface models, *Adv. Water Resour.*, 24, pp. 1173-1193, 2001.

Lohmann, D. E. Raschke, B. Nijssen and D. P. Lettenmaier, Regional scale hydrology: II. Application of the VIC-2L model to the Weser river, Germany, *Hydrol. Sci. J.*, 43(1), pp. 143-158, 1998a.

Lohmann, D., et al., The project for intercomparison of land-surface parameterization schemes (PILPS) phase 2(c) Red-Arkansas River Basin experiment: 3. Spatial and temporal analysis of water fluxes, *Global Planet. Change*, 19, pp. 161-179, 1998b.

Mallat, S., A theory for multiresolution signal decomposition: The wavelet representation, *IEEE T. Pattern Anal.*, 11(7), pp. 674-693, 1989a.

Mallat, S., Multiresolution approximations and wavelet orthonormal bases of $L^2(R)$. *Trans. Amer. Math. Soc.*, pp. 315, 69-87, 1989b.

Maurer, E. P., G. M. O'Donnell, D. P. Lettenmaier, and J. O. Roads, Evaluation of NCEP/NCAR Reanalysis Water and Energy Budgets using Macroscale Hydrologic Simulations, In: *Land Surface Hydrology, Meteorology, and Climate: Observations and Modeling, AGU series in Water Science and Applications*, AGU series in Water Science and Applications 3, V. Lakshmi, J. Albertson, and J. Schaake eds., pp. 137-158, 2001.

Nijssen, B., D. P. Lettenmaier, X. Liang, S. Wetzel, and E. F. Wood, Streamflow simulations for continental-scale river basins, *Water Resour. Res.*, 33(4), pp. 711-724, 1997.

Nijssen, B., G. M. O'Donnell, D. P. Lettenmaier, D. Lohmann, and E.F. Wood, Predicting the discharge of global rivers, *J. Clim.*, 14, pp. 3307-3323, 2001.

O'Donnell, G.M., K.P. Czajkowski, R.O. Dubayah, and D. P. Lettenmaier, Macroscale hydrological modeling using remotely sensed inputs: Application to the Ohio River basin, *J. Geophys. Res.*, 105(D10), pp. 12499-12516, 2000.

Saco, P., and P. Kumar, Coherent modes in multiscale variability of streamflow over the United States, *Water Resour. Res.*, 36(4), pp. 1049-1067, 2000.

Smith, L. C., D. L. Turcotte, and L. I. Bryan, Stream flow characterization and feature detection using a discrete wavelet transform, *Hydrol. Processes*, 12, pp. 233-249, 1998.

Sorooshian, S., and J. Dracup, Stochastic parameter estimation procedures for hydrologic rainfall-runoff models: correlated and heteroscedastic error cases, *Water Resour. Res.*, 19, 1, pp. 251-259, 1980.

Sorooshian, S., Q. Duan, and V. K. Gupta, Calibration of rainfall-runoff models: Application of global optimization to the Sacramento Soil Moisture Accounting model, *Water Resour. Res.*, 29(4), pp. 1185-1194, 1993.

Yapo, P. O., H. V. Gupta, and S. Sorooshian, Multi-objective global optimization for hydrologic models, *J. Hydrol.*, 204, pp. 83-97, 1998.

Wood, E. F., D. P. Lettenmaier, X. Liang, B. Nijssen, and S. W. Wetzel, Hydrological modeling of continental scale basins, *Annu. Rev. Earth Planet Sci., 25*, pp. 279-300, 1997.

Wood, E. F., et al., The project for intercomparison of land-surface parameterization schemes (PILPS) phase 2(c) Red-Arkansas river basin experiment: 1. Experiment description and summary intercomparisons, *Global Planet. Change*, 19(1-4), pp. 115-135, 1998.

Jonathan P. Fram, jfram@uclink.berkeley.ed

Xu Liang, liang@ce.berkeley.edu

Laura M. Parada, lparada@ce.berkeley.edu, Department of Civil and Environmental Engineering, University of California, Berkeley, CA 94720

Estimating Parameters and Structure of a Hydrochemical Model Using Multiple Criteria

Thomas Meixner

Department of Environmental Sciences, University of California, Riverside, California

Hoshin V. Gupta, Luis A. Bastidas, and Roger C. Bales

Department of Hydrology and Water Resources, The University of Arizona, Tucson, Arizona

Challenges in developing an accurate and precise parameter estimation method for catchment hydrochemical models have been a persistent roadblock to improving model performance. We investigate the use of generalized sensitivity analysis and multi-criteria calibration techniques to investigate parameter estimates, model structure, and natural processes using the Alpine Hydrochemical Model (AHM) of the Emerald Lake watershed (ELW), Sequoia National Park, California. A traditional generalized sensitivity analysis was conducted. The results of this sensitivity analysis were used to develop four subsets of criteria used to apply a multi-criteria parameter estimation algorithm to the AHM model of ELW. The sensitivity results revealed that mass flux measures of model error permitted focusing on the spring snowmelt period of time while concentration criteria focused on important processes throughout the year. This result led us to believe that a combination of mass flux and concentration criteria would be a productive approach in selecting the criteria to combine in a multi-criteria calibration of the model. In multi-criteria calibration we improved estimates of several hydrologic and biogeochemical processes in addition to identifying a flaw in the current representation of mineral weathering within the AHM, as applied to the Emerald Lake watershed. However, the calibration results also indicated that sensitivity analysis and model calibration evaluation procedures are useful in different ways for developing knowledge about watershed water quality models.

1. INTRODUCTION

For the past two decades there has been robust development of parameter estimation and calibration methodologies in conjunction with rainfall-runoff models. However, while improved parameter estimation and calibration methodologies as applied to catchment water quality models has been seen as a needed research goal [*Kirchner et al.*, 1996; *Christophersen et al.*, 1993; *Beck*, 1987] little progress has been made in applying the robust methods often applied to rainfall-runoff models to models of surface water quality. This inability to develop better parameter estimation, model calibration and model uncertainty methods for use with water quality models also limits the application of these models to important social issues, such as how to improve surface water quality [*National Academy of Sciences*, 2001]. We believe that one of the reasons for this lack of improvement is the absence of suitably developed methodologies to deal with the problems inherent in catchment water quality modeling.

Calibration of Watershed Models
Water Science and Application Volume 6

Past studies with watershed water quality models have shown the promise of using multi-criteria methods to better define model structure and estimated parameters [*Hooper et al.*, 1988; *De Grosbois et al.*, 1988; *Mroczkowski et al.*, 1997; *Kuczera and Mroczkowski*, 1998]. Still these studies have focused on using multiple criteria to better identify parameters and structure for the purely hydrologic components of watershed hydrologic and water quality models while ignoring the parameters that directly affect water quality (e.g. cation exchange and mineral weathering parameters). This approach presents a problem; on the one hand we know that chemical concentrations can give us information about watershed hydrologic properties, but on the other we know that chemical concentrations give information useful for parameterizing our models for watershed chemical properties. We present here a case study looking at the application of multi-criteria methods for parameter sensitivity analysis and parameter estimation of both chemical and hydrologic properties. We also consider how these methods can be used to investigate model structure and determine how the existing structure of a water quality model could be improved. This study is divided into two parts. For both parts we use the Alpine Hydrochemical Model (AHM) to simulate stream chemical composition for the Emerald Lake watershed (ELW). The first part deals with a general sensitivity analysis that we use to investigate the importance of certain model parameters to the model fluxes in our study [*Meixner et al.*, 1999]. The second part deals with using the results of the sensitivity analysis to investigate different combinations of criteria for use in developing parameter estimates and determining model structure [*Meixner et al.*, 2002]. We sought to answer two questions with the sensitivity analysis. First, what can differences in parameter sensitivity tell us about the information content of different objective functions? Second, what do the parameter sensitivity results tell us about the processes occurring in the real watershed?

In the second part of our study we addressed two main questions. First, what subset of criteria is necessary for AHM parameter estimation and model evaluation, and what methodology is best suited to selecting those criteria? Second, what do the calibration results imply about the hydrologic and hydrochemical processes that control stream chemical composition in the ELW, versus what the results imply about model structure? Also addressed are the broader implications of the results for the multi-criteria calibration of hydrologic and hydrochemical models. Finally, we address the implications of the differences and similarities in our sensitivity analysis and parameter estimation results.

2. METHODS

2.1 Site Description

ELW is a 120 ha headwater catchment located in the Sierra Nevada (36° 35' N, 118° 40' W), with elevation ranging from 2800 m at the lake to 3416 m at the summit of Alta Peak [*Wolford et al.*, 1996; *Tonnessen*, 1991] (Figure 1). The watershed is 48% covered by exposed granite and granodiorite, 23% by soil, and 23% by talus, and includes a 2 ha lake. On average, snowfall represents 95% of total annual precipitation. Streamflow is dilute with conductivity ranging from 2 to 10 µs cm^{-1} and alkalinity ranging from 15 to 50 µeq L^{-1}. The watershed is considered to be sensitive to changes in climate and atmospheric deposition due to thin soils, dilute waters, and snow dominated hydrology. The 1986 water year (October 1985 to September 1986) had significantly above normal snowfall and in 1987, significantly below normal snowfall. A simulation over the two years thus represents a robust test of the model since it must perform well in both wet and dry periods.

2.2 Model

AHM is a lumped conceptual model for alpine watersheds that requires the parameterization of the hydrologic and biogeochemical processes occurring in a watershed [*Wolford et al.*, 1996]. The model has been applied at several watersheds and has been found to perform well [*Meixner et al.*, 2000; *Meixner et al.*, 1998]. For ELW, the watershed was partitioned into three terrestrial subunits (rock, talus, and soil), a stream, and a lake (Figure 2). Each terrestrial subunit is made up of different compartments representing the snowpack, snowpack free water, snowmelt, surface runoff, and interception by trees and litter, and may contain multiple soil horizons. Stream and lake subunits have compartments for the snowpack, snowpack free water, snowmelt, stream/lake ice, and (as appropriate) streamflow or a stratified lake. Within this structure, a set of parameters defines the routing of flow from the rock subunit to the talus and soil subunits, and from there sequentially into the stream, the lake, and out of the watershed. AHM adjusts hydrologic and chemical inputs, outputs, and state variables for 13 separate compartments representing snow, vegetation, infiltration, and soil processes. AHM calculates chemical equilibrium and moves water and chemicals between compartments on a daily time step, with some processes calculated on a sub-daily time step. Model output can include detailed descriptions of all chemical calculations, tracking of both chemical and hydrologic storage and changes in storage within the watershed, soil chemical concentrations, and stream concentrations.

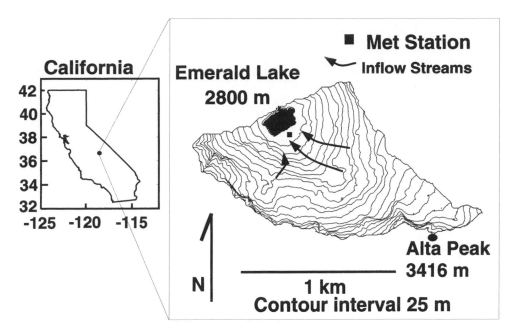

Figure 1. Elevation and location map of ELW.

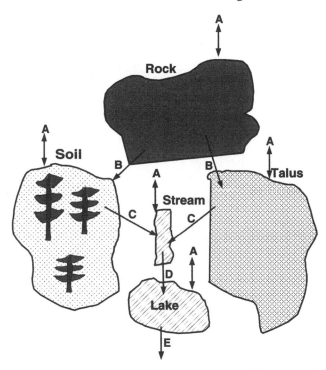

Figure 2. Schematic of AHM Model of the Emerald Lake watershed. A represents precipitation and evapotranspiration. B represents flow routed from rock subunit onto soil and talus subunits. C is surface and subsurface flow to stream from soil and talus. D is inflow to lake. E is lake outflow. Each subunit contains representations of major biogeochemical and hydrologic processes. For example, mineral weathering, cation exchange, unsaturated flow, and snowpack processes are represented in the soil subunit. Used with permission from *Meixner et al.* [2002].

2.3 Sensitivity Analysis Methods

A total of 20,000 Monte-Carlo simulations of ELW response were conducted for the sensitivity analysis. Each simulation was conducted by uniformly selecting values for the 24 model parameters from the ranges specified in Table 1: 10 of them are hydrologic parameters (five for the soil subunit and five for the talus subunit), and 14 are chemical parameters (4 cation exchange coefficients on each of the soil and talus subunits and 6 chemical parameters that are constrained by model structure to be the same for all subunits). For each simulation, 21 different objective functions were calculated, each being the sum of squared error (SSE) between a model simulated output and its associated value measured at the Emerald outflow (discharge and the 10 measured chemical species Ca^{2+}, Mg^{2+}, Na^+, K^+, Si, Cl^-, NO_3^-, SO_4^{2-}, pH, and acid neutralizing capacity (ANC)). For each of the 10 chemical species, 2 objective functions were calculated: i) the SSE between the measured and modeled concentration and ii) the SSE between the measured mass and the modeled mass flux. Differences between the concentration and mass flux objective functions can be understood qualitatively as concentration errors giving equal weight to errors throughout the year, and mass flux errors giving extra weight to model errors during the highest flow periods of the year during spring and summer snowmelt [*Meixner et al.*, 1999].

The 20,000 Monte-Carlo simulations were used as the basis for a multi-objective generalized sensitivity analysis (MOGSA) [*Bastidas et al.*, 1999; *Spear and Hornberger*,

Table 1. Parameters Varied and Range Relative to Values
by *Wolford et al.* [1996]

Parameter[a]	Range
ET[b]	0-1
Soil – depth[c]	0.5-1.5
N[c]	0.8-1.8
K[c]$_{sat}$	0.2-4
$_{sat}$[c]	0.5-1.5
Elution (D[d])	0.5-1.5
K-Ca^{2+e}	0.5-1.5
K-Mg^{2+e}	0.5-1.5
K-Na^{+e}	0.5-1.5
K-K^{+e}	0.5-1.5
K-SO$_4^2$ [f]	0.5-1.5
K-Si[f]	0.5-1.5
Soil P$_{CO_2}$[g]	0.5-1.5
K[h]	0.5-1.5
α[h]	0.5-1.5

a The hydrologic parameters and cation exchange coefficients were varied independently for the soil and talus subunits.

b ET represents the fraction of potential evapotranspiration permitted to occur.

c The parameter soil-depth represents total depth of soil on a subunit. Other hydrologic parameters are in the equation for unsaturated hydraulic conductivity:

$$K_u = K_{sat} \sqrt{W[1-(1-W^n)^{1/n}]^2}$$

where K_u is the unsaturated, and K_{sat} is saturated hydraulic conductivity, W is $(\theta - \theta_r)/\theta_{sat} - \theta_r)$, n is a drying curve coefficient that defines the relationship between θ and K_u, θ is soil water content, θ_r is the residual water content, and θ_{sat} is the saturated soil water content.

d The equation used to represent snowpack elution is:

$$\frac{C}{C_{ave}} = A\ B\ e^{(BX)} + (1 - A)\ D\ e^{(-DX)}$$

where C is the snowmelt solute concentration, C_{ave} is the initial concentration in the snowpack, coefficients A, B, and D define the magnitude and shape of the ionic pulse, and x is the fraction of already melted snowpack.

e The log K is used in the exchange reaction as shown for Ca^{2+}:
Ca^{2+} + 2(XH)   CaX$_2$ + 2H$^+$

f Log K for adsorption of SO$_4^2$ and H$_2$SiO$_4$ exchange as:

$$SO_4^2 + Y + 2H^+ \qquad YH_2SO_4$$

H$_2$SiO$_4$ + Z   Z - H$_2$SiO$_4$.

g The partial pressure of carbon dioxide PCO$_2$ was varied simultaneously for both subunits.

The weathering coefficients K and α were varied independently of each other but uniformly for all species. They contribute to weathering via the equation: Mol = A K[H$^+$]$^\alpha$ where Mol is moles added to the subunit, A is total area of the surfaces involved in reactions, [H$^+$] is hydrogen ion concentration, and K and α are constants. The total surface area is determined as the product of the soil depth, area, bulk density, and specific surface area for soil particles.

1980]. In MOGSA a subset of a population of 2000 samples was calculated using bootstrap methods. Sample populations smaller than 2000 resulted in fewer sensitive parameters while larger sample populations did not reveal additional sensitive parameters [*Meixner et al.*, 1999]. The sample population is then classified into behavioral (*B*) (i.e., having desired qualities) or non-behavioral (<u>B</u>) (i.e., not having desired qualities) sets; typically the desired quality is a low value for an objective function. The discrimination into behavioral and non-behavioral sets is entirely subjective and depends on the selection of a threshold "acceptable" value for the objective function. Here the 50% quantile was chosen as the threshold. For each parameter (θ_k) the empirical cumulative distribution function is computed for both behavioral, $F(\theta_k | B)$ and non-behavioral, $F(\theta_k | \underline{B})$ outcomes. The Komolgorov-Smirnov (K-S) statistic [*Stephens*, 1970] was used to discern whether the sampling distributions belong to the same underlying population distribution or not. If not, the parameter in question is deemed to be sensitive. A significance level α of 0.05 for the K-S test was used to discriminate sensitive from insensitive parameters, i.e., $\alpha < 0.05$ means that the behavioral and non-behavioral distributions of a parameter are not drawn from the same population and therefore the parameter being tested is sensitive.

2.4 Multi-Criteria Parameter Estimation Methodology

A thorough discussion of the application of multi-criteria theory to calibration of conceptual physically based models can be found in *Gupta et al.* [this volume, "Multiple…"]. The following is a brief summary of that methodology. Consider a model with parameters $\Theta = \{\Theta_1, ..., \Theta_n\}$ that is to be calibrated with observations (O_j) over m simulated model output variables. For each simulated response X_j, it is possible to define a criterion $f_j(\Theta)$ that represents the distance between the simulated value X_j and the observation O_j. The criterion f_j may be defined with any number of measures of model error or residual. The root mean squared error (RMSE) is a commonly used measure of model error. RMSE can be represented as:

$$RMSE(q) = \sqrt{\frac{1}{n} \sum_{t=1}^{n} (O_t - X_t(\theta))^2} \qquad (1)$$

where q is discharge, θ the set of model parameters, n the total number of observations O, and X the simulated value at time step t. The multi-criteria model calibration problem can formally be stated as:

Minimize $F(\theta) = \{f_1(\theta),...f_m(\theta)\}$ *subject to* $\theta \subset \Theta$ $\qquad (2)$

where the goal is to find values for θ within the feasible set Θ that minimize all of the criteria ($f_j(\Theta)$, $j=1,...,m$) simultaneously.

In practice it is not possible to minimize all criteria simultaneously. Instead, a set of solutions is commonly found, which have the property that it is necessary to deteriorate performance of one criterion in order to improve the performance of a second criterion within the set of solutions. The Pareto set represents the set of solutions that can objectively be considered better than all other possible solutions. However, objective comparisons within the set of solutions are not possible and therefore all of the solutions must be compared as a whole to other possible Pareto set solutions for a given model. The Pareto set represents the best solution available through model calibration without incorporating the subjective judgment stating that one or more of the criteria are more important than the others. The size of the Pareto set is related to errors in model structure and the calibration data set.

Yapo [1996] presented an efficient population-based optimization strategy that provides an approximate representation of the Pareto set with a single optimization run. This algorithm, Multi-Objective Complex Evolution (MOCOM-UA), is based on the Shuffled Complex Evolution (SCE-UA) optimization method [*Duan et al.*, 1993; *Duan et al.*, 1992]. The MOCOM-UA method begins by sampling the feasible space Θ at a number of locations. At each location the multi-criteria vector $F(\Theta)$ is computed, and the population is ranked and sorted using Pareto rank [*Goldberg*, 1989]. Simplexes of $n+1$ points are then selected using a rank-based method. Each simplex is evolved in an improvement direction using a multi-criteria extension of the downhill simplex method. Iterative application of the ranking and simplex evolution steps causes the population to converge towards the Pareto set. The algorithm terminates automatically when all points in the sample become mutually non-dominated [*Yapo et al.*, 1998].

2.5 Applying MOCOM-UA to AHM

The varied parameters and their feasible ranges were the same as the sensitivity analysis (Table 1). We developed four sets of complementary criteria and each set contained four criteria; using more than four criteria or fewer than four criteria gave inferior results [*Meixner*, 1999]. The first set was chosen on the basis of a correlation analysis of the observations of stream chemical composition at the Emerald outflow. The correlation analysis showed that the four least correlated time series, referred to as the data correlation case, were discharge (Q), and the concentrations of

H^+, Mg^{2+}, and Si (< 0.6 for each pair). A second set of four criteria was determined by using the four species with the least correlated RMSE values (< 0.05 for each pair). This set of four criteria, referred to as the criteria correlation case, were Q, ANC concentration, SO_4^{2-} mass, and Cl^- mass.

The remaining two sets of criteria were selected on the basis of information gained from the sensitivity analysis. The two sets were chosen based on which criteria shared the least number of sensitive model parameters. These two sets consisted of Q, Ca^{2+}, SO_4^{2-}, and Cl^- criteria. In the concentration case, the *concentration criteria* of SO_4^{2-} and Cl^- were used; in the other case, the mass flux case, the *mass flux criteria* for SO_4^{2-} and Cl^- were used. The concentration criterion for Ca^{2+} was used for both of these sets.

For each of these sets of criteria, the Pareto set was estimated using the MOCOM-UA algorithm. We selected a population size of 250 for the multi-criteria search; larger populations did lead to significantly better algorithm performance [*Meixner*, 1999]. Each calibration case thus has 250 different final parameter sets that are considered acceptable. Calibration success was compared using the Nash-Sutcliffe values [*Nash and Sutcliffe*, 1970] for all criteria. The Nash-Sutcliffe statistic is:

$$E = 1 - \frac{\sum_{i=1}^{N}(Q_i - P_i)^2}{\sum(Q_i - \bar{O})^2} \tag{3}$$

where O is the observation, $\bar{O}$ the mean observation, P the predicted value, O_i the observation or prediction in question and N the total number of observations.

A Nash-Sutcliffe value greater than zero indicates that the model is superior to the observational mean as a predictor of stream chemical composition [*Legates and McCabe*, 1999]. The calibration results for each of the four sets of criteria used in the MOCOM-UA algorithm will be viewed in four different ways: parameter space, criteria space, comparisons of the model results to the mean (using the Nash-Sutcliffe statistic) and comparisons between the calibration cases. Each model realization was given a rank from 1 to 1000 with a rank of 1 indicating the best possible simulation for a given model criteria. The ranks for each of the 4 cases were summed for each and every criterion and divided by 1000. In this situation a value of 31 for a given calibration case indicates that all 250 simulations for that case are superior to all simulations for other cases while a value of 219 indicates that all 250 simulations of a given case are inferior to all of the simulations from the other cases. The set of criteria that performs best is the one that has the smallest variation in estimated parameters, gives a calibrated model that is superi-

or to the mean observations, and is superior to the other available model calibrations.

3. RESULTS

3.1 Sensitivity Analysis Results

There were more sensitive hydrologic parameters than sensitive chemical parameters for concentration objective functions (Table 2). The PCO_2 in the subsurface, two weathering parameters, and the elution parameter, D, were sensitive for the largest number of concentration objectives. Exchange coefficients for the cations, Si and SO_4^{2-} were important for the species they are associated with and few other species.

The sensitivity analysis for the mass flux objectives tells a different story (Table 2). While more sensitive ion exchange parameters were found for mass flux objective functions, there were fewer sensitive hydrologic parameters. For example, only the elution parameter D was sensitive for Cl^- mass flux, while there were several sensitive hydrologic parameters for Cl^- concentration. In total, there were 40 sensitive chemical parameters for the 10 mass flux objective functions, while there were 41 sensitive chemical parameters for the concentration objective functions. However, the sensitive chemical parameters for concentration and mass flux were different. Fewer weathering parameters (15 weathering parameters for concentration, 7 for mass flux) but more exchange parameters (13 exchange parameters for concentration, 22 for mass flux) were found to be sensitive for mass flux objective functions.

For hydrologic parameters, using mass flux as opposed to concentration greatly decreased the number of sensitive parameters. There were 63 sensitive hydrologic parameters for chemical concentration as opposed to only 44 for mass flux. There were fewer sensitive hydrologic parameters for each of the chemical species except for Si, which saw no change, and SO_4^{2-}, which had 6 sensitive hydrologic parameters for mass flux as opposed to 2 for concentration. The evapotranspiration parameters (ET) for both of the subunits were important parameters for concentration objective functions, with a total of 10 sensitive ET parameters either on the soil or on the talus subunit; however, when mass flux was used, ET was only sensitive once.

3.2 MOCOM Results

Parameter space results indicate that the data correlation case contained little information about the processes represented by the model. The criteria correlation case, the concentration case, and the mass flux case resulted in relatively

Table 2. Parameter Sensitivity for Mass Flux and Concentration Criteria

Parameters	Objective Functions										
	ANC	Ca²⁺	Cl⁻	K⁺	Mg²⁺	Na⁺	NO₃⁻	H⁺	Q	Si	SO₄²⁻
ET (soil)	C¹			C		C	CM	C	M	C	
Soil-D (soil)	C	C	C	CM		CM	CM	C	M	C	C
N (soil)		C	C	CM	C	M	C	M	M		
K$_{sat}$ (soil)	CM¹	CM		CM	CM	CM	C	CM	M	CM	M
$_{sat}$ (soil)	C			CM		CM	CM	C	M	M	M
ET (talus)			C	C			C	C	M		
Soil-D (talus)	CM	CM	C	C	CM	CM	CM	CM	M	M	CM
N (talus)	C	CM	C		C	M	C	CM	M		M
K$_{sat}$ (talus)		M		C	M	M	CM	C	M	M	M
$_{sat}$ (talus)	CM	CM	C	C	CM		CM	CM	M	C	M
Elution (D)	C		CM		C	CM	CM	C			M
K-Ca²⁺ (soil)		CM		CM	CM			M			M
K-Mg²⁺ (soil)					CM						
K-K⁺ (soil)		M		CM							
K-Na⁺ (soil)						CM					
K-SO₄²⁻ (both)	M¹	M									CM
K-Ca²⁺ (talus)	M	CM			M	M					CM
K-Mg²⁺ (talus)		M			CM						
K-K⁺ (talus)				CM							
K-Na⁺ (talus)						CM					
K-Si (both)										CM	
K-H₂CO₃ (both)	CM	CM		CM	CM	CM		M			CM
(all species)	C	C		C		CM				C	C
all species	CM	CM		C	CM	CM		M		CM	C

₁ A C in a box indicates that the concentration version of that objective function was affected by changes in the parameter value, an M indicates that the mass flux objective function was affected and CM indicates that both the concentration and the mass flux criteria are affected.

small parameter spaces for the final 250 Pareto set results. The parameter values from the data correlation case spanned the entire parameter space (Figures 3 and 4), while the other three methodologies occupied a smaller fraction of the parameter space. The coefficient of variation (CV, the standard deviation of the Pareto set parameter values divided by the mean of the Pareto set parameter values) for all of the parameters can be used as a summary statistic of the precision of the parameter estimates for each case. The criteria correlation and concentration cases each had CVs of 0.077, the mass flux results registered an even lower 0.043, and the data correlation case had a CV of 0.2.

The criteria-space results for the four cases provide additional information on how each of the cases performed (Figure 5). The criteria correlation case results show improvement over the results of the initial parameters for ANC, Na⁺, Cl⁻ and Si, while this set of criteria was significantly worse for Mg²⁺ either as mass flux or concentration. There is little coherence for the data correlation case results and they are similar to what would result from using 250 sets of random parameter values [*Meixner*, 1999]. The results for both sensitivity analysis cases have some interesting contrasts. The case using only concentration criteria improved the simulations for several criteria, but with Mg²⁺,

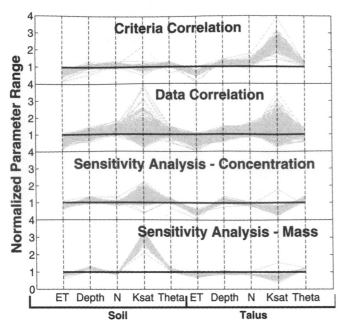

Figure 3. Hydrologic parameter ranges for four different calibration cases. See Table 1 for a description of the parameters. Soil and Talus labels refer to soil and talus subunits in the AHM model of Emerald Lake (Figure 2). Each panel of the figure is for a single criteria case that we investigated. A thick black line represents the initial parameter values [*Wolford et al.*, 1996]. Gray lines represent the 250 Pareto solutions for each calibration case. Y-Axis value is a multiple of the initial parameter value [*Wolford et al.*, 1996]. Used with permission from *Meixner et al.* [2002].

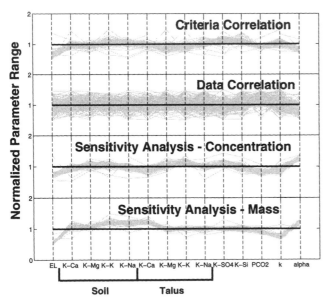

Figure 4. Chemical parameter ranges for four different calibration cases. Description of figure is the same as for Figure 3 except ranges for the 14 chemical parameters are shown. Used with permission from *Meixner et al.* [2002].

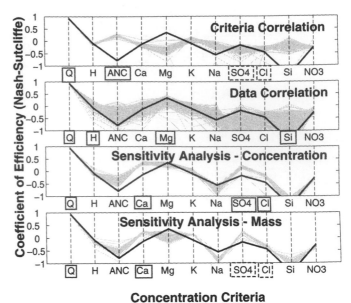

Concentration Criteria

Figure 5. Concentration criteria Nash-Sutcliffe values for the four calibration cases investigated. The thick black line indicates the Nash-Sutcliffe values for the parameter values determined by *Wolford et al.* [1996]. Boxed criteria indicate those criteria used to calibrate the model for each case. Criteria boxed with a dashed line were calibrated with mass flux form of criteria as opposed to the concentration criteria form. Used with permission from *Meixner et al.* [2002].

Na^+, NO_3^-, and H^+ predictions were not improved in any significant way over the initial parameter results. The results for the mass flux case were not nearly as good, with several criteria not being improved with respect to the initial parameter results. These results indicate that choosing criteria using the correlation or lack thereof of observations is not a good methodology for selecting the criteria used for multi-criteria calibration since results for the data correlation case were inferior to those of the other three cases.

First, comparing the overall results for each of the methodologies to the mean indicates that based on several criteria, model performance is inadequate at best (Table 3). In particular, the mean is a superior predictor for Cl^-, Si, Na^+, and NO_3^- for all but a few of the 1000 "best" simulations (250 parameter sets for each set of criteria) between the four sets of criteria used here. In general, the model does a good job of simulating the mass flux of all species except Cl^-. The model is expected to perform much better than the mean mass flux since mass flux varies by several orders of magnitude due to wide variations in flow while observed concentrations vary by a maximum of one order of magnitude. Among the four sets, the concentration case does the best on a mean basis with the average model prediction superior to the observational mean 178 out of 250 times. The worst of the four is the data correlation case, with the

Table 3. Number of Results Better Than the Mean

| Method | Q | Concentration Criteria | | | | | | | | | | Mass Flux Criteria | | | | | | | | | | Mean |
		H^+	ANC	Ca^{2+}	Mg^{2+}	K^+	Na^+	SO_4^{2-}	Cl^-	Si	NO_3^-	H^+	ANC	Ca^{2+}	Mg^{2+}	K^+	Na^+	SO_4^{2-}	Cl^-	Si	NO_3^-	
Criteria	250	0	250	1	2	241	0	0	0	11	1	250	250	250	250	249	250	249	247	250	250	155
Data	250	150	7	48	236	152	6	3	0	2	0	250	250	250	250	244	250	250	56	249	250	150
Sens. Conc.	250	3	0	250	250	242	0	248	0	0	0	250	250	250	250	250	250	250	249	249	250	178
Sens. Mass	250	0	0	249	98	250	0	248	0	0	0	250	250	250	250	250	250	250	249	250	250	171

model being better than the mean as a predictor of stream chemical composition only 150 out of 250 times.

Comparing the 4 different calibration cases with the sum of their ranks according to each criterion gives a different perspective on the results (Table 4). Of note is the fact that the data correlation case performs the worst of the four cases when looking at the ranking results of the criteria for which it was calibrated (average rank of 112). The mass flux case performs very poorly with respect to those objective functions for which it was not calibrated (average rank of 163) and thus performs relatively poorly overall (average rank of 132) when compared to the other criteria selection methods. These results indicate that the data correlation case performed worst overall (average rank of 135) with the mass flux case being the second worst overall. The criteria correlation case and concentration case both end up performing relatively well compared to the other two cases (average ranks of 119 and 116 respectively) with the concentration case performing marginally better mostly because in relative terms it does a better job in improving model performance for those criteria used to calibrate the model (average rank of 70).

4. DISCUSSION

4.1 Implications of Sensitivity Analysis Results

The results show systematic differences in parameter sensitivity for chemical concentration objective functions as opposed to mass flux objective functions. Using concentration as an objective function revealed parameter sensitivities that are important all year such as the fraction of potential evapotranspiration (ET) and the mineral weathering parameters (κ and α). Using mass flux as the objective function revealed parameters that can have a large effect for a short period of time, and especially during higher flows, like ion exchange and snowpack elution parameters.

Differences for hydrologic parameters other than ET are more mixed. Hydrologic parameters that determine soil pore-water volume θ_{sat}, soil depth, and residence time in that volume (K_{sat} and N) are important year round. These parameters determine the mixing volume of the subsurface and the rate of exchange between surface and subsurface water respectively.

These results indicate that mass flux and concentration objective functions contain different information about watershed processes. In particular, mass flux emphasizes parameters (e.g. exchange coefficients and snow elution parameters) that have a faster response within the model and possibly within the watershed. These processes are important in determining stream chemical composition dur-

ing spring snowmelt, the period of greatest sensitivity to acid deposition for alpine catchments [Williams et al., 1993]. Using concentration as an objective function highlighted sensitivities to evapotranspiration and mineral weathering parameters, which affect model output throughout the year and influence the mean model output. Thus, for investigating model error and structure with the goal of improving catchment models, it would be important to include measurements of model error that incorporate both mass flux and concentration objective functions.

Finally, our results indicate that the volume of talus and residence time of water in the soil subunit are among the most important hydrologic parameters in determining model output. Each subunit has two parameters that represent soil water holding capacity (θ_{sat}, and soil depth) and two parameters that represent rate of flow through that water holding volume (K_{sat} and N for unsaturated flow). A summary of the results from Table 2 shows that for soil, 22 flow rate parameters and 24 soil volume parameters were sensitive for chemical (concentration or mass flux) objective functions. For talus, 18 flow rate parameters and 32 talus volume parameters were sensitive for chemical objective functions. These results indicate that the AHM model of ELW is more sensitive to talus volume than it is flow rate through the talus. These results also indicate that flow rate through soil is more important than is flow rate through talus. Our results indicate that field efforts should focus on estimating the reactive volume of talus fields. This result confirms other previous studies identifying the importance of talus in alpine catchments [Campbell et al., 1995; Mast et al., 1995; Williams et al., 1997].

4.2 Implications of Multi-Criteria Calibration Results

According to the current analysis, the four criteria that are best suited for multi-criteria analysis are Q and the concentration of Ca^{2+}, SO_4^{2-}, and Cl^-. The companion set of criteria using the mass flux of SO_4^{2-} and Cl^- was inferior because it did not improve model results for all criteria as much as the concentration case did. Both the data correlation case and criteria correlation case were inferior to the concentration case (Figures 3, 4, and 5; Tables 3 and 4). The criteria correlation results had inferior simulations of the 21 criteria modeled using either the mean or the initial parameters as the benchmark for model comparison, but was only slightly inferior when comparing the 4 cases using the sum of ranks.

Criteria selected using multi-criteria sensitivity analysis resulted in the best-calibrated models, suggesting selection should consider the criteria that give the most independent parameter sensitivity results. Selecting criteria using sensitivity analysis results incorporates information contained in

Table 4. Sum of Rank Results For All Criteria [1]

		Concentration Criteria											Mass Flux Criteria												
Method	Q	H+	ANC	Ca²⁺	Mg²⁺	K⁺	Na⁺	SO₄²⁻	Cl⁻	Si	NO₃⁻	H⁺	ANC	Ca²⁺	Mg²⁺	K⁺	Na⁺	SO₄²⁻	Cl⁻	Si	NO₃⁻	Uncal.[4]	Cal.[5]	Mean
Criteria	123[2]	153	**31**	198	218	105	43	*164*[3]	*112*	35	104	218	32	217	218	57	43	**141**	**90**	4	163	129	97	119
Data	203	3	132	176	**87**	184	90	212	181	**114**	48	66	169	150	*105*	153	92	213	209	*41*	52	147	112	135
Sens. Conc.	139	90	135	**32**	42	165	153	**75**	**33**	211	132	67	167	*101*	46	111	179	*113*	*158*	09	84	128	70	116
Sens. Mass	35	213	202	**94**	153	47	215	*50*	*175*	140	216	150	132	**33**	131	180	187	**33**	**44**	17	201	163	51	132

1 Sum of rank refers to a bulk measure of calibration performance. Good performance means a low value. See methods section for more detail.
2 Numbers in bold indicate criteria that were included in the calibration.
3 Numbers in italics mean that the opposite pair of the concentration and mass flux criteria was used as a calibration criterion.
4 Uncal. refers to the average rank for the criteria not used in the calibration.
5 Cal. refers to the average rank of the criteria used in the calibration.

the data, model, and user's intuition. The sensitivity analysis incorporates information from the data since the data are used to determine which simulations are considered behavioral and which are considered non-behavioral. Information from the model is obviously incorporated since the purpose of the sensitivity analysis is to determine which criteria are sensitive to which model parameters [*Bastidas et al.*, 1999]. The user's intuition is involved since the user must determine which criteria contain information about the model form and the processes represented by the model. The user does this by using the sensitivity results and their physical understanding of the important processes to determine which set of criteria would be best.

Further improvements in criteria selection should be sought, since the current method of selecting criteria is still rather subjective. Preliminary results with a cluster analysis of the criteria values from the 20,000 Monte-Carlo simulations supported the four criteria selected using the sensitivity analysis. However, the cluster analysis did not provide a single unique result. Using other sets of criteria that would be acceptable according to the cluster analysis did not result in superior model simulations.

4.3 Natural Process

The calibration results reveal several things about the processes controlling stream chemical composition in the Emerald Lake watershed. First, the results indicate that evapotranspiration is less than currently modeled (Figure 3) using the initial parameters. Field and modeling exercises should focus on summertime estimates of ET, since model predictions of stream chemical composition are insensitive to ET parameters during spring snowmelt [*Meixner et al.*, 1999]. Second, the results indicate that the hydraulic conductivity (K_{sat}) of the soil should be higher than the current value, for the talus results were mixed as they related to K_{sat} (Figure 3). Finally, the results point to a deeper talus and a larger saturated water holding capacity (θ_{sat}) than in the initial parameters. These results indicate the need for a larger reactive volume for the talus subunit than currently used in the model. They also corroborate the earlier results from the sensitivity analysis that these two processes, rate of flow through soil and the volume of talus, are the key hydrologic processes in the AHM representation of ELW. For the chemical parameters, there are fewer clear-cut examples of the analyses' ability to limit the parameter space and to understand the processes controlling stream chemical composition. One exception is that all cases except the data correlation case indicated a lower elution parameter (EL).

Finally, the two correlation cases indicated a slower weathering rate while the two sensitivity analysis cases indicated a higher rate. These results might be caused by parameter interaction between α and κ. However, there is no discernible relationship for all four cases between the two parameters, for the 250 parameter sets for each case ($p < 0.02$). The mixed results may arise because the criteria behave differently. Silica, Na^+, and ANC would be better simulated by a lower value for the weathering rate parameter α while cations Ca^{2+}, Mg^{2+}, and K^+ need a higher α.

4.4 Model Structure

This conflict represents a problem within the current model structure for mineral weathering. Three criteria, ANC, Si, and Na^+ had mean, median, and at times maximum Nash-Sutcliffe values (out of the 250 Pareto results) less than zero for most of the cases; thus for these three criteria the mean is a better predictor of stream chemical composition than is the model. This result, together with the conflict between modeling ANC, Si, and Na^+ vs. Ca^{2+}, Mg^{2+}, and K^+, points to a flaw in model structure. This conflict might also be the reason for the relatively poor results using the criteria selected by correlation of the observed data. Three other sets of criteria simulated based on a cluster analysis of criteria values gave similar results. For these other calibration cases, the only good results were for those cases that included only non-weathering product criteria and either ANC, Na^+, and Si criteria or Ca^{2+}, Mg^{2+}, and K^+ criteria. For example, when calibrating on Q, Cl^-, Ca^{2+} and K^+, results were poor for ANC, Na^+, and Si.

These results indicate a mistake in model structure either in the temporal variability or stoichiometry of mineral weathering. Currently weathering is assumed to occur evenly throughout the year, which was confirmed by *Shaw* [1997]. However, a different representation of weathering, such as equilibrium dissolution of kaolinite into solution as suggested by *Campbell et al.* [1995] might be a better approach than the current simple kinetic approach adopted by *Wolford* [1992].

4.5 Broader Implications for Multi-Criteria Analysis

Our results illustrate the problems that need to be addressed to successfully apply multi-criteria analysis for water quality models. First, our method shows that criteria not used in the calibration can be used to determine the success of the calibration procedure. In our case, this methodology was successful in determining which combination of criteria was best suited for use in a multi-criteria analysis. Second, our results illustrate that in addition to the classic problem of parameter interaction and parameter correlation; there is now the problem of correlation of criteria and criteria interaction. This problem may explain some of the difficulties others have had in using multi-criteria analysis. For example, some of the problems *De Grosbois et al.* [1988] and *Hooper et al.* [1988] had in their analysis of the Birkenes model may have been solved if they had tools such as MOCOM-UA available and if they had investigated the interaction of criteria more fully (incorporating several geochemical criteria, e.g. ANC or Ca^{2+}). Such an analysis may have prevented the later conflicts between the improvements they made to model hydrologic structure and the deterioration in simulating stream chemical composition when their results were extended by others [*Stone and Seip*, 1989; *Lundquist et al.*, 1990]. Our results also indicate that scientific intuition played a role in several of the past successes with multi-criteria model analysis [*De Grosbois et al.*, 1988; *Hooper et al.*, 1988; *Uhlenbrook and Leibundgut*, 1999]. These previous investigations all used external information from the literature or from their personal experience with the data set and model at hand to select the multiple criteria they used for calibration.

Comparing the results of *Mroczkowski et al.* [1997] and *Kuczera and Mroczkowski* [1998] to ours we might infer that groundwater stage and stream discharge data contain conflicting information about watershed processes as represented by their model. As they stated, there is a danger in "assuming that augmenting streamflow data with other response time series data will significantly reduce parameter uncertainty." This warning should be expanded to include hydrochemical models and biogeochemical models. Due to the cost of acquiring time series other than streamflow, they recommended better studies a priori of the worth of additional time series of data. We would add to this caution that different combinations of criteria may improve results in different and possibly conflicting ways and that these combinations can be used to investigate model structure and the relative worth of data time series.

In addition, these multi-criteria calibration results can assess the validity of existing manual calibration methodologies that use median parameter values from multiple manual calibrations. To answer this question we compared simulation results using the median parameter values of the 250 parameter sets for the four cases we studied to the 250 criteria space results for each case. The results by case and criteria are summarized in Table 5, which shows the fraction (0.0-1.0) of Pareto results that performed worse than the median for each of the 21 criteria and each of the four cases we investigated. A value of 0 indicates that the median parameter values performed worse than every member of the Pareto set; a value of 1.0 means that the median parameter values performed better than all members of the Pareto set. On average, the median parameter values cases performed better than the Pareto solutions 43 percent of the time. This result indicates that the median of the parameter values for each case typically performed a little worse than the median of the Pareto results. The median does not perform equally well for all criteria. For example, the median is significantly inferior to the Pareto results for NO_3^- but significantly superior to the Pareto results for Si.

The two cases that had the best performance for the median parameter values were the data correlation case and the concentration case. The result for the data correlation case is expected since the parameter and criteria value results for that case were so poor. The concentration case results indicate that for a properly conducted multi-criteria calibration (either automatic or manual) the median represents a reasonable way to arrive at a preferred solution from the Pareto set results. Therefore, in circumstances where a competent multi-criteria calibration has been conducted (e.g. the concentration case in the current study), using the median is an acceptable means of determining a single preferred solution to be used for simulations of perturbed conditions.

4.6 Contrasts Between Sensitivity Analysis and Multi-Criteria Calibration

Sensitivity analysis techniques and multi-criteria calibration reveal different things about our model and more broadly indicate different aspects of the AHM. However, the contrasts between the two offer an additional lesson on the utility of the two methods. In our traditional approach to applying parameter sensitivity analysis we learned useful things about the model and about the watershed we were trying to model. For example, we found out that the volume of talus was important but the rate of flow through soil was also important. We learned about the relative seasonal sensitivity of mineral weathering and ion exchange.

However, we also used the sensitivity analysis to give us information as we went forward and investigated our model using multi-criteria calibration techniques. In this application we see interesting contrasts that might be helpful in the joint application of these methods. From the sensitivity results alone we argued that incorporation of mass flux and concentration objective functions would be the best approach. Our calibration results indicate that this was not the case. Looking at the sum of rank results (for the mass flux case) it is noticeable that incorporating mass flux criteria improved simulations of stream discharge (Q) resoundingly over the other calibration cases (average rank of 35). This result is not too surprising since improving simulations of discharge would improve all simulations of mass flux. However, improving simulations of discharge appears to have decreased model performance with respect to simulat-

Table 5. Fraction of Pareto Results Performing Worse Than Median Parameter Values

Method	Concentration Criteria											Mass Flux Criteria										
	Q	H^+	ANC	Ca^{2+}	Mg^{2+}	K^+	Na^+	SO_4^{2-}	Cl^-	Si	NO_3^-	H^+	ANC	Ca^{2+}	Mg^{2+}	K^+	Na^+	SO_4^{2-}	Cl^-	Si	NO_3^-	Mean
Criteria	0.57	0.36	0.16	0.00	0.21	0.60	0.91	0.74	0.00	0.22	0.00	0.32	0.16	0.11	0.26	0.69	0.78	0.83	0.42	0.01	0.00	0.35
Data	0.78	0.39	0.53	0.35	0.17	0.73	0.83	0.47	0.18	1.00	0.00	0.24	0.53	0.31	0.26	0.94	0.86	0.38	0.96	1.00	0.00	0.52
Sens. Conc.	0.44	0.01	0.07	0.20	0.17	1.00	0.80	0.83	0.00	1.00	0.00	0.09	0.10	0.13	0.09	0.86	0.43	0.98	1.00	1.00	0.00	0.46
Sens. Mass	0.44	0.03	0.14	0.11	0.31	0.98	0.89	0.96	0.00	1.00	0.00	0.30	0.06	0.02	0.45	0.30	0.47	0.98	0.01	1.00	0.00	0.40
Mean	0.56	0.20	0.22	0.16	0.21	0.83	0.86	0.75	0.05	0.81	0.00	0.24	0.21	0.14	0.27	0.70	0.64	0.79	0.60	0.75	0.00	0.43

ing non-calibrated concentrations and mass flux (average rank of 163, worst performance of any case) (Table 5). The reason for these poor simulations for the non-calibrated criteria may be that mass flux criteria put too much weight on getting discharge right. By weighting discharge more heavily, the hydrologic parameters in the model are optimized more for discharge than for chemical species.

This result indicates that an alternative weighting methodology should be developed for trying to better estimate appropriate chemical parameter values during the spring snowmelt period. This statement is supported by a combination of the sensitivity analysis results and the calibration results. The calibration results tell us that using mass flux measures is a poor way to estimate model parameters. However, the sensitivity analysis results indicate that focusing on the spring snowmelt period permits better identification of model parameters related to short time period properties. An alternative may be to use a weighting scheme focused on the period of snowmelt as opposed to mass flux error for each chemical species. Such an approach might remove the conflicts that appear to be present in our results.

5. CONCLUSIONS

The two investigative tools we used to dissect the AHM model of the Emerald lake watershed, MOGSA and MOCOM-UA, revealed different aspects of model structure, model parameters, and the natural processes controlling stream chemical composition in the Emerald Lake watershed. Each tool is suited to specific questions and with the combination of the two tools we were able to draw additional conclusions. Mass flux objectives proved more sensitive to snowpack elution and ion exchange parameters while concentration criteria were more sensitive to mineral weathering and evapotranspiration parameters. On the physical process side our results indicate that future field efforts for alpine basins should be concentrated on talus and soil hydrologic properties.

Using a multi-objective algorithm to estimate model parameters and investigate model structure enabled us to reach several conclusions. First, the model being used, the data, and the user's intuition (knowledge of each criteria's connection to processes represented by the model) must be combined to determine what set of criteria to use in a multi-criteria analysis. Second, fluxes not used in calibration are useful for evaluating calibration results and are an alternative to the traditional split sample method of evaluating calibration results. Third, multi-criteria calibration uncovered a flaw in the representation of mineral weathering. The ability of multi-criteria methods to improve model performance shows much promise, but caution should be used in pro-

ceeding as not all criteria contain information useful for improving model performance. Finally, our results confirm an existing simple practice for using a single compromise solution (the median parameter value of several manual calibrations) to make extrapolative model simulations.

The combination of the sensitivity analysis and multi-criteria parameter estimation results revealed some additional information about the model. While the sensitivity analysis results indicated that better estimates would be achieved by combining mass flux and concentration objective functions, the calibration results lead us to the opposite conclusion. The conflict is probably due to an overemphasis on getting stream discharge right when conducting parameter estimation including mass flux criteria. The results also indicate that sensitivity analysis results are only useful for revealing information as to whether a parameter is important or not in affecting model predictions, while model calibration is better at identifying conflicts between criteria and the information contained in them. It is recommended that similar sensitivity analysis and multi-criteria calibration methodologies be used to investigate model structure and parameterization, during model code verification, prior to broad use of a water quality model.

Acknowledgements. A Canon National Park Science Scholarship for the primary author made this work possible. An NSF CAREER award to the first author (EAR-0094312) provided support. NASA-EOS, NSF and NOAA research grants provided additional support for this work. Special thanks to J. Sickman, A. Leydecker and J. Melack for their help in interpreting field data and to the Surface Water Calibration Research Group at the University of Arizona for the many good conversations and quality insights into model calibration and parameter estimation. A debt of gratitude is owed to S. Sorooshian, whose Analysis of Hydrologic Systems course led the primary author to his current research path.

REFERENCES

Bastidas, L. A., H. V. Gupta, S. Sorooshian, W. J. Shuttleworth, and Z.-L. Yang, Sensitivity Analysis of a Land Surface Scheme using Multi-Criteria Methods, *J. Geophys. Res., 104* (D16):481-519, 1999.

Beck, M. B., Water Quality Modeling: A Review of the Analysis of Uncertainty, *Water. Resourc. Res., 23*:1393-1442, 1987.

Campbell, D. H., D. W. Clow, G. Ingersoll, A. Mast, N. E. Spahr, and J. T. Turk, Processes Controlling the Chemistry of Two Snowmelt Dominated Streams in the Rocky Mountains, *Water. Resourc. Res., 31*:2811-2821, 1995.

Christophersen, N., C. P. Neal, and R. P. Hooper, Modelling the hydrochemistry of catchments: a challenge for the scientific method, *J. Hydrol., 152*:1-12, 1993.

De Grosbois, E., R. P. Hooper, and N. Christophersen, A multisignal automatic calibration methodology for hydrochemical models: a case study of the Birkenes Model, *Water. Resourc. Res.,* 24:1299-1307, 1988.

Duan, Q., V. K. Gupta, and S. Sorooshian, Effective and efficient global optimization for conceptual rainfall-runoff models, *Water. Resourc. Res.,* 28:1015-1031, 1992.

Duan, Q., V. K. Gupta, and S. Sorooshian, A shuffled complex evolution approach for effective and efficient global minimization, *Journal of Optimization Theory Applications,* 76:501-521, 1993.

Goldberg, D. E., *Genetic Algorithms in Search, Optimization, and Machine Learning,* Addison-Wesley Publishing Co., Reading, MA, 1989.

Hooper, R. P., A. Stone, N. Christophersen, E. De Grosbois, and H. M. Seip, Assessing the Birkenes model of stream acidification, using a multisignal calibration methodology, *Water. Resourc. Res.,* 24:1308-1316, 1988.

Kirchner, J. W., R. P. Hooper, C. Kendall, C. Neal, and G. Leavesley, Testing and validating environmental models, *The Science of the Total Environment,* 183:33-47, 1996.

Kuczera, G. and M. Mroczkowski, Assessment of hydrologic parameter uncertainty and the worth of multiresponse data, *Water. Resourc. Res.,* 34:1481-1489, 1998.

Legates, D. R. and G. J. McCabe, Evaluating the use of "goodness-of-fit" measures in hydrologic and hydroclimatic model validation, *Water. Resourc. Res.,* 35:233-241, 1999.

Lundquist, D., N. Christophersen, and C. Neal, Towards developing a new short-term model for the Birkenes Catchment, *J. Hydrol.,* 116:391-401, 1990.

Mast, M. A., C. Kendall, D. H. Campbell, D. W. Clow, and J. Back, Determination of hydrologic pathways in an alpine-subalpine basin using isotopic and chemical tracers, in *Biogeochemistry of seasonally snow-covered catchments,* edited by K. Tonnessen, M. W. Williams, and M. Tranter, pp. 263-270, International Association of Hydrological Sciences, Boulder, Colorado, 1995.

Meixner, T., Alpine Biogeochemical Modeling: Case Studies, Improvements and Parameter Estimation, Ph.D. thesis, University of Arizona, 1999.

Meixner, T., R. C. Bales, M. W. Williams, D. H. Campbell, and J. S. Baron, Stream chemistry modeling of two watersheds in the Front Range, Colorado, *Water. Resourc. Res.,* 36:77-87, 2000.

Meixner, T., L. A. Bastidas, H. V. Gupta, and R. C. Bales, Multi-criteria parameter estimation for models of stream chemical composition, *Water. Resourc. Res.,* 38:9-1-9-9, 2002.

Meixner, T., A. D. Brown, and R. C. Bales, Importance of biogeochemical processes in modeling stream chemistry in two watersheds in the Sierra Nevada, California, *Water. Resourc. Res.,* 34(11):3121-3133, 1998.

Meixner, T., H. V. Gupta, L. A. Bastidas, and R. C. Bales, Sensitivity Analysis Using Mass Flux and Concentration, *Hydrol. Processes,* 13:2233-2244, 1999.

Mroczkowski, M., G. P. Raper, and G. Kuczera, The quest for more powerful validation of conceptual catchment models, *Water. Resourc. Res.,* 33:2325-2335, 1997.

Nash, J. E. and J. V. Sutcliffe, River flow forecasting through conceptual models, I, A discussion of principles, *J. Hydrol.,* 10:282-290, 1970.

National Academy of Sciences, *Assessing the TMDL Approach to Water Quality Management,* National Academy Press, Washington, D.C., 2001.

Shaw, J. R., Modeling of silicate mineral weathering reactions in an alpine basin of the southern Sierra Nevada, California, Ph.D. thesis, Department of Hydrology and Water Resources, University of Arizona, Tucson, AZ, 1997.

Spear, R. C. and G. M. Hornberger, Eutrophication of Peel Inlet ii, identification of critical uncertainties via generalized sensitivity analysis, *Water. Resourc. Res.,* 14:43-49, 1980.

Stephens, M. A., Use of the Komolgorov-Smirnov, Cramer-von Mises and related statistics without extensive tables, *Journal of Royal Statistical Society: Ser. B,* 33:115-122, 1970.

Stone, A. and H. M. Seip, Mathematical models and their role in understanding water acidification: An evaluation using Birkenes model as an example, *Ambio,* 18:192-199, 1989.

Tonnessen, K. A., The Emerald Lake watershed study: introduction and site description, *Water. Resourc. Res.,* 27:1537-1539, 1991.

Uhlenbrook, S. and Ch. Leibundgut, Development and validation of a process oriented catchment model based on dominating runoff generation processes, *Physics and Chemistry of the Earth,* 1999.

Williams, M. W., A. D. Brown, and J. M. Melack, Geochemical and hydrologic controls on the composition of surface water in a high-elevation basin, Sierra Nevada, California, *Limnol. Oceanogr.,* 38:775-797, 1993.

Williams, M. W., T. Davinroy, and P. D. Brooks, Organic and Inorganic Nitrogen Pools in Talus Fields and Subtalus Water, Green Lakes Valley, Colorado Front Range, *Hydrol. Processes,* 11:1747-1760, 1997.

Wolford, R.A. Integrated hydrogeochemical modeling of an alpine watershed: Sierra Nevada, California. Tucson, AZ: Department of Hydrology and Water Resources, University of Arizona, 1992.

Wolford, R. A., R. C. Bales, and S. Sorooshian, Development of a hydrochemical model for seasonally snow-covered alpine watersheds: Application to Emerald Lake Watershed, Sierra Nevada, California, *Water. Resourc. Res.,* 32(4):1061-1074, 1996.

Yapo, P. O., A multiobjective global optimization algorithm with application to calibration of hydrological models, Ph.D. thesis, Department of Hydrology and Water Resources, University of Arizona, Tucson, AZ, 1996.

Yapo, P. O., H. V. Gupta, and S. Sorooshian, Multi-objective global optimization for hydrologic models, *J. Hydrol.,* 204:83-97, 1998.

R. C. Bales, L. A. Bastidas, and H. V. Gupta, Department of Hydrology and Water Resources, College of Engineering and Mines, Harshbarger, Bldg. 11, PO Box 210011, University of Arizona, Tucson, AZ 85721-0011, U.S.A. (e-mail: roger@hwr.arizoan.edu

T. Meixner, Department of Environmental Sciences, University of California, Riverside, CA 92521 U.S.A. (e-mail: tmeixner@mail.ucr.edu)

Parameter, Structure, and Model Performance Evaluation for Land-Surface Schemes

Luis A. Bastidas[1], Hoshin V. Gupta, Kuo-lin Hsu, and Soroosh Sorooshian

Department of Hydrology and Water Resources, University of Arizona, Tucson, Arizona

We explore the potential of multi-criteria methods for identifying and quantifying the sources of error in land surface models (LSMs). Using observations of state variables (ground temperature and soil moisture) and heat fluxes (sensible and latent heat), we bound the parameters of land surface models to achieve optimal performance via multi-objective calibration. The optimizations are carried out using, simultaneously, the sensible heat, the skin temperature, and the near surface soil moisture content. The quantification of the data error is achieved via estimation of the different series using an artificial neural network approach. The parameter related error is identified using optimization and the model associated error is estimated by difference. The parameter related error for the heat fluxes is in the order of 15 to 20% of the total error and about 45% for the near surface soil temperatures. The model error is or the order of 45 to 50 % of the total error for the heat fluxes and 30% for the soil temperatures. We also explore the consistency in the model performance by using output series not included in the optimization processes. The additional series used are of soil temperatures and moisture content at different depths and the ground heat fluxes. We found that the model performance is consistent but the statistics of the model performance deteriorate with depth. This problem is relevant because of the possible use of remote sensing information on skin temperature and soil moisture for data assimilation and/or parameter estimation.

1. INTRODUCTION

Land surface modeling is a fundamental tool in the study of climate and hydrology. The reliance on this tool is increasing as hydrologists endeavor to examine large-scale phenomena, predict the hydrologic effects of climate variability, and/or examine land-surface-atmosphere hydrologic and energetic interactions. It is generally recognized that the validity of model simulations can be no better than model assumptions, and no more reliable than model inputs, initial conditions, and parameter values. Field measurements, prior information, and calibration are three techniques used in parameter estimation. With the availability of remote-sensing technology, the use of measurements is gaining importance. However, practical experience indicates that virtually all models will continue to require calibration of at least some parameters.

It is recognized that improving calibration techniques will not solve all of our hydrologic modeling problems. However, in most situations, there will be considerable value in being able to improve our estimates of the preferred parameter values (or region of the parameter space) for a given model given some observational data. If preferred parameters (within the constraints of the available data) can be identified, then there will be a clearer understanding of the uncertainties in the subsequent model simulations.

---

[1] Now at the Department of Civil and Environmental Engineering, Utah State University, Logan, Utah

Calibration of Watershed Models
Water Science and Application Volume 6
Copyright 2003 by the American Geophysical Union
10/1029/006WS17

A great deal of the current focus in hydrology is on the emerging generation of models designed to represent the hydrologic and energetic interactions between the land-surface and atmosphere. While traditional hydrologic models are typically characterized by a single output flux (streamflow at the watershed outlet), the new land-surface representations have multiple output fluxes and large numbers of parameters.

The last 30 years have witnessed the development of numerous models that attempt to represent the surface-atmosphere interactions. There is a wide variation in the complexity of the representation of the processes involved, from the simple bucket-type models [Manabe, 1969] to complex multilayered vertical representations such as BATS [Dickinson et al., 1986], OSU-LSM [Mahrt and Pan, 1984], SiB [Sellers et al., 1986], VIC [Wood et al., 1992], and many others. The models have been subjected repeatedly to improvements that include better representations of the vegetation physiology and attempts to represent surface heterogeneity at the GCM subgrid scale. Furthermore, new versions of some of the models have been developed, e.g., VIC-2L [Nijssen et al., 1997], SiB2 [Sellers et al., 1996], BATS2 [Dickinson et al., 1998], NOAH-LSM [Mitchell et al., 2000], and others. The increase in complexity of the process representation has resulted in large numbers of model parameters. However, the manner in which model parameter values are assigned has changed very little, namely, look-up tables based on literature review and ascribed to different vegetation and soil characteristics are still widely in use.

The large number of models currently in use led to the Project for Intercomparison of Land-Surface Parameterization Schemes (PILPS) [e.g., Henderson-Sellers and Brown, 1992; Henderson-Sellers et al., 1995; Pitman and Henderson-Sellers, 1998]. Originally, PILPS assumed that the parameters having the same physical interpretation should have the same value in all the models because some of them may be subject to measurement and/or estimation. It has been suggested, however, that this is not necessarily the case because LSMs are used on a GCM grid scale and, hence, effective values are required [e.g., Bastidas, 1998; Beven, 1995; Brewer and Wheatcraft, 1994; Gupta et al., 1999b; Sorooshian et al., 1999]. These effective values are, by their very nature, dependent on the particular parameterization. The fact that they might share the name and conceptual representation does not mean that they have the exact same meaning under different parameterizations. This difference in meaning and, thus, difference in the parameter value, should specifically be accounted for when carrying out a model intercomparison.

Remote sensing has the potential to provide information about the space-time variations of the land-surface processes. This is of particular relevance because this kind of information can be used to parameterize LSMs and to derive estimates of the latent heat flux [see, for example, Bastiaansen et al., 1994; Kustas and Humes, 1996; Lakshmi, 2000, Pelgrum and Bastiaanssen, 1996; Wood and Lakshmi, 1993].

Available observational data can be used to constrain the models, i.e., to bound the parameter values so that the model outputs are consistent with the field observations. This consistency with observations provides the means not only to evaluate and test the model performance but also to help in the identification of proper parameter values. The assignment of values to the model parameters should provide consistency between the model outputs and the observational data. Only when this consistency is achieved can the models be properly compared to each other. To attain this consistency, different parameter sets should be obtained for different environmental conditions, hence the need for multiple observational data sets from different environments.

Methodologies for a proper assignment of LSM parameter values by constraining the models with observational data, based on a multi-criteria calibration framework [Gupta et al., 1998], are being developed [e.g., Bastidas et al., 1999; Gupta et al., 1999b]. The multi-criteria methods are specially suited for the calibration of the LSM because of their multiple output nature.

In this chapter, we illustrate and discuss how multi-criteria methods can be used not only to constrain the models with observations but to evaluate the consistency between model outputs and observations. We also describe some of the limitations of the models associated with the quality and availability of the information.

2. MODELS AND DATA

For this work we used the NOAH Land Surface Model (National Centers for Environmental Prediction, Oregon State University, Air Force, National Weather Service-Office of Hydrology) version 2.1 [Mitchell et al., 2000], and the BATS (Biosphere-Atmosphere Transfer Scheme) version 1e [Dickinson et al., 1993] available from the BATS home page (www.atmo.arizona.edu/faculty/research/bats/batsmain.html).

Two data sets are used. One from station E13 (Central Location) of the Atmospheric Radiation Measurement Cloud and Radiation Testbeds (ARM-CART) program in the Southern Great Plains site (SGP) in Oklahoma. The data set covers the period April-July 1995. The second data set is from an agricultural site south of Champaign, Illinois, (40.01 N, 88.37 W) collected by T. Meyers of NOAA/ARL, and covers the entire year 1998. Both data sets have a time interval of 30 minutes and include all the necessary atmospheric forcings for the model and observational information on sensible heat (H in W/m²) and latent heat fluxes (λE W/m²). The ARM-CART site includes soil temperature (T_g in K) as the average of five sensors that integrate the temperature over the top 5 cm, and the average of five soil-moisture content measurements (S_w in weight of water per weight of dry soil) at a depth of 2.5 cm. The Champaign data has measurements of soil skin temperature and soil temperature and soil moisture at 5, 20, and 60 cm depths. The data are representative of the local- (small) scale hydrometeorology. All of this information is used to constrain the model parameters.

3. ESTIMATING LSM PARAMETERS USING MULTI-CRITERIA METHODS

3.1. Multi-Criteria Approach

Gupta et al. [1998] presented a framework for the application of the multi-criteria theory to the calibration of conceptual physically-based models. In *Gupta et al. [1999b]*, the methodology is extended to LSMs. The method is also described in *Gupta el al., Chapter 9 this volume*. In general, due to the multiple output nature of the LSMs the multi-criteria methodology is extremely well suited for the problem of calibration of this type of models. In fact, it is, arguably, the only appropriate way to carry out a calibration for such models.

To carry out the calibrations we have used the MOCOM-UA (Multi-Objective COMplex evolution) which is a general purpose global multi-objective optimization algorithm that provides an effective and efficient estimate of the Pareto solution space within a single optimization run and does not require the commonly used subjective weighting of the different objectives. MOCOM-UA is based on an extension of the SCE-UA population evolution method reported by *Duan et al. [1993]*. A detailed description and explanation of the method are given by *Yapo et al. [1997] and Gupta et al., Chapter 9 this volume*.

The implementation of the procedure outlined above requires the specification of a set of relatively unrelated objective functions F ("unrelated" in the sense that they measure different aspects of the differences between the observed data D and the model simulations $y(\theta)$) that extract the useful information contained in the data and transform it into estimates of the parameter set. In the systems theoretic sense, useful "information" can be viewed as that which enables one to test a hypothesis. There are two important issues to be considered here.

First, it should be noted that the hypothesis to be tested is always a subjective consequence of the interaction between the context of the problem and what the modeler considers to be important. In the context of SVATS modeling, the modeler must determine the important characteristics of the modeled behavior to be reproduced by the calibrated model and what constitutes an effective measure of that behavior. For example, during manual calibration, the modeler may examine the values of several, if not all, of the fluxes computed by the model. It might well be the case that two different fluxes provide redundant information. Another consideration could be the temporal scale at which the objective function is computed; for instance, yearly values of latent heat or the absolute temperature minimum over a year or the average monthly streamflow. This is important because a model could perform well on an annual or monthly basis but still be unable to match the high- frequency fluctuations (daily and/or hourly) of the observed variables. The final outcome of manual calibration is a result of attempting to strike a balance in optimizing (minimizing or maximizing, as appropriate) all of these measures. The hypothesis is that it is possible to find values for the model parameters that can achieve acceptable values of the measures under consideration.

Second and equally important is the fact that a hypothesis typically involves several underlying assumptions that must be tested as part of the hypothesis testing procedure. In the context of hydrological modeling, this might involve a rigorous analysis of the residuals to verify that they belong to some a priori assumed distribution, are unbiased, are homogeneous, and have non-systematic components. For example when applying the maximum likelihood theory to the calibration of a SVATS model the hypothesis might be that it is possible to find a set of parameters such that the variance of the residuals can be minimized and belong to some distribution, typically having zero mean and insignificant autocorrelation. When optimizing on different fluxes, several different measures can be ascribed to the different observed fluxes. Hence, the number of objectives can be significantly higher than the number of observed fluxes. *Gupta et al., [1999b]* and *Bastidas et al., [2001]* showed that the inclusion of series that contain information of the energy and water budgets is required to obtain a good calibration. In particular they suggest the use of sensible heat, soil temperature, and soil moisture, as the series to be included in any calibration process. This suggestion is used in the context of the present work and only the root mean square error is used as the objective function ascribed to each of the observed series.

Every optimization procedure requires the definition of the feasible parameter space, i.e., maximum and minimum allowable parameter values. In the present case, the values prescribed in the BATS model description *[Dickinson et al., 1993]* and in the NOAH model description *[Mitchell, 2000]* were used for the definition of the feasibility region. To preserve the physical soundness of the parameterization, additional constraints such as successively increasing thicknesses of the soil layers with depth were imposed.

4. CASE STUDIES

4.1. Error Estimation

A very important question to be addressed within the context of the model performance evaluation is the issue of the error of the simulation. The usual assumption is that the error is the difference between the observed and the computed series and that it has two sources: observation errors and model errors. The model error is the combined result of the model structural error and the error due to improper identification of the parameter set:

$$
\begin{matrix} \text{Total} \\ \text{error} \end{matrix} = \begin{matrix} \text{data} \\ \text{error} \end{matrix} + \begin{matrix} \text{model} \\ \text{structured} \\ \text{error} \end{matrix} + \begin{matrix} \text{parameter} \\ \text{specification} \\ \text{error} \end{matrix} \quad (1)
$$

The identification and elimination of the latter error is one of the goals of the current work. The reduction of the model structural error (the total elimination of this error would not be feasible) is a task for the developers of the models and has not been attempted here. However, suggestions regarding the areas of possible improvement are made. Some of the troubles in the model structure are identified as a result of a proper calibration procedure.

The data error is related to the way the information is collected and to the instrumentation used. For example, it is a well-known fact that rain gauges underestimate the actual amount of rainfall. An additional error is introduced due to the precision of the instrument. When using a model, the data error existing in the input data induces an error in the modeled output sequences. The observed output sequences, used for contrasting with the modeled outputs, also have an error. In general, the identification of these errors is a difficult task. One approach to try estimate the data error is the use of artificial neural networks (ANN) because they do not have a dependence on model structure.

The ability of the ANN for finding the relationship between output and input series is well established. In fact, unless there is an error in the observations, the ANNs are capable of matching the output series very closely. Therefore, after carrying out the training procedures for fitting the outputs from the inputs, the remaining error in the matching can be ascribed to an error in the data (input or output). A particular ANN model, named self-organized linear output (SOLO) model, developed by *Hsu et al., [2002]*, was applied to the input and output series of the ARM-CART site. An estimate of the RMSE smaller than that of the optimization procedures was found and assumed to be representative of the data error.

The equally-weighted compromise solution for $\{H, T_g, S_w\}$ (calibration performed on those variables) is used as representative of the optimization procedures for estimating the error due to the improper parameter specification. The error difference represents the possibility of improvement by enhancements in the model parameterization. It is worth noting that, for a compromise solution, the weighting coefficient assignment is important due to the unit dependency of the errors. Therefore, either a lot of weighting coefficients combinations need to be optimized or, as was done, a prior normalization of the error functions needs to be performed. The latter can be done only if there is available

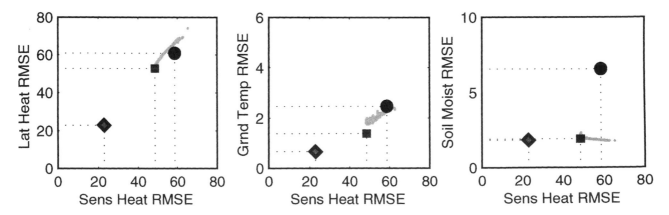

Figure 1. ARM-CART site. Error estimation. BATS default (a priori parameter values) - circle; $\{H, T, S_w\}$ multi-criteria calibration - gray dots; $\{H, T, S_w\}$ compromise solution - square; ANN estimated error data - diamond.

Table 1. Root Mean Square Error Distribution.
BATS Model at ARM-CART E13 site.

Series	Error			
	Data (ANN)	Model Struct.	Param. Ident.	Total (a priori)
Sensible Heat [w/m²]	22.86	25.67	10.47	59.00
Latent Heat [w/m²]	22.91	29.85	8.24	61.00
Ground Temp [K}	0.67	0.72	1.11	2.50
Soil moisture	0.0185	0.0008	0.0467	0.0660
Sensible Heat [w/m²]	39%	44%	18%	100%
Latent Heat [w/m²]	38%	49%	14%	100%
Ground Temp [K]	27%	29%	44%	100%
Soil moisture	28% ?	1% ?	71%	100%

knowledge about the ranges of the optimal values. This is not usually the case and could be done here only because a set of Pareto optimization results (obtained with MOCOM) were already available.

Figure 1 presents the results for the ARM-CART site using the BATS model. The circle represents the existing total error and corresponds to the error when the default parameter set (a priori information embedded in the model) is used. The square is the compromise solution. The distance between the circle and the square represents the error due to the improper parameter set. The diamond represents the error of the ANN fitting. As explained before, this estimates the error in the observation data. The distance between the diamond and the square represents the model structural error.

Table 1 presents the numerical values of the error estimations. The reduction in the total error due to the optimization procedures for the heat fluxes is between 15 and 30%. For the ground temperature the error reduction is on the order of 45%. The reduction in the soil moisture error is 70%. The BATS model structural error is therefore on the order of 30-50% of the total error for the fluxes, and 20-30% for the temperature. The model error for the soil moisture is uncertain. Generally, a high degree of error is associated with soil moisture observations.

4.2. Output Series Evaluation

In several previous studies, where the calibration approach was used, it has been a customary assumption to consider the latent heat time series as the most important [e.g., *Sellers et al., 1989; Rocha et al., 1996*] and in some others as the sole consideration for calibration [e.g., *Franks and Beven, 1997*].

In this work, however, we are focusing our attention on how uncertain are the outputs from the models even after a multi-criteria calibration process has been carried out. It should be noted that the uncertainty presented here is not associated with any probability value, it is associated with

the "trade-off uncertainty" of the optimal region, as defined in *Gupta et al.,* Chapter 9, this volume.

As before, we have used the MOCOM algorithm for the identification of the Pareto optimal region. The calibrations were carried out for the NOAH model at the Illinois site, according to *Bastidas et al., [2001]* recommendations, i.e. sensible heat, ground skin temperature, and soil moisture at 5 cm were used for the calibration purpose. The objective function used for minimization was the root mean squared error. To span the optimal surface and be sure that we are close enough to it, 250 points were used as the complex size, following *Bastidas, [1998]*. The whole set of 250 solutions was run, i.e. we have 250 trajectories for each of the time series analyzed. The gray areas observed in Figures 2-4 represent the set of all the different trajectories for a period of 30 days. It should be noted, however, that the entire period of 365 days was used for the error computation and the analysis. To have an idea of how different the trajectories are, and where the majority of them lie, the average value of those 250 trajectories, at each time step, is also plotted. The scatter plots shown are of the average values versus the observed ones, and correspond to the entire 365 days period. Table 2 presents the values of different error functions for different outputs.

The results for a chosen 50-day period of the heat fluxes are presented in Figure 2. The best results are arguably those for the sensible heat, despite the error statistics reported in Table 2, which are based on the average value of the range. This behavior has been observed at other sites and with other models, e.g. *Bastidas [1998], Bastidas et al., [2001]*, and confirms the assertion made there that sensible heat is the flux that provides the most consistent information for calibration purposes. The results for the latent heat are similar to those obtained for the sensible heat, both in terms of the statistics and the width of the trade-off uncertainty. The ground heat flux has the most trade-off uncertainty, and the biggest relative errors when compared with the magnitudes of the fluxes.

Table 2. Errors for different outputs.
NOAH model at Illinois site.

Series	Error		
	Cor. Coeff.	RMSE	Bias
Sensible Heat [w/m²]	0.85	31.95	1.09
Latent Heat [w/m²]	0.94	33.16	-6.66
Ground Heat [w/m²]	0.72	39.75	-1.16
Ground Skin Temp [K]	0.99	1.75	0.56
Ground Temp @ 5 cm[K]	0.96	2.77	0.11
Ground Temp @ 20 cm [K]	0.98	1.90	-0.35
Ground Temp @ 60 cm [K]	0.98	1.47	0.39
Soil Moisture @ 5 cm	0.60	0.05	-0.01
Soil Moisture @ 20 cm	-0.29	0.09	-0.08
Soil Moisture @ 60 cm	-0.21	0.07	-0.07

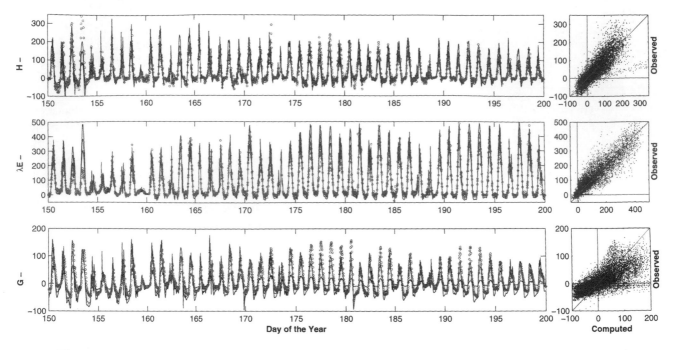

Figure 2. Observed (dots) and simulated (lines delimiting gray area) sensible, latent, and ground heat 50-day time series by using the parameter sets obtained from the $\{H, T_g, S_w\}$ calibration.

The soil moisture series for the entire year 1998 are presented in Figure 3. Only the soil moisture value for the 5 cm depth was used for calibration. Most of the observations at that particular depth are bracketed by the Pareto trajectories. However the trade-off uncertainty has an almost constant value throughout of around 20%. This same uncertainty is observed for the moisture at 20 and 60 cm depths but the observations are not bracketed by the model outputs. The soil moisture variation in the model outputs at 20 cm depth is somewhat similar to that at 5 cm depth. However that is not the case with the observational data, which remains almost constant for the whole year with values bigger than 40%. These values, close to saturation, for the entire year cast some doubts into the quality of the data. The same is observed for the depth of 60 cm.

The soil temperature values are presented in Figure 4. As stated before, only the skin temperature was used for calibration purposes. At the different depths the behavior of the model is consistent, the trade-off uncertainties also are consistent. However, we should note that the temperature observations at the depths of 5 and 20 cm between days 195 and 225 are of dubious quality because of a sudden drop in the amplitude of the daily temperatures (not shown here). The multi-criteria methods allow for an evaluation of the model and the calibration procedure, not only by using the standard split-sample test, which is not necessarily the best way to proceed (see for example *Gupta el al., 1999a*). In the present case, a trade-off uncertainty of around 5 K is observed

for all the depths. However, we can clearly see that the model has troubles tracking the daily amplitude variation in temperatures at the different depths.

5. DISCUSSION

The use of the multi-criteria approach implies that there is not a single or unique solution to the calibration problem. Rather, the best that one can obtain using multi-objective procedures is a model set, specifiable as a region of the parameter space. In the context of multiple measures of model performance, this model set defines the Pareto solution set (which is also a minimal estimate of the parameter uncertainty) in which it is not possible to objectively select a specific parameter set (model) as being superior to any other parameter set (model). This Pareto solution translates into a trade-off uncertainty in the model predictions. The size and properties of this model set and the sizes and properties of the trade-off uncertainty in the model predictions help in the evaluation of the adequacy or inadequacy of the model. The additional sources of information that are incorporated into the model identification by using multi-objective techniques allow for a better way of identifying difficulties associated with the model structure.

From the results presented in the previous section it is possible to say that the models are capable of reproducing the observed quantities, both fluxes and state variables, with a high degree of accuracy, if the proper parameter sets are cho-

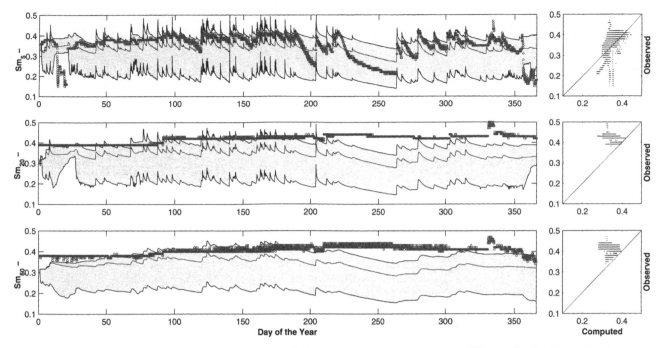

Figure 3. Observed (dots) and simulated (lines delimiting gray area) soil moisture at different depths time series by using the parameter sets obtained from the $\{H, T_g, S_w\}$ calibrations.

sen. This conclusion, however, is limited to the model outputs close to the surface. The performance is not as good when evaluating the modeled quantities against observed at different depths within the soil. This is important for the hydrology of the site. In general, the model has a better ability to reproduce the fluxes than it does to reproduce the state variables. A proper matching of the soil-moisture and soil temperature time series may be possible if those series are included in the optimization. That could not be done in the present work due to the dubious quality of the observations, particularly at deeper depths. However, the general pattern is good, which suggests that spatially distributed remotely sensed information on ground temperature can be used to identify parameter sets that will simulate the flux series reasonably well. Some attempts in this direction have already been made [e.g. *Franks and Beven, 1999*]. This is particularly relevant in the context of four-dimensional data assimilation (4DDA) techniques because the identification of proper parameter values will significantly enhance the usefulness of such techniques. One, has to be aware that, as suggested by the results and at the present stage, the predictive ability of the models decreases with the depth within the soil.

The default parameter set model error was reduced by 30%, on average, by using the automatic optimization procedures. This finding is of importance for the comparison of different models. A fair comparison is possible only if the best solutions from different models are compared. Only the elimination of the improper parameter identification-induced error allows for the comparison of the model structural error.

The flux measurements represent an integrated response to surface characteristics over a relatively large area, unlike the measurements of soil moisture and soil temperature that can be considered point measurements, or representative of an area of a few square meters. The deterioration of the model predictive ability with depth may be explained by a lack of sufficient correspondence between what is observed and what is actually being modeled. The quality of the observations seems to decrease also with the depth, and therefore more efforts have to be devoted to the improvement of such observations.

The multi-criteria calibration approach was found to be effective at bounding the parameter values within physically meaningful ranges. Calibration provides a way to gain insight into what the best values for the model parameters are. However, calibration requires observational data that are not available at many sites. There is a need to develop methodologies to transfer the knowledge gained by means of calibration to places where no data are available.

Acknowledgements. Partial financial support for this research was provided by the National Aeronautics and Space Administration (NASA-EOS grant NAGW2425), the National Oceanic and Atmospheric Administration (NOAA-GCIP grant NA86GP0324), and from SAHRA (Sustainability of Semi-Arid Hydrology and Riparian Areas) under the STC Program of the National Science Foundation Agreement No. EAR-9876800. Special thanks are due to Dr. Kenneth Mitchell of the NCEP for providing the code of the NOAH model and the

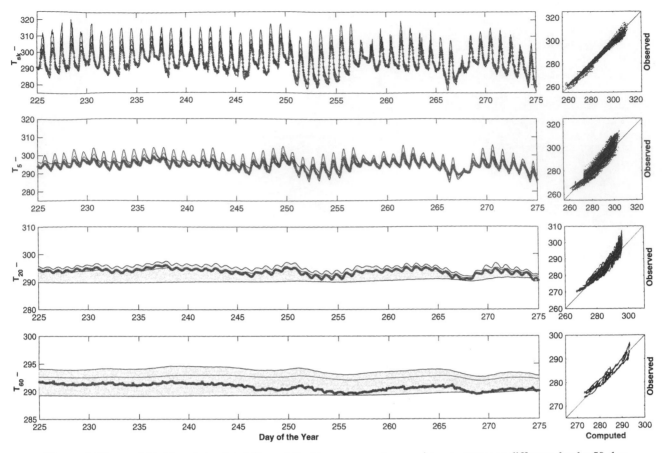

Figure 4. Observed (dots) and simulated (lines delimiting gray area) ground temperatures at different depths 50-day time series by using the parameter sets obtained from $\{H, T_g, S_w\}$ calibration.

ranges for the optimization of parameters. We also wish to thank Bart Nijssen and Qingyun Duan whose comments improved this manuscript.

REFERENCES

Bastiaanssen, W. G. M., D. H. Hoekman, and R. A. Roebling, A methodology for the assessment of surface resistance and soil water storage variability at mesoscale based on remote sensing measurements, a case study with the HAPEX-EFEDA data, *IAHS Special Publications*, 2, 66 p., 1994.

Bastidas, L. A., Parameter estimation for hydrometeorological models using multi-criteria methods, Ph.D. Dissertation, Department of Hydrology and Water Resources, The University of Arizona, Tucson, Arizona, 1998.

Bastidas, L.A., H. V. Guta, and S. Sorooshian, Bounding the parameters of land surface schemes using observational data, in *Land Surface Hydrology, Meteorology, and Climate: Observations and Modeling*, V. Lakshmi, J. Albertson, J. Schaake, Eds., p. 65-76, American Geophysical Union, 2001.

Bastidas, L. A., H. V. Gupta, S. Sorooshian, W. J. Shuttleworth, and Z. L. Yang, Sensitivity analysis of a land surface scheme using multi-criteria methods, *Journal of Geophysical Research-Atmospheres*, 104(D16), p. 19,481-19,490, 1999.

Beven, K. J., Linking parameters across scales: subgrid parameterizations and scale dependent hydrological models, *Hydrological Processes*, 9, p. 507-525, 1995.

Brewer, K. E. and S. W. Wheatcraft, Including multi-scale information in the characterization of hydraulic conductivity distributions, in *Wavelets in Geophysics*, Edited by E. Foufoula- Georgiou and P. Kumar, pp. 213-248, Academic Press, San Diego, CA, 1994.

Dickinson, R.E., A. Henderson-Sellers, P.J. Kennedy, and M. Wilson, Biosphere Atmosphere Transfer Scheme (BATS) for the NCAR Community Climate Model. NCAR Technical Note, NCAR/TN-275+STR, 1986.

Dickinson, R.E., A. Henderson-Sellers, and P.J. Kennedy, Biosphere Atmosphere Transfer Scheme (BATS) Version 1e as Coupled to the NCAR Community Climate Model. NCAR Technical Note, NCAR/TN-387+STR, 72 p., 1993.

Dickinson, R.E., M. Shaick, R. Bryant, and L. Graumlich, Interactive canopies for a climate model, *Journal of Climate*, 11, pp. 2823-2836, 1998.

Duan, Q., V. K. Gupta, and S. Sorooshian, A shuffled complex evolution approach for effective and efficient global minimization, *Journal of Optimization Theory and Applications*, 76(3), pp. 501-521, 1993.

Franks, S. W., and K. J. Beven, Bayesian estimation of uncertainty in land-surface-atmosphere flux predictions, *Journal of*

Geophysical Research - Atmospheres, 102(D20), pp. 23,991-23,999, 1997.

Franks, S. W., and K. J. Beven, Conditioning a multiple-patch SVAT model using uncertain time-space estimates of latent heat fluxes as inferred from remotely sensed data, *Water Resources Research,* 35(9), pp. 2,751-2,761, 1999.

Gupta, H.V., S. Sorooshian, and P.O. Yapo, Towards improved calibration of hydrologic models: multiple and non- commensurable measures of information, *Water Resources Research,* 34 (4), pp. 751-763, 1998.

Gupta, H.V., S. Sorooshian, and P. O. Yapo, Status of automatic calibration for hydrologic models: comparison with multilevel expert calibration, *J. of Hydrologic Engineering,* 4(2),pp. 135-143, 1999a.

Gupta, H.V., L. A. Bastidas, L., S. Sorooshian, W.J. Shuttleworth and Z.L. Yang, Parameter estimation of a land surface scheme using multi-criteria methods, *Journal of Geophysical Research,* Vol. 104, No. D16, p. 19,491-19,504, 1999b.

Henderson-Sellers, A., V. B. Brown, Project for intercomparison of landsurface parameterization schemes (PILPS): First science plan, GEWEX Technical Note, IGPO Publication Series No 5, 53 pp., 1992.

Henderson-Sellers, A., A. J. Pitman, P.K. Love, P. Irannejad, and T. H. Chen, The project for intercomparison of land surface parameterization schemes (PILPS): phases 2 and 3, *Bulletin of the American Meteorological Society,* 76, pp. 489-503, 1995.

Hsu, K., H.V. Gupta, X. Gao, S. Sorooshian, and B. Imam, SOLO-an artificial neural network suitable for hydrologic modeling and analysis, *Water Resources Research,* in press, 2002.

Imam, B., D. S. Yakowitz, and L. J. Lane, Effects of Optional Averaging Schemes on the Ranking of Alternatives by the Multiple Objective Component of a U.S. Department of Agriculture Decision Support System, in *Multiple Objective Decision Making for Land, Water, and Environmental Management,* S. A. El-Sawify and D.S. Yakowitz. Editors, Lewis Publishers, pp. 217-232., 1996.

Kustas, W. P., and K. S. Humes, Sensible heat flux from remotely sensed data at different resolutions, in *Scaling up in Hydrology Using Remote Sensing,* J. B. Stewart, et al., Editors, pp. 255, John Wiley and Sons, New York, 1996.

Lakshmi, V., A simple surface temperature assimilation scheme for use in land surface models, *Water Resources Research,* 36(12), pp. 3687-3700, 2000.

Liang, X., D. P. Lettenmaier, E. F. Wood, and S. J. Burges, A simple hydrologically based model of land surface water and energy fluxes for general circulation models, *Journal of Geophysical Research,* 99(D7), pp. 14,415-14,428, 1994.

Manabe, S., Climate and the ocean circulation. I. the atmospheric circulation and the hydrology of the Earth's surface. *Monthly Weather Review,* 97, pp. 739-774, 1969.

Mahrt L. and H. Pan, A two-layer model of soil hydrology. *Boundary Layer Meteorology,* 23, pp.1-20, 1984.

Mitchell, K., Y. Lin, E. Rogers, C. Marshall, M. Ek, D. Lohmann, J. Schaake, D. Tarpley, P. Grunmann, G. Maninkin, Q. Duan, and V. Koren, Recent GCIP-sponsored advancements in coupled land-surface modeling and data assimilation in the NCEP ETA mesoscale model, AMS 15th Conference on Hydrology, 9-14 January, Long Beach, CA, pp 180-183, 2000.

Nijssen, B., D. P. Lettenmaier, X. Liang, S. W. Wetzel, and W. F. Wood, Streamflow simulation for continental-scale river basins, *Water Resources Research,* 33(4), pp. 711-724, 1997.

Pelgrum, H., and W. G. M. Bastiaansen, An intercomparison of techniques to determine the area-averaged latent heat flux from individual in situ observations: A remote sensing approach using the European Field Experiment in a desertification-threatened area data, *Water Resources Research,* 32(9), pp. 2775-2786, 1996.

Pitman, A. J., A. Henderson-Sellers, Recent progress and results from the project for the intercomparison of landsurface parameterization schemes, *Journal of Hydrology,* (212), pp. 128-135, 1998.

Rocha, H. R., P. J. Sellers, G. J. Collatz, I. R. Wright, and J. Grace, Calibration and use of the SiB2 model to estimate water vapour and carbon exchange at the ABRACOS forest, in *Amazonian Deforestation and Climate,* J. H. Gash, C. A. Nobre, J. M. Roberts, and R. L. Victoria, Editors, pp. 459-471, 1996.

Sellers, P. J., Y. Mintz, Y. C. Sud, and A. Dalcher, A Simple Biosphere Model (SIB) for use within general circulation models, *Journal of the Atmospheric Sciences,* 43(6), pp. 505- 531, 1986.

Sellers, P. J., W. J. Shuttleworth, J. L. Dorman, A. Dalcher, and J. M. Roberts, Calibrating the simple biosphere model (SiB) for Amazonian tropical forest using field and remote sensing data, part 1, average calibration with field data, *Journal of Applied Meteorology,* 28, pp. 728-756, 1989.

Sellers, P. J., D. A. Randall, G. J. Collatz, J. A. Berry, C. B. Field, D. A. Dazlich, C. Zhang, G. D. Collelo, and L. Bounoua, A revised land surface parameterization (SiB2) for atmospheric GCMs. part I: model formulation, *Journal of Climate,* 9, pp. 676-705, 1996.

Sorooshian, S., L. A. Bastidas, and H. V. Gupta, Application of multi-objective optimization algorithms for hydrologic model identification and parameterization, 2nd International Conference on Multiple Objective Decision Support Systems for Land, Water, and Environmental Management, Brisbane, Australia, August 1-5, 1999.

Wood, E. F., and V. Lakshmi, Scaling water and energy fluxes in climate systems: three land-atmosphere modeling experiments, *Journal of Climate,* 6, pp. 839-857, 1993.

Wood, E. F., D. P. Lettenmaier, and V. G. Zartarian, A land surface hydrology parameterization with subgrid variability for general circulation models, *Journal of Geophysical Research,* 97(D3), pp. 2717-2728, 1992.

Yapo, P.O., H.V. Gupta, and S. Sorooshian, Multi-objective global optimization for hydrologic models, *Journal of Hydrology,* 204, pp. 83-97, 1997.

Luis Bastidas, Civil and Environmental Engineering, Utah State University, Logan, UT 84322-8200 (email: lucho@hwr.arizona.edu), Hoshin Gupta, Kuo-lin Hsu, Soroosh Sorooshian, Hydrology and Water Resources, University of Arizona, Tucson AZ 85721-0011.

Use of *a Priori* Parameter Estimates in the Derivation of Spatially Consistent Parameter Sets of Rainfall-runoff Models

Victor Koren, Michael Smith, and Qingyun Duan

Hydrology Laboratory, Office of Hydrologic Development, NOAA/NWS, Silver Spring, Maryland

It is well known that hydrologists rely much on trail-and-error process in estimating conceptual model parameters. While there has been a great deal of research into the development of automatic calibration methods, subjective expert judgment still plays a significant role in the selection of 'optimal' parameter sets. Any technique for calibrating rainfall-runoff model parameters requires many years of historical hydrometeorological data, and usually performs a single basin analysis. The quality and quantity of historical data can vary significantly for different regions, and even for different river basins in the same region. These inconsistencies can lead to non-optimal calibration results, and consequently significant and inappropriate randomness in the spatial patterns of model parameters. Therefore, an objective estimation procedure is needed that can produce spatially consistent and physically realistic parameter values. This paper investigates the possibility of using *a priori* parameter estimates to improve the calibration/estimation process. A set of physically based relationships between the Sacramento Soil Moisture Accounting model parameters and soil properties were developed to estimate *a priori* parameter values. Two tests, model parameter transferability to ungaged basins and constrained automatic calibration, were performed for a number of headwater watersheds in the Ohio river basin. The results suggest that the use of soil derived parameters can improve the spatial and physical consistency of parameter estimates while maintaining hydrological performance. Soil derived parameters provide a quantitative measure of possible differences between parameters of neighboring basins that allow one to 'rescale' calibrated parameters to ungaged watershed. Use of constrained calibration reduces inappropriate randomness in the spatial pattern of model parameters.

1. INTRODUCTION

The successful application of any rainfall-runoff model greatly depends on its parameterization. It is well known that hydrologists rely much on trail-and-error process in estimating of conceptual model parameters because their parameters generally are not directly observable [*Duan et*

al., 2001]. Even if model parameters are related to observable physical properties (e.g., parameters of so-called 'physically-based' models), some fine tuning or calibration of *a priori* parameters would still be required because basin-scale heterogeneities of physical properties and data uncertainties could significantly affect on an estimation process [*Carpenter et al.*, 2001]. While there has been a great deal of research into the development of automatic calibration methods (e.g., see *Rajaram and Georgakakos*, [1989]; *Sorooshian and Gupta*, [1995]; *Yapo et al.*, [1998]), subjective expert judgment still plays a significant role in the selection of 'optimal' parameter sets [*Hogue et al.*, 2000]. Existing calibration techniques tend to produce 'noisy'

Calibration of Watershed Models
Water Science and Application Volume 6
Copyright 2003 by the American Geophysical Union
10/1029/006WS18

parameter estimates. As stated by Burnash, [1995], "This could occur because the data set was not stable, the historic sequence did not include an adequate sequence of events to exercise some of the model's characteristics, or the optimization function was not sensitive to the discrete functions associated with the proper use of particular parameters." Implementation of fully distributed models increases requirements to the calibration system to preserve a physically reasonable spatial pattern of model parameters [*Refsgaard*, 1997].

Without a systematic approach, spatial inconsistencies can enter the calibration process at several points. For example, any technique for calibrating rainfall-runoff model parameters requires many years of historical hydrometeorological data, including precipitation, temperature, streamflow discharge, etc. The quality and quantity of these data can vary significantly for different regions, and even for different river basins in the same region. In addition, it is common for each river basin to be calibrated independently. Moreover, parameters can sometimes be given values that cause the process they represent to be simulated improperly, even though the overall statistical results indicate a good fit. These inconsistencies can lead to non-optimal calibration results, and consequently significant and inappropriate randomness in the spatial patterns of model parameters. Therefore, an objective estimation procedure is needed that can produce spatially consistent and physically realistic parameter values. The procedure should be constrained so that parameter adjustment takes place within a range of values which retains conceptual consistency. This paper investigates the possibility of using *a priori* parameter estimates to improve the calibration/estimation process. Section 2 contains a brief overview of the Sacramento Soil Moisture Accounting (SAC-SMA) model that was used in analysis, and an approach by Koren et al., [2000] to generate *a priori* estimates of the SAC-SMA model parameters from soil-vegetation data. Section 3 discusses a practical procedure of estimation of soil derived SAC-SMA model parameters, and the experimental design for testing the use of these estimates in the derivation of spatially consistent parameters. Test results are presented in Section 4. Section 5 contains a summary and recommendations for future work.

2. SOIL-BASED ESTIMATES OF SAC-SMA MODEL PARAMETERS

Parameters of conceptual models such as the SAC-SMA model are usually derived from input-output data analysis using automatic or manual calibration procedures, but are not readily derived from physical basin characteristics. This deficiency restricts the application of these models (e.g.,

limited use in ungaged basins, high spatial resolution applications, etc.) significantly. Improvements in quality and quantity of high resolution GIS data have stimulated developments of regional relationships between basin properties and model parameters which could be used in *a priori* parameter estimation. Abdulla et al., [1996] derived empirical equations which correlate the VIC-2L LSM parameters to easily determinable basin characteristics for the GCIP Large Scale Area-Southwest. Duan et al., [1996] correlated the parameters of the Simple Water Balance (SWB) model and basin characteristics for the southeast quadrant of the US. In both cases, model parameters were calibrated for selected basins prior to the derivation of regression equations. The disadvantage of this approach is that the calibration procedure can introduce significant uncertainties in the 'optimal' parameter set, and subsequently into the regression equations because the input/output data are noisy. Recently, soil/vegetation data were explicitly used to derive physically-based analytical relationships between soil properties and conceptual model parameters. Knowles, [2000] developed such relationships for the Bay-Delta Watershed Model (BDWM). The BDWM structure is similar to the conceptual structure of the SAC-SMA model. Koren et al., [2000] developed analytical relationships for the most SAC-SMA model parameters. In this study, we adopted an approach developed by Koren et al., [2000] that uses high resolution soil and vegetation data.

2.1. SAC-SMA Model Structure and Parameters

A detailed description of SAC-SMA can be found in Burnash et al., [1973] and Burnash, [1995]. The basic design of the SAC-SMA model centers on a two layer structure: a relatively thin upper layer, and usually a much thicker lower layer which supplies moisture to meet the evapotranspiration demands. Each layer consists of tension and free water storages that interact to generate soil moisture states and five runoff components. The free water storage of the lower layer is divided into two sub-storages: the *LZFSM* which controls supplemental (fast) base flow, and the *LZFPM* which controls primary (slow) ground water flow. Partitioning of rainfall into surface runoff and infiltration is constrained by the upper layer soil moisture conditions and the percolation potential of the lower layer. No surface runoff occurs before the tension water capacity of the upper layer, *UZTWM*, is filled. After that, surface runoff generation is controlled by the content of the upper layer free water storage, *UZFWM*, and the deficiency of lower layer tension water, *LZTWM*, and free water storages. Each free water reservoir can generate runoff depending on a depletion coefficient of the reservoir, namely the *UZK* coefficient for the upper layer, and *LZSK* and *LZPK* for the lower layer supplemental and primary free water storages, respectively. The percolation rate into the lower layer, I_{perc}, is a nonlinear function of the saturation of lower layer reservoirs, W_{LZ}, and the upper layer free water reservoir, W_{UZF}:

$$I_{perc} = I_o [1 + ZPERC \cdot (1 - \frac{W_{LZ}}{LZWM})^{REXP}] \frac{W_{UZF}}{UZFWM} \qquad (1)$$

where *ZPERC* is a ratio of maximum and minimum percolation rates, *REXP* is an exponent value that controls the shape of the percolation curve, *LZWM=LZTWM+LZFSM+LZFPM* is a total capacity of the lower layer, and I_o is the minimum percolation rate under fully saturated conditions in the upper and lower layers which equals the maximum rate of drainage from lower layer free water storages:

$$I_o = LZFSM \cdot LZSK + LZFPM \cdot LZPK \qquad (2)$$

Percolated water into the lower layer is divided among three storages of the layer. A parameter *PFREE* is used to express the fractional split of percolated water between tension and free water storages of the lower layer.

There are five minor parameters that control impermeable area runoff and riparian evapotranspiration. Table 1 lists all SAC-SMA model parameters.

Although there are strong physical arguments to support the model [*Burnash*, 1995], 16 model parameters can not be

measured. Some helpful rules were suggested for estimating of initial values of SAC-SMA parameters using hyetograph-hydrograph analysis [*Burnash*, 1995]. These initial estimates play a key role in the manual calibration procedure of the National Weather Service River Forecast System (NWS-RFS) [*Smith et al.*, this volume]. However, this procedure is based on trial-and-error approach and depends much on expert experience.

Recent developments by the University of Arizona research group [*Boyle et al.*, 2001; *Boyle et al.*, 2000; *Hogue et al.*, 2000; *Yapo et al.*, 1998; *Sorooshian and Gupta*, 1995] have significantly improved the automatic calibration process of the SAC-SMA model. However, limitations on the selection of an objective function, structural problems of the model, and uncertainties in input/output data reduce the ability of automatic calibration to obtain unique and conceptually realistic parameter estimates. On the other hand, a single basin calibration approach limits the analyses of the spatial pattern of model parameters, and can lead to inappropriate spatial randomness of calibration results.

2.2. Soil Texture and SAC-SMA Model Parameter Relationships

Koren et al., [2000] developed a physically based approach to quantify the relationships of 11 major parameters of the SAC-SMA model with soil properties (these parameters are highlighted in Table 1). As defined in Section

Table 1. SAC-SMA model parameters and their feasible ranges.

No	Parameter	Description	Ranges
1	**UZTWM**	The upper layer tension water capacity, *mm*	10-300
2	**UZFWM**	The upper layer free water capacity, *mm*	5-150
3	**UZK**	Interflow depletion rate from the upper layer free water storage, *day^{-1}*	0.10-0.75
4	**ZPERC**	Ratio of maximum and minimum percolation rates	5-350
5	**REXP**	Shape parameter of the percolation curve	1-5
6	**LZTWM**	The lower layer tension water capacity, *mm*	10-500
7	**LZFSM**	The lower layer supplemental free water capacity, *mm*	5-400
8	**LZFPM**	The lower layer primary free water capacity, *mm*	10-1000
9	**LZSK**	Depletion rate of the lower layer supplemental free water storage, *day^{-1}*	0.01-0.35
10	**LZPK**	Depletion rate of the lower layer primary free water storage, *day^{-1}*	0.001-0.05
11	**PFREE**	Percolation fraction that goes directly to the lower layer free water storages	0.0-0.8
12	**PCTIM**	Permanent impervious area fraction	
13	**ADIMP**	Maximum fraction of an additional impervious area due to saturation	
14	**RIVA**	Riparian vegetation area fraction	
15	**SIDE**	Ratio of deep percolation from lower layer free water storages	
16	**RSERV**	Fraction of lower layer free water not transferrable to lower layer tension water	

2.1, the SAC-SMA model is a typical storage type model that assumes that all rainfall losses are allocated in the upper and lower storages of a conceptual soil profile. Each layer consists of fast components (free water) driven mostly by gravitational forces, and slow components (tension water) driven by an evapotranspiration and diffusion. According the soil moisture property definition, Koren et al., [2000] assumed that slow component storages of the SAC-SMA model are related to available soil water, and that fast component storages are related to gravitational soil water. Available soil water and gravitational soil water were derived from soil properties such as the saturated moisture content, θ_s, field capacity, θ_{fld}, and wilting point, θ_{wlt}. These soil properties can be estimated from STATSGO dominant soil texture grids available for eleven soil layers (from ground surface to 2.5m beneath) for the conterminous United States [*Miller and White*, 1999]. The combined thickness of the upper and lower layers (as a water depth) was assumed to be equal to the total thickness of gravitational and available water storages to the soil profile depth, Z_{max}. A concept of an initial rain abstraction [*McCuen*, 1982] was used to split the soil profile into the upper and lower layers. The Natural Resources Conservation Service (NRCS) (formerly, Soil Conservation Service (SCS)) developed an approach to estimate the initial rain abstraction based on soil and vegetation type, as well as on soil moisture conditions [*McCuen*, 1982]. In the method by Koren et. al., [2001], it was assumed that under the average soil moisture condition stipulated by NRCS, the upper layer tension water storage is full and the free water storage is empty. In this case, the initial rain abstraction should satisfy the upper layer free water capacity. The upper layer thickness can then be calculated based on a SCS curve number, *CN*, for the soil profile. Under these assumptions all SAC-SMA storages (*UZTWM, UZFWM, LZTWM, LZFSM, LZFPM*) defined in water depth units can be estimated as functions of soil porosity, field capacity, wilting point, soil depth, and SCS curve number [*Koren et al.*, 2000].

A relationship for the depletion coefficient of the lower layer primary free water storage was obtained from the solution of Darcy's equation for an unconfined homogeneous aquifer [*Dingman*, 1993] that required estimation of the saturated hydraulic conductivity, K_s, and the specific yield of soil, μ. The percolation parameter *ZPERC* was estimated from other known SAC-SMA parameters as follows. From Eq. 1, it can be seen that the maximum percolation, I_{max}, occurs when the upper layer is fully saturated and the lower layer is dry:

$$I_{max} = I_o \cdot (1 + ZPERC) \tag{3}$$

It, therefore, was assumed that the maximum percolation rate is the maximum contents of the lower layer storages released per time interval Δt. Using these assumptions, an expression for *ZPERC* parameter can be obtained from Eq. 3:

$$ZPERC = \frac{(LZTWM + LZFSM + LZFPM)/\Delta t - I_o}{I_o} \tag{4}$$

Empirical relationships were suggested for other SAC-SMA parameters, *UZK, LZSK, REXP*, and *PFREE*. Ratios of field capacity (θ_{fld}/θ_s) and wilting point (θ_{wlt}/θ_s) were used as integrated indexes of soil properties. A few coefficients of these relationships were estimated using calibration results from a number of well calibrated headwater basins. Relationships for the 11 SAC-SMA parameters are presented in Appendix.

3. USE OF *A PRIORI* PARAMETERS FOR ESTIMATING SPATIALLY CONSISTENT PARAMETER SETS FOR HEADWATER BASINS

Limited tests of *a priori* parameters of the SAC-SMA model were presented in Koren et al., [2000] and Duan et al., [2001]. While overall statistics showed that *a priori* parameters compared well to carefully calibrated parameter sets for a few river basins, it was found that these derived relationships could not account for some specific local river basin conditions. Consequently, the accuracy of *a priori* parameters can vary for different regions. As an example, the estimated parameters of the lower layer free water storages may not be reliable in regions with deep ground water because the NRCS soils information is only defined to a depth of 2.5m. The split between the upper and lower layers based on the SCS curve number can also contribute to *a priori* parameter uncertainties. Other limitations arise because the approach is based on physical assumptions regarding relationships between model parameters and soil properties, and between soil properties themselves. Although most assumptions are obvious, some quantitative expressions were assigned empirically using SAC-SMA calibration results from a limited number of river basins. Another limitation of the approach relates to available soil and SCS curve number data. STATSGO data consist of soil texture data derived from 1:250000 scale soil maps and interpolated into 1x1 km grids for 11 soil layers. This introduces some limitations on the reliability of *a priori* parameters due to possible spatial sampling of soil texture over large areas (100-200 km^2 in some regions). Therefore, *a priori* parameters should be adjusted if there are observed rainfall-discharge data. The main objective of these relationships is to give reasonable initial values, and to reduce uncertainties in parameter ranges. Another benefit is that these relationships are based on available physical properties of soils and can be used on ungaged basins.

3.1. Estimation of Soil Based Parameters for Selected River Basins

STASGO dominant soil texture grids [Miller & White, 1999] for 11 soil layers were used in this analysis. Hydraulic soil properties θ_s, K_s, and Ψ_s (the saturation matrix potential) and b (the slope of the Campbell's, [1974] retention curve) for each USDA texture class were calculated using regression equations from Cosby et al., [1984]:

$$\theta_s = -0.00126\, F_{sand} + 0.489 \tag{5}$$

$$\Psi_s = -7.74e^{-0.0302F_{sand}},\ kPa \tag{6}$$

$$b = 0.159\, F_{clay} + 2.91 \tag{7}$$

The percentages of sand, F_{sand}, and clay, F_{clay}, were obtained from midpoint values of each textural class [Cosby et al., 1984] using the USDA textural triangle. Field capacity and wilting point estimates were calculated from the Campbell's matric water potential function using parameter values from Equations 5-7

$$\theta_{fld} = \theta_s\, (\Psi_{fld}/\Psi_s)^{-1/b} \tag{8}$$

$$\theta_{wlt} = \theta_s\, (\Psi_{wlt}/\Psi_s)^{-1/b} \tag{9}$$

Matric potential at the field capacity, Ψ_{fld}, was assumed to be -10 kPa for the 1-3 sandy soil classes (see Table 2), and -20 kPa for all other soil classes [ASCE, 1990]. Matric

potential at the wilting point, Ψ_{wlt}, was assumed to be -1500 kPa.

Saturated hydraulic conductivity, K_s, stream channel density, D_s, and specific yield, μ, values for each soil texture class are required to estimate the depletion rate of the lower layer primary free water (see Appendix, Eq. A8). Experimental data [Li et al., 1976] reported by Clapp and Hornberger, [1978] were adopted for the saturated hydraulic conductivity. Stream channel density does not vary much depending on soil properties, and a constant value of 2.5 was assumed in this analysis. Since there are no systematic data of the specific yield of different soils, an empirical relationship was developed for this analysis using limited data reported by Armstrong, [1978]:

$$\mu = 3.5\, (\theta_s - \theta_{fld})^{1.66} \tag{10}$$

Results from Eq. 10 for all soil texture classes and Armstrong's estimates are plotted in Figure 1. A 1.6 value of parameter n (see Appendix, Equations A3, A5-A7, A9) was used to maintain an average ratio between the supplemental and primary storage capacities close to 1/3 [Koren et al., 2000]. The values of physical soil properties used in this analysis are given in Table 2.

The NRCS developed a classification system to estimate a curve number, CN, based on soil type, land use, agricultural land treatment class, hydrologic condition, and antecedent soil moisture [McCuen, 1982]. To assess these factors, soil surveys and site investigations are recommended in addition to the use of soil-land use maps. Some of the factors could not be assessed in this study because only STATSGO grids were available for analyses. In light of this limitation, we utilized a simplified approach in which curve numbers were estimated based on USDA Hydrologic Soil

Table 2. Physical properties of different soil classes defined for this analysis.

No	Texture class	% sand	% clay	θ_{max}	θ_{fld}	θ_{wlt}	K_s mm/hr	μ
1	Sand	92	3	0.37	0.15	0.04	633.6	0.29
2	Loamy sand	82	6	0.39	0.19	0.05	562.6	0.23
3	Sandy loam	58	10	0.42	0.27	0.09	124.8	0.15
4	Silty loam	17	13	0.47	0.35	0.15	25.9	0.10
5	Silt	9	5	0.48	0.34	0.11	20.0	0.12
6	Loam	43	18	0.44	0.30	0.14	25.0	0.13
7	Sandy clay loam	58	27	0.42	0.29	0.16	22.7	0.12
8	Silty clay loam	10	34	0.48	0.41	0.24	6.1	0.04
9	Clay loam	32	34	0.45	0.36	0.21	8.8	0.07
10	Sandy clay	52	42	0.42	0.33	0.21	7.8	0.07
11	Silty clay	6	47	0.48	0.43	0.28	3.7	0.02
12	Clay	22	58	0.46	0.40	0.28	4.6	0.03

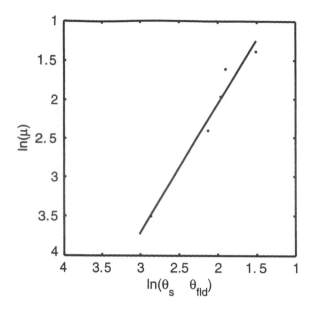

Figure 1. The specific yield (μ) as a function of the free water capacity ($\theta_s - \theta_{fld}$). Armstrong's estimates are shown with circles.

entire USA is displayed in Figure 2 as an example. These grids are now available in an ArcView application called the Calibration Assistance Program (CAP) [*Reed et al.*, 2001] that is designed to assist the calibrator in deriving initial parameter estimates. The CAP computes mean, maximum and minimum values of these parameters for basins and/or elevation zones of interest, and presents the results in a tabular format.

3.2. Tests Design and Data

Two tests, model parameter transferability to ungaged basins and constrained automatic calibration, were performed for a number of headwater watersheds in the Ohio river basin, Table 3. Rainfall-runoff simulations were generated in a lumped mode assuming that input data and model parameters were uniform over each basin. *A priori* SAC-SMA parameters for each watershed were estimated as an arithmetic averages from 1x1 km resolution parameter grids generated as described in Section 3.1.

Group grids (HSG) [*Miller & White*, 1999] assuming 'pasture or range land use' under 'fair' hydrologic conditions for the entire region [*McCuen*, 1982].

Eleven SAC-SMA parameter grids having a 1x1 km resolution were generated for the conterminous United States using data from Table 2 and HSG-based SCS curve numbers. The lower layer tension water capacity map for the

3.2.1. Parameter transferability test.
First, a test of SAC-SMA parameter transferability is performed on a number of neighboring headwater basins of the Upper Monongahela River, West Virginia (see Figure 3, watershed numbers 1-6). These watersheds are located in the southeastern portion of the Upper Monongahela basin. Slight differences in mean basin elevation exist. Comparison of observed hydrographs

Figure 2. The lower layer tension water capacity derived from soil data for the conterminous US.

Table 3. List of basins selected for the analysis.

No.	Watershed name	Latitude	Longitude	Basin area, ml^2	Elevation, ft.
	First group of basins				
1	Dry Fork at Hendricks, WV	39.072	-79.623	349.0	3240
2	Buckhannon R. at Hall, WV	39.051	-80.115	277.0	2060
3	Middle Fork R. at Audra, WV	39.040	-80.068	148.0	2850
4	Blackwater R. at Davis, WV	39.127	-79.469	85.9	3350
5	Tygart Valley R. at Dailey, WV	38.809	-79.882	185.0	2840
6	Shavers Fork below Bowden, WV	38.913	-79.771	151.0	3120
	Second group of basins				
7	Tygart Valley R. at Belington, WV	39.029	-79.936	408.0	1679
8	Middle Island C. at Little, WV	39.475	-80.997	458.0	631
9	Bluestone R. nr Pipestem, WV	37.544	-81.011	394.0	1527
10	Greenbrier R. at Buckeye, WV	38.186	-80.131	540.0	2086
11	Ohio Brush C. nr West Union, OH	38.804	-83.421	387.0	511
12	SF Licking R. at Cynthiana, KY	38.391	-84.303	621.0	689
13	Stillwater R. at Englewood, OH	39.869	-84.282	650.0	700
14	White R. at Noblesville, IN	40.047	-86.017	858.0	738
15	Big Blue R. at Shelbyville, IN	39.529	-85.782	421.0	738
16	Sugar C. nr Edinburgh, IN	39.361	-85.998	474.0	646
17	French Broad R. at Blantyre, NC	35.299	-82.624	296.0	2060
18	French Broad R. at Asheville, NC	35.609	-82.579	945.0	1950

shows much similarity in the response of the watersheds with some variations that are primarily related to elevation. The 'best' SAC-SMA parameter sets for all basins were available from the Ohio River Forecast Center (OHRFC). OHRFC hydrologists used the NWSRFS calibration procedure [*Smith et al.*, this volume] which is based on visual fitting of simulated and observed hydrographs, and comparing different statistics. While this procedure is rather subjective, it provides physically reliable and robust estimates of the SAC-SMA model parameters. Time series of mean areal six-hourly precipitation and air temperature values, and daily discharges were available from the OHRFC. 25-45 year time series were generated for most basins. However, only eight years of historical data were available for the Shavers Fork below Bowden.

Control simulations for the entire historical period (when input/output data were available) were first performed using two parameter sets for 6 selected watersheds: 1) 'best' OHRFC manually calibrated parameters for each basin, and 2) soil derived parameters for the same basins. These results provide an objective evaluation of the performance of soil derived parameters compared to OHRFC parameters. Because OHRFC parameters were derived using a subjective procedure, a simple comparison of just parameter values can not provide conclusive information.

To test parameter transferability, it was assumed that only one basin had historical time series to perform model calibration. The Dry Fork at Hendricks watershed was selected as representative and the best calibrated for the first group (based on the OHRFC expert judgment [*Tom Adams*, per-

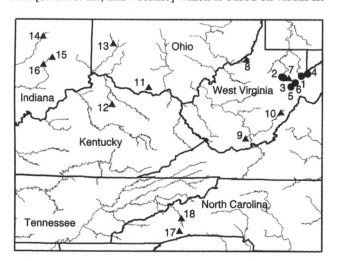

Figure 3. Location of outlets for the first (shown with circles) and second (shown in triangles) groups of test watersheds.

sonal communication]). All SAC-SMA parameters for five other watersheds of this group were assumed to be equal to calibrated parameters for the Dry Fork basin, $X_{DFH,calb}$. This approach is usually used when calibration is performed on a large river basin with a number of ungaged watersheds.

Other parameter sets for these watersheds were generated by scaling of soil derived parameters for each basin, $X_{i,soil}$, based on the ratio of calibrated, $X_{DFH,calb}$, and soil derived, $X_{DFH,soil}$, parameters for the Dry Fork basin:

$$X_{i,*} = \frac{X_{DFH,calb.}}{X_{DFH,soil}} X_{i,soil} \qquad (11)$$

where $Xi_{,*}$ are scaled parameters for a basin j. It assumes that soil information represents reasonably well the spatial pattern of model parameters while their magnitudes may be not optimal.

3.2.2. Use of soil derived parameters in an automatic calibration. This test involves automatic calibration of a larger number of basins representing different climatic and hydrological conditions. The second group of 12 headwater watersheds are spread through the Ohio-Tennessee River basin including Ohio, Indiana, Kentucky, West Virginia, and North Carolina states, see Figure 3, and represent different climatic and hydrological conditions. Annual precipitation varies from 500mm in the northwest portion of the region to 1500mm in the southeastern portion [*Schaake et al.*, 2000]. Potential evaporation varies much less throughout the region. Consequently, significant differences in annual runoff for the northwest (200mm) and southeastern (900mm) portions of the basin are present. Daily precipitation, air temperature, and discharges for 45-50 years period were obtained from the Model Parameter Estimation Experiment (MOPEX) [*Schaake et al.*, 2001] project databases.

First, automatic calibration was performed for all selected basins without the use of soil derived parameters. Parameters were allowed to vary in a broadly defined feasible space [*Brazil*, 1989; *Boyle et al.*, 2001]. Table 1 lists parameter ranges used in this study. The second set of calibration runs were conducted using soil derived parameters to define parameter ranges that are tied to basin physical characteristics. 25 percent bounds from soil derived parameters were used in these runs, i.e.

$$(1-0.25) X_{i,soil} < X_i < (1+0.25) X_{i,soil} \qquad (12)$$

The University of Arizona Shuffled Complex Evolution (SCE-UA) calibration technique [*Duan et al.*, 1992] was used in this test. The SCE-UA method is a global search pro-

cedure that uses concepts from random search algorithms, along with the strength of the downhill simplex method. It has been tested extensively in the last few years and is found to be efficient and consistent in finding the global optimum of multi-parametric nonlinear problems encountered in the calibration of conceptual watershed models. A weighted error function was selected as a minimization criterion:

$$F = \alpha \cdot MVRMS + (1-\alpha) DRMS \qquad (13)$$

where *MVRMS* is a mean square error of monthly runoff volumes, *DRMS* is a mean square error of daily discharges, and α is a weight parameter; 0.8 value was selected for this test. A 15 year period was used in the calibration process, and the rest of data (usually 25-28 years) were used for validation.

4. RESULTS AND DISCUSSION

4.1. Parameter Transferability Test Results

Some accuracy statistics of hydrographs simulated using calibrated and soil-derived parameters are shown in Table 4. These statistics include a daily discharge root mean square error, *DRMS*, a monthly volume root mean square error, *MVRMS*, a daily discharge root mean square error during flood events only, *FDRMS*, percent of total bias

Table 4. Accuracy statistics of hydrographs simulated using calibrated and soil derived parameters for the Upper Monongahela basin

Basin #	DRMS cms	MRMS, mm	BIAS, %	FDRMS, %	R^2
Calibrated parameters					
1	16.9	17.8	3.2	47	0.86
2	10.2	12.5	1.3	32	0.91
3	6.5	14.1	-0.9	40	0.89
4	4.7	20.8	2.0	49	0.84
5	7.9	12.5	1.6	43	0.89
6	8.1	18.3	-0.6	43	0.87
Avg.	9.1	16.0	1.6 [1]	42	0.88
Soil derived parameters					
1	18.2	17.6	2.3	50	0.84
2	11.3	12.9	-2.7	40	0.89
3	7.0	14.5	-4.6	40	0.88
4	5.1	20.5	-0.5	47	0.82
5	7.9	12.4	-5.3	44	0.89
6	9.9	18.7	-1.1	51	0.81
Avg.	9.9	16.1	2.7 [1]	45	0.86

1 - Estimated as the average of absolute biases for 6 watersheds.

of simulated and observed hydrographs, *BIAS*, and a correlation coefficient of daily discharges, *R*. As seen in Table 4, calibrated parameters usually produce higher accuracy although the gain is not as significant as compared to use of soil-derived parameters. As an example, simulated and observed hydrographs are plotted in Figure 4 for the Middle Fork River at Audra. Both parameter sets lead to good simulations of the observed hydrographs. The semi-log scale plot in Figure 4b suggests that base flow recessions are not well simulated by the soil-derived parameter sets. A possible reason for this was discussed earlier in Section 3.

Accuracy statistics from the transferability test simulations are shown in Table 5. The values suggest that scaled parameters improved simulation accuracies for 5 'ungaged' watersheds. Furthermore, the accuracy statistics are close to those obtained when each watershed was calibrated independently (compare Tables 5 and 4). While the overall statistics from a single watershed calibration (the Dry Fork at Hendricks watershed) are not greatly different from those derived from the scaled parameter version, there are significant degradations in bias (*BIAS*) and flood (*FDRMS*) statistics for some outlets, specifically for the Middle Fork River at Audra and the

Buckhannon River at Hall. The reason for this is that most soil derived parameters do not differ much for selected watersheds excluding the Middle Fork and Buckhannon river basins, see Figure 5, and, as a result, scaled parameters will produce similar results. However, the lower zone tension water (*LZTWM*) and supplemental free water (*LZFSM*) storages as well as the depletion rate of the primary free water storage are much lower for mentioned two watersheds (see Figure 5, thick lines). As a result, scaled parameters produced more runoff for these watersheds, and lead to improved bias and flood statistics compared to the constant parameter case.

4.2. Automatic Calibration Test Results

Calibration and validation results are presented in Figures 6 and 7. Daily runoff errors (*DRMS*) and monthly volume errors (*MVRMS*) from unconstrained and constrained calibration/validation, and soil derived parameter simulations are plotted for 10 watersheds in the Ohio basin. Results from two watersheds in the Tennessee basin were excluded from these plots, and will be discussed later. As seen from Figures 6 and 7, unconstrained calibration leads to slightly better sta-

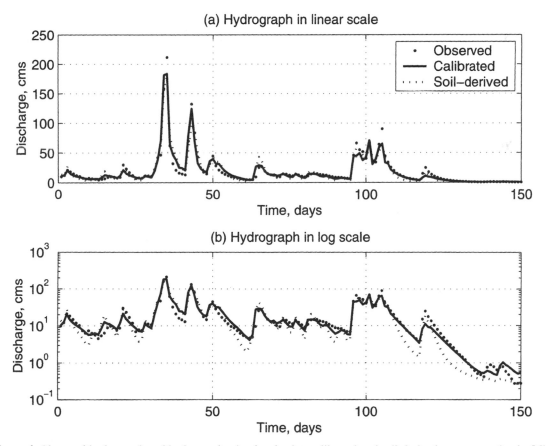

Figure 4. Observed hydrograph and hydrographs simulated using calibrated and soil derived parameters for the Middle Fork River at Audra, February - June 1967.

Table 5. Statistics for the parameter transferability test, the Upper Monongahela river basin.

Basin #	DRMS, cms	MRMS, mm	BIAS, %	FDRMS, %	R^2
Parameters calibrated for the Dry Fork at Hendricks (basin #1)					
2	10.8	13.4	-2.0	38	0.90
3	7.2	14.9	-5.3	42	0.87
4	4.6	20.8	1.1	45	0.85
5	7.9	12.2	-3.5	43	0.88
6	8.8	18.6	-0.3	47	0.85
Avg	7.9	16.0	2.4 [1]	43	0.87
'Scaled' soil derived parameters					
2	10.5	13.2	-0.5	36	0.91
3	7.6	14.1	-2.5	34	0.86
4	4.6	21.0	0.4	46	0.85
5	8.1	12.5	-3.5	45	0.88
6	9.0	18.8	-0.1	44	0.85
Avg	8.0	15.9	1.4 [1]	41	0.87

1 - Estimated as the average of absolute biases for 5 watersheds.

tistics compared to constrained calibration on the calibration data sets, however, this gain was practically eliminated on the validation data sets. While the use of soil-derived parameters alone provides reasonable simulation results, minor parameter adjustments can improve the overall performance.

The major benefit of use of soil derived parameters as calibration constraints is in generating spatially consistent parameter sets. The spatial variability of one SAC-SMA parameter, *UZFWM*, derived from unconstrained and constrained (values in parentheses) calibration can be seen in

Figure 8. It can be seen that overall, constrained and unconstrained results are consistent for most outlets. However, *UZFWM* values from unconstrained calibration can differ by 3-5 times for neighboring watersheds (highlighted values in italic). Figure 9c shows that the same behavior can be seen for most of the other parameters, which vary over the entire feasible parameter ranges. On the other hand, constrained calibration generates more consistent parameter sets while maintaining hydrological performance as shown in Figure 9b. Comparison of Figures 9a and 9b confirms that the spatial pattern of parameters derived by constrained calibration is consistent with soil derived patterns slightly adjusted to local physical properties and possibly data uncertainties.

Figure 10 shows that the most affected parameters from unconstrained calibration were the percolation parameters *ZPERC* and *REXP*, the upper layer free water storage *UZFWM*, and the lower layer tension water storage *LZTWM*. Deviations of these parameters from soil-based parameters were more than 60%. At the same time, deviations of constrained calibration parameters were much less than the allowed constraints of 25%. Overall constrained calibration results suggest that only 12% of the final parameter values were constrained by the specified search boundaries. Of these, 65% were values of the least identifiable from soil data parameters *LZPK* and *LZTWM*.

5. SUMMARY AND FUTURE WORK

This study illustrates the benefit of using soil-derived parameters to estimate conceptual model parameters for ungaged watersheds, and to improve results of automatic calibration. The results suggest that the use of soil derived parameters can improve the spatial and physical consisten-

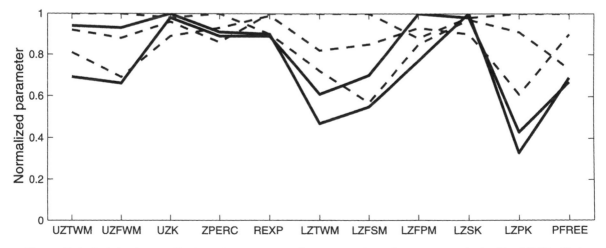

Figure 5. Soil derived normalized model parameters for the transferability test watersheds. The Middle Fork and Buckhannon river watersheds are shown with thick lines. In this Figure and later on, parameters were normalized based on their ranges for the unconstrained optimization.

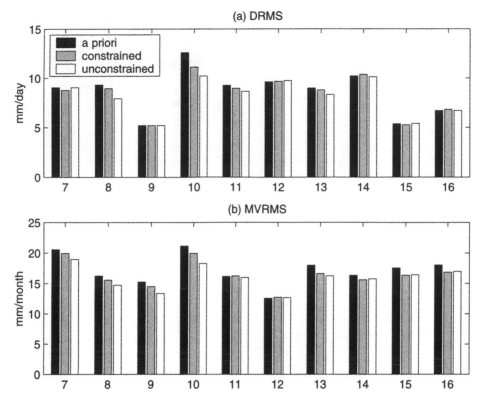

Figure 6. Constrained and unconstrained calibration results for the second group of watersheds: Calibration period.

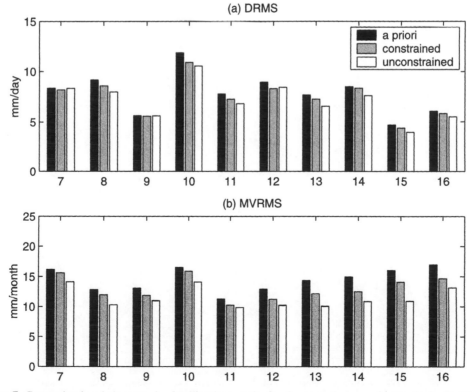

Figure 7. Constrained and unconstrained calibration results for the second group of watersheds: Verification period.

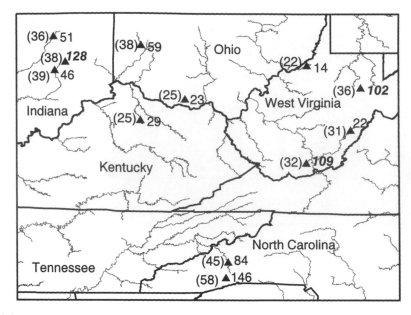

Figure 8. Spatial variability of the upper layer free water capacity derived from unconstrained and constrained (values in parentheses) calibration for the second group of watersheds.

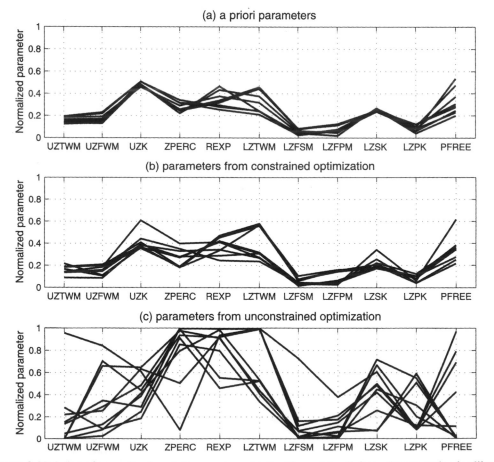

Figure 9. Variation of normalized model parameters obtained from (a) soil data, (b) constrained calibration, and (c) unconstrained calibration for the second group of watersheds.

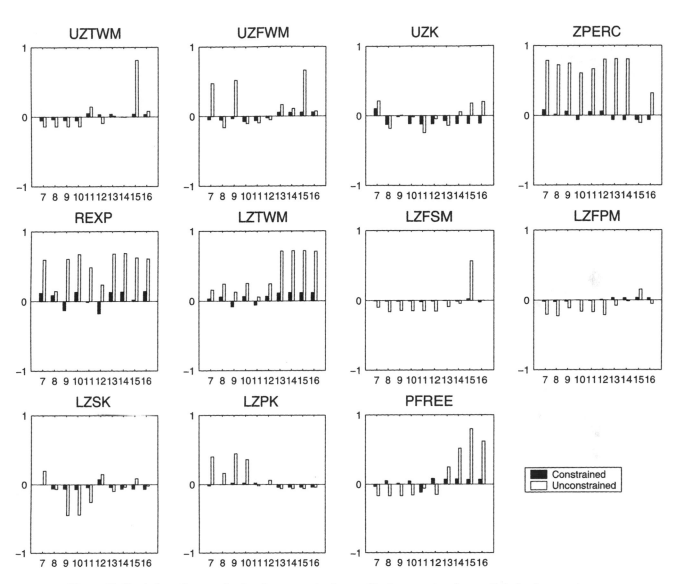

Figure 10. Deviation of constrained and unconstrained normalized parameters from soil derived parameters for the second group of watersheds.

cy of estimated model parameters while maintaining hydrological performance. When transferring model parameters from well-calibrated watersheds to ungaged watersheds, RFC experts rely on qualitative information such as soil, vegetation, etc. Soil derived parameters provide a quantitative measure of possible differences between parameters that allow one to 'rescale' calibrated parameters to ungaged watershed.

Use of constrained calibration reduces non-regularities in the spatial pattern of model parameters. RFC forecasters routinely make run-time adjustments to the hydrologic model states and certain parameters in order to keep the forecast models in close agreement with observed streamflow data. On a practical level, spatially consistent hydrologic model parameters should allow RFC forecasters to more efficiently make these run-time modifications, especially in the case of non-standard conditions during a rainfall event. With spatially consistent parameter sets, the forecaster can expect to use similar adjustments throughout a basin, thus saving time and allowing for the evaluation of more scenarios.

A simple approach was used to incorporate soil derived parameters into the automatic calibration procedure. Search regions of parameters were constrained by some percentage of soil-derived parameters. However, the percentage can vary for different regions depending on the accuracy of soil derived parameters and the quality of input/output data. Large uncertainties of soil derived parameters complicate the calibration procedure and in some cases can eliminate the benefit of using constrained calibration. This problem was encountered when constrained calibration was performed on two watersheds in the Tennessee basin, the French Broad River at Blantyre and Asheville. Daily and monthly statistics from constrained calibration were degraded significantly for both calibration and validation data sets as shown in Table 6. Large uncertainties in the soil derived parameters of the lower layer free water storages are the main reason for this. As discussed above, the soil-based approach does not account for a deep ground water aquifer, and as a result, underestimates the lower layer free water storages. Unconstrained calibration generated much higher values of the lower layer free water parameters for these two basins, *LZFSM*=270mm and *LZFPM*=950mm compared to 25mm and 124mm respectively for the soil derived parameters. To deal with this problem, constrained calibration should account for large uncertainties of soil derived parameters. One possibility would be to use an explicit measure, D_X, of the deviation of calibrated parameters, X_i, from soil derived parameters, X_i^*, in the automatic calibration procedure:

$$D_x = \left(\sum_{i=1}^{N} \left(\frac{X_i^* - X_i}{X_{max,i} - X_{min,i}} \right)^2 \right)^{0.5} \quad (14)$$

where $X_{max,i}$ and $X_{min,i}$ are maximum and minimum parameter values in a feasible space, and N is the number of calibrated parameters. A single objective function can be selected that will weight the gain in simulation accuracy versus the increase in D_X. However, estimation of a weight function may be a real challenge of this approach. Another possibility would be to incorporate a parameter deviation measure into multi-criteria calibration [*Boyle et al.*, 2000].

As stated in Section 3, there are weaknesses in the derivation of soil-based SAC-SMA parameters. Future research should be conducted to include more data sources in the estimation technique. Specifically, ground water information and hydrograph analysis may be helpful in estimating lower layer free water storages and depletion coefficients. New developments in generating more consistent SCS curve number grids can also lead to better estimates of fast runoff parameters.

APPENDIX:
SOIL BASED RELATIONSHIPS FOR ESTIMATING A PRIORI PARAMETERS OF THE SAC-SMA MODEL

Below there are SAC-SMA parameter and soil property relationships as they appeared in Koren et al., [2000]. Two printing errors in the original paper were fixed here: a coefficient 4 in the denominator was removed and a basic time step, Δt, (in the SAC-SMA model it equals 24 hours) was added in Eq. A8, and a coefficient 50.8 in Eq. A12 was replaced by 5.08. Parameter and soil property notations are consistent with Table 1 and Section 3.

Upper layer parameters:

$$UZTWM = (\theta_{fld} - \theta_{wlt}) \cdot Z_{up} \quad (A1)$$

$$UZFWM = (\theta_s - \theta_{fld}) \cdot Z_{up} \quad (A2)$$

$$UZK = 1 - (\theta_{fld}/\theta_s)^n \quad (A3)$$

Lower layer parameters:

$$LZTWM = (\theta_{fld} - \theta_{wlt}) \cdot (Z_{max} - Z_{up}) \quad (A4)$$

$$LZFSM = (\theta_s - \theta_{fld}) \cdot (Z_{max} - Z_{up}) \cdot (\theta_{wlt}/\theta_s)^n \quad (A5)$$

$$LZFPM = (\theta_s - \theta_{fld}) \cdot (Z_{max} - Z_{up}) \cdot [1-(\theta_{wlt}/\theta_s)^n] \quad (A6)$$

$$LZSK = \frac{1-(\theta_{fld}/\theta_s)^n}{1 + 2(1 - \theta_{wlt})} \quad (A7)$$

Table 6. Daily, DRMS, and monthly, MVRMS, statistics for two watersheds of the French Broad river from automatic calibration test

Basin #	DRMS, cms			MVRMS, mm		
	Uncon-strained	Con-strained	Soil derived	Uncon-strained	Con-strained	Soil derived
Calibration period						
17	4.35	7.39	13.61	16.73	31.66	36.95
18	10.79	15.39	24.67	10.88	21.97	26.37
Validation period						
17	4.08	5.65	10.30	18.62	31.11	38.21
18	10.20	13.47	20.75	9.73	19.47	25.11

$$LZPK = 1 - e^{-\frac{\pi K_s D_s (Z_{max} - Z_{up})\Delta t}{\mu}} \quad (A8)$$

$$PFREE = (\theta_{wlt}/\theta_s)^n \quad (A9)$$

$$ZPERC = \frac{LZTWM + LZFSM \cdot (1-LZSK)}{LZFSM \cdot LZSK + LZFPM \cdot LZPK} + \frac{LZFPM \cdot (1-LZPK)}{LZFSM \cdot LZSK + LZFPM \cdot LZPK} \quad (A10)$$

Percolation parameters:

$$REXP = [\theta_{wlt}/(\theta_{wlt,sand} - 0.001)]^{0.5} \quad (A11)$$

Upper layer thickness:

$$Z_{up} = 5.08 \frac{1000/CN - 10}{\theta_s - \theta_{fld}}, \quad mm \quad (A12)$$

REFERENCES

Abdulla, F. A., D. P. Lettenmaier, E. F. Wood, and J. A. Smith, 1996. Application of a macroscale hydrologic model to estimate the water balance of the Arkansas-red river basin, *JGR, 101(D3)*, 7,449-7,459.

Armstrong, B. L., 1978. Derivation of initial soil moisture accounting parameters from soil properties for the national weather service river forecast system, *NOAA Technical Memorandum, NWS HYDRO 37*, 53p.

ASCE, 1990. Evapotranspiration and Irrigation water requirements, *ASCE Manuals and Reports on Engineering Practice*, M. E. Jensen et al. (Ed.), No 70, ASCE.

Boyle, D. P., H. V. Gupta, and S. Sorooshian, V. Koren, Z. Zhang, and M. Smith, 2001. Toward improved streamflow forecasts: Value of semidistributed modeling, *WRR, Vol. 37, No. 11*, 2749-2759.

Boyle, D. P., H. V. Gupta, and S. Sorooshian, 2000. Toward improved calibration of hydrologic models: Combining the strengths of manual and automatic methods, *WRR, Vol. 36, No. 12*, 3663-3674.

Brazil, L., 1989. Multilevel calibration strategy for complex hydrologic simulation models, *NOAA Tech. Rep., NWS 42*, 178p.

Burnash, R. J. C., R. L. Ferral, and R. A. McGuire, 1973, A generalized streamflow simulation system - Conceptual modeling for digital computers, *Technical Report, Joint Federal and State River Forecast Center, U.S. National Weather Service and California Department of Water Resources, Sacramento, California*, 204p.

Burnash, R. J. C., 1995. The NWS river forecast system - catchment modeling, *In Computer Models of Watershed Hydrology, V. P. Singh (Ed.), Water Resources Publications, Littleton, Colorado*, 311-366.

Carpenter, T. M., K. P. Georgakakos, J. A. Sperfslagea, 2001. On the parametric and NEXRAD-radar sensitivities of a distributed hydrologic model suitable for operational use, *J. of Hydrology, 253*, 169-193.

Clapp, R. B. and G. M. Hornberger, 1978. Empirical Equations for Some Soil Hydraulic Properties, *WRR, Vol. 14, No. 4*, 601-604.

Cosby, B. J., G. M. Hornberger, R. B. Clapp, and T. R. Ginn, 1984. A Statistical Extrapolation of the Relationships of Soil Moisture Characteristics to the Physical Properties of Soil, *WRR, Vol. 20, No. 6*, 682-690.

Dingman, S. L., 1993. Physical Hydrology, *Prentice Hall, Englewood Cliffs, New Jersey 07632*, 575p.

Duan, Q., S. Sorooshian, and V.K. Gupta, 1992, Effective and Efficient Global Optimization for Conceptual Rainfall-Runoff Models, *Water Resour. Res., 28(4)*, 1015-1031.

Duan, Q., V. Koren, P. Koch, and J. Schaake, 1996. Use of NDVI and soil characteristics data for regional parameter estimation of hydrologic models. *EOS, AGU, 77(17), Spring Meet. Suppl.*, 138.

Duan, Q., J. Schaake, and V. Koren, 2001. *A Priori* Estimation of Land Surface Model Parameters, *In Land Surface Hydrology, Meteorology, and Climate: Observation and Modeling, V. Lakshmi, et al. (Ed.), Water Science and Application, Vol. 3, AGU, Washington, DC*, 77-94.

Hogue, T. S., S. Sorooshian, H. Gupta, A. Holz, and D. Braatz, 2000. A multistep automatic calibration scheme for river forecasting models, *J. of Hydrometeorology, Vol. 1*, 524-542.

Knowles, N, 2000. Modeling the Hydroclimate of the San Francisco Bay-Delta Estuary and Watershed, *Ph.D., University of California, San Diego*.

Koren, V. I., M. Smith, D. Wang, and Z. Zhang, 2000. Use of soil property data in the derivation of conceptual rainfall-runoff model parameters, *Preprints, 15th Conference on Hydrology, Long Beach, CA, Amer. Meteor. Soc., 10-14 January 2000*, 103-106.

Li, E. A., V. O. Shanholtz, and E. W. Carson, 1976. Estimating saturated hydraulic conductivity and capillary potential at the wetting front, *Dep. of Agr. Eng., Va. Polytech. Inst. And State Univ., Blacksburg*.

McCuen, R. H., 1982. A guide to hydrologic analysis using SCS methods, *Prentice Hall, Englewood Cliffs, New Jersey 07632*, 145p.

Miller, D. A. and R. A. White, 1999. A Conterminous United States multi-layer soil characteristics data set for regional climate and hydrology modeling, *Earth Interactions, 2, (available at http://EarthInteractions.org)*.

Rajaram, H., and K. Georgakakos, 1989. Recursive Parameter Estimation of Hydrologic Models, *WRR, Vol. 25, No. 2*, 281-294.

Reed, S., S. King, V. Koren, M. B. Smith, Z. Zhang, and D. Wang, 2001. Parameterization Assistance for NWS Hydrology Models Using ArcView, *Proc. of the 21st Annual ESRI International User Conference, San Diego, CA, July 2001, http://www.esri.com/library/userconf/proc01/professional/papers/pap1082/p1082.htm.*

Refsgaard, J. C., 1997. Parameterization, calibration and validation of distributed hydrological models, *J. of Hydrology, 198,* 69-97.

Schaake, J., Q. Duan, V. Koren, and A. Hall, 2001, Toward improved parameter estimation of land surface hydrology models through the Model Parameter Estimation Experiment (MOPEX), *in Soil-Vegetation-Atmosphere Transfer Schemes and Large-Scale Hydrological Models,*in *Dolman A.J., A.J. Hall, M.L. Kavvas, T. Oki, & J.W. Pomeroy* (Ed), *IAHS Pub. No. 270,* 91-97.

Schaake, J. C., Q. Duan, S. Cong, 2000. Retrospective analysis of water budget for the Mississippi river basin, *Preprints, 15th Conference on Hydrology, Long Beach, CA, Amer. Meteor. Soc.,* 10-14 January 2000, 91-94.

Sorooshian, S., and V. K. Gupta, 1995. Model calibration, *In Computer Models of Watershed Hydrology, V. P. Singh (Ed.), Water Resources Publications, Littleton, Colorado,* 23-68.

Yapo, P. O., H. V. Gupta, S. Sorooshian, 1998. Multi-objective global optimization for hydrologic models, *J. of Hydrol., 204,* 83-97.

_____

Victor Koren, Michael Smith, and Qingyun Duan, Hydrology Laboratory, Office of Hydrologic Development, NOAA/National Weather Service, 1325 East-West Highway, Silver Spring, MD 20910, USA.

Use of *a Priori* Parameter-Estimation Methods to Constrain Calibration of Distributed-Parameter Models

G. H. Leavesley, L. E. Hay, R.J. Viger, and S. L. Markstrom

U.S. Geological Survey, Denver, Colorado

Over-parameterization, ungauged basins, and the assessment of the impact of land-use and climate change are a few of the problems that limit the use of calibration techniques for distributed-parameter models. One approach to addressing these problems is the use of *a priori* parameter-estimation procedures to minimizing the number of parameters to be calibrated, or to obtain parameter values where calibration is not possible. A set of modeling and analytical tools is being developed using the US Geological Survey's Modular Modeling System to facilitate the development and evaluation of objective *a priori* methods. Initial testing of these tools was conducted on basins in the Rocky, Sierra-Nevada, and Cascade Mountain Regions of the United States. *a priori* parameter estimates were made for the USGS distributed-parameter model PRMS using available digital datasets of terrain, soils, vegetation type and density, and climatological data. Only the Rocky Mountain basin had an acceptable uncalibrated performance. Model performance for all basins improved as the parameters calibrated were increased incrementally from none, to those affecting the water balance, then hydrograph timing, and then all soils and vegetation related parameters. Problems were identified in the use of the forest-density dataset as a surrogate for forest cover density. A full evaluation of the soils dataset for determining available water-holding capacity was not possible due to the insensitivity of the model to this parameter in these snowmelt basins. Key issues in *a priori* parameter estimation for this limited application were identified to include regional climatic and physiographic differences, dataset limitations, and selection of measures of parameter and model performance. These will be addressed in an expanded research effort using tens of basins in different climatic and physiographic regions of the United States and the world.

INTRODUCTION

A major difficulty in the use of distributed-parameter models is the general lack of objective methods to estimate the values of distributed parameters. Calibration techniques are typically used to compensate for various sources of uncertainty in these estimates. However, the transferability of calibrated results to other basins is often an issue due to the over-parameterization of many distributed-parameter models and the incorporation of model and data errors in fitted parameter values. The application of calibration techniques to problems such as ungauged basins, or assessing the impact of land-use and climate change, is further limited because there are typically no measures of system response against which to calibrate. Estimating parameters where calibration is not possible, and addressing the over-parameterization problem by minimizing the number of parameters to be fitted, requires the development of methods that relate parameter values to measurable climatic and basin characteristics.

The development of methodologies to relate selected model parameters to climatic and basin characteristics has been conducted by a number of disciplines in the field of hydrology.

Calibration of Watershed Models
Water Science and Application Volume 6
Copyright 2003 by the American Geophysical Union
10/1029/006WS19

Studies at the point and plot scale have typically been used to define these relations. For example, in the area of soil physics, Rawls and Brackensiek (1983) developed a methodology to estimate soil water-holding capacities and Green-Ampt infiltration-model parameters using soil-texture information. More recent work by Schaap et al. (1998) has focused on the use of soil-properties data to develop pedotransfer functions for the estimation of water-retention and hydraulic properties. Similar efforts are being conducted in other areas with regard to a variety of hydrologic and climatic processes (e.g. Koren, et al., this volume). However, the application and evaluation of such techniques over larger areas have been limited. The ability to define the most appropriate parameter-estimation methods for use with different models in different climatic and physiographic regions, and to specify the robustness and reliability of these methods and their associated datasets, are major knowledge gaps.

The increasing availability of high-resolution spatial and temporal datasets of climatic and basin characteristics now provides the opportunity to investigate and develop *a priori* estimation procedures for distributed model parameters. To facilitate the development, testing, and evaluation of *a priori* parameter-estimation methods for a variety of models and datasets, a set of tools is being developed using the US Geological Survey's Modular Modeling System (MMS) (Leavesley et al., 1996; 2002). MMS is an integrated system of computer software that provides a common framework for multidisciplinary research and operational efforts to develop, evaluate, and apply a wide range of modeling capabilities and analytical tools. The long-term objectives of this research effort are to (1) develop and evaluate objective *a priori* parameter-estimation methods using available spatial and temporal datasets, and (2) evaluate and identify the most robust process-model conceptualizations and parameters for both uncalibrated and calibrated applications in different climatic and physiographic regions.

This paper focuses on the first objective. It describes the initial development and testing of a set of methodologies and tools for use with available digital datasets in three snowmelt regions of the western United States. The effort was limited to applications in mountainous regions where snow accumulation and melt processes dominate the hydrological cycle. This provided a focus on common hydrologic processes but in different climatic regimes. As the first step in a larger, more comprehensive effort, the study was further limited to only one model and a single basin in each of the three snowmelt regions. The next steps in this research will be the application of the tools and knowledge developed in this study to ten's of basins in the study regions and the development of a fully integrated set of models, methods, and tools to address both research objectives given above.

STUDY BASINS

Snow-dominated, mountain basins were chosen in the Rocky, Sierra Nevada, and Cascade Mountain Ranges in the United States. The basins selected (Fig. 1) were (1) the Animas River basin, which has a drainage area of 1820 km[2], and elevation that ranges from approximately 2000 to 4000m; (2) the East Fork of the Carson River basin (hereafter referred to as the Carson River basin), which has a drainage area of 920 km[2] and elevations that range from approximately 1600 to 3000m; and (3) the Cle Elum River basin, which has a drainage area of 526 km[2] and elevations that range from 600 to 2000m. Vegetation on all the basins is predominantly coniferous forest with a mix of alpine tundra and bare rock occurring on areas above timberline.

MODEL

The USGS Precipitation-Runoff Modeling System (PRMS) (Leavesley et al., 1983; Leavesley and Stannard, 1995) is a distributed-parameter, physical-process watershed model. Distributed-parameter capabilities are provided by partitioning a watershed into units, using characteristics such as slope, aspect, elevation, vegetation type, soil type, and precipitation distribution. Each unit is assumed to be homogeneous with respect to its hydrologic response and to the characteristics listed above. Each unit is termed a hydrologic response unit (HRU). A water balance and an energy balance are computed daily for each HRU. The sum of the

Figure 1. Study basin locations.

responses of all HRUs, weighted on a unit- area basis, produces the daily watershed response.

Snow is the major form of precipitation in the Animas, Carson, and Cle Elum River basins, and the major source of streamflow. The snow components of PRMS simulate the accumulation and depletion of a snowpack on each HRU. A snowpack is maintained and modified both as a water reservoir and as a dynamic heat reservoir. A water balance is computed each day and an energy balance is computed for two 12-hr periods each day. The energy-balance computations include estimates of net shortwave and longwave radiation, the heat content of precipitation, and approximations of convection and condensation terms.

PRMS uses daily inputs of solar radiation and the variables precipitation (PRCP), maximum air temperature (TMAX), and minimum air temperature (TMIN). Solar radiation was distributed to each HRU as a function of HRU slope and aspect. Solar radiation data were not available on a daily basis and so were computed using existing algorithms in PRMS. Estimates of daily shortwave radiation received on a horizontal surface were computed using air temperature, precipitation, and potential solar radiation. A list of PRMS parameters referred to in this paper and their definitions are provided in Table 1.

TOOLS

The GIS Weasel

The GIS Weasel is a geographic information system (GIS) interface for applying tools to delineate, characterize, and parameterize topographical, hydrological, and biological basin features for use in a variety of lumped- and distributed-modeling approaches. It is composed of Workstation ArcInfo (ESRI, 1992) GIS software, C language programs, and shell scripts.

Parameter-estimation methods are implemented using ARC Macro Language (AML) functions applied to available digital datasets. A library of parameter-estimation methods is maintained in a similar fashion to the library of process modules in MMS. For a given model, a recipe file of AML functions can be created and executed to estimate a selected set of spatial parameters. This recipe file can also be modified to change the parameter-estimation method associated with a selected parameter, thus enabling the evaluation of alternative parameter-estimation methods.

Table 1. Definition of selected PRMS parameters.

Group	Parameter	Definition
GIS Weasel	covden_win	Winter vegetation cover density for the major vegetation type on an HRU
	jh_coef_hru	Air temperature coefficient used in Jensen-Haise potential evapotranspiration computations for each HRU.
	rad_trncf	Transmission coefficient for short-wave radiation through the winter vegetation canopy
	soil_moist_max	Maximum available water holding capacity of soil profile
Meteorological	adjmix_rain	Monthly factor to adjust rain proportion in a mixed rain/snow event
	bias	Precipitation adjustment factor to account for gage catch efficiency and other sources of measurement error
	tmax_allrain	Maximum daily temperature above which all precipitation is assumed to be rain
	tmax_allsnow	Maximum daily temperature equal to or below which all precipitation is assumed to be snow
Runoff Timing	emis_noppt	Average emissivity of air on days without precipitation
	gwflow_coef	Groundwater reservoir routing coefficient
	soil2gw_max	Amount of the soil water excess for an HRU that is routed directly to the associated ground-water reservoir
	ssrcoef_sq	Non-linear subsurface -reservoir routing coefficient

XYZ Precipitation and Temperature Distribution

Recent research has resulted in the development of a new distribution methodology for daily values of PRCP, TMAX, and TMIN (Hay et al., 2000a,b). Significant geographic factors affecting the spatial distribution of PRCP, TMAX, and TMIN within a river basin are latitude (x), longitude (y), and elevation (z). Multiple linear regression (MLR) equations are developed for each dependent climate variable (PRCP, TMAX, TMIN) using the independent variables of x, y, and z from available climate stations. The general form of the MLR equation for precipitation at a given HRU is

$$PRCP_{(HRU)} = b_0 + b_1 x_{(HRU)} + b_2 y_{(HRU)} + b_3 z_{(HRU)} \quad (1)$$

The resulting fit from equation 1 describes a plane in three-dimensional space with slopes b_1, b_2, and b_3 intersecting the PRCP axis at b_0. Similar equations are used for TMAX and TMIN. Use of the station x and y coordinates in the MLR provides information on the local-scale influences on the climate variables that are not related to elevation (for example, the distance to a topographic barrier). To account for seasonal climate variations, MLR equations are developed for each month using mean values of PRCP, TMAX, and TMIN (dependent variables) and station x, y, and z (independent variables) from a set of stations selected from regional National Weather Service and Snow Telemetry (SNOTEL) stations that fall within and outside the selected basins. The monthly MLRs are computed to determine the regression surface that describes the spatial relations between the monthly dependent variables and the independent variables. Note that for each month the best MLR relation does not always include all the independent variables.

Estimates of daily PRCP, TMAX, and TMIN for each HRU are computed using the following procedure: (1) mean daily values of PRCP (TMAX and TMIN) and corresponding mean x, y, and z values from a selected station set (described in the Exhaustive Search analysis below) are used with the slopes of the monthly MLRs to compute a unique b_0 for that day and (2) the MLR equation is then solved using the x, y, and z values of the HRUs.

The regional MLR equations, typically developed for areas thousands to tens-of-thousands of square kilometers in size, may under- or over-estimate the mean precipitation (or temperature) in smaller basins typically used for hydrologic simulations. These smaller basins often range in size from a few hundred to a few thousand square kilometers. Also, measurement errors associated with precipitation, particularly precipitation gage under-catch of snow, may lead to significant errors in hydrologic simulations. To address these issues, an Exhaustive Search (ES) analysis is used to

(1) determine the optimal precipitation- and temperature-station sets to anchor the xyz distribution methodology; (2) provide an estimate of bias associated with the selected precipitation stations; and (3) define a separate precipitation-station set to determine daily precipitation frequency.

Precipitation and temperature stations are selected independently since the best precipitation station choice generally differs from the best choice for temperature distribution in a basin. For every combination of these precipitation- and temperature-station sets, a precipitation bias and a station set to indicate precipitation frequency are also tested. The range of the bias correction is from 0 to 50 percent and the correction is applied only to snowfall events. The correction actually compensates for the net effect of a number of biases related to precipitation measurement, such as gauge under-catch, gage location, and/or lack of gauges at high elevation. It may also correct for other sources of bias in PRMS.

The ES analysis is run to test all single stations and possible combinations of two, three, and four station groups comprising the xyz-station sets. For each ES analysis, the best station sets for temperature and precipitation, along with an associated precipitation bias and frequency, are determined by comparing the sum of the absolute value of the difference between measured and simulated runoff. The ES analysis ends when the sum of the absolute errors associated with the above combinations shows no significant improvement from one station group to the next.

Analysis Tools

Optimization and sensitivity analysis tools are provided in MMS to analyze model parameters and evaluate the extent to which uncertainty in model parameters affects uncertainty in simulation results. Two optimization procedures are available to fit user-selected parameters. One is the Rosenbrock technique (Rosenbrock, 1960), as it is implemented in PRMS. The second is a hyper-tunnel method (Restrepo and Bras, 1982).

Several methods for parameter sensitivity analysis are also provided. One is the method described in the PRMS user's manual (Leavesley et al., 1983), which allows the evaluation of a variety of measures including relative parameter sensitivity, error propagation, and parameter correlation. A second method evaluates the sensitivity of any pair of parameters and develops the objective function surface for a selected range of these two parameters. To address the question of parameter uncertainty, a Monte Carlo procedure is available to evaluate alternative combinations of model parameters.

The basic measures of model and parameter performance used in this study were the comparisons of measured to simulated daily streamflows. One measure was expressed in terms of the sum of the absolute values of the differences between measured and simulated daily streamflow. This measure was used as the objective function in all model calibrations and was used for comparison of individual parameter sensitivity and performance. A second measure was the Nash-Sutcliffe coefficient of efficiency (CE) (Nash and Sutcliffe, 1970). It was used as a measure of model performance for alternative parameter sets.

METHODOLOGY

Parameter Estimation

PRMS HRUs were delineated and characterized using the GIS Weasel. The Animas, Carson, and Cle Elum River basins were divided into 121, 96, and 124 HRUs, respectively. Spatially distributed topographic, vegetation, and soils parameters on each HRU were estimated using available digital datasets. The datasets used were: (1) USGS 3-arc second digital elevation models (DEMs); (2) 1-km gridded version of the State Soils Geographic (STATSGO) soils data (U.S. Department of Agriculture, 1994); (3) Forest Service 1-km gridded vegetation type and forest-density data (U.S. Department of Agriculture, 1992): and the USGS GIRAS land-use/land-cover gridded coverage. A composite of GIRAS Land Cover and the Forest Type Groups was created. In this composite, the GIRAS data was only used where the Forest Type Group data described "non-forest". The resulting Land Cover dataset has a total of 44 classes.

Topographic parameters such as elevation, slope, and aspect were computed for each HRU using the USGS 3-arc second DEM. Elevation was calculated as the median of the distribution of the DEM grid-cell elevations. Slope was calculated as the mean of the distribution of the grid-cell slopes. Grid-cell aspect values were reclassified into one of eight aspect classes that represent the eight cardinal points of the compass. HRU aspect was calculated as the aspect class having the dominant number of grid cells.

The vegetation type and density datasets were used to estimate vegetation-type and vegetation-cover-density parameters on each HRU, as well as the associated parameters of interception-storage capacity and the transmission coefficient for solar radiation. HRU vegetation type was reclassified into one of four classes defined as forest, shrub, grass, and bare. HRU vegetation type was then determined as the dominant reclassified vegetation type in an HRU. Vegetation cover density was defined as the percentage of the area of the HRU covered by the dominant vegetation type canopy. For forest vegetation types, the canopy density was assumed to be equal to the mean of the forest-density values for the forest types found in the HRU, expressed

as a percentage of the entire HRU area. For example, an HRU with 60 percent of its area in forest having a mean forest density of 50 percent would have a vegetation cover density of 30 percent.

Interception-storage capacity was calculated by multiplying the computed HRU vegetation cover density times the average depth of precipitation storage per unit area of the cover type. A table of interception-storage-capacity values for all vegetation types was created using values from the literature. For deciduous vegetation types, cover density and its associated interception-storage capacity were calculated for two periods, one with and one without leaves. The transmission coefficient was calculated using the "cover density - transmission coefficient" relation provided in the PRMS user's manual (Leavesley et al., 1983). For deciduous vegetation, the cover density for the period without leaves was used in this computation.

Soil type in PRMS is categorized as sand, loam, or clay, and was calculated for each HRU using the soil texture data in the STATSGO soil dataset. Soil type was calculated as the dominant soil type on the HRU. The available water-holding capacity of the soil on each HRU is a function of the water-holding characteristics of the soil and the average rooting depth of the dominant vegetation. Calculation began with the identification of the dominant vegetation type. Average rooting depth for each vegetation type was estimated from the literature and a table of rooting-depth values was linked to the vegetation-type dataset. Available water-holding capacity values in the STATSGO dataset were processed to provide an average water-holding capacity value per unit depth of soil. This value was multiplied times the average rooting depth of the HRU vegetation type to calculate the HRU water-holding capacity.

Climate-related parameters were estimated using daily data obtained from the National Weather Service and the SNOTEL data network. Precipitation- and temperature-distribution relations were computed using the xyz methodology. A threshold temperature parameter (tmax_allsnow) is used to determine precipitation form (rain, snow, or a combination of both). The estimate of tmax_allsnow was based on the assumptions that a precipitation event will have a cloud base 305 m above the ground and that a temperature at the cloud base of $0°$ C will produce snow. The $0°$ C cloud-base temperature was assumed to provide a near-surface air temperature of $1.7°$ C for tmax_allsnow.

PRMS parameters related to the partitioning of water among processes related to surface, subsurface, and groundwater flow, as well as all remaining parameters, were estimated from the results of other model applications in these mountainous regions and were provided as a common set for all the basins.

Parameter Evaluation

Evaluation of the *a priori* parameter-estimation methods and their use in constraining parameter calibration was conducted by examining model performance at four levels of parameter fitting. At the first level, the model was run using all estimated parameters. No calibration was conducted. HRU and meteorological parameters were estimated using the GIS Weasel and the regional MLR relations of the xyz methodology. The subset of precipitation and temperature stations used consisted of all stations within or immediately adjacent to the modeled basin. No exhaustive search was conducted.

The second level focused on the calibration of those parameters related to obtaining reasonable monthly and annual water-balances. The parameters fitted were those controlling the magnitude and distribution of precipitation and potential evapotranspiration (PET). PET was computed using a modified Jensen-Haise method (Jensen et al., 1969) with a parameter value that was varied by month. The monthly PET parameter values were calibrated to monthly estimates of PET for each region based on values obtained from the literature. Then the exhaustive search was applied in the xyz method to determine the optimal station set, bias correction, and precipitation frequency station set.

At the third level, the fitted values obtained at level two were retained and the additional parameters that affect the timing of streamflow were optimized. These included parameters affecting precipitation form, snowmelt rates, and the rates and timing of surface, subsurface, and groundwater flow processes.

At the fourth level, the fitted values obtained at level three were retained and the most sensitive spatial and non-spatial PRMS parameters were calibrated. This included several of the parameters estimated using the GIS Weasel. Level 4 was considered comparable to a full model calibration of all sensitive parameters.

With the exception of the xyz methodology, all parameter calibrations were conducted using the Rosenbrock optimization technique. The procedure used to fit distributed parameters was to adjust all values of a specific parameter simultaneously. The assumption was made that all the values of a distributed parameter were correct relative to each other and to their spatial location. The mean value of the parameter and the deviation from the mean for each HRU was computed. The mean was then adjusted and the deviations were used to recompute the individual HRU values. In the recomputation procedure, the HRU values can be increased or decreased by the same magnitude or by the same percentage of their initial value. An upper and lower bound were specified for each parameter, and individual HRU values were reset to the boundary value if they exceeded the specified bound.

Meteorological and streamflow time-series data were available for the period 1978-1996 in the Animas and Carson basins and 1978-1994 in the Cle Elum basin. Parameter calibration was conducted using the period 1978-1988 and all analyses of parameter sensitivity and model performance were conducted using the period 1989 to the end of the record or selected years within this period.

RESULTS

Parameter Sensitivity

A first step in the evaluation of parameter-estimation methods was to determine if the model had any sensitivity to the parameters being estimated. Parameter sensitivities were determined using the PRMS sensitivity analysis procedures. For comparative purposes the parameters were grouped into the general categories of (1) GIS Weasel computed, (2) meteorological process, and (3) runoff timing process. A selection of the most sensitive parameters in each group is shown in Table 2.

Comparing differences among the parameter groups for all basins showed that the meteorological parameters, in most cases, were about an order of magnitude more sensitive than those in the other two groups. Tmax_allsnow is a scalar value but was applied to each HRU and thus affected the spatial distribution of rain and snow through the effect of the temperature-distribution relations determined using the xyz-method.

Differences among the basins within parameter groups reflected the differences in the climatic and physiographic characteristics of the three mountain regions. The most sensitive GIS Weasel estimated parameters were (1) winter cover density (covden_win) and solar radiation transmission coefficient (rad_trncf) which affect the snowpack energy balance relations, and (2) the Jensen Haise HRU ET coefficient (jh_coef_hru) and available soil-water storage (soil_moist_max) which affect the water-balance relations. The higher sensitivity of rad_trncf in the Animas basin reflects the somewhat larger effect of shortwave radiation on the snowpack energy balance in the Rocky Mountains as compared to the Sierra Nevada and Cascade ranges. The most sensitive parameter in the meteorological group in all basins was tmax_allsnow, which delineates precipitation form between snow and rain. It was most sensitive in the Carson and Cle Elum basins where rain-snow combinations and rain-on-snow events are much more common than in the Animas basin.

The most sensitive runoff-timing parameters were the emissivity term in the longwave energy equation (emis_noppt) and the daily flux rate of water movement

Table 2. Percent change in model standard error for a 10 percent increase in selected parameter values.

Group	Parameter	Animas	E Fk Carson	Cle Elum
GIS Weasel	covden_win	.02	1.01	.02
	jh_coef_hru	.58	.28	.07
	rad_trncf	2.66	.82	.17
	soil_moist_max	.41	.15	.05
Meteorological	adjmix_rain	2.80	.95	.39
	bias	.61	.02	.01
	tmax_allrain	7.31	2.13	.02
	tmax_allsnow	8.51	30.45	21.67
Runoff Timing	emis_noppt	.78	3.14	.59
	gwflow_coef	.09	.18	.01
	soil2gw_max	.63	.38	.05
	ssrcoef_sq	.05	.09	.10

from the soil zone to the ground-water reservoir (soil2gw_max). The emis_noppt parameter affects the long-wave energy balance computation for days with no precipitation. Soil2gw_max affects the distribution of runoff between more rapid subsurface and slower ground-water flow sources. Comparing the sensitivity of rad_trncf to emis_noppt within each basin indicates the relative effects of shortwave and longwave energy on the total energy balance for snowmelt computations in each basin. Rad_trncf is more sensitive than emis_noppt in the Animas basin but less sensitive in the Cle Elum basin, which is again indicative of a greater effect of shortwave energy in the Animas basin. The smaller sensitivity of rad_trncf in the Carson basin may be anomalous and is related to a problem of the underestimation of rad_trncf discussed in the next section.

Uncalibrated Parameters

Evaluation of the performance of selected *a priori* parameter estimates was conducted using a Monte Carlo analysis procedure. A test case was constructed to evaluate a selected set of sensitive parameters that were estimated from the spatial and climatic datasets. One thousand model runs were made using parameter sets with randomly generated values for the four parameters estimated by the GIS Weasel (Table 2) and the tmax_allsnow and the bias parameters. The results for the rad_trncf, soil_moist_max, and tmax_allsnow parameters on each basin are shown as dotty plots in Figure 2. These plots reflect the concept of equifinality where a number of different parameter sets may be suitable for reproducing observed basin streamflow (Beven and Freer, 2001). The arrows indicate the dots that represent the uncalibrated

model parameter set and objective function values of the initial uncalibrated run. For a distributed parameter, the x-axis value is the mean of all HRU values weighted by HRU area. The objective-function values for the uncalibrated runs are larger than the objective-function values for the best-fit runs by about 45 percent in the Animas basin, 85 percent in the Carson basin, and 30 percent in the Cle Elum basin.

The parameter sets containing the best-fit values for rad_trncf were reasonably well constrained on the Animas and Carson basins but less so on the Cle Elum. This reflects in part the sensitivity of each basin to shortwave energy input. The mean of the *a priori* estimate of rad_trncf was overestimated in the Animas basin and underestimated in the Carson basin but had a value with a less clearly defined error in the Cle Elum basin. The higher estimate of rad_trncf in the Animas basin produced an overestimate of shortwave energy available for snowmelt while the lower estimate in the Carson and Cle Elum produced an underestimate of shortwave energy available for snowmelt.

The *a priori* estimates of rad_trncf were computed using the HRU winter vegetation cover densities computed from the forest-density dataset. The estimated mean value of the winter cover-density parameter covden_win was about 35 percent in the Animas basin and about 71 percent in the Carson and Cle Elum basins. The mean value of 71 percent for covden_win in the Carson basin appeared high. An examination of the forest density dataset for the Carson basin showed a large number of grid cells with the value of 100 percent forest density. The forest density values were based on the coregistration of Advanced Very High Resolution Radiometry (AVHRR) data and Landsat Thematic Mapper (TM) and on regression analysis of sta-

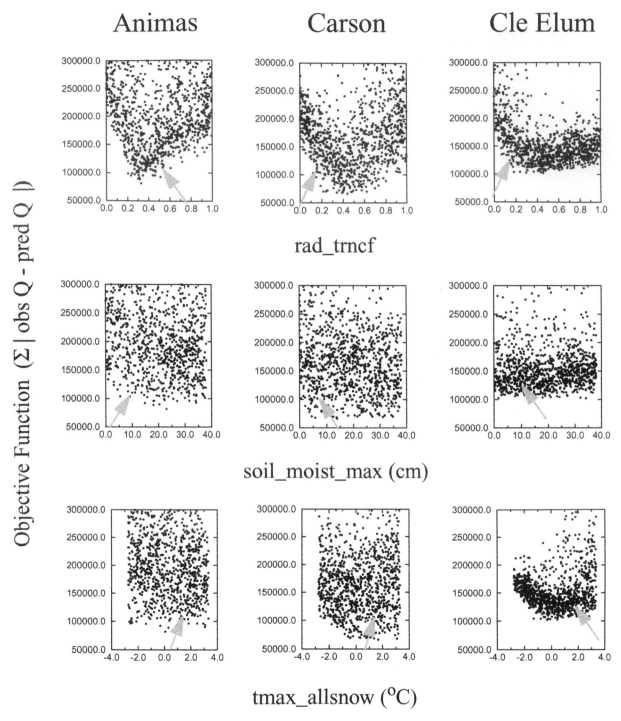

Figure 2. Monte Carlo analyses for the parameters rad_trncf, soil_moist_max, and tmax_allsnow on the Animas, Carson, and Cle Elum basins. (Arrows indicate uncalibrated parameter set values and results.)

tistical relations between the two data types (Zhu, 1994). The forest density was the percentage of forested TM cells within one AVHRR cell. There were about 1,225 TM cells in an AVHRR cell and a TM cell was considered forested if it contained a classified forest type.

Thus, 100 percent forest density from the dataset does not necessarily mean 100 percent cover of the land surface, only that 100 percent of the TM cells had some forest cover. However, covden_win is defined as one minus the percentage of the sky visible from the land or snow surface. A value of 100 percent forest cover effectively eliminates shortwave energy from the snowpack energy-balance computation. The forest-density dataset was used as an index of cov-den_win in this application and some adjustment for the difference in interpretation will be needed to better estimate covden_win and rad_trncf.

The soil_moist_max parameter was computed from the STATSGO soils dataset. It shows a much larger degree of uncertainty when compared to the other two parameters shown (Figure 2). This reflects the fact that in the snowmelt regions selected, the soil typically remains at or near field capacity during most or all of the snowmelt-runoff period and thus its storage capacity has only a small effect on streamflow. The snowmelt period is the major source of the annual streamflow in these basins. Consequently, the value of the STATSGO dataset for estimating soil_moist_max cannot be fully evaluated in these basins.

The relative insensitivity of tmax_allsnow in the Animas basin reflects the fact that there are few winter rain or rain-snow mixed events and that most snow events occur at temperatures well below $1.7°$ C. In the Carson and Cle Elum basins, winter rain and rain-snow mixed events are more common and the threshold effect of tmax_allsnow is more evident. The large increase in the objective function at tmax_allsnow values less than about $0°$ C results from the associated increase in the proportion of rain versus snow. This response provides some confidence that the model responds correctly to physically unrealistic values of tmax_allsnow. Smaller increases in the objective function for tmax_allsnow values greater than $0°$ C may indicate less sensitivity to decreases in the proportion of rain versus snow, or that rain events occur at temperatures much warmer than $0°$ C.

Constrained Calibration Performance

Calibrated model performance at each level of parameter fitting was measured using the Nash-Sutcliffe CE (Table 3). The uncalibrated parameters produced the poorest performance of all the levels of fitting. At this level, the best simulation results were obtained in the Animas basin with less satisfactory results in the other basins.

At the second level, calibrating the PET parameter and applying the exhaustive search procedure in the xyz methodology provided improved results in all basins. The increases in the CE were 0.26 in the Carson basin and 0.05 in the Animas and Cle Elum basins. The smaller levels of improvement in the Animas basin implies that the initial selection of climate stations in and near the basin was reasonable but that improvement in model performance is possible. A similar statement can be made for the Cle Elum

basin but only after observing the results from level 3 which indicated that timing was a larger source of error than the water balance.

The fitted bias corrections, associated with gauge-catch error for snow, were 30, 0, and 10 percent for the Animas, Carson, and Cle Elum basins respectively. These values appear reasonable, with the exception of the Carson basin. As with any parameter fitting exercise, the station-set selection and bias correction may be adjusting for biases in the data as well as biases in other parameters or model conceptualizations.

Calibrating the runoff-timing parameters at level three increased the CE by 0.17 for the Cle Elum basin but only 0.05 and 0.02 for the Carson and Animas basins respectively. Calibrating the GIS Weasel estimated parameters at level four increased the CE an additional 0.05 in the Animas and Carson basins but only 0.01 in the Cle Elum basin.

The stepwise fitting provided a mixed picture of the value of the estimated parameters in each group. Improvement in model performance varied among the basins with the fitting of each group. Improvement in the Animas was about the same for each fitting step, while the meteorological parameter fitting on the Carson and timing parameter fitting on the Cle Elum provided the greatest improvement in model performance.

DISCUSSION AND CONCLUSIONS

The results presented have provided an overview of the initial methods and tools that are being integrated into a modeling framework for use in the development, testing, and application of *a priori* parameter-estimation methodologies. While limited in scope, the results raise a number of issues that need to be addressed in the continued development and enhancement of the methods and tools. These issues relate to the general categories of datasets, parameter-

Table 3. Nash-Sutcliffe coefficient of efficiency for four levels of parameter fitting.

Level	Animas	East Fork Carson	Cle Elum
1. Uncalibrated	.73	.47	.52
2. Exhaustive search xyz	.78	.73	.57
3. Optimize timing	.80	.78	.74
4. Full optimization	.85	.83	.75

estimation methods, and measures of model performance.

Dataset issues include concerns of dataset consistency, areal extent, and value. The datasets used in this study were selected because they were available for the entire United States and each was produced with a consistent methodology. This enabled the comparison of parameter estimation methods and the information content of the dataset among different regions of the United States. The value of a dataset relates to measures that include the accuracy of the data and the effect of the data on improved model performance. The value of the STATSGO soils dataset was shown to be limited in the basins tested because of the insensitivity of the model to the computed available water-holding capacity parameters. To get a more complete measure of the value of the STATSGO dataset, basins with rain as a more dominant precipitation form will need to be included in the study. Expanding the study to all regions of the United States will enable a comparable evaluation of the value of other available datasets for a range of models and process parameterizations.

Alternatively, parameters calculated using the vegetation type and density datasets had a much higher sensitivity but raised another concern that leads into the issue of appropriate parameter-estimation methodology. An attempt was made to use the forest density dataset as a surrogate for the forest cover density parameter in PRMS. However, inconsistencies in the interpretation of these two physical measures were a major source of parameter and model error.

An alternative approach to using dataset values directly in parameter computation is to first calibrate spatially distributed parameters on a typical set of basins within a region of interest. Then regression equations, relating the calibrated parameter values to values in the dataset, are developed. The resulting regression equations and original datasets are then used to estimate distributed parameters on other basins in the region (Abdulla and Lettenmaier, 1997; Xu, 1999). While the referenced examples used modeling approaches different from PRMS, the method should be applicable to a wide of variety of lumped and distributed models. One concern, however, is that fitted parameters may be biased by other sources of error, thus the value of this approach compared to other approaches needs more evaluation. It also requires a reasonable number of gauged basins with sufficient spatial variability to address the full range of related basin characteristics and parameter values.

Methods to define the distribution of precipitation and temperature are key to being able to accurately simulate distributed hydrologic processes and streamflow. While this is a problem in all regions, it is most pronounced in areas of complex and mountainous terrain. In the Animas basin, the uncalibrated xyz methodology using precipitation and tem-

perature stations in and near the basin produced reasonable model performance. Application of the xyz exhaustive search procedure brought model performance in the Carson basin to an acceptable level and improved performance in the Animas basin further. The testing and development of the xyz methodology began in mountainous regions and has only recently begun testing in other regions of the United States. It should be applicable for temperature distribution in all climatic and physiographic regions. For precipitation distribution, it is most applicable for frontal type precipitation events that occur over an entire watershed. Modifications to the current xyz method as well as other techniques are being evaluated for use with more localized, convective-type storms.

In the Cle Elum basin the effects of the xyz methodology on model performance was masked to some degree by the errors associated with poor estimates of the runoff-timing parameters. This observation identifies two issues. One is the general lack of a regional or global dataset for geological and hydrogeological characteristics that could be used to assist in the estimation of runoff-timing parameters related to the apportioning of the surface and surface components of streamflow generation. These parameters were estimated from model application to other basins in this region and the results reflect some of the potential difficulties in transferring parameters from one basin to another.

The second issue is the question of how to best identify and measure the sources of error, such as data, parameter, and model error, and what measures are most appropriate for objectively defining parameter and model performance. The Animas and Carson model results were described above as being acceptable. The level of acceptability is typically a subjective judgment and needs to be more clearly defined in terms of what specific measures are most appropriate and what are acceptable magnitudes of those measures.

Measures of model performance are also needed to compare uncalibrated performance versus the calibrated model. Appropriate measures could be used to provide confidence limits for simulation results on ungauged basins. Such measures would also provide a consistent way to compare other methodologies and models. Defining the appropriate measures of performance is a question that still needs to be addressed.

Historically the measure most typically used for calibrating and evaluating distributed parameters has been streamflow. However, streamflow integrates the spatial variability of the process parameters being fitted. Thus it is possible to obtain a reasonable simulation for the wrong reason. A more appropriate measure of distributed parameter performance would be spatial measures related to the process being simulated. Increasing availability of remotely sensed data is now making it possible to begin to develop some independent measures of distributed parameter model performance. One such measure currently available is snow-covered area.

The ability of PRMS to adequately simulate streamflow and the spatial and temporal distribution of snow-covered area has been demonstrated in the Carson basin (Leavesley and Stannard, 1990) and in five basins adjacent to the Animas basin (Leavesley et al., 2002). A comparison of the model simulated snow-covered area with that measured by satellite, throughout the melt season, showed that the spatial and temporal distribution of snowpack accumulation and melt agreed well with the satellite data for the basins in both regions. Concurrently, simulated streamflow agreed well with the volume and timing of measured streamflow. The agreement in simulated snow-covered area and streamflow volume and timing infer a measure of confidence in the parameter estimation methods applied and in the transferability of the methods to ungauged basins. The Carson and Animas basins were selected for this study in part based on the results of these previous studies.

To address the issues raised in this study and to build on its results, the parameter-estimation methodologies will be tested and enhanced using tens of basins in a number of climatic and physiographic regions of the United States using a variety of process conceptualizations. Test basins provided by the Model Parameter Estimation Experiment (MOPEX) project (http://www.nws.noaa.gov/oh/mopex/), which is a cooperative activity of the international scientific community, will be used to expand the study to basins in other regions the world.

To facilitate this, the GIS Weasel and xyz methods will be enhanced to operate in a batch mode. User-specified recipe files in the GIS Weasel will define the delineation, characterization, and parameterization procedures to be applied to the select basins in each region. Alternative recipe files will be developed for each model and each set of parameter estimation methods and datasets to be evaluated. The Shuffle Complex Evolution Optimization algorithm (Duan et al., 1993) and the Multi-Objective COMplex Evolution algorithm (Yapo et al., 1998; Gupta et al., this volume, "Multiple Criteria Global Optimization for Watershed Model Calibration"), which is capable of solving multi-objective optimization problems, are also being incorporated in to MMS. The Monte Carlo methodology is being expanded to incorporate the Generalized Likelihood Uncertainty Estimation (GLUE) procedure (Beven and Binley, 1992; Beven and Freer, 2001; Freer et al., this volume).

This research effort is not unique. A variety of systems and tools to address the issues of parameter estimation and uncertainty analysis are being developed by other investigators using approaches that include multi-criteria optimization, sensitivity analysis, and generalized likelihood uncertainty analysis techniques (Beven and Binley, 1992; Beven and Freer, 2001; Yapo et. al., 1998; Wagener et. al., 1999; Wheater and Lees, 1999). What separates MMS from these other systems is the Open Source software system approach in which all members of the scientific community can participate in the design and development of the system framework, process modules, and analysis and support tools. The resulting toolbox will facilitate the multidisciplinary, systematic approach that is needed to 1) identify the most appropriate estimation methods for use with different models in different climatic and physiographic regions, and 2) define the robustness and reliability of these methods and their associated datasets.

Further information on MMS and the GIS Weasel can be found at:

http://wwwbrr.cr.usgs.gov/mms
http://wwwbrr.cr.usgs.gov/weasel

REFERENCES

Abdulla, F.A., and Lettenmaier, D.P., Development of regional parameter estimation equations for a macroscale hydrologic model, *J. Hydrol.,* 197(1-4), 230-257, 1997.

Beven, K.J., and Binley, A. The future of distributed models: Model calibration and uncertainty prediction, *Hydrol. Process.* 279-298, 1992.

Beven, K., and Freer, J., Equifinality, data assimilation, and uncertainty estimation in mechanistic modelling of complex environmental systems using the GLUE methodology, *J. Hydrol.,* 249, 11-29, 2001.

Beven, K.J., Lamb, R., Quinn, P.F., Romanowicz, R., and Freer, J., TOPMODEL, in Singh, V.P. (Ed). *Computer Models of Watershed Hydrology,* Water Resources Publications, Highlands Ranch, CO, 627-668, 1995.

Duan, Q., Gupta, V.K., and Sorooshian, S., A shuffled complex evolution approach for effective and efficient global optimization, *J. Opt. Theory and Its App.,* (3), 501- 521, 1993.

Environmental Systems Research Institute, *ARC/INFO 6.1 user's guide.* Redlands, CA., 1992.

Hay, L.E., Clark, M.P, and Leavesley, G.H., Use of Atmospheric Forecasts in Hydrologic Models,. Part two: Case Study, *Proceedings of the American Water Resources Association's Spring Specialty Conference on Water Resources in Extreme Environments,* April, 2000, Anchorage, Alaska, 2000a.

Hay, L.E., Wilby, R.L., and Leavesley, G.H., A comparison of delta change and downscaled GCM scenarios for three mountainous basins in the United States', *J. Am. Wat. Resour.,* 36(2), 387-397, 2000b.

Jensen, M.E.,Rob, D.C.N., and Franzoy, C.E., Scheduling irrigation using climate-crop-soil data: *National Conference on Water Resources Engineering,* Am. Soc. Civil Eng., New Orleans, LA, 20 p., 1969

Leavesley, G.H., Lichty, R. W., Troutman, B. M., and Saindon, L. G., Precipitation-Runoff Modeling System: User's Manual. *U.S. Geol. Surv. Wat. Res. Invest. Report 83-4238*, 1983.

Leavesley, G.H., Markstrom, S.L., Restrepo, P.J., and Viger, R.J., A modular approach to addressing model design, scale, and parameter estimation issues in distributed hydrological modeling, *Hydrol. Process.*, 16, 173-187, 2002.

Leavesley, G.H., Restrepo, P.J., Markstrom, S.L., Dixon, M., and Stannard, L.G., The modular modeling system - MMS: User's manual: *U.S. Geol. Surv. Open File Report 96-151*, 142 p, 1996.

Leavesley, G.H., and Stannard, L.G., Application of remotely sensed data in a distributed-parameter watershed model, in Kite, G.W. and Wankiewicz, eds., *Proceedings of Workshop on Applications of Remote Sensing in Hydrology*, National Hydrologic Research Centre, Environment Canada, 47-64, 1990.

Leavesley, G.H., and Stannard, L.G., The precipitation-runoff modeling system - PRMS Chap 9 of Singh, V.P., ed., *Computer models of watershed hydrology*, Water Resources Publications, Highlands Ranch, Colorado, p. 281-310, 1995.

Nash, J.E., and Sutcliffe, J.V., River flow forecasting through conceptual models, Part I, A discussion of principles: *J. Hydrol.* 10, 282- 290, 1970.

Rawls, W.J., and Brackensiek, D.L., A procedure to predict Green and Ampt infiltration parameters, *Proceedings of the National Conference on Advances in Infiltration*, 102-112, 1983

Restrepo, P.J. and Bras, R.L., Automatic parameter estimation of a large conceptual rainfall-runoff model: A maximum-likelihood approach, *Ralph M. Parsons Laboratory Report No. 2*, Massachusetts Institute of Technology, Department of Civil Engineering, Cambridge, 1982.

Rosenbrock, H.H., An automatic method of finding the greatest or least value of a function, *Computer J.*, 3, 175-184, 1960.

Schaap, M.G., Leij F.J. and van Genuchten M.Th., Neural network analysis for hierarchical prediction of soil water retention and saturated hydraulic conductivity. *Soil Sci. Soc. Am. J.* 62,847-855, 1998.

U.S. Department of Agriculture, State Soil Geographic (STATSGO) database - Data use information: Natural Resources Conservation Service, *Misc. Pub. No. 1492*, 107p, 1994.

Department of Agriculture, Forest land distribution data for the United States: Forest Service, Accessed August 10, 1998 at URL http://www.epa.gov/docs/grd/forest_inventory, 1992.

Wagener, T., Lees, M., and Wheater, H.S., A generic rainfall-runoff modeling toolbox, *Eos Trans. AGU, Fall Mtg. Suppl.* 80, F409, 1999.

Wheater, H.S., and Lees, M.J., A framework for development and application of coupled hydrological models', *Eos Trans. AGU, Fall Mtg. Suppl.* 80, F405, 1999.

Yapo, P.O., Gupta, H.V., and Sorooshian, S., Multi-objective global optimization for hydrologic models, *J. Hydrol.*, 83-97, 1998.

Xu, C-Y, Estimation of parameters of a conceptual water balance model for ungauged catchments, *Water Resour. Manag.* 13(5), 353- 368, 1999.

Zhu, Z., Forest Density Mapping in the lower 48 States: A regression procedure, Southwest Forest Experiment Station, *Res. Pap. SO- 280*, 1994

Ordered Physics-Based Parameter Adjustment of a Distributed Model

Baxter E. Vieux and Fekadu G. Moreda

School of Civil Engineering and Environmental Science University of Oklahoma, Norman, Oklahoma

Distributed hydrologic models based on conservation laws have identifiable optimal values and expected behavior and interaction during calibration. This paper describes a calibration method that exploits these model characteristics and presents results for two river basins: the Illinois River (2300 km²) and Blue River basins (1142 km²) in Oklahoma. Distributed parameter watershed models that are physics-based offer distinct advantages over conceptual rainfall-runoff models. Spatially distributed parameters derived from soil properties, land use/cover, topography, and input from radar rainfall require new methods for adjustment in order to minimize differences between simulated and observed hydrographs. The scheme presented is an ordered physics-based parameter adjustment (OPPA) method. Two river basins are simulated with volume errors with good agreement in volume for a series of eight storm events over each basin. The calibrated simulations for the Blue are within 1.5 mm average difference, 6.9% average error, with a root mean square error (RMSE) of 9.7 mm. The calibrated simulations for the Illinois are within 2.5 mm average difference, −12.5% average error, and RMSE=3.6 mm. Realistic parameter values are obtained for both basins based on soil properties and land use/cover maps. Equivalent calibration factors are obtained for the two basins even though they are simulated at differing resolutions and are located in different geographic-climatic regions.

INTRODUCTION

The goal of distributed modeling of streamflow is to better represent the spatio-temporal characteristics of a watershed governing the transformation of rainfall into runoff. Thus, the motivation for development of distributed hydrologic modeling. A number of modeling approaches exist that rely on conservation equations for the routing of runoff through a distributed representation of a watershed. Such models, termed physics-based or physically-based distributed models (PBD), include *r.water.fea* (*Vieux and Gauer* 1994; *Vieux* 2001), a parallel version of r.water.fea called the distributed hydrologic model (DHM) ; CASC2D [*Julien and Saghafian,* 1991; *Ogden and Julien* 1994; *Julien et al.*

Calibration of Watershed Models
Water Science and Application Volume 6
Copyright 2003 by the American Geophysical Union
10/1029/006WS20

1995], Systeme Hydrologique European (SHE) (*Abbott et al.* 1986a; b) and the Distributed Hydrology Soil Vegetation Model (DHSVM) [*Wigmosta et al.* 1994]. These models formulate runoff generation and routing based on conservation equations to various degrees.

Because the parameters are derived from physical properties, prior knowledge exists for starting points in the calibration process, and may be applied to ungauged watersheds. PBD models are well suited to simulating specific events at locations where streamflow records may not exist or are relatively short. Conceptual rainfall-runoff (CRR) models include Precipitation-Runoff Modeling System (PRMS) by *Leavesley et al.* [1983], the Sacramento Soil Moisture Accounting Model (SAC-SMA) [*Burnash et al.* 1973]. The SAC-SMA model simulates runoff generation with 16 conceptual parameters and routes the runoff using unit hydrographs to an outlet. Deriving these conceptual parameters from soil properties may help extend the application of the SAC-SMA model to ungauged watersheds

[*Koren et al.* 2000]. Both CRR and PBD models require initial parameter estimates, which are then refined through calibration (see *Koren et al.* this volume).

CRR models are not physics-based by definition, but may be considered semi-distributed by subdividing the watershed into sub-basins. Assessment of performance improvements associated with subdivision of the Blue River basin from a single lumped basin to 8-subbasins is described by *Boyle et al.* [2001]. This approach was tested along with various combinations of lumped versus distributed parameters, soil moisture, routing, and precipitation inputs. Improved model predictions were found by using semidistributed parameters with three subbasins, but little improvement when the subdivision was extended to eight. The semidistributed model performed significantly better than the lumped representation of parameters. Current efforts to transform the SAC-SMA model into a semi-distributed representation are underway at the US National Weather Service (NWS) [*Koren et al.,* 1999; and *Boyle et al.* 2001]. Motivation for this stems from a desire to improve performance of the lumped SAC-SMA by subdividing it into subbasins or grids.

Development of automated computer-based calibration methods has focused mainly on the development of 1) mulitcriteria objective functions [*Boyle et al.,* 2001; *Bolye et al.* this volume], and 2) optimal value search algorithms [*Sorooshian and Dracup,* 1980; *Duan et al.,* 1992, 1994]. *Yapo et al.* [1998] extended the single-objective function method to a multi-objective complex evolution (MOCOM-UA) capable of exploiting the observed time series. *Boyle et al.* [2000] compared automatic with manual methods of calibration to develop a multicriteria approach to optimization that combines features from both. Regardless of the search algorithms used to calibrate CRR models, parameter interaction, convergence and interstorm/interannual stability are still problematic.

PBD models are parameterized by deriving estimates of parameters from physical properties, viz.; databases of soil properties used to derive infiltration parameters. Besides uncertainties in parameters and inputs, model prediction accuracy depends on how well the model structure represents physical conditions. Balance between model complexity and number of parameters, given limited streamflow observations is a significant concern [*Jakeman and Hornberger,* 1993; *Hornberger and Spear,* 1981; and Freer, et al. this volume]. Because of the common experience with CRR models that a number of parameter sets give equal performance, the concept of multiple models with no optimal parameter set (equifinality) has been advanced by *Beven and Freer* [2001].

Vieux [2001] used a distributed model to investigate the hydrologic worth of distributed data. Considerable sensitiv-

ity to spatially averaged (lumped) parameters in a fully distributed model is found. Two effects of lumping calibrated parameters were noted: 1) a bias in simulated results caused by delayed and attenuated peaks and reduced volume, and 2) degraded prediction accuracy resulting in poorer performance in volume and peak discharge. Scale effects introduced through discretization can introduce bias requiring re-calibration [*Finnerty et al.,* 1997; *Obled et al.* 1994]. Better parameter representation of watershed characteristics generally improves model prediction accuracy whether it is a PBD or CRR model. The degree or extent of improvement likely depends on the particular model structure used to investigate the importance of the spatial variability affecting the process.

Distributed Model Calibration

PBD model calibration differs from CRR calibration in two important ways. First, some scheme must be devised to adjust the grid cell parameters affecting the output. Second, as a result of the governing equations derived from the physics of conservation of mass and momentum, the parameters should exhibit expected behavior. Because of the known behavior of the model, a sequence of adjustment is possible that identifies the optimal parameter set. This method is termed as ordered physics-based parameter adjustment (OPPA). The OPPA method described herein capitalizes on the expected behavior of a physics-based model.

Automated retrieval of the optimal set can be implemented using a cost surface to search out the optimal parameter values. Manual methods involve making the adjustment to model parameters and inspecting each hydrograph individually, or inspecting the cost surface formed from many events and identifying the optimal parameters. Automatic retrieval involves search algorithms that find the minimum or optimal values on a cost surface. One automated technique is the adjoint equation, which is the inverse of a partial differential equation subject to minimization of a cost function. Finding the solution that minimizes the objective function constitutes an optimal control problem using scalars to multiply maps of parameters [*Vieux et al.,* 1998; *Vieux,* 2001; *White et al.,* 2001a,b]. Because there are underlying differential equations in PBD models, we can form an adjoint of the forward model and invert in the presence of data to find if there is a unique solution. A unique solution exists if the model is invertible. Existence of identifiable optimal parameter sets can be demonstrated mathematically through the adjoint method, or through direct computation of a cost surface as described in this chapter.

Several aspects of the PBD model calibration are of particular importance: 1) Maps of parameters derived from geographic information system or remote sensing (GIS/RS) data provide spatial distribution, 2) Parameters may be scale dependent because of sampling characteristics of the GIS/RS source, 3) Slope and drainage length are dependent on DEM resolution, and 4) Calibration is used to adjust initial parameter estimates from soil properties, DEM, and land use/cover.

The agreement between the observed and simulated volume and peak flow may be expressed in terms of bias and departures. The bias indicates systematic over or under prediction. The departure, whether expressed as an average difference, percentage error, coefficient of determination, or as a root-mean-square error, serves as a measure of the prediction accuracy. Three objective functions may be considered:

1) Square of errors between observed and simulated volume;
2) Square of errors between observed and simulated peak flow;
3) Sum of the normalized errors of volume and peak flow.

A physics-based model has the advantage of having expected parameter response and interaction. Calibrating such a model profits from the physical relationship, physical significance, and expected response to adjustment of parameters derived from physical properties. To summarize the approach taken for distributed rainfall-runoff calibration, the following list may be enumerated.

1) Estimate the spatially distributed parameters from physical properties;
2) assign channel hydraulic properties based on measured cross-sections where available;
3) study the sensitivity of each parameter:
   a. Identify response sensitivity to each parameter;
   b. run the model for a range of storms from small, medium, to large events;
   c. observe the characteristics of the hydrograph over the range of storm sizes;
   d. observe any consistent volume bias;
   e. identify seasonal effects that may influence radar estimation of rainfall, land use/cover, or other factors;
   f. identify any systematic bias due to radar, soil moisture, or hydraulic conductivity;
   g. derive range of response for a given change in a parameter, e.g., soil moisture;
   h. categorize parameters according to response magnitude.
4) the optimum parameter is that set which minimizes the respective objective function;

5) volume should be adjusted first, followed by parameters affecting timing and peak;
6) re-adjust hydraulic conductivity if necessary to account for changes due to parameter interaction.

Table 1 shows the expected response and the parameter affecting the model response. This table can be read as: *increasing the volume of the hydrograph (+)* is achieved by *decreasing hydraulic conductivity (-)*. Similarly, *increasing peak flow (+)* is achieved by *decreasing hydraulic roughness (-)*. Channel parameterization should be applied according to measured cross-sections, and measured or visual estimates of hydraulic roughness. If this is unavailable, then channel hydraulic characteristics may be estimated from similar channels, geomorphic relationships or local knowledge and then adjusted. Channel hydraulics primarily affect the timing, and to some degree the peak discharge. Multiple gauging stations within a river basin helps resolve timing problems associated with channel hydraulics and aids in the adjustment process. Consistent bias in timing may be related to the channel or the overland flow hydraulics, or both. In either case, these parameters are estimated, and then adjusted to minimize the objective function.

The OPPA method is described and demonstrated for two watersheds located in Oklahoma and Arkansas. The remaining sections describe the model structure, study watershed locations and data, results and discussion showing the results of the OPPA method, and conclusions.

MODEL DESCRIPTION

The model used herein was first developed by *Vieux* [1988] and applied to a small agricultural watershed without channel routing, which was later added. The solution using linear, one-dimensional elements presented by *Vieux et al.* [1990] uses a single chain of finite elements for solving overland flow. *Vieux and Gauer* [1994] extended this finite element solution to a network of elements representing a watershed domain with channels within a GIS. The resulting

Table 1. OPPA Method Parameter and Response.

Response	Parameter
Runoff Volume (+)	Hydraulic conductivity (-)
Runoff Volume (+)	Initial degree of saturation (+)
Runoff Volume (+)	Radar rainfall bias, G/R (+)
Peak flow (+)	Hydraulic roughness (-)
Time to peak (+)	Hydraulic roughness (+)

model was the distributed hydrologic model *r.water.fea*, developed in 1993 for the U.S. Army Corps of Engineers, Construction Engineering Research Laboratory, Champaign, Illinois (USA-CERL). The initial development of the model is a part of the public domain GIS called GRASS (Geographic Resource Analysis Support System). Several derivative models now exist. The *r.water.fea* model was ported from Unix to a Windows version that runs as an ArcView Extension, called *Arc.water.fea* [*Vieux*, 2001]. The *r.water.fea* model accesses a map database for various parameters controlling the hydrologic process. For example, rainfall rates derived from radar or other sources are sampled over each grid cell as input to the model. Integration of the r.water.fea model with GIS routines requires that the model be run within either the GRASS or ArcView GIS. Other versions are modifications to run in special environments. The DRUM (distributed runoff model) is coupled with the SHEELS model for soil moisture modeling [*Crosson et al.* 1999]. A parallel version is also being developed as a part of the Environmental Hydrology Applications Team (EHAT) project at the National Center for Supercomputing Applications [*NCSA-EHAT*, 2001].

A recent version of the finite element approach written in Java™, takes advantage of client-server applications within a real-time operational context. Due to the computational efficiency of the finite element method (described below), large watershed domains may be simulated in real-time using precipitation from radar updated every 5 or 6 minutes. This model, called Vflo™, is a new implementation of the finite element approach with many improvements related to ease of operation and stream routing options. It is currently operational for real-time hydrologic prediction for the Central Weather Bureau, Republic of China (Taiwan), in the U.S. over the coastal Carolinas, and the intermountain Southwest in the Salt and Verde watersheds in Arizona for the Salt River Project. The family of models: *r.water.fea, arc.water.fea,* and Vflo™ are network models capable of utilizing widely available digital datasets describing topography, land use/cover, soils, and radar rainfall. In the following sections, the mathematical analogy, numerical solution, and implementation are described.

The connectivity between grid cells in a digital elevation model is used to develop a system of equations for solving the kinematic wave analogy (KWA). Figure 1 shows the grid cell scheme used by *r.water.fea* to define the finite elements connecting overland flow and channel elements. Because overland flow in the KWA is in the direction of principal landsurface slope, it relies on a drainage direction map derived from a digital elevation model (DEM). This scheme is efficient since cross-drainage gradients are not present in the KWA and therefore, need not be computed.

The mathematical formulation and computer implementation are not burdened from computing unneeded terms provided the model is correctly applied in situations where the land surface gradient dominates. The KWA assumption is appropriate whenever backwater effects are not important, or with some error in hydraulically mild slopes.

Mathematical Formulation

The KWA for overland flow is a simplification of the conservation of mass and momentum equations wherein the dominating principle gradient is the land surface slope. The conservative form of the full dynamic equations relates the temporal and x-direction gradients of flow depth, y and velocity, V as:

$$\frac{\partial V}{\partial t} + V\frac{\partial v}{\partial x} + g\frac{\partial y}{\partial x} - \left(S_o + S_f\right) = 0 \quad (1)$$

If all other terms are small or an order of magnitude less than the bed slope, S_o, or friction gradient, S_f, the KWA is an appropriate representation of the wave movement downstream in many practical watershed applications [*Singh* 2002; *Chow et al.,* 1988]. The simplified momentum equation and the continuity equation comprise the KWA. The one-dimensional continuity equation for overland flow resulting from rainfall excess is expressed by:

$$\frac{\partial h}{\partial t} + \frac{\partial (uh)}{\partial x} = R - I \quad (2)$$

where R is rainfall rate; I is infiltration rate; h is flow depth; and u is overland flow velocity. In the KWA, we equate the bed slope with the friction gradient, which amounts to the *uniform flow* assumption. Using this fact together with an appropriate relationship between velocity, u, and flow depth, h, such as the Manning equation, we obtain:

$$u = \frac{S_o^{1/2}}{n} h^{2/3} \quad (3)$$

where S_o is the bed slope or principal land surface slope, and n is the hydraulic roughness. Velocity and flow depth depend on the land surface slope and the friction induced by the hydraulic roughness. Hydraulic roughness is derived from land use/cover or remotely sensed vegetation maps. Both overland and channel flow are represented by Eqs. 2 and 3 with suitable adaptation for channel characteristics. This eliminates the need to assign time of travel explicitly for each cell as in DHSVM or the HEC-HMS ModClark method, because timing and the effects of drainage network configuration are accounted for implicitly as defined by the

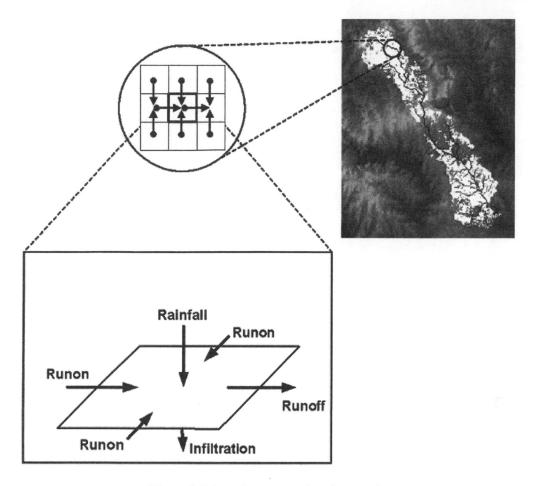

Figure 1. Schematic representation of *r.water.fea.*

drainage network, slope, and hydraulic roughness in the drainage network comprised of overland and channel cells.

Two of the most important parameters in this model are the saturated hydraulic conductivity, k, controlling infiltration, I, in Eq. 3 and the roughness, n. Hydraulic conductivity controls the total amount of water that will be partitioned into the surface runoff and the subsurface, whereas the hydraulic roughness mainly affects the peak flow and the time to peak. Model results obtained from Eqs. 2 and 3 are adjusted by scalars applied to spatially distributed parameters:

$$\frac{\partial h}{\partial t} + \beta \frac{s^{1/2}}{n} \frac{\partial h^{5/3}}{\partial x} = \gamma R - \alpha I \qquad (4)$$

where the three scalars α, γ, β and are multipliers controlling the infiltration rate, I, rainfall rate, R, and hydraulic roughness, n, respectively. The flow depth, h and slope, s is the principal land-surface slope at the center of each grid cell. The slope and hydraulic roughness are spatially variable, while rainfall, infiltration, and flow depth are spatially

and temporally variable. If we consider the rainfall as accurately known, i.e., no bias, then γ is equal to one.

The OPPA method relies on the scalars in Eq. 4 as adjustments of the PBD model. When viewed as controls of the solution in the adjoint formulation, an optimal control scheme results [*Vieux et al.,* 1998; *Vieux,* 2001; *White et al.,* 2001a,b]. The fact that Eq. 4 may be inverted indicates that there is a unique solution and that the parameters are identifiable. For a distributed-parameter model, existence of a set of unique controls (scalar multipliers) is predicated on knowing the spatial pattern of the parameter. Knowing the spatial pattern reduces the distributed parameter in two-dimensions to simple adjustment of the scalars. Without knowledge of the spatial pattern, a uniform value may be used, which is a special case of spatial variability. Whether manual or adjoint methods are used, an optimal parameter set is found where the cost surface is a minimum.

Finite element solution. The finite element solution to the KWA for a watershed is described in detail by *Vieux* [2001]. Linear one-dimensional elements are laid out in the eight principal directions of slope connecting each grid cell.

There is no reason that the lengths must be of equal length except that the drainage network connectivity is derived from GIS maps of drainage direction, which are generally of constant resolution. Using 1-D linear elements to solve the KWA equations is described by *Vieux,* [1988], *Vieux et al.,* [1990], *Vieux and Segerlind* [1989], *Vieux and Gauer* [1994] and *Vieux* [2001]. The finite element method is based on forming the elemental residual, R(e) and then minimizing this over the domain represented by a system of equations. For the overland flow case, and replacing *uh* with the unit-width flow rate, q in Eq. 2 becomes:

$$R^{(e)} = \int_\Omega N^T \left[\frac{\partial h}{\partial t} + \frac{\partial q}{\partial x} - (R - I) \right] = 0 \quad (5)$$

Using weighting functions N^T that are the same as linear shape function approximations for $h^{(e)}$ and $q^{(e)}$ in Eq. 5, the elemental residual is obtained. This forms the basis for assembling the elemental contribution into a system of ordinary differential equations in time, which are then solved with an explicit finite difference scheme.

Computational efficiency of the finite element method. The limitation to widespread use of a fully distributed hydrologic model was thought to be the computation time that limits model utility for real-time flood forecasting. *Johnson* [2000] tested the finite difference model CASC2D for the 130 km² Buffalo Creek watershed in Colorado, and found a 1:1 simulation:rainfall computational time. This means that for a CASC2D simulation for a basin receiving 20 hours of rainfall takes 20 hours of computational time. It was also found that the CASC2D model was sensitive to, and produced variable results depending on the resolution and timestep used.

The efficiency in the finite element method of diagonalizing the time dependent matrix permits sufficient computational efficiency to solve large watersheds over days of response in just minutes or seconds on a single processor. Simulation time on a PC, 500 mHz, Intel Pentium-3 processor and 1-Gb RAM takes 5-6 minutes to simulate 8 days of runoff at a 100-second time step for the storm events tested over the 2400 km² Illinois River basin at 1-km resolution. Simulation time from event to event depends on the variability of the precipitation in each subbasin. Because of the efficiency of the finite element model, it is possible to perform many sensitivity studies by direct computation of the parameter combination permutations.

The prospects for using a distributed-parameter model in operational flood forecasting are possible because of this efficiency. Computational efficiency of the finite element method implemented in the Vflo™ model is the main reason that real-time applications are practical. If the model cannot complete the simulation before the next input from the radar scan, say every 5 or 6 minutes, then the model never arrives at a solution until the rainfall is over, and quite possibly, not until after the flood has occurred.

Surface Runoff Generation

Infiltration excess (IE) is treated by the model as the source of runoff. The model represents overland flow as a uniform depth over a computational element. From hillslope to stream channel, there may be areas of IE and Saturation Excess (SE), however, the model treats runoff generation as solely IE. Simulation of IE requires soil properties and initial soil moisture conditions. The well-known Green-Ampt equation is used to account for the effects of initial degree of saturation on infiltration rate. The rate form of the Green-Ampt equation for the one-stage case of initially ponded conditions and assuming a shallow ponded water depth is

$$f(t) = K_e \left[1 + \frac{\Psi_f \Delta\theta}{F(t)} \right] \quad (6)$$

where $\phi(t) = dF(t)/dt$ infiltration rate; K_e = effective saturated hydraulic conductivity, estimated from soil properties or measured in the laboratory; ψ_f = average capillary potential (wetting front soil suction head); $\Delta\theta$ = moisture deficit; and $F(t)$ = cumulative infiltration depth. The soil-moisture deficit can be computed as:

$$\Delta\theta = \phi_{total} - \theta_i \quad (7)$$

where ϕ_{total} is the total porosity; and θ_i is initial volumetric water content. Eq. 6 is solved for cumulative infiltration depth, F for successive increments of time using a Newton iteration to obtain the instantaneous infiltration rate which when subtracted from the rainfall rate, R, becomes rainfall excess for routing through the finite element drainage network. Run-on from upslope is added to the rainfall vector making it possible for downslope areas to infiltrate runoff even though rainfall has ceased over any particular cell. This concludes the description of the relevant model components. Following sections describe the study area and results of the model calibration for the Illinois and Blue River basins.

STUDY AREA AND DATA SOURCE

Watershed Location

The Illinois River watershed straddles the Oklahoma/Arkansas border with approximately 54% of the

2,400 km² drainage area at Tahlequah located in Oklahoma as shown in Figure 2. The basin spans Delaware, Adair, Cherokee, and Sequoyah counties on the Oklahoma side. The average annual flow of the Illinois River as it enters Oklahoma near Watts is 20 m³/s, which increases at Tahlequah to 29 m³/s, after which it flows into Tenkiller Ferry Lake. Both rivers are designated by the Oklahoma Scenic Rivers Commission as Scenic Rivers, and are not controlled by reservoirs. Rocky soils and outcrops tend to remain in either forest or pasture, whereas cropland is concentrated in the more fertile soil in the lowlands or flatter slopes in the watershed. The dominant industry in the basin is agriculture in both the Blue and Illinois.

Hydroclimatology of the Blue and Illinois Rivers

The Oklahoma Climate Survey reports that over the counties encompassing the Blue River, between 70-74 days are expected to have measurable precipitation. Normal annual precipitation ranges from 914 mm to over 1016 mm with fewer than 4 days of snow expected. The mean annual temperature is 16 C and between 80-89 days >32C are expected. The South Central region of Oklahoma only has a mean annual runoff estimated to range from 152-203 mm [*Ryder*, 1996]. The Illinois River typically receives measurable precipitation between 80-90 days annually. Normal annual precipitation ranges from 1066 to 1219 mm with 10-14 days of snow expected. The mean annual temperature is 14-15 C and less than 49 to 59 days >32C are expected. Mean annual runoff in the East Central region of Oklahoma is estimated to range from 304 to 508 mm [*Ryder*, 1996]. Mean annu-

al runoff in the region encompassing the Illinois River is roughly twice that of the Blue River.

The USGS 07332500 Blue River near Blue, OK has been in operation since 1936. It has a drainage areas 1142 km². Hourly and daily streamflow hydrographs reveal that the hydrograph peaks increase rapidly and then return to flow around 5-10 m³/s. The maximum discharge is 1300 cms with a daily mean of 9 cms. Baseflow is a minor component compared to peaks of individual hydrographs.

Since 1935, The USGS 07196500 Illinois River near Tahlequah, OK has been in operation. At this location, the drainage area is 2300 km². Hourly and daily streamflow hydrographs reveal the hydrograph peaks increase rapidly and then return to low flow around 20-30 m³/s. The maximum discharge is 2583 cms with a mean of 27 cms and 90% of being equal to or less than 4 cms. Baseflow though higher than the in the Blue River, is small compared to hydrograph peaks. Milder temperatures along with higher annual precipitation produce more annual runoff and consistently higher baseflow in the Illinois River than in the South Central region of Oklahoma where the Blue River is located.

Digital Elevation Model (DEM)

In the Illinois basin, the 1:24,000 USGS 30-m resolution DEM is re-sampled to obtain a 960-m resolution map of elevation. In the case of the Blue River basin, the 1:250,000 USGS 3-arc second DEM is resampled to 270-meter resolution. The Blue DEM consists of 16,900 grid cells (270 meter), whereas, the Illinois DEM consists of 2,700 grid cells (960 meter). The finite element representation is a network of elements and does not require subwatersheds. The DEM is used to delineate the watershed into three sub-basins in the Illinois and 16 sub-basins in the Blue as listed in Table 2.

Subdivision of the river basins into subwatersheds is done for organizing the simulations. This has particular importance in achieving load balance in a parallel computing environment. The number of basins affects the number of channels and the efficiency at which runoff arrives at the outlet. The coarser resolution chosen for the Illinois River is consistent with a 1-km resolution DEM used by the NWS River Forecast Center in Tulsa, Oklahoma to delineate river basins and forecast locations. The finer resolution chosen for the Blue was based on earlier studies on resampling of the 3-arc second DEM. The 270-meter resolution is simply a 3x3 kernel aggregation used to smooth irregularities, pits/peaks present in the USGS DEM.

Choosing differing resolutions is also a test of the robustness of the model. That is, can comparable results be achieved even though widely differing resolutions are chosen for each basin? Grid cell resolution may change model

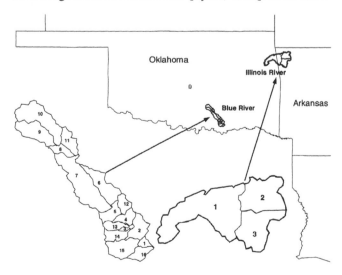

Figure 2. Location of the 2,400 km₂ Illinois River Basin, subbasins and USGS stream gauging stations in Oklahoma and Arkansas.

Table 2. Sub-Basin Properties and Channel Hydraulic Characteristics for the Two Basins

Basin	Area (km^2)	Channel properties		
		Stream width (m)	Channel slope (m/m)	Channel roughness
		Illinois		
1	950	66	0.5	0.04
2	670	50	0.15	0.04
3	700	50	0.15	0.04
		Blue		
1	19	18	0.0008	0.05
2	72	15	0.0009	0.05
3	11	12	0.0017	0.05
4	43	10	0.0004	0.05
5	51	8	0.0006	0.05
6	112	5	0.0021	0.05
7	254	8	0.0026	0.05
8	35	8	0.005	0.05
9	117	5	0.0023	0.05
10	172	5	0.0015	0.05
11	58	5	0.0018	0.05
12	38	5	0.0047	0.05
13	35	5	0.0047	0.05
14	47	5	0.0025	0.05
15	98	5	0.0011	0.05
16	44	5	0.0063	0.05

results due to resampling because slope decreases with increasing grid cell size as described by *Vieux* [1993] and *Vieux* [2001]. Drainage length also shortens with increase grid cell size and can compensate for the effect of slope flattening. Thus, calibrations may be resolution specific as other researchers have found [*Obled et al.*, 1994]. Ongoing studies are addressing this dependency.

Channel Characteristics

Data for Illinois River channel cross-sections were taken from those surveyed and reported by *Harmel* [1997] in the lower reaches above Tahlequah. USGS stream gauging measurements supplemented surveyed data. For the Blue River, cross-sections were estimated from USGS measurements at the outlet then proportionally distributed upstream. Trapezoidal cross-sections are interpreted from the surveyed cross sections to obtain average values for the three streams corresponding to the three basins. These were further adjusted to achieve a better fit between observed hydrograph peaks timing and travel between Watts and Tahlequah. It should be noted that each stream reach is represented using a single finite element in *r.water.fea* and *arc.water.fea*. Whereas, the Vflo™ model has as many channel finite elements as there are grid cells traversed by the channel reach. Different calibrated channel characteristics are expected to result given different levels of discretization.

Rainfall

Hourly rainfall maps as input are obtaiqned from Stage III NEXRAD rainfall at a nominal 4 x 4 km resolution. As listed in Table 3, eight storms are used for the Illinois and Blue River basins. The rainfall is re-sampled to obtain a resolution corresponding to the DEM resolution. Comparison with rain gauge accumulations show a mean field bias averaged over eight storms to be within 10% for the Illinois. Further, since the implementation of the so-called P1 adjustment at the Arkansas-Red River Basin Forecast Center (ABRFC) in December 1996, the bias is close to 1.0 for most events over the Illinois and Blue basins compared to gauge. In the results reported herein, no attempt is made to adjust the Stage III estimates of rainfall.

Discharge

The simulated hydrographs for each storm are compared to the USGS observed hydrographs at USGS stream gage 07196500, located along the Illinois River near Tahlequah, OK, and gauging station 078332500 located at Blue, Oklahoma. Currently, the model simulates the direct runoff component requiring base flow separation by the straight-line method [*Chow et. al.*, 1988]. Because any of the models at present only simulate the direct surface runoff components, simulation results are compared to the runoff separated from the base flow. Because we are focusing on the storms that have generated substantial flow, the base flow components are normally less than 10 percent of the flow during these events. Table 3 shows the runoff coefficient, *RC,* computed as the ratio of the Stage III radar rainfall and observed USGS runoff, after baseflow separation, averages RC=0.36 for the Blue and 0.39 for the Illinois. The range of *RC* is from 0.23 to 0.56 for the Illinois and 0.21 to 0.60 for the Blue basin. Thus, the two basins have relatively similar partitioning of rainfall into runoff.

Infiltration and Roughness Parameters

The Green and Ampt soil parameter maps are calculated using the *Brooks and Corey* [1964] equations that relate to estimate the infiltration parameters to based on soil properties such as clay, sand content, bulk density, and pore size distribution and others. These soil properties are then used in Rawls and Brakenseik relationships to estimate the Green and Ampt infiltration parameters [*Rawls et al.* 1983a,b.]. The soil properties of for soils over in the Illinois River basin are obtained from the Map Information Assembly and Display System (MIADS). MIADS is a soils database compiled at 200-meter resolution by the USDA-Natural Resources Conservation Service for the State of Oklahoma from county-level soil surveys. The Manning coefficient maps are obtained by relating LULC maps to the corresponding roughness coefficients. Coefficients of roughness are estimated based on the dominant land use/cover classification based on the Anderson Classification system and Manning hydraulic roughness

[*Vieux,* 2001]. Table 4 shows the range of values for the parameter maps input to the *r.water.fea* model.

Soil Moisture

For the Illinois and Blue events, antecedent soil moisture is modeled using the Green and Ampt equations and is entered into the model as a spatially variable map or as a lumped degree of saturation. Except for the Illinois November-December 1996 and February-1997 events, a prolonged dry spell resulted in dry conditions, which are modeled with 20% initial degree of saturation. This assumption was further supported by prolonged low flow in the river for a month or more antecedent to the event. Inspection of rainfall records for the November-December 1996 and February-1997 events revealed that while there was minimal stream flow, there had been precipitation. The initial degree of saturation was increased to 70%. Simulations using the model, SHEELS, reported by *Crosson, et al.* [2001], and applied to the Blue River basin by *Martinez and Duchon* [2001] confirmed this value as 71% [personal communication]. Sensitivity studies for this basin show little variation in rainfall-runoff response at initial degrees of saturation below 50%. Sensitivity to soil moisture is part of the research supported herein using the DRUM version of *r.water.fea* coupled with SHEELS.

RESULTS AND DISCUSSION

The OPPA method may be applied in both automatic and manual modes. OPPA capitalizes on known behavior of the

Table 3. Storm Event Radar Rainfall and Observed Runoff for the Illinois and Blue Basis

Illinois			
Event	Rainfall (mm)	Runoff (mm)	RC
4-Mar-95	30.4	9.7	0.32
10-May-96	32.5	12.2	0.38
20-Apr-96	80.4	18.2	0.23
4-Nov-96	75.7	21.0	0.28
8-Jun-95	68.8	23.0	0.33
13-Jan-95	71.4	26.4	0.37
25-Nov-Dec-1996	53.3	29.6	0.56
19-Feb-97	76.1	33.1	0.43
Mean	61.1	21.7	0.36
Blue			
Event	Rainfall (mm)	Runoff (mm)	RC
26-Sep-1996	52.74	10.9	0.21
21-Apr-1996	64.81	13.7	0.21
12-Nov-1994	103.37	31.0	0.30
19-Feb-1997	95.25	35.4	0.37
14-Mar-1998	83.84	35.4	0.42
06-Nov-1996	67.42	40.5	0.60
19-Oct-1996	89.83	48.1	0.54
06-May-1995	100.52	49.2	0.49
Mean	82.2	33.0	0.39

physics-based model and parameters. Certain features of the cost surface and of PBD models make the order of parameter identification important. First, we address the overall performance achieved for the two basins. Secondly, we address the features of the cost surfaces obtained.

Calibration Performance

The model performance in terms of reproducing the total volume of flow is excellent for the eight storm events shown in Figure 3. The result of calibration is encouraging given that several storms during different seasons were well reproduced with high accuracy with a single adjustment to applied to all storms. There is good agreement between simulated and observed for the February 1997 and April 1996 hydrographs shown in Figures 4a and b, respectively. In this case, the rising limb and peak flow would have been acceptable for flood forecasting without adjustment ($\alpha=1$ and $\beta=1$). Similar results are observed for the other storms except for the March 1995 storm, which is not reproduced as well. It is noted that the March 1995 storm has the lowest rainfall, 30 mm, among the storms considered producing only 10 mm of runoff averaged over the watershed. By taking the best-fit-line slope to the simulated and observed volume and peak discharge, the prediction accuracy achieved is within 11% in volume and 20% in peak flow for the 8 storm events simulated. Peak discharge performance could be improved with better channel parameterization. Some of the *flashy* model behavior can be attributed to the runoff generation method (infiltration excess), the channel routing method (kinematic wave), and because only three finite elements are used to represent the channels in *r.water.fea* for the Illinois River basin. In the Blue basin, 16 elements (as many as the number of sub-basins) are used to represent the channels. In both cases, trapezoidal cross-sections are used to represent natural channels, which may not properly represent out-of-bank flow at high discharge rates.

Cost Surface Shape

Typical approaches to calibration of CRR models are to apply a wide range of parameter values and then contour the objective function or otherwise search out the minima. Applying the same methodology to this event-based model, the objective function may be comprised of volume, peak flow, or a normalized composition of the two. Following the OPPA scheme of adjusting for volume, then for peak magnitude, we present the results for the Illinois River Basin and the Blue River basin. Figures 5 and 6 show the contours of the square of volume differences for the two basins simulated as a function of parameter scalars multiplying hydraulic conductivity and hydraulic roughness.

Forming an objective function in terms of volume, the cost surface is computed for 64 parameter pairs. Figures 5 and 6 show such a surface for a range of scalar multipliers $\alpha=0.5$ to 4 and $\beta=0.5$ to 4 for the Illinois and Blue, respectively. The contours are the squares of the volume differences in cubic meters summed over the eight storm events. In the case of the Illinois River basin, Figure 5 reveals the elongated trough traced by the 1e+15 contour extending from $\beta=0.75$ to 2.75 with its axis centered over $\alpha=0.5$. Given a reasonable estimate of β and initial estimates of hydraulic roughness, n, the optimal value for α may be retrieved. Optimal solutions will be retrieved more efficiently when starting with hydraulically smooth and impervious scalar values. The shape of the cost surface has a single minimum where the hydraulic roughness multiplier causes the cost surface to be minimized. For peak flow, we see that the surface is complex with a relatively flat area that is sickle shaped and has a minimum in and around $\beta=2$. Given the flatness of the response, $\beta=1.5$ to 2.5 yield roughly equivalent error. Because the optimal parameter set may be indistinguishable at small distances around a minimum, inspection of hydrographs may be used to resolve which parameter set to use. From inspection of individual hydrographs, a value of $\beta=1.5$ is selected together with $\alpha=0.5$ yielding good agreement.

The shape of the cost surface is interesting because it shows that the optimal solution may be more difficult to automatically retrieve depending on the direction of search. Starting at large multipliers, say $\alpha=3$ and $\beta=3$, would place the solution in a relatively flat region making it difficult for any automatic search algorithm to retrieve the optimal set. If the starting point is at say $\alpha=0.1$ and $\beta=0.1$, then the

Table 4. Green and Ampt Infiltration and Hydraulic Roughness Parameters for Illinois and Blue River Basins

Parameter (map)	Illinois	Blue	Units
Manning roughness coefficient	0.01-0.09	0.01-0.09	$m^{1/6}$
Soil saturated hydraulic conductivity	0.0-6	0.0-6	cm/hr
Soil suction at wetting front	32-50	32-50	cm
Soil porosity	0-0.35	0-0.35	-
Soil degree of saturation	20-70	20-70	%

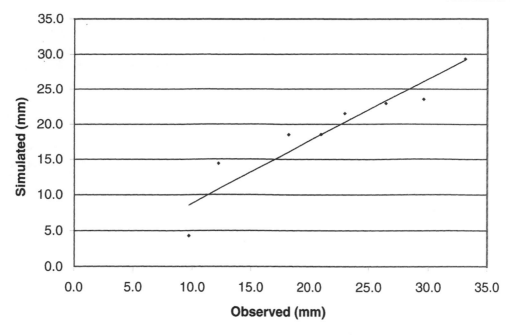

Figure 3. Calibrated volume for the eight storms for the Illinois basin. (With permission, Kluwer Academic Press, *Vieux* [2001]).

search algorithm will find a region of steep descent. This results in a faster retrieval of the optimal set and is more important to automatic retrieval.

The shape of the cost surface is not uniform in all directions. In fact, it is *ill-conditioned* from a mathematical viewpoint which can slow automatic retrieval. The trough shape indicates that there is more sensitivity to hydraulic conductivity than roughness when considering volume, as would be expected. This ill-conditioned aspect has the advantage in that hydraulic conductivity may be easily retrieved with any close value of hydraulic roughness. Once hydraulic conduc-

tivity is retrieved, hydraulic roughness is retrieved using a peak flow objective function. Some interdependency between the two parameters is expected for pervious watersheds. If at the region of minimal errors, individual hydrograph inspection can be used to resolve which parameters sets to use.

The Blue basin behaves similarly in terms of the shape of the cost surface as shown in Figure 6. The volume cost surface presents an elongated trough traced by the 2e+15 contour that extends between $\beta = 0.5$ to 2.75, with its axis centered over $\alpha = 0.5$. Turning to the cost surface for peak

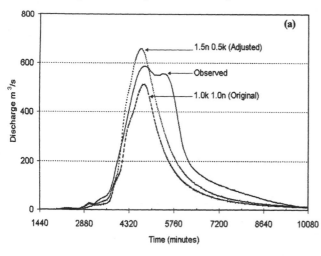

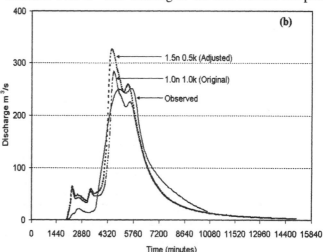

Figures 4a,b. Comparison of observed stream flow with simulated initial and adjusted estimates of ßn and αk, (a) February 1997, and (b) April 1996. (With permission, Kluwer Academic Press, *Vieux* [2001]).

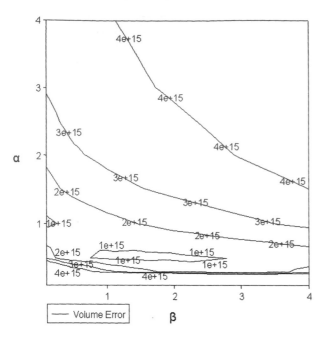

Figure 5. Cost surface for the volume-only objective function and the hydraulic conductivity and roughness multipliers, and for the Illinois River.

flow, we again find a relatively flat area that is sickle shaped and a minimum traced by the 1e+5 contour focused on β =2. From these simulations, β =2.0 is selected together with α=0.3 yielding excellent agreement of an average difference of simulated to observed volume difference of -1.5 mm. Tables 5a and b summarize errors between simulated and observed volumes for the Blue and Illinois River basins and storms tested. The calibrated simulations for the Blue are within 1.5 mm average difference, 6.9% average error, with a root mean square error (RMSE) of 9.7 mm. The calibrated volume simulations for the Illinois are within 2.5 mm average difference, -12.5% average error, and RMSE=3.6 mm. Figure 7 shows the cost surface for the combined objective function for the Illinois basin. The viewing angle reveals a small dip or minimum at the optimal parameter set of α=0.5 and β=2 where both volume and peak flow errors are a minimum. Note that there is an identifiable optimal parameter set for the storms tested.

The larger errors are associated with smaller storm events. The 26-Sep-1996 event over the Blue has a simulated volume of 18 mm compared with 10 mm observed runoff, which is a 69% overestimate. Compared to the Blue River, the Illinois River basin has a better RMSE of 3.6 mm. Experiments with radar calibration, scaling issues, and other parametric studies can be accomplished with such a watershed model given the accuracy and computational efficiency.

Reasons for underestimation may be due to the location of the basin in relation to the radars used to generate the

Stage III rainfall estimates. The distal end of the Blue is at ~200 km from KTLX and lies along a radial of the radar. KFRD is at the same distance around ~150 km to the midpoint of the basin. Stage III is a mosaic of these two radars. Given the distance of the Blue River basin from KTLX and KFRD, low-level precipitation processes may be below the beam of the radar. Larger convective events typically have higher cumulus cloud tops with improved detection by the radar at these distances. The largest underestimate is for the 14-Mar-1998 event over the Blue with -45.91% error in volume, which is a -16.24 mm difference. This event falls during the winter and may not be adequately sampled due to the height of radar in relation to stratiform precipitation processes. It is also noteworthy that the smaller magnitude runoff events have the largest errors when they are on the order of the errors associated with calibration of 9.7 mm RMSE for the Blue. The same holds true for the Illinois basin that has an RMSE of 3.6 mm. Resolution of the possible causes of under/over estimation is part of on-going studies involving radar corrections and model improvements.

The relatively similar behavior between the Illinois and Blue River basins is striking. Both present a relatively flat, sickle-shaped area in the peak flow cost surface with optimal scalar values adjusting hydraulic roughness of β =1.5 and 2.0, for the Illinois and Blue, respectively. Similarly for volume, a trough is traced with optimal scalar values adjusting volume of α =0.5 (Illinois) and of α =0.3 (Blue) yielding excellent

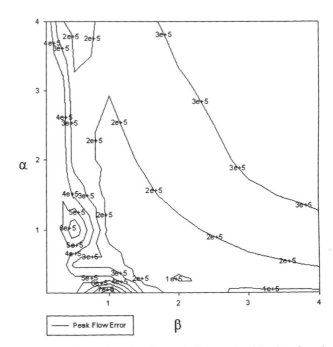

Figure 6. Cost surface for the peak flow-only objective function and the hydraulic conductivity and roughness multipliers, and for the Blue River.

Table 5a. Calibrated Volume for the Blue River Basin α = 0.3 β = 2.0

Event	Simulated (mm)	Observed (mm)	Diff (mm)	Error (%)	RMSE (mm)
26-Sep-1996	18.47	10.88	7.6	69.7	
21-Apr-1996	19.66	13.74	5.9	43.1	
12-Nov-1994	40.58	31.02	9.6	30.8	
19-Feb-1997	26.05	35.36	-9.3	-26.3	
14-Mar-1998	19.13	35.37	-16.2	-45.9	
6-Nov-1996	48.52	40.54	8.0	19.7	
19-Oct-1996	39.33	48.08	-8.8	-18.2	
6-May-1995	40.47	49.22	-8.8	-17.8	
			-1.5	6.89%	9.7

Table 5b. Calibrated Volume for the Illinois River Basin α = 0.5 β = 1.5

Event	Simulated (mm)	Observed (mm)	Diff (mm)	Error (%)	RMSE (mm)
4-Mar-95	4.3	9.7	-5.4	-55.7	
10-May-96	14.4	12.2	2.2	18.0	
20-Apr-96	18.5	18.2	0.3	1.6	
4-Nov-96	18.5	21.0	-2.5	-11.9	
8-Jun-95	21.5	23.0	-1.5	-6.5	
13-Jan-95	22.9	26.4	-3.5	-13.3	
25-Nov-Dec-1996	23.5	29.6	-6.1	-20.6	
19-Feb-97	29.3	33.1	-3.8	-11.5	
			-2.5	-12.5	3.6

agreement in both cases. Stability of these values with the addition of more storm events are being tested.

The model is robust because roughly equivalent calibration factors are achieved for both basins regardless of resolution and differing geographic-climatic regions where the basins are located. Comparable results are achieved, even though widely differing resolutions are chosen. In both cases, volumes are estimated to within 1.5-2.5 mm for the suites of eight storms over each basin, and peak flow is estimated to within 20% for each basin. These are encouraging results especially when considered in the light of CRR model results that require considerably more historical data to calibrate.

CONCLUSIONS

Physics-based models have important characteristics distinguishing themselves from CRR models. Whether automatic or manual search algorithms are used, the order of parameter adjustment is important to ensure that timing and peak discharge are retrieved for the correct volume of runoff water in the watershed. Hydraulic conductivity is retrieved from the volume-only objective function. Hydraulic roughness is retrieved from the peak-only objective function and is sensitive to hydraulic conductivity demonstrating the advantage of for an ordered search in the OPPA method.

Computational efficiency permits retrieval of optimal parameters with reasonable expenditure of time using direct computational approaches. Beginning with parameters derived on a physical-basis realistic results requiring only minor adjustment is obtained. Compared with CRR models which rely on empirical equations, parameters may take on a wide range of values making search algorithms particularly inefficient compared to the minor adjustments necessary in the physics-based model presented herein.

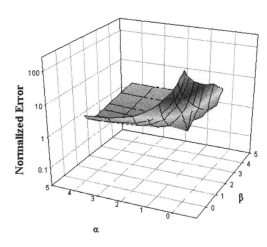

Figure 7. Illinois combined cost surface, peak flow and volume.

Calibrated volume estimates simulated for the Blue basin are within -1.5 mm average difference, 6.9% average absolute error, and with an RMSE of 9.7 mm. Small storms on the order of 10 to 20 mm total runoff are more difficult to simulate given that the total runoff volume is on the order of the simulation RMS error. Though such errors have importance in longer-term water resources applications, they are of less interest from the flood-forecasting viewpoint. Volume simulated for the Illinois basin has similar agreement to within 11%. Both errors are comparable even though the Illinois is roughly twice the size of the Blue and was simulated with 16,900 grid cells (270 meter), whereas, the Illinois simulation employed 2,700 grid cells at 960-meter resolution. In both cases, the excellent agreement in volume was achieved with a scalar multiplier α =0.5 and 0.3 for the Illinois and Blue, respectively. Peak flow was simulated to within 20% for the Illinois and the Blue for β =2.0. The eight storms were chosen for each basin independently (not the same storms) and represent different seasons over different geographical regions in Oklahoma.

Improved rainfall-runoff prediction benefits a wide range of users within governmental and private sectors. The PBD model presented herein shows promise given the accuracy achieved through calibration with relatively few storm events compared with CRR models. Capitalizing on the unique characteristics of a PBD model, the OPPA method is proven to be an effective means for adjusting physics-based models using spatially distributed initial estimates based on soil properties, land use/cover, and digital elevation data. Performance of *r.water.fea* shows exceptionally good prediction accuracy for the two basins simulated. Considering accuracy and computation efficiency of the finite element network model approach, real-time flood forecasting using a PBD model coupled with radar inputs is achievable.

Acknowledgements. This research was supported through funding by: *Estimating Surface Water and Energy Fluxes Using Models and Observations from RADAR and the Oklahoma Mesonet,* NASA-EPSCoR, Award No. NCC5-171; NASA-EPSCOR Grant and the NSF/EPA Water and Watersheds Program, *Ecological Risks, Stakeholder Values and River Basins: Testing Management Alternatives for the Illinois River,* Award No. R825791.

REFERENCES

Abbott, M.B., J.C. Bathurst, J.A. Cunge, P.E O'Connell, and J. Rasmussen, An introduction to European Hydrological System -Systeme Hydrologique Europeen, "SHE", 1 History and philosophy of a physically-based distributed modeling system., *J. Hydrol., 87,* 45-59, 1986a.

Abbott, M.B., J.C. Bathurst, J.A. Cunge, P.E O'Connell, and J. Rasmussen, An introduction to European Hydrological System -Systeme Hydrologique Europeen, "SHE", 1 Structure of a physically-based distributed modeling system, *J. Hydrol., 87,* 61-77, 1986b.

Beven, K., and J. Freer, Equifinality, data assimilation, and uncertainty estimation in mechanistic modeling of complex environmental systems using the GLUE *methodology, J. Hydrol., 249,* 11-29, 2001.

Boyle, D.P., H.V. Gupta, and S. Sorooshian, Toward improved calibration of hydrologic models: Combining the strengths of manual and automatic methods, *Water Resour. Res., 36*(12), 3663-3674, 2000.

Boyle, D.P., H.V. Gupta, and S. Soorooshian, Toward improved streamflow forecasts: Value of semidistributed modeling, *Water Resour. Res. 37*(11), 2749-2759, 2001.

Brooks, R. H. and A.T. Corey, Hydraulic properties of porous media, *Hydrology Paper No. 3, Colorado State University, Fort Collins, Colorado,* 1964.

Burnash, R.J.C., R.L. Ferral, and R.A. McGuire, A general stream flow simulation system - Conceptual modeling for digital computers, *Report by the Joint Federal State River Forecasts Center, Sacramento, California. 1973.*

Chow, V. T., D.R. Maidment, and L.W. Mays, *Applied Hydrology.* McGraw-Hill, New York 1988.

Crosson, W.L., C.A. Lymon, R. Inguva, and M. Schamshula, Assimilating remote sensing data in a surface flux-soil moisture model, *J. Hydrol. Processes, 16*(8), 1645-1662, 2002.

Finnerty B.D, M.B. Smith, V. Koren, D.J. Seo, and G. Moglen, Space-Time Scale Sensitivity of the Sacramento Model to Radar-Gage Precipitation Inputs, *J. Hydrol., 203,* 21-38, 1997.

Harmel, D.R., *Analysis of bank erosion on the Illinois River in northeast Oklahoma,* Ph.D. dissertation, Oklahoma State University, Oklahoma, 1997.

Hornberger G. M., and R. C. Spear, "An approach to the preliminary analysis of environmental systems", *Journal of Environmental Management,* 12: 7-18, 1981.

Jakeman, A.J., and G.M. Hornberger, How much complexity is warranted in a rainfall-runoff model?, *Water Resour. Res., 29*(8), 2637-2649, 1993.

James, W.P. and K.W. Kim, A distributed dynamic watershed model. *Water Resources Bulletin* 24(4), 587-596. 1990.

Johnson, L.E., Assessment of flash flood warning procedures. *J. Geophys. Res., 105*(2), 2299-2314, 2000.

Julien, P.Y. and B. Saghafian, CASC2D user manual - a two dimensional watershed rainfall-runoff model. *Civil Engineering Rep. CER90-91PYJ-BS-12, Colorado State University, Fort Collins,* 66p, 1991.

Koren, V.I., B.D. Finnerty, J.C. Schaake, M.B. Smith, D.-J. Seo, and Q,-Y. Duan, Scale dependencies of hydrologic models to spatial variability of precipitation, , *J. Hydrol., 217,* 285-302, 1999.

Koren V.I., M.B. Smith, D. Wang, Z. Zhang, Use of soil property data in the derivation of conceptual rainfall-runoff model parameters. Reprint from the AMS 15th Conference on Hydrology, 9-14 January, Long Beach, CA, pp. 103-106, 2000.

Maidment, D.R., J.F. Olivera, A. Calver, A. Eatherall, and W. Fraczeck, A unit hydrograph derived from a spatially distributed velocity field, J. *Hydrological Processes, 10* (6), 1996.

Martinez, J.E., and C.E. Duchon, Effect of the number of soil layers on a modeled surface water budget, *Water Resour. Res. 37*(2), 2001.

Perkins, W.A., M.S. Wigmosta, and B. Nijssen, Development and testing of road and stream drainage network simulation within a distributed hydrologic model, Eos Trans. AGU, 77 (46), F232, Fall Meet. Suppl., 1996.

Ogden, F.L., and P.Y. Julien, Runoff model sensitivity to radar rainfall resolution, *J. Hydrol., 158,* 1-18, 1994.

Rawls, W. J., D.L. Brakensiek, and B. Soni, Agricultural management effects on soil water processes, Part I: Soil water retention and Green and Ampt infiltration parameters. *Transactions of the American Society of Agricultural Engineers, 26(6),* .1747-1752, 1983b.

Rawls, W. J., D.L. Brakensiek, and N. Miller, Predicting Green and Ampt infiltration parameters from soils data, *ASCE, J. of Hydraulic Engineering, 109(1),* 62-70, 1983a.

Ryder, P.D., Ground Water Atlas of the United States: Oklahoma, Texas, *U.S. Geological Survey, HA 730-E,* 1996.

Shaake J.C., V.I. Koren, Q,-Y. Duan, K. Mitchell, and F. Chen, Simple water balance model for estimating runoff at different spatial and temporal scales. *J. Geophys. Res. 101 (D3),* 7461-7475, 1996.

Shah, S.M.S, P.E. O'Connell, and J.R.M. Hosking, Modeling the effects of spatial variability in rainfall catchment response - 2. Experiments with distributed and lumped models, *J. Hydrol., 175,* 89-111, 1996.

Storck, P., D.P. Lettenmaier, B.A. Connelly, and T.W. Cundy, Implications of forest practices on downstream flooding, phase II final report, University of Washington, Seattle, 1995.

Vieux, B.E. and N.S. Farajalla, Temporal and spatial aggregation of NEXRAD rainfall estimates on distributed hydrologic modeling, *Proceedings of Third International Conference on GIS and Environmental Modeling, NCGIA, Jan. 21-25,* 199-208, 1996.

Vieux, B. E., V. F. Bralts, L.J. Segerlind, and R. B. Wallace, Finite element watershed modeling: One-dimensional elements, *ASCE*

J. *Water Resources Planning and Management, 116*(6), 803-819, 1990.

Vieux, B. E., N. S. Farajalla, and N. Gauer, Integrated GIS and Distributed Stormwater Runoff Modeling. Edited by Goodchild, M. F., Parks, B O., and Steyaert, L. T. Oxford University Press, New York, 1994.

Vieux, B.E. and N. Gauer, Finite Element Modeling of Storm Water Runoff Using GRASS GIS, *Microcomputers in Civil Engineering, Vol. 9:4, pp.* 263-270, 1994.

Vieux, B.E., Finite-element analysis of hydrologic response areas using geographical information system. *Ph.D. dissertation, Department of Agricultural Engineering, Michigan State University, East Lansing, Mich.,* 1988.

Vieux, B.E., F. LeDimet, D. Armand, "Optimal Control and Adjoint Methods Applied to Distributed Hydrologic Model Calibration." *Proceedings of Int. Assoc. for Computational Mechanics, IV World Congress on Computational Mechanics,* 29 June-2 July, Buenos Aires, Argentina. p. II1050, 1998.

Vieux, B.E., *Distributed Hydrologic modeling using GIS.* Kluwer Academic Press, Water and Science Technology Series, 38, 2001.

White, L.W., B.E. Vieux, D. Armand, Surface flow model: inverse problems and predictions, *J. Advances in Water Resources, 25,* 317-324, 2001a.

White, L.W., B.E. Vieux, D. Armand, Estimation of Optimal Parameters for a Surface Hydrology Model, *J. Advances in Water Resources,* accepted, 2001b.

Wigmosta, M.S., L.W. Vail, and D.P. Lettenmaier, "A distributed hydrology-vegetation model for complex terrain", *Water Resources Research* 30(6), 1665-1669, June, 1994.

Zhang, W., and T.W. Cundy, Modeling of two-dimensional overland flow. *Water Resources Research, 25(9):* 2019-2035, 1989.

Baxter E. Vieux, School of Civil Engineering and Environmental Science, University of Oklahoma Room CEC-334, 202 West Boyd Street, Norman Oklahoma 73019.

Fekadu G. Moreda, Present address: National Weather Service, National Oceanic and Atmospheric Administration, 1325 East West Highway: w/OHD1, Silver Spring Maryland 20910.

Process Representation, Measurements, Data Quality, and Criteria for Parameter Estimation of Watershed Models

Stephen J. Burges

Department of Civil and Environmental Engineering, University of Washington, Seattle

A short review of procedures used to calibrate lumped continuous simulation hydrologic models is given to provide a starting point for considering the hydrologic features to include, how those features can be represented, and data required for calibrating distributed continuous simulation models. Most calibration procedures have concentrated only on attempting to match simulated and recorded streamflow time series without any explicit recognition of data errors. Model calibration procedures should be designed to accommodate data uncertainty and errors, and the nature and amount of the various signals that are used to effect calibration. Spatial models necessitate use of multiple objectives. Examples ranging in scale from the hillslope, to a small zero order basin, to a larger basin are used to illustrate the need for different data and objective measures to calibrate the fitted model, assess parameter uncertainty, and to provide model predictions that incorporate explicitly parameter uncertainty.

INTRODUCTION

The major focus of this paper is on the quality of long-term continuous simulations to describe existing or prior conditions, and to predict the water balance for changed catchment conditions or climatology.

I start by providing seven fundamental requirements of hydrologic modeling, introduced by *James and Burges* [1982], and provide general guidelines for calibrating and testing continuous hydrologic simulation models. Issues of data quality assurance and quality control are then discussed briefly. This is followed by a discussion of hillslope and channel features that need to be included in models. I provide a short chronology of the development of hydrologic models and model components starting with processes and leading to complete continuous simulation models. This leads to a discussion of the additional needs of spatial hydrologic models and is followed by data needs for all continuous simulation models with emphasis on systematic errors in data and the need for nested measurements within catchments. Model parameter estimation schemes are dis-

Calibration of Watershed Models
Water Science and Application Volume 6
Copyright 2003 by the American Geophysical Union
10/1029/006WS21

cussed as well as objective measures and summary performance statistics and time series displays. Finally, the propagation of data, model, and parameter estimation errors into model predictions is discussed and several potentially productive approaches are identified.

HYDROLOGIC MODELING

Much has been written about how to build models of the hydrologic cycle at scales that range from a few square meters to continental scale in plan area. *James and Burges* [1982] discussed in detail the basic requirements of a hydrologic model as well as how to select, calibrate and test a model that is suitable for a particular application.

Modeling requires [*James and Burges*, 1982]:

- Identification of the hydrologic quantities important to the user
- Identification of the hydrologic processes that need to be modeled
- Selection of equations to represent each process
- Synthesizing equations into a computational framework
- Determining model parameters that best represent the catchment hydrologic response
- Testing the adequacy of model estimates, and
- Communicating results to decision-makers.

Communication of results to decision-makers is receiving much needed attention. There are many opportunities here for using sophisticated geographic information systems (GIS) and all forms of moving film and computer based animations.

Guidelines to Continuous Simulation Model Calibration and Testing

Crawford and Linsley [1966], particularly in Chapter 5, offered some of the earliest (known to me) guidelines for model calibration and testing. One of the earliest comprehensive tests of unit hydrograph based "event models" was done by *Hoyt et al* [1936], soon after *Sherman* [1932, 1949] developed the unit hydrograph approach for describing the basin outflow hydrograph. *Clarke* [1973] provided one of the most complete published categorizations of hydrologic models. There have been specific guidelines provided for calibrating particular models e.g., *Peck* [1976], for the "Sacramento Model", and *Sugawara et. al.* [1984], for the "Tank Model".

Model calibration Guidelines have been summarized by *James and Burges* [1982]:

1) Simulated and recorded flow values should agree for the:
  * Annual flow volume for each water year
  * Seasonal flow volumes for each water year
  * Weekly and daily volumes
2) Simulated and recorded hydrographs for a given storm should have:
  * Similar shapes
  * The same peak values
  * The same time of peak flow
3) Predicted evapotranspiration (ET) should be less than or equal to potential ET for the region.
4) Modeled hillslope stored water should fluctuate with precipitation patterns
5) Model parameters must be consistent with observed catchment properties
6) The relative amounts of "surface" and "base" flow must be consistent with soil and geological conditions.
7) All comparisons must be consistent with the accuracy and errors of the recorded data!

The seven requirements for model calibration must hold at:
  * a small spatial element, and
  * the catchment as a whole

The requirements dictate the spatial scale for modeling in any particular application.

DATA QUALITY ASSURANCE QUALITY CONTROL (QAQC)

The profession has evolved to a state where most of those who use data are not involved in the collection of the data or in quality assurance and quality control. Three illustrations of data QAQC follow.

Streamflow Data

It is essential that the data user know all shortcomings of the data that are being used. In the United States, the U.S. Geological Survey provides quantitative descriptions for the quality of the reported streamflow (excellent, good, etc.,). For the best stream gauge stations (excellent) there is an approximately 95% chance that the streamflow rate that has been reported is within + or - 5% of the reported quantity. (This is valid for a stable "stage-discharge" rating curve. Rating curves usually change during a flood when the riverbed form or the channel-section changes or both change. It is difficult to determine the actual time when the bed change occurs unless river gauging was ongoing at the time of change). Any model that attempts to reproduce the measured hydrograph should include explicitly uncertainty bounds on the modeled streamflow hydrograph. More hydrologists attempt to show such information now that we have enormous computing power available (see, e.g., *Kuczera and Parent* [1998]).

Precipitation Data

Documentation of QAQC for precipitation data is less complete than documentation for streamflow. It is not always clear when a record has been "filled in" when a recording device was inoperable or if there were other problems in recording precipitation. In most situations there is no way for correcting for variable systematic bias created by wind influences on precipitation gauges. *Duchon and Essenberg* [2001] provide a sharp reminder of this ubiquitous problem. We have to document our precipitation data more completely and provide QAQC flags for each reported value.

Radiant Energy Data

An example of the documentation for some of the most carefully collected environmental data in the US is given by Augustine et al (2000) in which they describe their efforts to collect and make available the data for six SURFRAD sites across the US. In this work the authors state the accuracy for a pyranometer as +or - 2 to 5%, pyrgeometer as + or - 9 Wm^{-2}, and a pyrheliometer as + or - 2 to 3%. This has considerable implications for global climate change. These data are the best available; measurements were started in 1996. *Baker* (2001) indicates that the "change in flux due to CO2 increase over the past 200 years is less than 2 Wm-2". Various assumptions used in global climate models for the presence of ice in the upper-troposphere clouds is as much as 17 Wm-2 of flux entering or leaving the earth. These quantities are within the measurement errors of some of the earth-based devices that are needed to document them.

LAND SURFACE MECHANISMS TO INCLUDE IN MODELS

Figure 1 (Figure 3.1 from Hillslope Hydrology, M. J. Kirkby © 1978 John Wiley & Sons Limited. Reproduced with permission.) shows schematically what has to be represented at the hillslope hydrology scale. The various features shown in the figure are absolutely crucial for getting flow

paths correct. Representing hillslope elements such that the main flow paths that are influenced by the indicated geological, soil, root, burrowing animals, and vegetation features are included, is necessary to get the hillslope hydrology approximately correct. It will be of even greater importance when hillslope hydrologic processes are coupled more closely with hillslope bio-geochemical processes.

Flow Paths and Fluxes

Figure 1 shows that water reaches the receiving stream by a few ribbon paths on a seepage face. Of all the ways that water moves through, over, and below the soil, few measurements of them are made or are available. Our measurements of the input (precipitation — more realistically, throughput) are sparse. Even when radar measurements are at their best, they will be available at a relatively crude scale of 1 km2 or more. The smallest time increments will likely be about 15 minutes. A dense network of "good" rain

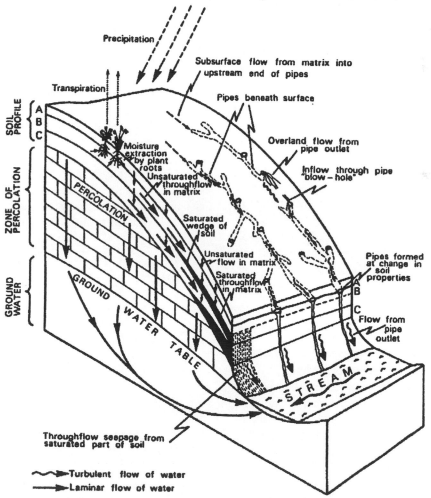

Figure 1. Flow routes followed by subsurface runoff on hillslopes [Figure 3.1 from Atkinson (1978)].

gauges will likely be more useful in many situations. It is rare that there are any measurements of the vertical state of soil water or perched ground water. Few measurements are available to permit locating relatively impermeable subsurface layers that define the flow field. In short, the boundaries are poorly defined (if at all) for the "boundary valued" hillslope flow situation. It is extremely unusual for streamflow to be measured at locations along a channel at distances closer than 10 km or so. How then will it be possible to get the flow paths and fluxes approximately correct except in a gross, areally averaged way? There are many challenges for the future.

Spatial Variation in Hillslope Properties and Fluxes

We should also consider the orientation of the hillslope in Figure 1 to incoming sunshine and to the direction of the prevailing wind fields. Consider the simplest situation of rainfall and non-freezing soils. If the hillslope is exposed to much of the prevailing wind, there will be greater evaporation from intercepted rainfall than from a sheltered hillslope. For the same rainfall input more water will reach the ground for the protected hillslope than for the exposed hillslope. This means that the permeability and nature of the underlying soil and rock in the "wetter" location is likely to be higher than in the "dryer" location. The biogeochemistries and "hydroecologies" [*Rodriguez-Iturbe*, 2000] are likely to differ. A similar situation exists for solar driven evaporation. This means that our models, even when using crude hillslope averages, need to have different "hydraulic" properties to reflect the actual state. *Seyfried and Wilcox* [1995] report on the influence of snowdrifts and the resulting spatial differences in land-surface infiltration properties. *Crawford and Linsley* [1966] approximated these features explicitly by using uniform probability distributions across the landscape for infiltration and evaporation.

Influence of Temperature on Soil Water Movement

Water temperature as it moves over and through the landscape is largely ignored in most models. Figure 2, from *Musgrave* [1955, p158], shows the relationship between the infiltration rate and near surface soil temperature. This classic figure for a 72-hour long infiltration test demonstrates the importance of the viscosity of water on infiltration. The fluctuations in infiltration rate are diurnal and the rate increased by more than a factor of 2 as the soil temperature changed from about 2°C to about 20°C. (There are other "apparent" viscosity influences with air-water flow in soil, particularly with flood irrigation practice).

Flow Over Hillslopes

How does the water move across the landscape? Water movement has been approximated as shallow water wave flow across assumed plane surface hillslopes in some fluid mechanics based schemes. This might give approximately the right answer for the timing and bulk delivery of water to the channel but is incorrect everywhere on the hillslope. There is much evidence of the movement of roll waves on hillslopes that produce partial to complete overland flow. There are transitions from viscous to turbulent flow as the flow deepens. The infiltration rate for overland flow increases as the flow rate increases in many situations. This has been documented and explained by *Dunne et al* [1991] and qualitatively by *Seyfried and Wilcox* [1995]. The importance of representing the detailed hydraulic geometry properties of overland flow in models has been discussed by many. The need to include the influence of centimeter-scale topography in the representation of hillslope surface flow characteristics was demonstrated by *Zhang* [1990]. *Grayson et. al.* [1992] demonstrated the critical importance of representing small-scale surface hydraulic geometry and roughness features when modeling overland flow.

Energy Transfers to the Atmosphere

Water (vapor) and energy transfers from and to the atmosphere pose many challenges of measurement and representation. The physics and bio-physics are fairly well understood. The difficulty is in implementation. I am hopeful that we will eventually develop some form of "observation mechanism" that will augment and enhance what we now do with expensive to install, maintain, and operate instru-

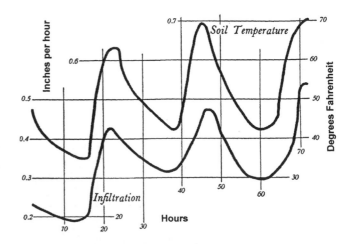

Figure 2. Temperature and infiltration in a 72-hour continuous test near Colorado Springs, where infiltration is proportion to viscosity of water [from Musgrave (1955)].

ment towers that sample far too little of the domain. I mentioned this hope and challenge in *Burges* [1986].

Approximations of Hydrologic Processes

How do we approximate the various hydrologic processes in our models? We have learned to include the mechanisms and mass balance dynamic schemes associated with the names Horton, Hursh, Betson, Hewlett, Dunne, Darcy, Richards, Thornthwaite, Dalton, Penman, and Bowen and a few others. Excellent summaries and illustrations of these approaches are given in the collection of papers in the book "Hillslope Hydrology", [*Kirkby*, 1978]. I particularly appreciate the papers by *Betson and Ardis* [1978], *Chorley* [1978], *Dunne* [1978], and *Freeze* [1974], and the book by *Montieth and Unsworth* [1990]. *Brutsaert's* [1982] coverage of evaporation into the atmosphere is essential reading.

Channel Flow

One part of modeling that is done relatively well is movement of water in the main stream and river channels in the absence of channel losses to infiltration or river aquifer interactions. There is still room for considerable work, however, in this part of hydrologic modeling. Much needs to be done concerning the interaction of river flow with the flood plain and flood plain vegetation. Much greater coupling of river flow with sediment action is needed. In many situations, river water temperature is not given much consideration. In others it is essential to consider water temperature to represent diurnal fluctuations in river flow rates associated with temperature influenced infiltration of water to the riverbed. Ronan et. al. (1998) describe careful field measurements and a sophisticated approach to thermal influences on riverbed infiltration.

MODELS AND MODEL COMPONENTS

Surface water hydrologic models have components and approaches that have evolved from ideas and principles that have been introduced over about the last one hundred and fifty years. The principle of runoff equilibrium was introduced by *Mulvaney* [1851]. His thoughtful work formed the basis for the widely misused "rational method". *Sherman* [1932] introduced the unit hydrograph concept of the time distribution of runoff from a landscape. Comprehensive tests of the unit hydrograph approach were done by *Hoyt et al* [1936]. Their work showed that the use of the then available daily rainfall amounts for hydrologic modeling had numerous limitations.

Quantitative description of the movement of unsteady flow through a reach of a river was provided by *McCarthy*

[1938] whose work in the Muskingum River basin gave rise to what we know as the "Muskingum Method" for hydrologic flow routing. Important understanding about the "time-contributing area" response of a catchment was provided by *Clark* [1945] with the introduction of the "time-area" diagram. His approach has been adopted in part or in whole (often without attribution) in large-scale hydrologic models. The dynamics of water movement across a relatively plane surface was quantified in careful laboratory experiments by *Izzard* [1946]. Izzard provided a key representation of outflow rate indexed against the volume of water that remained on the land surface. *Dooge* [1992] reported that Izzard's indexing scheme was based on principles introduced by Robert Horton for the description of the movement of water across a landscape. *Woolhiser and Liggett* [1967] formalized the approach for modeling water movement across an assumed plane surface using shallow water wave approximations to the 1-D equations for mass and momentum.

The movement of water into the soil is usually modeled with some approximation to *Richards* [1931] formulation of flow in partially saturated soil. The most common approximations are the *Horton* [1933], and *Philip* [1954], [see complete details in *Philip*, 1969], 1-Dimensional vertical infiltration models. Saturated flow movement is represented by means of approximate solutions to "Darcy" saturated flow dynamics. Subsurface flow is represented in many models as input to and release from conceptual linear or non-linear reservoirs.

Empirical models based on a co-axial correlation for "runoff depth" production in humid areas were developed by the US Weather Service [see, e.g., *Kohler and Linsley*, 1951]. The variables, "antecedent precipitation index", "week of year", "rainfall duration", and "rainfall amount", were used to estimate "basin recharge" from which "runoff depth" could be obtained by difference from the rainfall depth. These main variables captured, to first order, the important variations in catchment vegetation, soil-water state, solar radiation input, and the influences of agricultural and silvicultural practice and the influence of burrowing animals. The best modern attempts to model the hydrologic cycle do not capture all of these variables and states explicitly.

CONTINUOUS HYDROGRAPH SIMULATION MODELS

The earliest developments of continuous hydrologic modeling that I know about in the pre- digital computer era were by *Linsley and Ackerman* [1942], where they approached modeling of runoff production for the Valley River Basin in North Carolina by a procedure of moisture accounting. *Sugawara* [1961] published a description of what is now known as the "Tank Model". *Linsley and Ackerman* [1942] examined numerous storms and noted that they were sufficiently uniform in intensity that they could approximate the rainfall that way. Linsley knew that there were many other situations where runoff response was related to rainfall intensity but had to wait until hourly data and digital computers were available before he could take the procedure forward.

With the advent of digital computers the earliest models were what *Clarke* [1973] described as "lumped". The development that most influenced the field was made by *Linsley and Crawford* [1960] with the forerunner to the moisture accounting "Stanford Watershed Model", and *Crawford and Linsley* [1962, 1963, and 1966], the last being the "Stanford Watershed Model IV". *Burnash et al* [1973] worked along an independent path to produce the well known "Sacramento Model" used by the US National Weather Service. Professor Sugawara continued to refine the Tank Model and a full description of the model and approaches to its calibration and use was presented in English by *Sugawara et al* [1984]. There have been numerous other models developed and used.

There have been many spatially distributed models developed. One example of a Soil-Vegetation-Atmosphere-Transfer (SVAT) type of continuous simulation model is that developed by *Wigmosta et al* [1994]. The SVAT models explicitly include both atmospheric vapor fluxes and energy accounting at the land surface. In distributed models, the landscape is usually broken into pixels that range in size from 5 m to 10 Km depending on the desired model application. *Beven* [2001] has posed the critical question: how far can we go in distributed hydrologic modeling?

All of these models need some form of calibration to match up recorded time series with modeled time series, usually daily streamflow volumes. Distributed models can in principle be calibrated against measured and pixel predicted states. The candidate states are those that are observed from remote platforms and typically include some pixel averaged near surface moisture depth and fraction of snow covered area. There are many challenges in the development, calibration, and use of such models.

DATA CONSIDERATIONS FOR SPATIAL MODELS

Most developers and users of spatial hydrologic models have relied on streamflow measurements at a few locations along rivers to check parts of their models for plausibility. This practice is fraught with difficulties. There is an infinity of ways that a distributed model could be set up and "calibrated" to emulate approximately a streamflow time series. It is essential that data be available at a range of nested spatial scales to support development, calibration, and testing of spatial models. Three examples are provided to emphasize the need for nested hydrological measurements at a range of spatial and temporal scales.

Example 1

Figure 3, which is Figure 10.3 from *Dunne and Leopold* [1978], emphasizes the importance of, and need for, flow measurements at a range of nested scales. The figure shows the 15-hour duration hydrograph for locations along the Sleepers River in Vermont corresponding to nested basins having drainage areas of 0.2, 3.2, 16.6, and 43 square miles for three storm pulses that occurred over the first six hours. The data for the 0.2 square mile catchment reflect the strong rain signal and permit the greatest opportunity for representing major flow producing hillslope processes. It would be a more difficult problem to represent the processes well if data were available at the 3.2 square mile catchment scale. I do not know of any way to identify the processes if data were only available at the 16.6 square mile scale.

The channel attenuation influences on the hydrograph shown in Figure 3 mask the hillslope input patterns to the channels. Few streams are gauged at even the largest catchment scale of 43 square miles. It is rare to have information at this level of detail, but this is representative of the level of

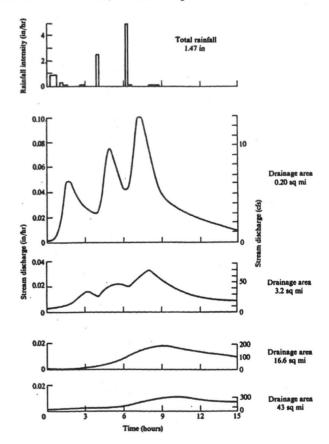

Figure 3. Changes in hydrograph shape at a series of stations along the Sleepers River near Danville, Vermont over a 15-hour period–Figure 10-3 from Dunne and Leopold (1978).

nested catchment measurements that are needed to support distributed modeling.

Example 2

A second example is for nested catchments in the Evans Creek basin of Western Washington. Hydrologic data were recorded for four water years for the small (0.37 km²) Novelty Hill Catchment at location A in Figure 4. A longer record is available for Evans Creek, location B in Figure 4. A detailed description of the hydrologic setting and monitoring (soil depth, rainfall, and piezometer depth, and weir flow measurements) for the 0.37 km² till-plateau Novelty Hill is given in *Wigmosta and Burges* [1997] and *Burges et al* [1998].

Figure 5 shows the four years of measured data. The low flow patterns in Figure 5 are completely different. The small upland catchment recharges ground water through a till layer and yields less hillslope flow to channels than occurs

at the larger 37 km² scale. The delayed flow from the upland recharge zones, similar to location A, is crucial to baseflow production at the larger scale and for the ecological health of the larger creek. The ability to represent the hydrologic processes at the scale of Novelty Hill is particularly important when land use changes are planned or have occurred.

Figure 6 further emphasizes the differences in flow production patterns. Evans Creek drains to Bear Creek (area 123 km²). The accumulated runoff, expressed as mm over the catchment, is shown for Novelty Hill, Evans Creek, and Bear Creek. The cumulative runoff depths for Bear and Evans Creeks are almost indistinguishable. Their cumulative flow patterns are substantially different from Novelty Hill. If the gauged data were available for Bear Creek, it might be possible to model the gross hydrologic features for the smaller Evans Creek. There is no known, or scientifically testable, way that the hydrologic fluxes from Novelty Hill could be elucidated from the Evans Creek gauge data alone. It is unusual to have data recorded at the relatively small

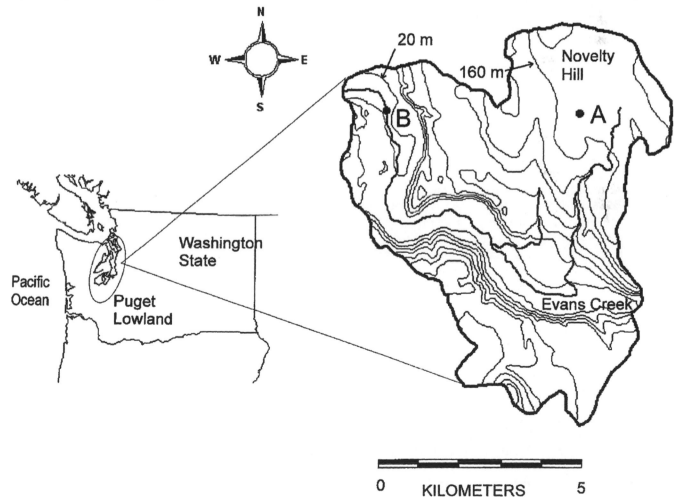

Figure 4. Location map for Novelty Hill catchment (A) nested within the Evans Creek (B) basin in Washington State.

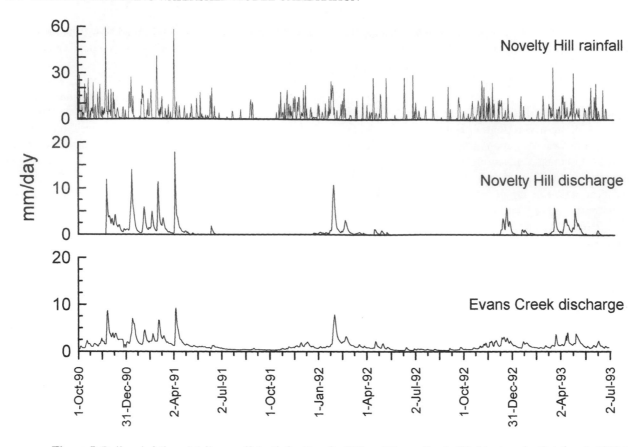

Figure 5. Daily rainfall, and daily runoff depth for Novelty Hill and Evans Creek, Washington, for October 1, 1990 to July 2, 1993.

scale of Evans Creek (37 km²), let alone the rare case of Novelty Hill (0.37 km²).

The Novelty Hill and Evans Creek data provide an important illustration of the need for nested data over scales ranging from a fraction of a square kilometer to tens and hundreds of square kilometers if we are to model hydrologic balances properly. Such data and additional measures that permit closure of the water and energy balances at the hillslope scale provide the level of detail needed to determine the flow paths and fluxes that are critical for both sharpening estimates of the water balance and modeling bio-geochemical states and fluxes.

Example 3

A third example is included to indicate the importance of soil and vegetation at the scale of a sub pixel used in the finest resolution distributed hydrological models (less than 10m by 10 m). Figure 7, from *Kolsti et al* [1995], shows hydrographs for two plots, (Plot 1 and Plot 2), at an experimental site, located at the Urban Horticulture Center (Latitude 47° 39' 29" N, Longitude 122° 17' 31" W) at the University of Washington, Seattle, campus.

Each plot has dimensions: length, L = 9.75 m, width, W = 2.44 m, and depth, D = 0.30 m. The slope is 5%. The length and depth were chosen to be representative of typical suburban lawns in the Seattle region. The plots have impermeable liners and subsurface and surface flow rates are measured using tipping bucket gauges. The gauges are calibrated

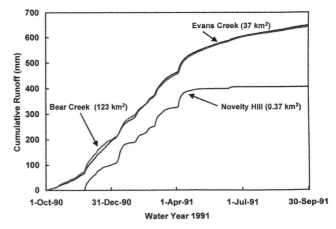

Figure 6. Accumulated runoff depth (mm) for Novelty Hill (0.37 km²), Evans Creek (37 km²), and Bear Creek (123 km²), Washington, for October 1, 1990 to September 30, 1991.

at least annually; the measured flow data are accurate to approximately 1%. The choice of 0.3 m for depth was based on advice from landscape architects of the feasible depth of soil development after houses had been built. Plot 1 consists of till with grass. Plot 2 consists of till amended with compost and grass. The grass root structures differ significantly in Plots 1 and 2. Both surfaces are planar, but there is extensive centimeter-scale variation in surface topography and roughness created by the grass. Details of plot construction and soil properties are given in *Kolsti et al* (1995).

Figure 7, shows, at one-hour time increment, the four-day runoff response to two storms in December 1994. We show the equivalent rainfall input rate (without convolution) as rainfall rate multiplied by plot area (to illustrate when water is being stored in the plots) and the outflow rate. It is likely that the rainfall amounts are understated because the gauge used is wind influenced. This is discussed in more detail below.

The soil in Plot 1 is less permeable than in Plot 2. There is a more rapid response of plot water outflow to rainfall and at a higher rate for Plot 1 than for Plot 2 during the first storm (22-mm rain). The compost amended soil and denser

grass root structure for Plot 2 is more effective initially at storing water than the unamended plot. (Do we know how to model with fidelity such a soil-grass root system using soil physics methods with some kind of a Richards equation approach? The earthworms are active at different levels at different times of year too). The attenuated patterns are similar for the second storm, but as time progresses and the plots have exhausted their soil water storage capacities, the hydrographs become similar. This is the case even though the outflow mechanisms are quite different.

If one had the rare opportunity to have carefully measured data at the small scale of Plot 1, would there be any possibility of predicting the hydrologic response of Plot 2, even if the geometry was known completely? I do not know of any soil-vegetation classification scheme that would permit accurate modeling of this situation. If measurements were made that collected the combined outflow from Plots 1 and 2, I do not know of any scheme that would permit deconvoluting the signal to yield the component hydrographs. I included this small-scale example so we can think constructively about issues of spatial modeling.

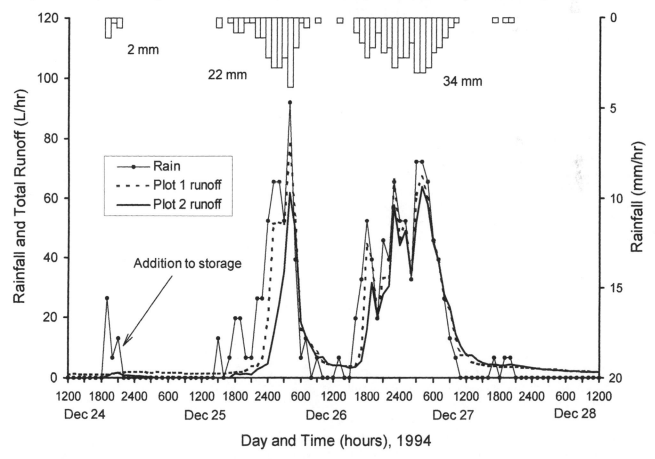

Figure 7. Hourly rainfall and runoff from Plots 1 and 2, Center for Urban Horticulture, University of Washington, Seattle for December 24-28, 1994.

A Note on the "Rational Method"

For those who use the "Rational Method" (it is still widely used), Figure 7 shows, at the most favorable small scale, that the fundamental requirement of the method -- local equilibrium—is not satisfied. What would one use for a "runoff coefficient"? We have not observed any significant "overland" flow. This is fortunate because the centimeter- and sub-centimeter-scale surface roughness variations would pose modeling problems for simplistic solutions to continuity and momentum equations for assumed properties of shallow surface flow.

RAINFALL DATA MEASUREMENT UNCERTAINTIES

I have spent a considerable amount of time since mid 1995 concentrating on measurement of rainfall with the aim of sharpening and describing the uncertainties of measurements of this major input to any rainfall-runoff model. I have chosen to illustrate here some issues of determining point- and area-rainfall for use in models, or in simple mass balance calculations based on measurements, that are needed to close the water balance for a given catchment. If the rain input has variable systematic errors, the model outputs lack credibility, and it is impossible to get the water balance correct.

Many thoughtful investigators have addressed the issue of errors and biases in rainfall measurement and how those measurements influence rainfall-runoff model results. Three examples illustrate the range of approaches that have been taken. *Crawford and Linsley* [1966] built into the Stanford Watershed Model a rain gauge scaling factor as a calibration parameter. In many calibrations of that model, modelers have elected to multiply the recorded rainfall time series by factors that range up to about 1.1. This largely accounts for wind influences on gauge catch, but simple scaling does not hold over all storms. For cases where rain is recorded at low elevation, larger upward adjustments are made to approximate orographic influences.

Crawford and Linsley [1966] include illustrative examples on "weather modification" and show sensitivity results for runoff production for three "model calibrated" basins to uniform increases of 10% in precipitation and 10% potential evapotranspiration. *Dawdy and Bergmann* [1969] presented one of the pioneering investigations to assess the effects of data errors on rainfall-runoff model simulation results. More recently *Faures et al* [1995] demonstrated the modeling consequences of using a selection of gauges from a small basin. Their work is representative of some of the most careful ground based rainfall measurements. They used a dense network of rain gauges; the catch of each gauge, however, had an unknowable wind influence. Their

work demonstrated the critical nature of representing spatial and temporal rain patterns correctly in a small arid basin that experiences convective rainfall.

POINT RAINFALL MEASUREMENT

Figure 8 shows daily measured rainfall for a 203 mm Belfort rain gauge (funnel rim height 30 inches (0.76 m)) and a 193 mm buried rain gauge located at the University of Washington Center for Urban Horticulture site. The data were recorded in the autumn and winter months of 1999 to 2001 when daily maximum air temperatures were typically below 12° C, so any evaporation that might occur from the collector bucket in the surface Belfort gauge would be negligible. On days when winds were light to negligible, the two gauges indicated almost the same rainfall depth. In all cases the buried gauge had a higher catch than the wind exposed gauge.

Is the buried gauge representative of rainfall? There is a second 193-mm buried gauge located approximately 7 m to the west. The storm rainfall depths measured in the two buried gauges seldom differ by more than 0.2 mm.

The surface Belfort rain gauge "under catch" is a function of wind pulse patterns and raindrop size. For a given turbulent burst pattern (or steady uniform flow pattern), the under catch for small drops is greater than for larger drops. I do not have disdrometer data at the site, but drops are usually on the smaller size. Figure 8 indicates that the rain recorded in the wind-exposed gauge has a variable systemic error. There is no known way to correct the data.

I recommend use of buried rain gauges to measure point rainfall. The experience in Seattle has been reinforced by our observations in Mississippi where the rainfall is much

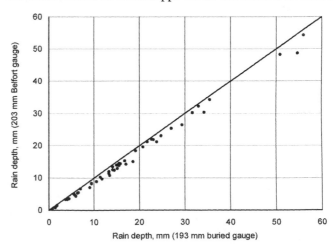

Figure 8. Daily rain depth recorded by a wind-influenced Belfort gauge and a buried gauge, Center for Urban Horticulture, University of Washington, Seattle, winter 1999 and autumn-winter 2001-2002.

more intense and larger raindrops are more usual. *Duchon and Essenberg* [2001] provide data for buried and above ground rain gauges in Oklahoma. They found that even with 1-minute time increment wind data at the gauge rim height they had no way to correct for wind under catch of exposed rain gauges because they did not have drop size data. The largest difference they recorded for a squall line storm was 15% under catch for wind exposed gauges relative to buried "pit" gauges.

SPATIAL RAINFALL MEASUREMENT

What are our prospects for spatial rainfall measurement? Based on my experience, I have concluded that a "point rain fall" measurement can best be determined from measurements from at least three rain gauges located approximately a few to ten meters apart. My preference is for two appropriately drained buried "pit" gauges and one surface gauge in the rare case that a "pit" gauge floods. All gauges must have the capability to record accumulated rainfall with the accumulated rainfall stored to permit an independent volume or weight measurement. At least one gauge has to have the capability of recording both rate and accumulated amount.

Accurate point measurements can be used with remotely sensed spatial rainfall to provide improved spatial rainfall depth estimates. *Steiner et al* [1999] reported on a detailed investigation of the comparison of NEXRAD Radar (WSR 88 radar system) measured rainfall against a dense network of surface rain gauges at the 21.4 km^2 Goodwin Creek, Mississippi, experimental watershed. (The experimental facilities at Goodwin Creek are maintained by staff from the US Department of Agriculture, National Sedimentation Laboratory, located in Oxford, Mississippi). The radar used was located at Memphis, Tennessee, (35°20'41"N, 89°52'24"W), and 121.2 km to the north of the centrally located Goodwin Creek climate station, (34°15'16" N, 89°52'26" W).

Spatial rainfall patterns are illustrated using information from the storm of May 27, 1997 that moved across the Goodwin Creek basin. Figure 9, which is parts b and c of Figure 2 from *Steiner et al* [1999], shows the rainfall intensity and accumulated rain for the storm as measured at the central climate station. The rainfall rate is given every minute from data recorded by a Joss-Waldvogel RD-69 type disdrometer and every six minutes from processed radar reflectivity data from the Memphis, Tennessee WSR 88 radar. The original radar reflectivity data (1-km range and 1 degree azimuth) were used to create rainfall pixels at a scale of 1 km by 1 km. Figure 9 b shows the rainfall rate for the approximately 5-hour duration storm. The accumulated rainfall is shown in Figure 9c for the radar, disdrometer, and

surface tipping bucket rain gauge. Disdrometer information is essential for complete interpretation of the radar derived rainfall information. Inspection of Figure 9c indicates that the surface tipping bucket gauge and the disdrometer recorded almost identical rainfall totals. The radar under-reported rainfall for the heaviest part of the storm, but the storm total from all three measuring systems was quite close. That was not the situation for other storms analyzed in *Steiner et al* [1999].

Figure 10 shows radar-estimated rainfall for a 10 km by 10 km region centered around the climate station for the May 27, 1997 storm. Two numbers are shown for each of five rain gauges located throughout the catchment. The upper number is the accumulated rain gauge measured rainfall in mm. The lower number is the radar determined accumulated rainfall in mm for the 1 km by 1 km pixel that contains the rain gauge. Three numbers are shown for the climate station. They are the accumulated disdrometer, rain gauge, and radar rainfall depths, respectively, in mm.

There are numerous striking features in Figure 10. The rain gauge in the SW corner, located at the outlet of Goodwin Creek, recorded 36.8 mm and the radar estimate was 52.5 mm. The gauge catch is likely low. The radar estimated depth is likely high because a single Z-R relationship was used to convert radar reflectivity to rainfall rate. Figure 3 in *Steiner et al* [1999] shows how the rainfall intensity-radar reflectivity relationship varied during the storm for the pixel centered over the climate station. There is a considerable difference in the gauge measured depth and radar depth for the eastern most gauge location. In this situation the gauge funnel became clogged during the storm; this problem was identified when subsequent storms were analyzed.

Figure 10 shows the radar estimated spatial variation to be significant. There are variable and systematic spatial measurement errors. No single ground based measuring station

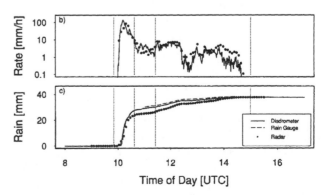

Figure 9. Rainfall rate (b) and accumulated rainfall depth (c) for the storm of May 27, 1997, at the climate station, Goodwin Creek, Mississippi. Figure extracted from Steiner et al (1999), Figure 2.

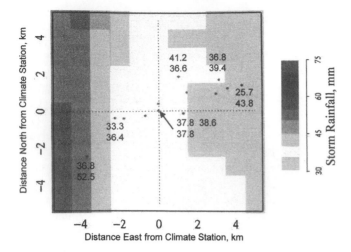

Figure 10. Spatial radar determined rainfall depth in mm for a 10 km by 10 km region centered about the Goodwin Creek, Mississippi climate station. Numbers show measured point rainfall (above) and radar estimated rainfall (below) for the associated 1 km by 1 km radar pixel.

would be sufficient to represent the average depth for the storm let alone variations in intensity. No three gauges located across the basin describe the spatial variability adequately. It is essential to capture the degree of variability in spatial rainfall-runoff modeling. There are many opportunities for combining radar-based measurements with a network of ground based recording rain gauges and disdrometers to sharpen the estimate of rainfall. I have made this case less completely in *Burges* [1998].

MODEL PARAMETER ESTIMATION SCHEMES

Single Objective

Calibration can be done manually where a skilled hydrologist adjusts parameters until the model outputs agree with recorded information to some desired level or by using organized computer search optimization schemes. *Nelder and Mead* [1965] developed a Simplex scheme, one of the most widely used approaches. This scheme is effective for situations when up to five to seven parameters need to be estimated. *Gupta and Sorooshian* [1985] showed derivative based Maximum Likelihood (MLE) schemes to be effective for calibrating models of limited complexity. *Duan et. al.* [1992] developed the most effective tool currently available for deterministic optimization, the Shuffled Complex Evolution (SCE-UA) method. This algorithm makes clever use of different starting locations in parameter space and a Nelder-Mead scheme. *Tanakamaru and Burges* [1996] demonstrated the power and utility of the SCE-UA algo-

rithm with the Tank Model where they calibrated twelve model parameters and four initial states. The obvious extension is to use multiple objectives in model fitting. *Madsen* [2000] presented an application of the Multi-Objective version (Pareto Optimality combined with the SCE-UA algorithm). A discussion of multi-objective calibration issues is given in the following section.

What is the best that can be achieved with an automatic calibration scheme and "error free" data? *Gan and Burges* [1990a] posed this question and devised tests for a situation when all hydrologic states and fluxes were postulated to be error free. Table 1 shows geometric, hydraulic properties, and the hydrologic response to two years of 6-minute time increment rainfall for two hypothetical small hillslopes. *Gan and Burges* [1990b] produced hillslope like hydrology using a plausible soil-physics based model [*Smith and Hebbert*, 1978] (S-H). The S-H model created time series of evapotranspiration, soil moisture distribution throughout the hillside, and surface and subsurface flow that were "hydrologic like". They treated these series as if they had been observed with an extensive array of error free instruments. The error free rainfall and resulting hydrographs shown in Figures 11 and 12 correspond to Cases 1 and 2, respectively, in Table 1. These hydrographs are not unlike natural systems. These time series could have been provided to any third party with the challenge to fit any hydrologic model of choice to these realistic appearing "data". A complete discussion of the approach is given in *Gan and Burges* [1990a].

Both hillslopes in Table 1 have slope of 10%, length 500 m, and width 100m. Both overly a relatively impermeable layer having saturated hydraulic conductivity, $K_L = 5x10^{-6}$ m/hr. The hillslopes are postulated to be homogenous and isotropic and have uniform root structures such that exfiltration can occur from the full depth of the soil column.

The hydrologic response for the shallow hillslope in Case 1 (depth 0.8 m, saturated hydraulic conductivity $K_U = 0.02$ m/hr, porosity 0.40), is shown in Figure 11. This hillslope produces principally saturated overland and Horton overland flow. The subsurface flow is 12% of the total flow. The modeled hydrologic response for the deeper and highly permeable hillslope in Case 2 (depth 1.6 m, saturated hydraulic conductivity K_U = 0. 2 m/hr, porosity 0.44) is shown in Figure 12. This hillslope produces principally subsurface flow (93% of the total flow). The evapotranspiration differs between the two cases with different initial- and end-state soil moisture accounting for the apparent mass balance differences.

The Sacramento model [*Burnash et. al*, 1973] (SMA) was calibrated to the streamflow signals in Figures 11 and 12 using automatic and manual calibration. The Nelder-Mead Algorithm was used with comparisons of "recorded", (S-H) and "simulated", (SMA) runoff made at a time step of 1 day.

Table 1. Geometry, hydraulic properties, and hydrologic responses of two hypothetical hillslopes [extracted from Gan and Burges 1990b)

	Width	Depth	Slope	Length	K_L	K_U	Porosity	Rain	ET	ET %Rain	Total Flow	Sub Surface Flow	%Total Flow
	m	*m*		*m*	*m/hr*	*m/hr*		*mm*	*mm*		*mm*	*mm*	
Case1	*100*	*0.8*	*0.1*	*500*	*$5x10^{-6}$*	*0.02*	*0.40*	*5790*	*996*	*17%*	*4850*	*580*	*12%*
Case2	*100*	*1.6*	*0.1*	*500*	*$5x10^{-6}$*	*0.2*	*0.44*	*5790*	*1088*	*19%*	*4874*	*4556*	*93%*

The quality of the calibrations for the two water years of data is evident in the residual time series plots shown in Figures 11 and 12. In the less permeable, shallower hillslope (Figure 11), the maximum daily flow volume was slightly less than 0.16 cmsd. The maximum error shown is approximately 0.02 cmsd. The maximum errors occur when the hillslope is in a relatively dry state when a storm arrives. For the deeper and highly permeable hillslope (Figure 12), the maximum daily flow volume was approximately 0.08 cmsd. The maximum residual error was approximately 0.03 cmsd.

Summary statistics do not illustrate the strengths and weaknesses of simulated hillslope responses as completely as residual time series plots. The *Nash and Sutcliffe* [1970] coefficient of efficiency, equation 7 in Gan and Burges (1990a), was calculated at the model time step increment of six minutes as well as at the daily summary level. For Case 1 (Figure 11), the corresponding efficiencies were 0.825, and 0.976, respectively. For the deeper more permeable situation (Figure 12), the efficiencies were 0.527, and 0.869, respectively. Several large simulation errors caused the decreased efficiencies in Case 2.

We deliberately plotted the time series of rainfall, hillslope output flow, and the residual daily flow volume error time series to highlight any apparent systematic errors as well as to show how well or poorly the calibrated model performed. We also did this to encourage others to display their modeling results as plainly and clearly as possible. *James and Burges* [1982] had recommended displaying error time series as well as providing several summary statistics for them. I am pleased to see more colleagues have since adopted this approach and would be delighted to see all time series results displayed this way.

Multiple Objectives

For many applications there will remain a major need to have models represent aspects of measured time series, particularly streamflow. Given timing errors in instruments and variable systematic bias in rainfall measurements, it appears that attaining a "good" solution for several objectives might be a profitable approach. One such illustration is provided in Figure 13.

Figure 13 shows the measured hydrograph for Plot 2 (shown previously in Figures 7) as well as a "simulated" hydrograph. The simulated hydrograph was produced by making a time shift of minus one hour for the first storm hydrograph and a plus one hour time shift for the second storm. The "simulated" hydrograph time series looks identical to the measured time series with the exception of the time shifts. Most modelers have chosen to be "slaves to time" and

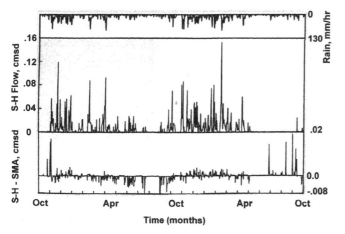

Figure 11. Time series of "observed error free" daily rainfall and runoff and the residual time series of simulated and "error free" runoff for a shallow hillslope (case 1: properties are given in Table 1 and the text). The figure was extracted from Figure 2a, *Gan and Burges* [1990b].

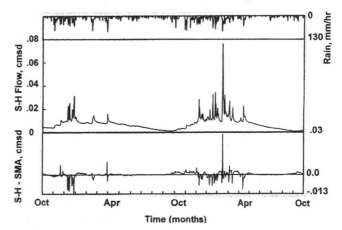

Figure 12. Time series of "observed error free" daily rainfall and runoff and the residual time series of simulated and "error free" runoff for a shallow hillslope (case 2: properties are given in Table 1 and the text). The figure was extracted from Figure 2b, *Gan and Burges* [1990b].

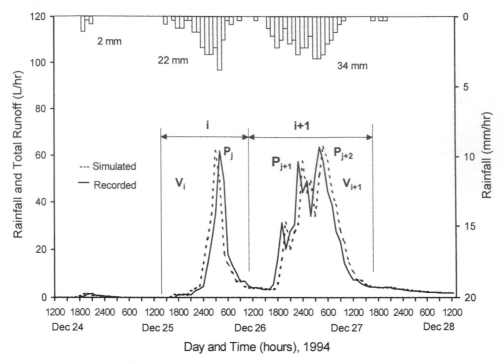

Figure 13. Measured hourly rainfall and runoff and hypothetical "simulated" runoff from Plot 2, Center for Urban Horticulture, University of Washington, Seattle.

compare modeled and recorded information for a specified time increment. In the situation in Figure 13, this approach leads to huge differences between modeled and recorded flow rates (as a function of time) for the rising and falling limbs. An alternative approach would bypass this problem, a problem identified immediately by the human eye.

I suggest that we should create appropriate time series for comparing "modeled" and "recorded" hydrographs. Two time series would suffice: series "i" contains the flow volume (or melt volume) associated with principal storms; series "j" contains peak flow rates above some specified threshold. The series "i", and "j" need not be of equal length. In the illustration in Figure 13 there are two volumes and three peak flow rates for the two principal storms. This approach permits identifying storms where measurement errors are thought to be serious. The corresponding data could be readily removed from the comparison scheme. The simplest automatic scheme to partition the recorded and modeled time series into volume and peak series would be one volume and one peak per storm or melt episode. It is important in hydrologic modeling to have peak flow rates and volumes correct. Sub series can be constructed from the volume and peak series. For example we could choose mid level peaks and corresponding volumes, lower magnitude and corresponding volumes, and so on, to permit testing features of models subject to threshold flow production conditions where greatest sensitivity to measurement errors might be expected.

Gupta et al [1998] provided an extensive discussion of issues in multi-objective calibration. They developed a multi-objective complex evolution scheme (MOCOM-UA) that built on the earlier shuffled complex evolution approach (SCE-UA). They provided in their Table 1 a list of calibration schemes that they and colleagues have used at the National Weather Service. Monthly flow volumes were one of the "volume" series that have been used. Given their observations, I suggest that the time series of volumes and peaks from Figure 13 be used in a two-objective (Pareto optimal) approach. This approach should overcome many existing limitations of data, particularly measurement and timing errors of rapidly changing states or fluxes.

TREATMENT OF PARAMETER AND INPUT UNCERTAINTY

Dawdy and Bergmann [1969] were among the earliest investigators to address explicitly parameter and input uncertainty effects on simulated streamflow time series. It is in relatively recent times that the computational demands for complete parameter and input uncertainty propagation could be accommodated. A formal Kalman Filter approach was used by *Kitanidis and Bras* [1980 a, b, c] to quantify model and parameter uncertainty in forecasts made with the Sacramento model. They assumed in their pioneering work aspects of parameter error structure and the form of the error distribution.

Garen and Burges [1981] sought the simplest scheme possible to propagate assumed parameter error through a calibrated Stanford Watershed model. They used both simple first-order uncertainty propagation and Monte Carlo uncertainty propagation and showed that the first-order approach was a good approximation to the more complete and computationally intensive Monte Carlo approach. They demonstrated the relative influence of parameter uncertainty on modeled hydrograph response for a given storm input for a hydrologically dry and a hydrologically wet initial state. The relatively uncertainties in hydrograph ordinates were completely different for the two cases. A much more comprehensive approach needs to take into account explicitly model and parameter errors, including bias, and data uncertainties.

The work of *Kuczera and Parent* [1998] is representative of the most complete schemes for calibrating a model with erroneous data and providing error bound predictions for model simulations. The direction of their work is particularly appealing because they predict uncertainties for a variety of states and variables. I think their general approach using the Metropolis algorithm, has much to recommend it. They used a daily hydrologic model with its attendant limitations, but they provide a format for using multiple signals for calibrating and testing any general model. The approach offers a powerful tool to demonstrate model structure inadequacy. For example, if measurements fall outside of prediction bounds there is need to question both data adequacy and model structure.

The general approach of *Kuczera and Parent* [1998] needs to be explored in considerably more detail. *Bates and Campbell* [2001] have introduced a fully Bayesian parameter estimation scheme that has additional potentially appealing features that also needs further exploration. An alternative parameter estimation scheme that explicitly considers uncertainties has been introduced by *Freer et al* [1996]. Works contained in this book that address these and related issues are the chapters by *Kavetski et al, Freer et al, Vrugt et al,* and *Misirli et al.*

SUMMARY

I have emphasized the need to focus on process details in our models. For most catchments, where there are no lakes, approximately 98 to 99% of the landform is the "hillslope". We have to get the details of water and energy balance right at the individual hillslope element before we can use any model for serious prediction applications. We would all hope to use a calibrated model to make useful predictions for all components of the hydrologic cycle for changes in precipitation and evaporation patterns and land use change. We need to include more information that would satisfy the flow paths and fluxes in multiple elements of the form shown schematically in Figure 1. We need to establish more "natural laboratories" and make appropriate nested measurements that address issues that I have raised with the material displayed in Figures 3 to 7. We need to address biases in point rainfall. (I have dodged the thorny issue of measuring snowfall, melt, and ablation). We need to establish rain measuring networks that reduce wind influenced under catch in whatever point measurement clusters we adopt to gain the most complete information we can from radar (or other remote observation devices) to describe spatial precipitation patterns.

Modeling has to be consistent with the measurement scale. There is little support for distributed modeling if only one "rain gauge" and one "streamflow measuring location" are available for the land unit of interest if it exceeds a small area. We need increased emphasis on measuring the vapor exchange with the atmosphere and with measuring recharge to ground water. This is essential to close the water budget at whatever spatial scale we model. We also need a renewed emphasis on data quality assurance and quality control so we can propagate errors appropriately and with confidence. When we have such information we can address algorithms for obtaining model parameters and parameter error structures. It is likely that multi-objective optimization schemes that blend features of the work of *Gupta et al* [1998] and *Kuczera and Parent* [1998] and *Bates and Campbell* [2001] will prove to be effective.

The overall objective in all this work is to sharpen considerably the measured and modeled mass and energy balances for catchments of all sizes. The data networks and models, that have been adequate for most prediction of hydrologic extremes and forecasting and water and land use decision making when there were fewer pressures on these resources, are no longer adequate for the needs of modern hydrology. Modern needs include hydrologically- and ecologically-based decision making and hydrologic hazard prediction for increasingly populated regions subject to flooding and drought.

Acknowledgments. This work was supported in part by grant EAR-9909579 from the National Science Foundation.

REFERENCES

Atkinson, T. C., Techniques for measuring subsurface flow on hillslopes, *in Hillslope Hydrology*, edited by M.J. Kirkby, Wiley-Interscience, New York, pp. 73-120, 1978.

Augustine, J. A., DeLuisi, J. J., and Long, C. N., SURFRAD -- A national surface radiation budget netw!ork for atmospheric research, *Bulletin of the American Meteorological Society*, 81(10), 2341-2357, 2000.

Baker, M., Inside history on droplets, *Nature*, 413, 586-587, 11 October, 2001.

Bates, B. C., and Campbell, E. P., A Markov chain Monte Carlo scheme for parameter estimation and inference in conceptual rainfall-runoff modeling, *Water Resources Research*, 37(4), 937-947, 2001.

Betson, R. P., and Ardis, C. V., Implications for modeling surface-water hydrology, in *Hillslope Hydrology*, edited by M. J. Kirkby, Wiley-Interscience, New York, 295-323, 1978.

Beven, K., How far can we go in distributed hydrologic modeling?, *Hydrology and Earth System Science*, 5(1), 1-12, 2001.

Brutsaert, W., Evaporation into the Atmosphere: Theory, History and Application, Reidel Publishing Co., Boston, p299, 1982.

Burnash, R. J. C., Ferral, R. L., and McGuire, R. A., A generalized streamflow simulation system, Conceptual Modeling for Digital Computers, U.S. National Weather Service, Sacramento, CA, 1973.

Burges, S. J., Trends and directions in hydrology, *Water Resources Research*, 22(9), 1s-5s, 1986.

Burges, S. J., Streamflow Prediction -- Capabilities, Opportunities, and Challenges", in *Hydrologic Sciences – Taking Stock and Looking Ahead*, Water Science and Technology Board, National Academy Press, pp. 101-134, 1998.

Burges, S. J. Wigmosta, M. S., and Meena, J. M., "Hydrologic effects of land-use change in a zero-order catchment", *Journal of Hydrologic Engineering*, 3(2), 86-97, 1998.

Chorley, R. J., The hillslope hydrological cycle, *in Hillslope Hydrology*, edited by M. J. Kirkby, Wiley-Interscience, New York, pp. 1-42, 1978.

Clark, C. O., Storage and the unit hydrograph, *Transactions*, ASCE, 110, 1419-1488, 1945.

Clarke, R. T., A review of some mathematical models used in hydrology, with observations on their calibration and use, *Journal of Hydrology*, 19, 1-20, 1973.

Crawford, N. H., and Linsley, R. K., The Synthesis of Continuous Streamflow Hydrographs on a Digital Computer, *Technical Report 12*, Department of Civil Engineering, Stanford University, Stanford, CA, 1962.

Crawford, N. H. and Linsley, R. K., Conceptual model of the hydrologic cycle, *International Association of Scientific Hydrology*, Publication No. 62, 1963.

Crawford, N. H. and Linsley R. K., Digital Simulation in Hydrology: Stanford Watershed Model IV, *Technical Report No. 39*, Department of Civil Engineering, Stanford University, p210, 1966.

Dawdy, D. R., and Bergmann, J. M., Effect of rainfall variability on streamflow simulation, *Water Resources Research*, 5(5), 958-966, 1969.

Dooge, J. C.I., Sensitivity of runoff to climate change: a Hortonian approach", *Bulletin American Meteorological Society*, 73(12), 2013-2024, 1992.

Duchon, C. E., and Essenberg, G. R., Comparative rainfall observations from pit and aboveground rain gauges with and without wind shields, *Water Resources Research*, 37(12), 3253-3263, 2001.

Duan, Q., Sorooshian, S., and Gupta, V. K., Effective and efficient global optimization for conceptual rainfall-runoff models, *Water Resources Research*, 28(4), 1015-1031, 1992.

Dunne, T., Field studies of hillslope flow processes, *in Hillslope Hydrology*, edited by M. J. Kirkby, Wiley-Interscience, New York, pp. 227-293, 1978.

Dunne, T., Zhang, W., and Aubrey, B. F., Effects of rainfall, vegetation, microtopography on infiltration and runoff, *Water Resources Research*, 27(9), 2271-2285, 1991.

Dunne, T., and Leopold, L. B., Water in Environmental Planning, Freeman, p818, 1978.

Faures, J.-M., Goodrich, D. C., Woolhiser, D. A., and Sorooshian, S, Impact of small-scale spatial rainfall variability on runoff modeling. *Journal of Hydrology*, 173:309-326, 1995.

Freer, J., Beven, K., and Ambroise, B. Bayesian estimation of uncertainty in runoff prediction and the value of data: an application of the GLUE approach, *Water Resources Research*, 32(7), 2161-2173, 1996.

Freeze, R. A., Streamflow generation, *Reviews of Geophysics and Space Physics*, 12(4), 627-647, 1974.

Gan, T. Y., and Burges, S. J., An assessment of a conceptual rainfall-runoff model's ability to represent the dynamics of small hypothetical catchments, 1: models, model properties, and experimental design, *Water Resources Research*, 26(7), 1595-1604, 1990.

Gan T. Y., and Burges, S. J., An assessment of a conceptual rainfall-runoff model's ability to represent the dynamics of small hypothetical catchments, 2: hydrologic responses for normal and extreme rainfall, *Water Resources Research*, 26(7), 1605-1619, 1990.

Garen, D. C., and Burges, S. J., Approximate error bounds for simulated hydrographs, *Journal of the Hydraulics Division*, ASCE, 107(HY11), 1519-1534, 1981.

Grayson, R. B., Moore, I. D., and McMahon, T. A., Physically based hydrologic modeling 1. A terrain-based model for investigative purposes, *Water Resources Research*, 28(10), 2639-2658, 1992.

Gupta, V. K., and Sorooshian, S., The automatic calibration of conceptual catchment models using derivative-based optimization algorithms, *Water Resources Research*, 21(4), 473-485, 1985.

Gupta, H. V., Sorooshian, S., and Yapo, P. O., Toward improved calibration of hydrologic models: Multiple and noncommensurable measures of information, *Water Resources Research*, 34(4), 751-763, 1998.

Horton, R. E., The role of infiltration in the hydrologic cycle, *Transactions*, American Geophysical Union, 14, 446-460, 1933.

Hoyt, W. G., et al., Studies of Relations of Rainfall and Run-off in the United States, U.S. Geological Survey, *Water-Supply Paper 772*, U.S. Government. Printing Office, pp. 301, 1936.

Izzard, C. F., Hydraulics of runoff from developed surfaces, *Proceedings*, Highway Research Board, twenty-sixth Annual Meeting, 1946.

James, L. D., and Burges, S. J., Selection, calibration, and testing of hydrologic models, *in Hydrologic Modeling of Small Watersheds*, edited by C. T. Haan, H. P. Johnson, and D. L. Brakensiek, American Society of Agricultural Engineers Monograph, 5, 437-472, 1982.

Kirkby, M. J., Hillslope Hydrology, John Wiley & Sons Limited, New York, p389, 1978.

Kitanidis, P. K., and Bras, R. L., Adaptive filtering through detection of isolated transient errors in rainfall-runoff models, *Water Resources Research*, 16, 740–748, 1980a.

Kitanidis, P. K., and Bras, R. L., Real-time forecasting with a conceptual hydrological model, 1, Analysis of uncertainty, *Water Resources Research*, 16, 1025–1033, 1980b.

Kitanidis, P. K., and Bras, R. L., Real-time forecasting with a conceptual hydrological model, 2, Applications and results, *Water Resources Research*, 16, 1034–1044, 1980c.

Kohler, M. A., and Linsley, R. K., Predicting the runoff from storm rainfall, *U.S. Weather Bureau Research Paper 34*, 1951.

Kolsti, K. F., Burges, S. J. and Jensen, B. W., Hydrologic Response of Residential-Scale Lawns on Till Containing Various Amounts of Compost, *Water Resources Series, Technical Report No.147*, Department of Civil Engineering, University of Washington, p143, 1995.

Kuczera, G., and Parent, E., Monte Carlo assessment of parameter uncertainty in conceptual catchment models: the Metropolis algorithm, *Journal of Hydrology*, 211, 69-85, 1998.

Linsley, R. K., and Ackerman, W. C., Method of predicting the runoff from rainfall, *Transactions*, American Society of Civil Engineers, Paper No. 2147, 825, 1942.

Linsley, R. K., and Crawford, N. H., Computation of a synthetic streamflow record on a digital computer, *Publication No. 51*, International Association of Scientific Hydrology, 526-538, 1960.

Madsen, H., Automatic calibration of a conceptual rainfall-runoff model using multiple objectives, *Journal of Hydrology*, 235, 276-288, 2000.

McCarthy, G. T., The unit hydrograph and flood routing, presented at Conference North Atlantic Division, US Corps of Engineers, June 1938

Montieth, J. L., and Unsworth, M. H., Principles of Environmental Physics, (2 ed.), Edward Arnold, p291, 1990.

Mulvaney, T. J., On the use of self-registering rain and flood gauges in making observations of the relations of rainfall and flood discharges in a given catchment, *Proceedings Institution of Civil Engineers Ireland*, 4 18-31, 1851.

Musgrave, G. W., How much of the rain enters the soil?, in *Water the Yearbook of Agriculture 1955*, edited by A. Stefferud, US Government Printing Office, pp. 151-159, 1955

Nash, J. E., and Sutcliffe, J. V., River flow forecasting through conceptual models, 1. a discussion of principles, *Journal of Hydrology*, 10 282-290, 1970.

Nelder, J. A., and Mead, R., A simplex method for functional minimization, *The Computer Journal*, 9, 308-313, 1965.

Peck, E. L., Catchment Modeling and Initial Parameter Estimation for the National Weather Service River Forecast System, NOAA, Tech. Memo, NWS HYDRO-31, 1976.

Philip, J. R., An infiltration equation with physical significance, *Soil Science*, 77, 153-157, 1954.

Philip, J. R., Theory of infiltration, in *Advances in Hydroscience*, edited by V. T. Chow, 5, 215-296, Academic Press, New York, 1969.

Richards, L. A., Capillary conduction of liquids through porous mediums, *Physics*, 1, 318-333, 1931.

Rodriguez-Iturbe, I., Ecohydrology: a hydrologic perspective of climate-soil-vegetation dynamics", *Water Resources Research*, 36(1), 3-9, 2000.

Ronan, R. D., Prudic, D. E., Thodal, C. E., and Constantz, J., Field study and simulation of diurnal temperature effects on infiltration and variably saturated flow beneath an ephemeral stream, *Water Resources Research*, 34(9), 2137-2153, 1998.

Seyfried, M. S., and Wilcox, B. P., Scale and the nature of spatial variability: field examples having implications for hydrologic modeling", *Water Resources Research*, 31(1), 173-184, 1995.

Sherman, L. K., The relation of hydrographs of runoff to size and character of drainage basins, *Transactions*, American Geophysical Union, 13, 332-339, 1932.

Sherman, L. K., The unit hydrograph, *in Hydrology*, edited by O. E. Meinzer, McGraw-Hill, New York, Chapter 11E, 1942; reprinted Dover, New York, 1949.

Smith, R. E. and Hebbert, R. H. B., Mathematical simulation of interdependent surface and subsurface hydrologic processes, *Water Resources Research*, 19(4), 987-1001, 1983.

Steiner, M., Smith, J. A., Burges, S. J., Alonso, C. V., and Darden, R. W., Effect of bias adjustment and rain gauge data quality control on radar rainfall estimation, *Water Resources Research*, 35(8), 2487-2503, 1999.

Sugawara, M. On the analysis of runoff structure about several Japanese rivers, *Japanese Journal of Geophysics*, 2, March 1961.

Sugawara, M., Watanabe, I., Ozaki, E., and Katsuyama, Y., Tank Model With Snow Component, *Research Notes No. 65*, National Research Center for Disaster Prevention, Japan, 1984.

Tanakamaru, H., and Burges, S. J., Application of global optimization to parameter estimation of the tank model, *Proceedings International Conference on Water Resources and Environmental Research* Vol. II, pp. 39-46, Kyoto, Japan, October 29-31, 1996

Wigmosta, M. S., Vail, L. W., and Lettenmaier, D. P., A distributed hydrology-vegetation model for complex terrain, *Water Resources Research*, 30(6), 1665-1679, 1994.

Wigmosta, M., and Burges, S. J., An adaptive modeling and monitoring approach to describe the hydrologic behavior of small catchments, *Journal of Hydrology*, 202, 48-77, 1997.

Woolhiser, D. A. and Liggett, J. A., Unsteady one-dimensional flow over a plane the rising hydrograph, *Water Resources Research*, 3(3), 753-771, 1967.

Zhang, W., Numerical simulation of the hydrodynamics of overland flow with spatial variation in its physical characteristics, Ph.D. dissertation, University of Washington, Seattle, 231p, 1990.

Burges, Stephen J., Department of Civil and Environmental Engineering, University of Washington, Seattle, WA 98195-2700

The Quest for an Improved Dialog Between Modeler and Experimentalist

Jan Seibert

Swedish University of Agricultural Sciences, Department of Environmental Assessment, Uppsala, Sweden

Jeffrey J. McDonnell

Department of Forest Engineering, Oregon State University, Corvallis, Oregon

Multi-criteria calibration of runoff models using additional data, such as groundwater levels or soil moisture, has been proposed as a way to constrain parameter values and to ensure the realistic simulation of internal variables. Nevertheless, in many cases the availability of such 'hard data' is limited. We argue that experimentalists working in a catchment often have much more knowledge of catchment behavior than is currently used for model calibration and testing. While potentially highly useful, this information is difficult to use directly as exact numbers in the calibration process. We present a framework whereby these 'soft' data from the experimentalist are made useful through fuzzy measures of model-simulation and parameter-value acceptability. The use of soft data is an approach to formalize the exchange of information and calibration measures between experimentalist and modeler. This dialog may also greatly augment the traditional and few 'hard' data measures available. We illustrate the value of 'soft data' with the application of a three-box conceptual model for the Maimai catchment in New Zealand. The model was calibrated against hard data (runoff and groundwater-levels) as well as a number of criteria derived from the soft data (e.g., percent new water, reservoir volume). While very good fits were obtained when calibrating against runoff only (model efficiency = 0.93), parameter sets obtained in this way showed, in general, poor internal consistency. Inclusion of soft-data criteria in the model calibration process resulted in lower model-efficiency values (around 0.84 when including all criteria) but led to better overall performance, as interpreted by the experimentalist's view of catchment runoff dynamics.

INTRODUCTION

Many different conceptual models of catchment hydrology have been developed during the last few decades [*Singh*, 1995]. These models have become valuable tools for water management problems (*e.g.*, flood forecasting, water balance studies and computation of design floods). The increasing awareness of environmental problems has given additional impetus to hydrological modeling. Runoff models have to meet new requirements when they are intended to deal with problems such as acidification, soil erosion and land degradation, leaching of pollutants, irrigation, sustainable water-resource management or possible consequences of land-use or climatic changes. Linkages to geochemistry, ecology, meteorology and other sciences must be considered explicitly and realistic simulations of internal processes become essential.

Calibration of Watershed Models
Water Science and Application Volume 6
Copyright 2003 by the American Geophysical Union
10/1029/006WS22

Despite much effort [*Hornberger and Boyer*, 1995], hydrological modeling is faced by fundamental problems such as the need for calibration and the equifinality of different model structures and parameter sets (*i.e.*, the phenomenon that equally good model simulations might be obtained in many different ways, *Beven*, 1993). These problems are linked to the limited data availability and the natural heterogeneity of watersheds [*e.g.*, *Beven*, 1993; *O'Connell* and *Todini*, 1996; *Bronstert*, 1999]. Problems also can be related to the procedures used for model testing. Traditional tests such as split-sample tests are often not sufficient to evaluate model validity or to assess the pros and cons of different model approaches. More powerful and rigorous methods for model calibration and testing are clearly required [*Kirchner et al.*, 1996; *Mroczkowski et al.*, 1997, *Kavetski et al.*, this issue].

Multi-criteria Model Calibration

Manual calibration of a model by trial and error is a time-consuming method and results may be subjective. This is particularly true when calibrating against more than one hydrological variable. Therefore, various automatic calibration methods have been developed [*Sorooshian* and *Gupta*, 1995; *Gupta et al.*, this volume, *Duan*, this volume]. In general, these methods allow for a quick and 'objective' calibration. On the other hand there is the danger that model calibration becomes a 'dumb' curve fitting exercise. By this we mean that unlike the manual calibration process where the hydrologist will implicitly make use of his/her process knowledge (e.g. by examining different aspects of the hydrograph or the simulation of internal variables), in the automatic approach, only explicitly stated criteria are considered. Thus, there appears to be a need for methods to infuse hydrological reasoning into the automatic calibration process.

Two 'ways forward' on the equifinality issue include: (1) making more detailed use out of the comparison between simulated and observed runoff series [e.g., *Boyle et al.*, 2000; this volume; *Burges*, this volume, *Freer*, this volume] or (2) incorporating additional data into the model calibration procedure. Boyle *et al.* [2000; this volume], followed the first approach and proposed a method to combine the strengths of manual and automatic calibration methods. Recognizing that one goodness-of-fit measure is not sufficient to judge the fit of observed and simulated runoff series, they examined different parts of the hydrograph separately. Our work, and this chapter, complements the work of Boyle *et al.* [2000; this volume] by exploring the second approach: *i.e.*, the utilization of additional data in the model calibration process.

The need to utilize additional data for model calibration and testing has been emphasized by others in the recent years [*de Grosbois et al.*, 1988; *Ambroise et al.*, 1995; *Refsgaard*, 1997; *Kuczera* and *Mroczkowski*, 1998; *Seibert*, 1999; *Meixner* and *Bastidas*, this volume]. Testing runoff models against variables other than simply catchment-outlet runoff is important for two main reasons: (1) in many hydrological questions, and for other sciences such as ecology, it may be of much more interest to know what happens *within* a catchment than at the outlet, and (2) to have confidence in model predictions, which are often extrapolations beyond the testable conditions, it must be ensured that the model not only works, but also does so for the right reasons.

Most parameters of conceptual runoff models need to be determined by calibration. Some parameters may have a physical basis but they are *effective* parameters on the catchment or subcatchment scale. The typical problem is that the information contained in the rainfall-runoff relationship usually does not allow the identification of one unique parameter set. Reducing the number of parameters is an unattractive option because it might transform the conceptual gray-box representation of the rainfall-runoff process into a pure black-box description. Another more attractive way to reduce parameter uncertainty is the use of additional data. Franks *et al.* [1998] demonstrated that the known percentage of saturated areas in the catchment helped to constrain calibrated parameter values and model predictions in an application of TOPMODEL. Seibert [2000] found for an application of the HBV model, that groundwater-level data helped to constrain the parameters of the groundwater routine. However, the worth of additional data varies depending on the kind of data, but also on the structure of the applied model. For instance, Kuczera and Mroczkowski [1998] found that groundwater levels helped little to reduce the parameter uncertainty in a hydrosalinity model, whereas stream salinity data more substantially reduced the uncertainties. Blazkova *et al.* [2002] mapped saturated areas and found that this information influenced optimized parameter values for TOPMODEL, but also that the additional information had only limited effect on constraining prediction bounds for stream discharge.

The Concept of Soft Data

In many cases the amount of available additional data is limited. However, a hydrologist might have a perceptual model [*Beven*, 1993], which is a highly detailed yet qualitative understanding of dominant runoff processes even in situations with limited field measurements. Thus, there exists in addition to hard data (streamflow hydrograph, well record) 'soft data' about catchment hydrology and its internal 'behavior'. While some groups have used the perceptual model to guide the construction of the model elements, little has been done to use this kind of data in the model calibration. The few to do this include Franks *et al.* [1998] who used maps of surface saturated area to constrain parameter ranges for TOPMODEL runs and Franks and Beven [1997] who used related fuzzy measures for evapotranspiration. Soft data can be defined as qualitative knowledge from the experimentalist that cannot be used directly as exact numbers but that can be made useful when transformed into quantitative data through fuzzy measures of model-simulation and parameter-value acceptability. Soft data may be based on 'hard' measurements but these measurements require some interpretation or manipulation by a hydrologist before being useful in model testing. While fuzzy, these soft measures can be exceedingly valuable for indicating 'how a catchment works'. Fuzzy measures, which implement the concept of partial truth with values between completely true and completely false, have been found to be useful in hydrological model calibration [*Seibert*, 1997; *Aronica et al.*, 1998; *Franks et al.*, 1998; *Hankin* and *Beven*, 1998]. *Aronica et al.* [1998], for instance, used a fuzzy-rule based calibration for a system containing highly uncertain flood information. A fuzzy measure varies between zero and one and describes the degree to which the statement 'x is a member of Y' or, in our case, 'this parameter set is the best possible set' is true.

Different methods are available for automatic optimization. Evolution-based optimization methods have been found to be suitable tools for the calibration of conceptual runoff models [*Wang*, 1991; *Duan et al.* 1992; *Franchini*, 1996; *Kuczera*, 1997; *Yapo et al.*, 1998, *Duan*, this volume]. Genetic algorithms are one class of these methods. The goal of genetic algorithms, originally suggested by Holland [1975; 1992], is to mimic evolution. Parameter sets are encoded to chromosome-like strings and different recombination operators are used to generate new parameter sets. The optimization starts with a population of randomly generated parameter sets. These are evaluated by running the model; those sets that give a better simulation according to some objective function, are given more chances to generate new sets than those sets that gave poorer results. Seibert [2000] used a genetic algorithm to find the true parameter values for a theoretical, error-free test case with synthetic data. For a real-world case, with calibration against observed runoff, he found that parameter values varied considerably for different calibration trials. However, approximately the same model efficiency was achieved in almost every trial. This possibility for different parameter sets in the case of a flat goodness-of-fit surface allows one to utilize the genetic algorithm to evaluate parameter uncertainty

using the variation of calibrated parameter values as a measure of parameter identifiability [*Seibert*, 2000]. The genetic algorithm can, thus, provide an indication of parameter uncertainty and serve as an alternative to Monte Carlo approaches like, the Generalized Likelihood Uncertainty Estimation (GLUE) techniques of Freer *et al.* [1996].

In this chapter we present a method for how to use the additional data that often exists in experimental catchments for the calibration of conceptual runoff models. We present a number of 'soft data' measures as means to improve the dialog between modeler and experimentalist. We describe and use the implementation of a genetic algorithm for calibration, as proposed by Seibert [2000], and illustrate these methods for the Maimai watershed in New Zealand. Our main message in this chapter is that additional soft data may be a useful way to ensure that a model of catchment hydrology not only works (for runoff simulation), but also does so for the right process reasons.

MATERIAL AND METHODS

Soft Data

We define soft data as knowledge from the experimentalist that cannot be used directly for model calibration and testing but that can be made useful through fuzzy measures of model-simulation and parameter-value acceptability. It is important to note that soft data may be based on 'hard' measurements that require some interpretation or manipulation by a hydrologist before being useful in model testing. Model simulations may be judged in more process-based, ways when soft data is used compared to when only the hard data is considered For instance, the experimentalist might have some observations concerning the range in which groundwater levels fluctuate within a given zone of the catchment, or conceptual model box (based on field campaign information or observations made over some irregular time periods) or the contribution of rainfall or 'new' water [*McDonnell et al.*, 1991] to peak flow (from event-based isotope tracing studies). Soft data can be used to constrain the calibration by: (1) evaluating the model with regard to simulations for which there might be no hard data available for comparison, and (2) assessing how reasonable the parameter values are, based on field experience. This range of 'reasonable' parameter values might be wide, especially when the parameter values are effective values at some larger scale.

When comparing model simulations or parameter values with soft data, there may be a relatively wide range of acceptable simulations or values. Furthermore, there might be a range of values that fall between 'fully acceptable' and 'not acceptable', based on the experimentalist's experience in the field and other synoptic measurements. Fuzzy measures of acceptance can be used to consider these ranges [*Franks et al.*, 1998]. For each soft data type, a trapezoidal function (Eq. 1), where the experimentalist is asked to assign values to the variables a_i, is used to compute the degree of acceptance, μ, from the corresponding simulated quantity or parameter value x. This trapezoidal function is a simple way to map experimentalist experience into a quantity, which then can be optimized (Fig. 1). Other functions with different shapes might be used instead of the trapezoidal function.

$$\mu(x) = \begin{cases} 0 & \text{if } x \leq a_1 \\ \dfrac{x - a_1}{a_2 - a_1} & \text{if } a_1 \leq x < a_2 \\ 1 & \text{if } a_2 \leq x < a_3 \\ \dfrac{a_4 - x}{a_4 - a_3} & \text{if } a_3 \leq x < a_4 \\ 0 & \text{if } x > a_4 \end{cases} \quad (1)$$

An important point is that that uncertainty exists in the experimentalist's view of the catchment and that data collected in the field have their related uncertainties [*Sherlock et al.*, 2002]. Thus, the trapezoidal function provides a way for the experimentalist to also provide his or her uncertainty bounds on the delivered rules to the modeler.

The general acceptability of a parameter set was defined by three components: (1) the goodness-of-fit measures for the hard data such as the model efficiency [*Nash and Sutcliffe*, 1970] for runoff (A_1), (2) the goodness of the model simulations with regard to soft data (*e.g.*, maximum groundwater levels) as quantified using Eq. 1 (A_2), and (3) the acceptability of the parameter values based on the experimentalist's experience (A_3). For all three components, a value of one for A_i corresponds to a perfect fit (or complete acceptability).

The overall acceptability, A, of a parameter set is computed as a weighted geometric mean with the weights n_1, n_2, and n_3 (Eq. 2). A can then be used as optimization criterion.

$$A = A_1^{n1} A_2^{n2} A_3^{n3} \quad \text{with } n_1 + n_2 + n_3 = 1 \quad (2)$$

The selection of the weights in Eq. 2 n_1, n_2, and n_3 determines which solution along the pareto-optimality sub-space will be found. The weights allow placement of more (or less) emphasis on the different types of data. A higher value for n_1, for instance, might be justified if there is much useful and accurate hard data, whereas a smaller value might be appropriate if the hard data consists of only runoff.

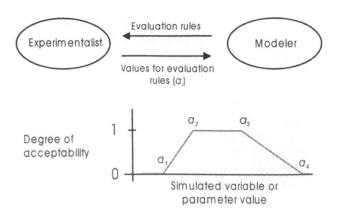

Figure 1. Framework for formalized dialog between experimentalist and modeler using a trapezoidal function as a means of assigning values to the soft data.

Description of the Genetic Algorithm

A genetic algorithm utilizes an evolution of parameter sets with elements of selection and recombination to find optimized parameter sets [*Duan*, this volume]. An initial population of n (set to 50) parameter sets is selected randomly within the parameter space. The 'fitness' of an individual set is quantified as the value of an objective function. A new population (generation) is generated from this population by n times combining two parameter sets, which are chosen randomly but with a higher chance of being picked for sets with a higher 'fitness' (*i.e.*, objective function). From the two parent sets (sets A and B) the new parameter set is generated by applying for each parameter randomly (with some probability, p_i), each of the following four rules: (1) value of set A ($p_1=0.41$), (2) value of set B ($p_2=0.41$), (3) random value between the values of set A and set B (alternatively, if both values were equal, a random value close to this value) ($p_3=0.16$), or, (4) random value within the limits given for the parameter (mutation) ($p_4=0.02$). The first two rules preserve the values of the preceding generation, whereas the other two rules provide an amount of random search. A balance between these rules is important for the success of the algorithm. However, within reasonable ranges adjustments to the probabilities for the different rules have only minor effects on the performance of the algorithm. Finally the fitness of each set in the new population is evaluated and the new generation replaces the old one. However, the best set is retained if there is no better set in the proceeding generation. This process is repeated for a number of generations.

The results of a genetic algorithm can be improved by combination with a local search method [*Wang*, 1991]. For instance the parameter set found by a genetic algorithm can be used as starting point for a local optimization [*Franchini*,

1996]. In addition to this form of subsequent 'fine-tuning', a local search approach can also be implemented during the 'evolution' process [Seibert, 2000]. At a small probability (p=0.02), the new parameter set is not found by the parameter-by-parameter combinations as described above; instead the new parameter set is the result of a one-dimensional optimization along the line determined by the two parameter sets using Brent's method [*Press et al.*, 1992]. In this chapter we divide the total number of 2500 model runs into 2000 runs for the genetic algorithm and 500 runs for the subsequent local optimization. We use Powell's quadratically convergent method for this multidimensional, local optimization, as described in Press *et al.* [1992].

Our genetic algorithm includes stochastic elements such as the randomly generated initial set of parameter sets and the partly random generation of offsprings during the 'evolution' of parameter sets. Thus, the calibrated parameter values may vary for different calibration trials, when different parameter sets result in similarly good simulations according to the goodness-of-fit measure. This makes this optimization algorithm suitable to address parameter uncertainty using the variation of calibrated parameter values as a measure of parameter identifiability. For the results presented in this study, sixty calibration trials were performed for each goodness-of-fit measure and the best 50 parameter sets were used for further analysis of model performance and parameter identifiability.

The Maimai Watershed

Maimai M8 is a small 3.8 ha headwater catchment located to the east of the Paparoa Mountain Range on the South Island of New Zealand. Slopes are short (<300 m) and steep (average 34°) with local relief of 100-150 m. Stream channels are deeply incised and lower portions of the slope profiles are strongly convex. Areas that could contribute to storm response by saturation overland flow are small and limited to 4-7 % [*Mosley*, 1979; *Pearce et al.*, 1986]. Mean annual precipitation is approximately 2600 mm, producing an estimated 1550 mm of runoff. There were 11 major runoff events during the period of record used for model simulation in this study (August-December, 1987) with a maximum runoff of 6 mm/h. Additional to rainfall and runoff data, groundwater levels extracted from the tensiometer data in McDonnell [1989, 1990], were available for two locations (one in the riparian and one in the hollow zone). Mean monthly values of potential evaporation estimated by L. Rowe [1992, pers.comm.] were distributed using a sine curve for each day [*J. Freer*, 2000, pers. comm.].

The Maimai M8 watershed is a well-studied watershed with ongoing hillslope research by several research teams since the late 1970s. During these studies a very detailed yet qualitative perceptual model of hillslope hydrology evolved (for review see McGlynn *et al.* [2002]).

Conceptual Three-box Model

While this chapter focuses on soft data for multi-criteria calibration, the soft data first helped guide the box-model construction. Our conceptual model is based on the three reservoirs identified from the experimental studies at M8: riparian, hollow and hillslope zones (Fig. 2, Table 1). These zones (or model boxes) display very different groundwater dynamics [*McDonnell*, 1990] and group clearly based on their isotopic characteristics [*McDonnell et al.*, 1991]. Water is simulated to flow from the hillslope zone into the hollow zone and from the hollow zone into the riparian zone. Outflow from the riparian zone forms the flow in the stream. Most importantly, and most novel for this model, is the formulation used to model the unsaturated and saturated storage. Due to the shallow groundwater (groundwater levels 0 – 1.5 m below the ground surface) growth of the (transient) saturated zone occurs at the expense of the unsaturated zone thickness. Thus, a coupled formulation of the saturated and unsaturated storage was used, as proposed by Seibert *et al.* [2002]. In this formulation, the amount of saturated storage determines the maximum space for unsaturated storage. For a more detailed description and equations of the three-box model the reader is referred to [*Seibert* and *McDonnell*, 2002].

Table 1. List of parameters used in the three-box model.

Parameter	Description	Unit
z_{max}	Soil depth [a] [mm]	
c	Parameter corresponding to water content at field capacity divided by porosity	[-]
d	Parameter corresponding to water content at wilting point divided by porosity	[-]
B	Shape coefficient determining groundwater recharge	[-]
$k_{1,riparian}$	Outflow coefficient, riparian box	[h^{-1}]
$k_{1,hollow}$	Outflow coefficient, hollow box, lower outflow	[h^{-1}]
$k_{2,hollow}$	Outflow coefficient, hollow box, upper outflow	[h^{-1}]
$k_{1,hillslope}$	Outflow coefficient, hillslope box	[h^{-1}]
$z_{threshold}$	Threshold storage for contribution from upper outflow in the hollow box	[mm]
p	Porosity [a]	[-]
$f_{riparian}$	Areal fraction of the riparian zone	[-]
f_{hollow}	Areal fraction of the hollow zone	[-]

[a] Different values were allowed for riparian, hollow and hillslope box

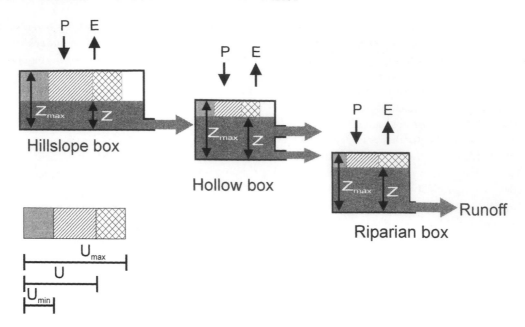

Figure 2. Structure of the three-box model developed for the Maimai M8 watershed including hillslope, hollow and riparian zone reservoirs. (P: precipitation, E: evaporation, z: groundwater level above bedrock, U: unsaturated storage). See also Table 1.

As for any model, several simplifications and assumptions are made to derive this conceptual three-box model [*Seibert* and *McDonnell*, 2002]. The model structure is guided by experimental findings at Maimai. Obviously these simplifications and assumptions are not universally applicable; for other watersheds, a different model structure may be more appropriate (perhaps different box configurations, different number of boxes or different sizes and connections of boxes). The dialogue between experimentalist and modeler using the soft-data framework might guide this construction of conceptual models for particular catchments.

Application of the Soft-Data Framework

For presentation in this chapter we include a subset of the available soft data for demonstration purposes: groundwater levels in the three boxes, the new-water contribution to peak runoff, and some other parameter values. Evaluation rules were developed using Eq. 1 to judge model performance with regard to minimum and maximum groundwater levels as well as the frequency of levels being above a specified level (Table 2). The values for these rules were motivated by field studies reported in McDonnell [1990], McDonnell et al. [1991] and Stewart and McDonnell [1991] for the same August-December 1987 period where groundwater response in the riparian and hollow zones were quantified with recording tensiometers that show distinctly different wetting, filling, draining behavior. Riparian zones were characterized by rapid conversion of tension to pressure potential

(*i.e.*, rapid conversion of unsaturated zone to a saturated zone by storage filling and water table rise from below). Water tables were sustained in this zone for 1-2 days following the cessation of rainfall. These data provided the soft data measures for minimum and maximum groundwater levels and frequency of levels above a specified level (listed in Table 2). The hollow zone response was much more sensitive to rainfall inputs: conversion of unsaturated zone to transient saturation occurred within the few hours of the hydrograph rising limb and pore pressure recession rates closely matched stream and subsurface-trench hydrograph recession rates. Soft data for the hillslope positions were gathered from previous throughflow pit analysis by Mosley [1979] including continuously recorded pit outflow from a number of distinct linear hillslope segments. Hillslope sections (unlike hollows and riparian zones) show very infrequent water table development—when water tables were present, they were restricted vis-à-vis the soft data measure trapezoidal function classification (see numbers in Table 2). The soil catena sequences in the Maimai catchment as mapped by McKie [1978] confirm these interpretations.

Hillslope soils show no evidence of any gleying whereas gleying appears in the hollow zone and is most dominant in the riparian zone. We view this as a long-term expression of the spatial delineation of boxes and water table longevity applied in this study. Table 2 includes also a number of soft-data rules including isotope hydrograph separation-derived new-water estimates (at peakflow). Values for these rules were based on results from hydrograph separations reported

Table 2. Evaluation rules based on soft data used for model calibration (the values for a_i define the trapezoidal function used to compute the degree of acceptance, see Eq. 1).

Type of soft information	Specific soft information	a_1	a_2	a_3	a_4	Motivation
New water contribution to peak runoff [-]	870930 18.00	0.03	0.06	0.12	0.15	McDonnell *et al.* [1991]
	871008 3.00	0.05	0.13	0.31	0.40	"
	871010 17.00	-	0	0.03	0.06	"
	871013 11.00	0.17	0.23	0.35	0.41	"
	871113 19.00	-	0	0.03	0.06	"
	871127 8.00	0.04	0.07	0.13	0.16	"
Range of groundwater levels, min./max. fraction of saturated part of the soil [-]	Maximum hillslope	0	0.2	0.5	0.7	Mosley [1979]
	Maximum hollow	0	0.5	0.75	1	McDonnell [1990]
	Minimum hollow	0	0.05	0.1	0.2	"
	Minimum riparian	0.05	0.1	0.3	0.5	"
Frequency of groundwater levels above a certain level (as fraction of soil) [-]	Hillslope, above 0.5 during events	-	0	0.1	0.3	Mosley [1979]
	Hollow above 0.7 during events	-	0	0.1	0.2	McDonnell [1990]
	Hollow above 0.9 during events	-	-	0	0.1	"
	Riparian above 0.2	0.6	0.8	1	1	"
	Riparian above 0.9 during events	0	0.25	0.75	1	"
Parameter values	Fraction of riparian zone [-]	0.01	0.03	0.07	0.10	Mosley [1979]
	Fraction of hollow zone [-]	0.05	0.10	0.15	0.20	McDonnell [1990]
	Porosity in hillslope zone [-]	0.45	0.6	0.7	0.75	McDonnell [1989]
	Porosity in hollow zone [-]	0.45	0.55	0.65	0.75	"
	Porosity in riparian zone [-]	0.45	0.5	0.6	0.75	"
	Soil depth for hillslope zone [m]	0.1	0.3	0.8	1.5	McDonnell *et al.* [1998]
	Soil depth for hollow zone [m]	0.5	1	2	2.5	"
	Soil depth for riparian zone [m]	0.15	0.4	0.75	1	"
	Threshold level in hollow zone, fraction of soil depth [-]	0	0.1	0.4	1	McDonnell [1990] McDonnell *et al.* [1991]

in McDonnell [1989] and McDonnell *et al.* [1991]. These evaluation rules allowed computation of degree of acceptance with respect to the simulated new-water. New water percentage is a very useful integrated measure of the relative contribution of rainfall versus displaced stored water contributions at various times through the storm hydrograph. Unlike the point-based water level measures and rules, the new water percentage subsumes point scale variability into an integrated measure of catchment runoff dynamics. In our dataset, the new-water percentages varied, from event to event, and some storms did not have rain isotopic concentration suitable for application of the two-component mass balance separation technique. The flexibility of the soft data is such that even for isolated measures from field campaigns or experiments (or when hydrograph separation was possible) rules may be developed to guide the model calibration process, even if this information is derived from periods outside the simulated calibration period.

We computed degrees of acceptance for a number of parameters using the soft data evaluation rules. Acceptance in this instance is defined as the degree to which model parameter values agree with the field experience and the perceptual model of the catchment runoff process. These acceptance values varied from one, if the value was within the desirable range and decreased towards zero with increasing deviations from this range (Table 2). For example, we allowed values from 1 to 10 percent for the areal fraction of the riparian zone (*i.e.*, the variable source area in this case), but the degree of acceptance was one, only for values between 3 and 7 percent (based on mapped saturated areas in the M8 catchment reported in Mosley [1979]). Based on the individual parameters the acceptability of a certain parameter set was computed as the geometric mean of the respective degrees of acceptance.

We quantified the acceptability of calibrations using hard data (A_1) using a combination of the efficiency measure, R_{eff}, and the relative volume error, V_E, (=accumulated difference divided by sum of observed runoff) for the runoff simulations as proposed by Lindström [1997] (Eq. 3). Following Lindström [1997], a value of 0.1 was used for the weighing coefficient, ω, which determines the relative emphasis on the volume error. The coefficient of determination, r^2, was used to assess the performance of the simulations for the groundwater levels in the riparian and the hollow zone, and A_1 is computed as average of these different goodness-of-fit measures (Eq. 3).

$$A_1 = \frac{1}{2}\left(R_{eff} - \omega|V_E| + \sqrt{r^2_{gw\,hollow}\; r^2_{gw\,riparian}}\right) \quad (3)$$

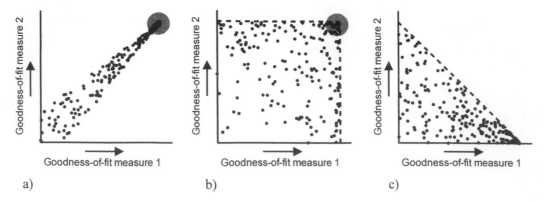

Figure 3. Three different types of relations between goodness-of-fit measures for the best realizations: a) a strong positive correlation, b) no correlation, and c) a negative correlation. Each dot represents one realization (or parameter set), the dashed line represents the pareto-optimality and the gray circle indicates the region in which the 'best' parameter sets are found.

Using the coefficient of determination, r^2, we did not force the model to *exactly fit* the observations, but allowed for an offset and a different amplitude. We argue that it is the dynamics, rather than the exact levels, that should be used from this kind of data where we compare the point observation from the field with a simulated average behavior of an entire zone (i.e., box within the model). By also utilizing soft data, there is no need to 'over fit' the model to the levels obtained from tensiometer observations at a few observation locations – in our case, one point in the hollow zone and another mid-way up the main valley bottom in the riparian zone (see McDonnell [1990] for field details).

Acceptability of the model simulations using soft data (A_2) was computed as the arithmetic mean of 15 evaluation rules of the soft data for groundwater levels and contribution of new water (Table 2). The arithmetic mean was used in this instance since the geometric mean is less suitable when values can become zero. Acceptability of the parameter values based on soft data (A_3) was computed as the geometric mean of nine evaluation rules of the different parameters (Table 2).

When plotting two different goodness-of-fit measures against each other for a number of realizations (parameter sets), the relations for the best realizations can be grouped into three basic cases: (1) a strong positive correlation, (2) no correlation, and (3) a negative correlation (Fig. 3). In case 1 the second criterion does not contribute with additional information and only one of the goodness-of-fit measures needs to be calculated. The situation is different for the case 2, where the both criteria provide different information. However, in both cases it is quite apparent from which region one would choose parameter sets to achieve optimal model performance, i.e., from a region where one can find realizations that are optimal for both criteria (see gray circle in Fig.3). In case 3 the two criteria also provide different information, but here the two criteria are

not unrelated and "conflict' one another. In other words, a good solution according to one criterion can only be obtained at the price of a poor performance according to the second criterion. It is therefore not possible to find a solution that is optimal according to the two criteria simultaneously, since the best values for the two criteria are negatively correlated. The best solutions are found along a pareto-optimality line (*i.e.*, 'compromise-solutions'). If the 'compromise-solutions' are too poor compared to the individual best solutions, this might indicate a problem with the model structure [*Seibert*, 2000]. As mentioned above, the selection of the weights n_1, n_2, and n_3 in Eq. 2 determines which solution along the pareto-optimality sub-space (lines in Fig. 3) will be found.

We tested different combinations to examine the relations between the different criteria. We quantified the value of the soft data by testing how the measures helped in ensuring internal model consistency and reducing parameter uncertainty. First we examined how model performance, as judged by the various criteria, varied when the model was calibrated considering different sets of criteria. Second, we compared the magnitude of parameter uncertainty when calibrating against runoff only and when calibrating against different combinations of criteria. For this part of the analysis we used values of 0.4, 0.4 and 0.2 for the weights in Eq. 2 n_1, n_2, and n_3 respectively to place more emphasize on the acceptability with regard to the simulations (both hard and soft data) and less weight on the acceptability of the parameter values.

RESULTS

Model Performance

The model was able to reproduce observed runoff very well. When calibrated with runoff data only, the model was able to simulate the observed runoff with values of 0.93 for the model efficiency [*Nash* and *Sutcliffe*, 1970]. Notwithstanding, while high model efficiency was obtained with the runoff-only (hard data) calibration, goodness-of-fit statistics for percent new water and soft groundwater measures for example, were very poor (Fig.4, shaded area). If one examines the simulated groundwater levels for each of the three boxes for the runoff-only calibration, several different response patterns are produced—each with a high model efficiency for runoff (Fig. 5a-c). In Fig. 5a, the riparian and hollow box fail to behave like observed reservoir dynamics reported in McDonnell [1990], with too much water remaining in the hollow box, especially between events. Fig. 5b is an example where each of the three boxes filled and drained too quickly during events. Fig. 5c shows an appropriate riparian box response but poor representation of the hollow zone where the zone is drained too quickly. This is a compelling example of how relying only on the traditional single-criterion, hard-data model calibration, can produce 'right answers for the wrong reasons'. It each case, without the insight of soft data, one may have been tempted to assume that the model 'worked well' given the high model efficiency for any of the very similar runoff simulations.

As additional hard and soft data were entered into the model calibration, the model efficiency for runoff decreased (from the 0.93 value to 0.84) but goodness-of-fit for the process description (*i.e.*, soft data on groundwater, percent-

new-water and parameter values) increased dramatically (Fig. 4 and 6). The combined objective function A (Eq. 2) increased from 0.46 to 0.79 when adding A_2 and A_3 to the optimization criterion. In general, the variability in the various goodness-of-fit measures decreased when more criteria were included into the calibration. Most importantly perhaps, the groundwater dynamics simulated with a parameter set obtained by this multi-criteria calibration are in keeping with experimental observations on reservoir response. The goodness-of-fit of the groundwater level simulations increased from 0.53 to 0.82 for the hard data and from 0.34 to 0.60 for the soft data, for parameter sets optimized using the combination of all criteria compared to the simulations using parameter sets calibrated to only runoff. Furthermore, the range of objective-function values generally decreased when a criterion was considered during calibration.

The simulation with the best overall performance caused a somewhat reduced model efficiency for runoff but displayed more 'realistic' internal dynamics (Fig. 6). Fig. 6 also shows the decrease of unsaturated storage through the event, indicative of the coupled formulation of saturated and unsaturated storage. We argue that this formulation is an important and new feature of the three-box approach because it is a more realistic conceptualization of the unsaturated-saturated storage interactions given the shallow groundwater. While application of the model to other catchments might involve different arrangements and numbers of boxes, the

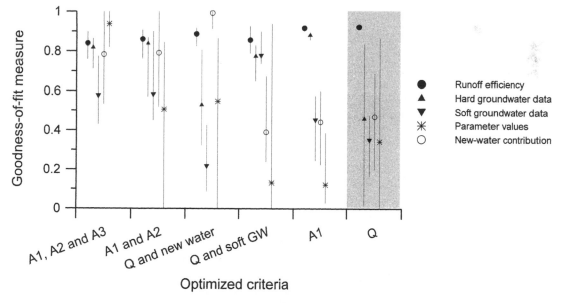

Figure 4. Goodness-of-fit measures for runoff, groundwater levels, new water ratios, soft groundwater measures, and parameter-value acceptability for calibrations against various combinations hard and soft information (see text for definition of the different optimization criteria). The symbol shows the median of 50 calibration trials and the vertical lines indicate the range of these trials. The shaded area relates to the traditional calibration approach using only runoff data and highlights the problem of internal consistency when calibrating against only runoff.

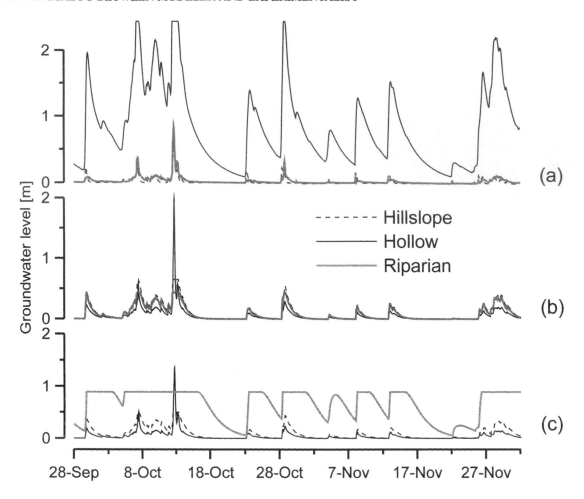

Figure 5. Three model runs with different parameter sets resulting in different groundwater dynamics (levels in [m] above bedrock). All three parameter sets had been calibrated to observed runoff and gave an almost similar goodness-of-fit (model efficiency ~0.93). None of the three sets of groundwater time series agrees with the perceptual model of the watershed.

coupled saturated-unsaturated zone formulation is one that is common to many headwater catchment conditions.

Relation between Optimization Criteria

Different parameter sets will be found through calibration if different weights (n_i) are used for the overall acceptability in Eq. 2. Using different combinations of A_1 and A_2 as well as A_1 and A_3 demonstrated that both soft-data criteria (A_2 and A_3) gave different information than the hard data (A_1) (Fig. 7). There is no conflict between the hard data and the soft data on parameter values (A_3) (Fig. 7b), *i.e.*, the calibrated solutions all follow the 'no-correlation'-pattern (compare Fig. 3b). On the other hand, there is a trade-off between the hard data and the soft data on model simulations (A_2) (Fig. 7a), *i.e.*, it is not possible to find a solution that is optimal according to both criteria simultaneously. The solutions form a curve that lies in between the 'nega-tive-correlation' and the 'no-correlation'-patterns (compare Fig. 3 b,c) indicating that there is some conflict between the criteria, but not total disagreement.

Parameter Uncertainty

For each parameter, 50 different values were obtained by the different calibration trials. The range between the 0.1 and 0.9 percentile divided by the median was computed for each parameter as measure of parameter uncertainty. The ratio between the values obtained from multi-criteria soft data calibrations and those derived from runoff-only hard data calibrations indicated a general reduction of parameter uncertainty (*i.e.*, the variation of calibrated parameter values decreased) when adding different criteria, but results varied from model parameter to model parameter. When optimizing the combination of all criteria (A_1, A_2 and A_3) the ratio varied between 0.03 and 0.65. The median was

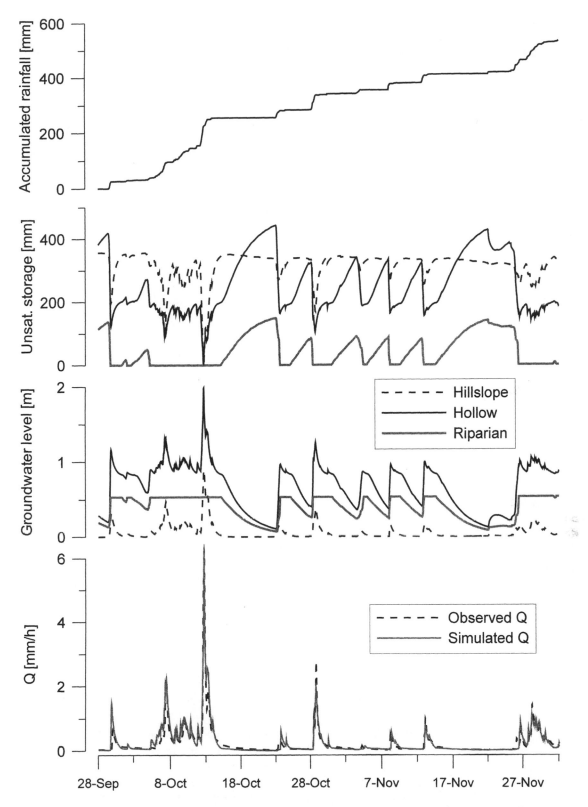

Figure 6. Simulation with best overall performance. Accumulated rainfall, simulated unsaturated storage and simulated groundwater levels (m above bedrock), as well as observed and simulated runoff. The model efficiency for runoff is 0.84 and the simulated groundwater dynamics agree in general with the perceptual model.

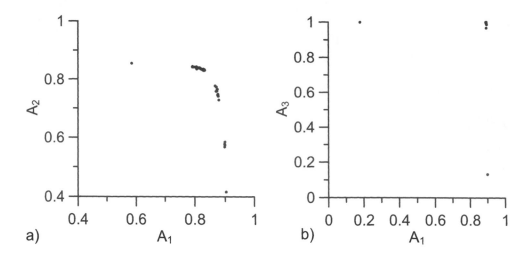

Figure 7. Relations between model performance according to a) A_1 and A_2 as well as b) A_1 and A_3. Each point represents the results with a parameter sets which was calibrated using different combinations of the two respective criteria as objective function (i.e., the combined acceptability measure with different weights n_i (using a value of zero for n_3 (a) and n_2 (b)).

0.4, implying that using all criteria helped to reduce parameter uncertainty on average by 60% relative to the single criterion calibration against only runoff. The reduction of parameter uncertainty was most obvious for the coefficients of the linear outflow equations, despite the fact that no 'desirable' parameter ranges were specified for these parameters. Including hard groundwater data or soft data for new-water contribution to peak runoff also reduced parameter uncertainty, but not as significantly as for the combination of all criteria.

DISCUSSION

Soft Data to Improve Model Performance

When a model is calibrated against different criteria, the overall 'best' parameter set often is a compromise between the different criteria. In other words, when the model is calibrated against several criteria, the value of an individual goodness-of-fit measure will be lower than when the model is calibrated against only this criterion. If this decrease in goodness-of-fit is large, then one might have to reject or reconsider the model structure. Seibert [2000] presents an example where the difficulty in simulating both runoff and groundwater levels with the same parameter set indicated a major problem in the model structure. With a modified model structure, it was less problematic to optimize the model against the two criteria.

In addition to the reduced parameter uncertainty, the multi-criteria calibration is assumed to provide parameter sets that are a more appropriate representation of the catchment, than a calibration against only runoff. Runoff will be simulated slightly worse during the calibration period, but the internal variables come into much better agreement with the conditions in the catchment. It seems reasonable that this improved internal consistency is associated with more reliable predictions outside the calibration domain. This assertion has to be tested in future studies using validation periods during which the hydrological conditions differ from those during calibration.

There exists a trade-off between model complexity and parameter uncertainty. It is difficult to test very parsimonious models with, for instance, only 3-6 parameters against data other than runoff, since measurable quantities have no clear counterparts in the model. In general, the testability of models increase with increasing model complexity. On the other hand, incorporation of additional variables used for calibration and validation often require extending the model, and the number of parameters may increase faster than the amount of additional information. Additional information may help improve the identifiability of parameter values, as demonstrated in this study, but if the only aim is to improve parameter identifiability, reducing the number of parameters might be a more efficient method. However, too parsimonious a model might be of limited usefulness if one intends to use the model for more than simply the simulation of runoff.

Relation between Optimization Criteria

The fact that the model performance decreases when the model is also calibrated against soft data shows that there is some conflict between the criteria. This was also indicated by the results of the calibrations with different weights (Fig. 7). This conflict might be caused by errors in the hard or soft data. More probably, however, it reflects the fact that the model structure is not perfect. Nevertheless, in our study the disagreement between hard and soft data was not tremendous and one might conclude that the model structure thus is an appropriate approximation. We are now implementing this approach in other well-studied experimental catchments to better understand these relations.

Types of Soft Data

The soft data measures used in this paper vary from static measures (*e.g.*, the spatial extent of the riparian zone) to data on groundwater level variations and highly integrated measures like the percent of new water at peakflow. The results of isotopic hydrograph separations have the advantage that the new-water contribution is an integrated measure of catchment response and offers much constraint on the preceptual model of runoff generation. Few studies to date have used isotope data in model calibration—despite the now common use of this in watershed analysis [*Kendall* and *McDonnell*, 1998]. Hooper *et al.* [1988] used continuous stream O-18 to calibrate the Birkenes model—another simple conceptual box model of runoff response. Similarly, Seibert *et al* [2001] have used continuous stream O-18 for model testing. In the present study, we use the new water *ratio* for discrete events rather than a continuous time series of O-18. Unlike higher latitude Scandinavia where previous attempts have been made, the Maimai catchment shows several periods of rainfall 'cross-over' with stream baseflow and ground water because of the lower amplitude of the seasonal O-18 variations (due primarily to lower annual temperature range)—making continuous time series modeling less valuable. Nevertheless, the new-water soft-data measure is an example of *making the most of data available* for a given situation. We advocate that in many catchment studies, additional (soft) data may be available that can, and should, be used to constrain model simulations. In snow-dominated environments, for instance, snow cover information may be used. In cases where the expansion and contraction of surface-saturated areas is important (and considered in the model), knowledge of the maximal portion of the catchment that might become saturated can be used. Franks *et al.* [1998] derived information on the extent of saturated areas at a certain time step from remote sensing and this information helped to constrain parameter values of TOPMODEL. In most cases measurements on the extent of saturated areas are not available, but hydrological reasoning and field experience might allow specifying a range of reasonable values (*e.g.* based on topography or vegetation

types). Mapped subsurface moisture distribution form one-the-ground remote sensing using non-invasive geophysical techniques may be another for useful soft data in the future. Sherlock and McDonnell [2002] showed that groundwater levels and soil water content could be mapped using electromagnetic induction at the hillslope scale, such techniques become applicable at the catchment scale, such pattern data may be a useful constraint on model parameters. At larger watershed scales, residence time of water in different boxes might be a useful soft data measure [*Uhlenbrook et al.* 2000].

CONCLUDING REMARKS

Today, obtaining some 'acceptable' fit between observed and simulated runoff is not such a difficult task, even in cases where the model structure is not necessary physically reasonable. Such models abound in the literature and in practice [Singh and Frevert, 2002]. By using one simple goodness-of-fit measure, such as the model efficiency for runoff, the calibration of a runoff model often becomes nothing more than a curve fitting exercise. Given the number of experimental watersheds around the world, the data and perceptual understanding of catchment hydrology gathered by experimentalists should be utilized much more in catchment modeling than it is done today. Given that additional data might allow for assessing internal model consistency, we advocate that this represents an important way forward towards more realistic conceptual models. We argue that the use of soft data may be a useful philosophy and approach in this regard, as an important complement to the use of traditional hard data measures, normally are used in model calibration. The concept of soft data together with a multi-criteria calibration, is a way to mimic hydrological reasoning (which exists implicitly in manual calibration approaches) in automatic calibration procedures. Obviously the exact numbers for the fuzzy evaluations (Eq. 1) and the weighing of the three components of the overall acceptibility that we describe (Eq. 2) are, to some degree, subjective decisions. However, these decisions, even if they are subjective, are more reasonable than ignoring all the qualitative process understanding that exists for most small research catchments. The soft-data framework might lead towards more reasonable model calibrations and more realistic model simulations. This dialog, that links the experimentalist and the modeler might, thus, be the needed catalyst for new progress in watershed modeling.

Acknowledgments. We thank the two reviewers for valuable comments and Jim Freer for compiling the data. This research was partly funded by the Swedish Research Council (grant 620-20001065/2001) and NSF grant EAR 9805475.

REFERENCES

Ambroise, B., Perrin, J.L, and Reutenauer, D.,. Multicriterion validations of a semidistributed conceptual model of the water cycle in the Fecht Catchment (Vosges, Massif, France), *Water Resources Research,* 31: 1467-1481, 1995.

Aronica, G., B. Hankin and K. Beven, Uncertainty and equafinity in calibrating distributed roughness coefficients in a flood propagation model with limited data. *Advances in Water Resources,* 22:349-365, 1998.

Beven, K., Prophecy, reality and uncertainty in distributed hydrological modeling, *Advances in Water Resources,* 16: 41-51, 1993.

Blazkova, S., K.J. Beven and A. Kulasova, On constraining TOPMODEL hydrograph simulations using partial saturated area information. *Hydrological Processes* 16: 441-458, 2002.

Boyle, DP, H.V. Gupta and S. Sorooshian, Towards improved calibration of hydrological models: combining the strengths of manual and automatic methods. *Water Resources Research,* 36: 3663-3674, 2000.

Bronstert, A., Capabilities and limitations of physically base hydrological modelling on the hillslope scale. *Hydrological Processes* 13: 21-48, 1999.

de Groisbois, E., Hooper, R.P., and Christophersen, N., 1988. A multisignal automatic calibration methodology for hydrochemical models: a case study of the Birkenes model, *Water Resources Research,* 24: 1299-1307

Duan, Q., Sorooshian, S., and Gupta, V.K., Effective and efficient global optimization for conceptual rainfall-runoff models, *Water Resources Research,* 28: 1015-1031, 1992.

Franchini, M., Using a genetic algorithm combined with a local search method for the automatic calibration of conceptual rainfall-runoff models. *Hydrological Sciences - Journal des Sciences Hydrologiques,* 41: 21-40, 1996.

Franks, S., Gineste, Ph., Beven, K.J. and Merot, Ph., On constraining the predictions of a distributed model: The incorporation of fuzzy estimates of saturated areas into the calibration process. *Water Resources Research* 34: 787-797, 1998.

Freer, J., Beven, K.J. and Ambroise, B., Bayesian estimation of uncertainty in runoff prediction and the value of data: An application of the GLUE approach. *Water Resources Research* 32: 2161-2173, 1996.

Hankin, B.G., and K.Beven, Modelling dispersion in complex open channel flows: Fuzzy calibration (2), *Stochastic Hydrology and Hydraulics,* 12: 397-412, 1998.

Holland, J., *Adaptation in natural and artificial systems.* Cambridge, MA: MIT Press, 211 pp. (First edition 1975, Ann Arbor: University of Michigan Press), 1975/1992.

Hooper, R.P., Stone, A., Christophersen, N., de Grosbois, E., and Seip, H.M., Assessing the Birkenes model of stream acidification using a multisignal calibration methodology. *Water Resources Research* 24: 1308-1316, 1988.

Hooper, R., B. Aulenbach, D. Burns, J.J. McDonnell, J. Freer, C. Kendall and K. Beven, Riparian control of streamwater chemistry: Implications for hydrochemical basin models. *International Association of Hydrological Sciences,* Publication 248: 451-458, 1998.

Hornberger, G.M. and Boyer, E.W., Recent advances in watershed modelling. *Reviews of Geophysics* supplement: 949-957, 1995.

Kendall, C. and J.J. McDonnell (eds.), Isotope tracers in catchment Hydrology, Elsevier Science Publishers, 816 pp., 1998.

Kirchner, J.W., Hooper, R.P., Kendall, C., Neal, C. and Leavesley, G., Testing and validating environmental models. *The Science of the Total Environment* 183:33-47, 1996.

Kuczera, G, Efficient subspace probabilistic parameter optimization for catchment models. *Water Resources Research,* 22: 177-185, 1997.

Kuczera, G. and M. Mroczkowski, Assessment of hydrological parameter uncertainty and the worth of multiresponse data. *Water Resources Research,* 34: 1481-1489, 1998.

Lindström, G., A simple automatic calibration routine for the HBV model. *Nordic Hydrology,* 28: 153-168, 1997.

McDonnell, J.J., The age, origin and pathway of subsurface stormflow. PhD Thesis, *University of Canterbury, Christchurch,* 270 pp., 1989.

McDonnell, J.J., A rationale for old water discharge through macropores in a steep, humid catchment. *Water Resources Research,* 26: 2821-2832, 1990.

McDonnell, J.J., M.K. Stewart and I.F. Owens, Effects of catchment-scale subsurface watershed mixing on stream isotopic response. *Water Resources Research,* 26: 3065-3073, 1991.

McDonnell, J.J., D. Brammer, C. Kendall, N. Hjerdt, L. Rowe, M. Stewart and R. Woods, Flow pathways on steep forested hillslopes: The tracer, tensiometer and trough approach, In Tani *et al.* (eds). *Environmental Forest Science,* Kluwer Academic Publishers, pp. 463-474, 1998.

McGlynn, B, J.J. McDonnell and D. Brammer, An evolving perceptual model of hillslope flow in a steep forested humid catchment: A review of the Maimai catchment. *Journal of Hydrology,* 257, 1-26, 2002.

McKie, D.A., A study of soil variability within the Blackball Hill Soils, Reefton, NewZealand. M.Ag.Sc. Thesis, University of Canterbury, 180 pp., 1978.

Mosley, M., Streamflow generation in a forested watershed, New Zealand. *Water Resources Research,* 15: 795-806, 1979.

Mroczkowski, M., Raper, G.P., and Kuczera, G., The quest for more powerful validation of conceptual catchment models. *Water Resources Research,* 33: 2325-2335, 1997.

Nash, J.E. and Sutcliffe, J.V., River flow forecasting through conceptual models, part 1 - a discussion of principles. *Journal of Hydrology* 10: 282-290, 1970.

O´Connell, P.E. and Todini, E., Modelling of rainfall, flow and mass transport in hydrological systems: an overview. *Journal of Hydrology* 175: 3-16, 1996.

Pearce, A., M Stewart and M. Sklash, Storm runoff generation in humid headwater catchments, 1: Where does the water come from? *Water Resources Research,* 22: 1263-1272, 1986.

Press, W.H., Flannery, B.P., Teukolsky, S.A., and Vetterling, W.T., *Numerical recipes in FORTRAN: The art of scientific computing.* 2nd edition, Cambridge University Press, Cambridge, Great Britain, 963 pp., 1992.

Refsgaard, J.C., Parameterisation, calibration and validation of distributed hydrological models. *Journal of Hydrology,* 198: 69-97, 1997.

Seibert, J., Estimation of parameter uncertainty in the HBV model. *Nordic Hydrology*, 28: 247-262, 1997.

Seibert, J., Conceptual runoff models - fiction or representation of reality? Acta Univ. Ups., *Comprehensive Summaries of Uppsala Dissertations from the Faculty of Science and Technology* 436. 52 pp. Uppsala. ISBN 91-554-4402-4, 1999.

Seibert, J., Multi-criteria calibration of a conceptual runoff model using a genetic algorithm. *Hydrology and Earth System Sciences*, 4: 215-224, 2000.

Seibert, J. and J. McDonnell, On the dialog between experimentalist and modeler in catchment hydrology: Use of soft data for multi-criteria model calibration. *Water Resources Research*, in press, 2002.

Seibert, J., A.Rodhe, K.Bishop, Simulating interactions between saturated and unsaturated storage in a conceptual runoff model, *Hydrological Processes*, in press, 2002.

Singh, V.P. (ed.), *Computer models of watershed hydrology*. Water Resources Publications, Highlands Ranch, Colorado, U.S.A., 1130 pp., 1995.

Singh, V.P. and D. Frevert (eds). Mathematical models of small watershed hydrology and applications. *Water Resources Publications*, Highland Ranch, Colorado, USA, *950p,* 2002.

Sherlock, M., N. A. Chappell, and J. J. McDonnell (2000). Effects of experimental uncertainty of hillslope flow paths. *Hydrological Processes,* 14, 2457-71.

Sherlock, M. and J.J. McDonnell, A new tool for hillslope hydrologists: Spatially distributed groundwater level and soil water content measured using electromagnetic induction. *Hydrological Processes*, in press 2002

Sorooshian, S. and Gupta, V.K., Model Calibration, in: Singh, V. (Ed.): *Computer models for watershed hydrology, Water Resources Publications*, Highland Ranch, Colorado, USA, *pp. 23-68,* 1995.

Uhlenbrook, S., Leibundgut, Ch., Maloszewski, P., Natural tracers for investigating residence times, runoff components and validation of a rainfall-runoff model Proceedings of the TraM'2000 Conference, Liège, Belgium, May 2000. IAHS Publ. No. 262, p.465-471, 2000.

Wang, Q.J., The genetic algorithm and its application to calibrating conceptual rainfall-runoff models, *Water Resources Research*, 27: 2467-2471, 1991.

Yapo, P.O., Gupta, V.K., and Sorooshian, S., Multi-objective global optimization for hydrologic models. *Journal of Hydrology*, 204: 83-97, 1998.

Jeffrey J. McDonnell, Oregon State University, Department of Forest Engineering, Corvallis OR 97331, U.S.A., e-mail: Jeffrey.McDonnell@orst.edu

Jan Seibert, Swedish University of Agricultural Sciences, Department of Environmental Assessment, Box 7050, S-750 07 Uppsala, Sweden, e-mail: jan.seibert@ma.slu.se

Effects of Model Complexity and Structure, Parameter Interactions and Data on Watershed Modeling

Thian Yew Gan and Getu Fana Biftu

Department of Civil and Environmental Engineering, University of Alberta, Edmonton, Alberta, Canada

On the basis of the calibration and validation results obtained from 37 sets of CRR model-data experiments formulated out of 5 CRR (conceptual rainfall-runoff) models and 5 catchments of wet, semi-wet and dry catchments, it is clear that parameters optimized automatically are data dependent. Global optimum parameters are impossible to obtain but conceptually sound parameters are possible if adequate calibration data is available, even for dry catchments based on standard CRR of 10 to 20 parameters. On a whole, more dependable results are expected from wet than dry catchments. Further, model performance depends more on the model structure and data quality than model complexity or data length. Parameters explicitly "coupled" to other parameters generally exhibit stronger interactions (which likely mean more identifiability problem), but calibration data could also cause some of the parameter interactions observed.

1. INTRODUCTION

Dry and mountainous catchments are generally more difficult to model than, say, temperate or wet catchments because their hydrologic processes are more complex and variable. Jakeman and Hornberger (1993) found that wet catchments could be well represented by a four-parameter, linear model of two components, which respectively represent a "slow" and a "fast" response modes. This simple approach is likely inappropriate for dry catchments with extended dry spells that often change drastically to wet seasons when torrential rain pours. Furthermore, the hydrologic processes of dry catchments should be distinct between high grounds where patchy and sparse vegetation dominate and valleys or areas covered with denser vegetation. The more abrupt and heterogeneous changes of dry catchments cause their hydrologic data to be generally noisy, unrepresentative or even erroneous.

Majority of the deterministic, lumped-parameter conceptual rainfall-runoff models (CRR) have been built for temperate or wet catchments where the hydrologic responses only involve a subset of the processes that occur in dry catchments. CRR models conceptualize hydrologic processes in inter-related conceptual storages defined by model parameters. Rates of recharge to or withdrawal of water from these storages are a function of exponents, state variables, storage capacities and water balance principles, rather than a combination of energy and water balances. Even then, because the rainfall-runoff transformation process is highly complex, CRR models that incorporate major hydrologic processes often have more than half a dozen of parameters (see Tables 2, 5, 9), which could result in an over-parameterization problem. This is partly because of the limited information contained in most of our standard hydrologic data available for calibrating (manual or automatic or both) non-measurable model parameters that govern the approximate mathematical functions for the soil moisture accounting phase of basin hydrology.

As long as the conceptual base of a model 'captures' the essential hydrologic processes of a catchment, past studies show that complex models do not necessarily perform better than simpler models (e.g., Loague and Freeze, 1985). With an ever increasing computing power, and the availability of powerful optimization algorithms, such as the SCE-UA (Duan et al., 1992), the danger of over-parameterization may be growing ever bigger.

What should be the recommended level of model complexity for dry catchments, given that dry catchments under-

Calibration of Watershed Models
Water Science and Application Volume 6
Copyright 2003 by the American Geophysical Union
10/1029/006WS23

go a wider range and more complex hydrologic process-es than temperate or wet catchments, yet usually have only worse data sets for model calibrations? Will the majority of traditional CRR models built for temperate catchments suffer from over-parameterization if applied to dry catchments? Will CRR models suffer from more identifiability problems and parameter interactions when applied to dry catchments? What impact will the choice of calibration data have on calibrating CRR models to dry catchments, given that the standard rainfall-runoff data are point measurements which contain errors and only the input-output, instead of the rainfall-runoff trans-formation information? Research objectives are outlined in Section 2, CRR models in Section 3, research proce-dures in Section 4, discussions of results in Section 5, and conclusions in Section 6.

2. RESEARCH OBJECTIVES

The primary objectives are to study the effects of model complexity and structure of five major CRR models, the wetness and climatic conditions of catchments selected from different continents, data length and variability on conceptual hydrologic modeling. The influence of param-eter identifiability and parameter interactions on the opti-mized parameters obtained via automatic calibration is also addressed. Input data to the five CRR models are rain-fall and potential evapotranspiration (ET). The basis of model calibration is the streamflow at the basin outlet (Table 1). The credibility of the optimized parameters is validated using data sets independent of the calibration experience.

3. CONCEPTUAL RAINFALL-RUNOFF MODELS

The five deterministic, lumped-parameter, conceptual rainfall-runoff (CRR) models used were: (i) the soil mois-ture accounting and routing model (SMAR) model of Ireland, (ii) the Xinanjiang model (XNJ) of China, (iii) the Nedbor-Afstromnings Model (NAM) model of Denmark, (iv) the Sacramento model (SMA) of the US, and (v) the Pitman (PTM) of South Africa. SMAR was developed in the University of Galway (O' Connell et al., 1970), while XNJ was developed in 1973 by the East China College of Hydraulic Engineering (now the Hohai University) to fore-cast floods in large humid basins where the infiltration rate is high, making surface runoff small and interflow or sub-surface flow high (Zhao, 1992). NAM, developed at the Technical University of Denmark (Danish Hydraulic Institute, 1982), operates by continuously accounting for moisture content in five mutually interrelated storages. The

Sacramento model (SMA) is the US National Weather Service model for operational river forecast (Burnash et al., 1973). The Pitman model (Pitman, 1976) is widely used throughout southern Africa.

These five deterministic CRR models have similar model structure but they differ in model complexity and the functions used to represent various hydrological sub-processes (see Table 1). SMAR is the simplest (9 parame-ters), SMA is the most complicated (21 parameters), while PTM (16 parameters), XNJ (15 parameters) and NAM (15 parameters) are comparable to each other. The actual num-ber of parameters calibrated for each model is shown in Table 1.

3.1. Calibration of CRR Models

Model calibration is a major aspect of hydrologic modeling. We could obtain erroneous results out of conceptually realis-tic models if they are not properly calibrated. Models are either calibrated manually, automatically or a combination of manual and automatic procedures. Manual calibration includes assigning parameter values through past experience and field data, and guidelines given in the literature. Model parameters obtained from an automatic procedure will depend on six elements: (i) optimization algorithm, (ii) objective function, (iii) calibration data, (iv) model structure and complexity, (v) parameter identifiability, and (vi) param-eter interactions. The paper's focus is on the third to the sixth elements.

The Shuffle Complex Evolution Method (SCE-UA) of Duan et al. (1992) is a global, probabilistic optimization method that evaluates the objective function at randomly spaced points in a feasible parameter space. SCE-UA was chosen for this study because it represents a synthesis of the best features of several methods. It combines the strength of Simplex (Nelder and Mead, 1965) and the concept of a con-trolled random search, a systematic evolution of points in the direction of global improvement, competitive evolution (Holland, 1975), and complex shuffling. Duan et al. (1992) compared the independent global search in the feasible space without sharing information as giving a number of competent people a difficult problem to solve without conferring with each other. The idea is for people to first work independently (individually or in small groups called complexes), and later gets together now and then to share information about their progress (shuffling). For the limited point measurements we have for these test catchments, we do not see any advantage in using the multi-objective extension of SCE-UA, the MOCOM-UA based on the Pareto solution space (Yapo et al., 1997). SCE-UA is likely the most advanced optimization algorithm we can justifiably use.

Table 1. Main Model Features of SMAR, Xinanjiang (XNJ), NAM, Sacramento (SMA), and the Pitman (PTM) models

Classification	SM AR	XNJ	NAM	SMA	PTM
Origin	Ireland(Europe)	China(Asia)	Denmark(Europe)	USA(North America)	South Africa
Input	Rainfall Pot. Evap.	Rainfall Pot. Evap.	Rainfall Pot. Evap.	Rainfall Pot. Evap	Rainfall Pot/Pan Evap
Output	DOF SOF ITF GWF	DOF SOF ITF GWF	DOF SOF ITF GWF	DOF SOF ITF GWF	DOF SOF Horton OF GWF
Time steps	Daily	Daily	Daily	1 hr, 6 hr, daily	Daily, monthly
No parameter/ No parameters optimized	9/6	15/15	15/13	21/13	16/9
No. of soil moisture zone	5 Maximum	3	2	2	1
Conceptual storage types	Surface storage Maximum 5 soil moisture Groundwater	Upper, lower & deep tension zones Free water	Snow storage (optional) Surface water Lower zone soil moisture Upper & lower groundwater	Upper zone tension & free water Lower zone tension, primary & secondary free water	Interception Soil moisture Groundwater
Actual Evaporation	Potential rate from upper storage & a fraction of potential rate from lower storage	Potential rate from upper storage & a fraction of potential rate from lower storage	Potential rate multiplied by the relative water content in the lower zone	PET x linear ratio of state vs. capacity of tension storages & upper zone free water storage	Linear function of soil moisture
Sub-basin routing	Nash's method of cascade of reservoirs	Unit hydrograph for surface flow & linear reservoir for baseflow	Linear reservoirs for overland flow, interflow, upper and lower baseflow	Unit hydrograph for all flows	Simple lag functions for groundwater and surface flow
Channel Routing	Linear reservoir	Muskingum or Nash Method	Linear reservoir	Muskingum and variable lag methods	Muskingum method for surface flow

The objective function used for the generation of the response surface was a simple daily root mean square (DRMS) objective function ($= 1/n\sqrt{SLS}$, where n is the number of data, and SLS = simple least square) that assumes the presence of Gaussian, independent homogeneous variance error. Unlike the objective function based on the maximum likelihood for the heteroscedastic error (HMLE), this approach does not involve data transformation and so it places more weight on high flows than on low flows. Using DRMS may not be desirable when most of the data comprise of low flows with a few large events because the latter tend to exert excessive influence on the calibration, leading to unrepresentative parameters estimated. The search of SCE-UA was either stopped after 20 shuffling iterations, or if the change in objective function and that in parameter values were both less than 0.0001, or if the number of iterations was greater than 50,000.

4. RESEARCH PROCEDURE

The research procedure adopted to fulfill the objectives outlined in Section 2 is given below.

4.1 Test Catchments

As one of the primary emphases of this study is to compare CRR's performance between dry and wet catchments, five catchments of different climatic conditions are chosen, representing one wet (Sunkosi 2), one semi-wet (Shiquan 2) and three dry catchments (Ihimbu, Bird Creek and Great Usuthu). In terms of rainfall, the mean rainfall of these catchments ranges from about 2.5 to 5 mm/day (Table 2). However, the streamflow/rainfall ratios of the dry catchments are relatively low (about 0.2) compared to that of Sunkosi2 of 0.86 and

Shiquan2 of 0.42. Figure 1 that shows the daily averages of rainfall, discharge and pan evaporation also reveals significant differences in climatic and hydrologic conditions among these catchments. Other than a wide range in climate, these catchments also represent a wide range in size, from about 14,000 km² (Shiquan2) to 2,300 km² (Bird Creek).

4.2. Effects of Data Variability and Length on Model Calibration

Generally for model calibration it is recommended to use 3 to 5 years of data that include average, wet and dry years so that the data encompass a sufficient range of hydrologic

events to activate all the model parameters during calibration. Other than the Great Usuthu catchment, all catchments were calibrated with 5 to 6 years of data and validated with 2-year (Sunkosi2, Shiquan2 and Bird Creek) or 3-year (Ihimbu) of data. Since the Great Usuthu catchment has 21 years of data, the effects of data variability and length on model calibration are separately investigated (Section 5.2).

4.3. Effects of Parameter Interactions on Model Calibration (Xinanjiang Model)

To study the effects of parameter interactions on the automatic calibration of CRR models, the strategy adopted is to

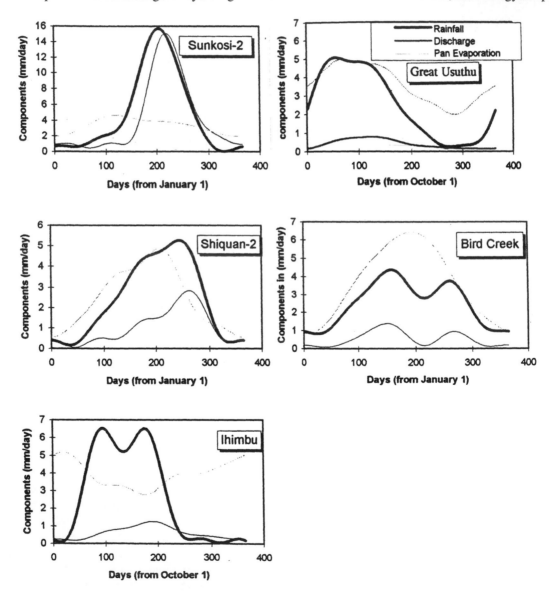

Figure 1. Plots showing daily averages of multi-year rainfall, discharge, and the pan evaporation in mm/day for the five test catchments.

Table 2. General Information About 5 Test Catchments.

Catchment - Symbol	Catchment Area (km²)	Data Length (days)	Mean Rainfall (mm/day)	Mean Evaporation (mm/day)	Mean discharge (mm/day)	Country	Wetness (Discharge/ Rainfall)
Sunkosi 2 -Sun2	10,336	2922	4.91	3.35	4.21	Nepal	Wet (0.86)
Shiquan 2 -Shi2	14,192	2922	2.43	2.50	1.03	China	Med (0.42)
Ihimbu -Ihim	2,480	3287	2.66	4.06	0.61	Tanzania	Dry (0.23)
Bird Creek -BC	2,344	2922	2.66	3.58	0.61	USA	Dry (0.23)
Great Usuthu-GU	2682	7670	2.53	3.5	0.41	Swaziland	Dry (0.16)

first "contaminate" the streamflow data of a river basin with white noise of mean zero and variance equal to the sample variance of observed streamflow data by Monte Carlo simulation. Over 100 sequences of such "contaminated" streamflow data were simulated, and each data set differs from others only slightly. For each "contaminated" streamflow data set, we obtained the optimized parameters of XNJ by SCE-UA. We then computed the correlation matrix of over 100 sets of optimized parameters for each test case to assess the parameter interaction problem. From the mean/median and standard deviation of each parameter, we also study the range of optimized parameter values obtained from the "contaminated" streamflow.

5. DISCUSSIONS OF RESULTS

Three statistical indices selected to compare the performance of the five models applied to the five catchments are the root mean square error (RMSE), bias (BIAS) and the coefficient of Efficiency (E_f) (Nash-Sutcliffe, 1970). With each case representing one model-catchment combination, 37 test cases were conducted. The discussion of results is mainly given in terms of E_f instead of all three statistics, partly because the three statistics gave similar results.

5.1. Comparisons of Model Performance

At the calibration stage, the difference in the performance between the 5 models for wet, semi-wet and dry catchments are mostly marginal, except for the Great Usuthu where the effect of data variability and length were tested (Figure 2). Since PTM was applied only to Bird Creek and the Great Usuthu, its overall performance is harder to assess. To realistically assess a calibration, validation runs were based on data sets independent of those used during calibration. As expected, model performance at the validation stage is generally lower than at the calibration stage. The drop in model performance at the validation stage is relatively modest for the wet and semi-wet catchments than for the 3 dry catchments, especially for Ihimbu whose data is deemed unreliable since all 5 models performed poorly at the validation stage.

On a whole, XNJ seems to be slightly but consistently more versatile than other models in handling a wide range of catchment conditions. Among the eight test cases (Figure 2) the only poor performance with XNJ was with the validation stage for Ihimbu ($E_f \approx 45\%$). It is believed that XNJ did better than other models likely because it is the only model that considers the uneven distribution of runoff producing area to simulate the runoff. In addition to Ihimbu, SMA also performed poorly for Bird Creek at the validation stage ($E_f \approx 30\%$). SMAR did badly for the Ihimbu catchment ($E_f \approx 38\%$) but did reasonably well with Bird Creek. NAM did not model well Ihimbu and Bird Creek ($E_f \approx 41\%$ and 56%) at the validation stage.

By comparing the performance of the five CRR under wet versus dry catchments, we could offer some possible explanations to the results obtained in terms of catchment conditions, model structure and complexity, parameter identifiability, and parameter interactions.

As the simplest model, SMAR (9 parameters) has a more restricted model structure. For example, even though it can operate up to five soil moisture zones, the capacity for each soil layer is set at 25 mm except for the lowest layer (lower zone groundwater). Further, it is only when all the soil layer zones are saturated will there be any runoff from the soil layers. This runoff, divided into surface runoff and groundwater by only one parameter G, has a ratio of surface runoff to groundwater that is more or less fixed by G. This relatively simple approach may work well with wet catchments but it is probably too restrictive for dry catchments, which experience a wide range of flow scenarios. However, its performance is comparable and sometimes better than that of NAM, PTM and SMA.

Surprisingly XNJ has been doing marginally better than other models even though XNJ was built for humid and semi-humid regions with rich vegetation, well-developed soil zone, low surface runoff and high interflow in China. This is likely because while other models assume a uniform distribution in soil moisture in the catchment, XNJ considers a nonuniform spatial distribution of soil moisture deficit and tension storage capacity in the catchment. By so doing, the runoff producing area is also simulated in terms of a *nonuniform* distribution. Further, the total runoff is separat-

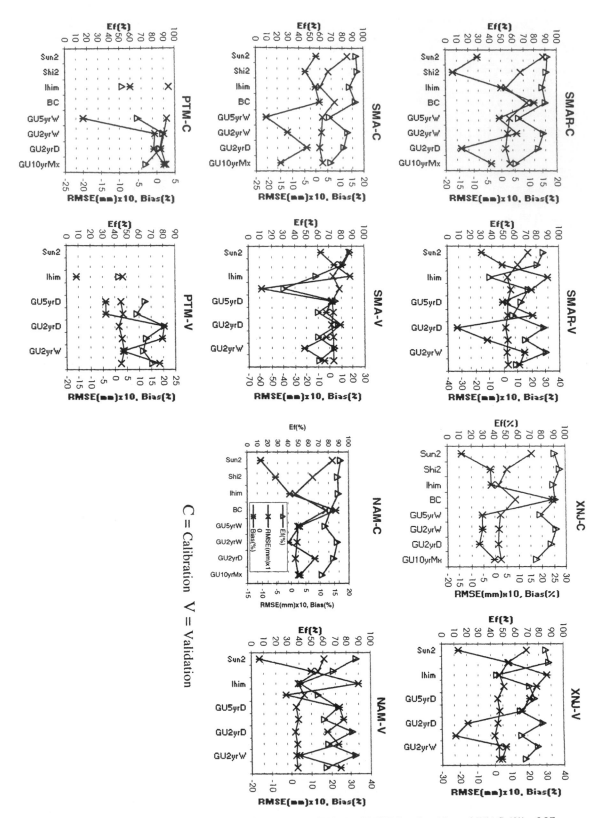

Figure 2. Calibration (C) and validation (V) results in terms of Ef(%), RMSE (mm) x 10, and BIAS (%) of 37 CRR/data test cases.

Table 3. Test Strategies Based on Calibration and Validation Data Types and Length for the Great Usuthu Catchment (Section 5.2).

Test No.	Data Length & Type	Test Period C=Calibration V=Validation	Plot Symbol of Figure 2
I	5- yr wet	1953-1958 (C)	GU5yrW
	5- yr dry	1963-1968 (V)	GU5yrD
	11-yr mixed	1963-1974 (V)	GU11yrMx
II	2- yr wet	1954-1956 (C)	GU2yrW
	2- yr dry	1963-1965 (C)	GU2yrD
	11-yr mixed	1963-1974 (V)	GU11yrMx
III	2- yr dry	1963-1965 (C)	GU2yrD
	2- yr wet	1954-1956 (V)	GU2yrW
IV	10-yr mixed	1953-1963 (C)	GU10yrMx
	11-yr mixed	1963-1974 (V)	GU11yrMx

ed into three components based on free water storage distributed in a parabolic manner.

Since the runoff producing areas of dry catchments are probably more unevenly distributed than humid catchments, and as the only model that takes this into consideration, XNJ generally did better than other models in all five catchments, particularly at the calibration stage. At the validation stage, it still did well but at times marginally inferior to other models. This could be partly attributed to XNJ assuming the soil moisture deficit and the storage capacity to be uniformly distributed in simulating the evaporation, which is inconsistent with the assumption used in simulating runoff.

The performance of NAM is better than SMA and PTM but slightly behind that of XNJ and SMAR. Despite of not considering tension and free water like SMA and XNJ, it accounts for soil moisture through 5 mutually inter-related storages. The snow storage was not used since there was no snowfall in the data. It did better than SMA partly because it uses separate linear reservoirs for overland flow, interflow and baseflow, instead of a unitgraph like SMA. It did not do as well as XNJ probably because it did not account for spatial variability in soil moisture and in its conceptual storages, even though it also has 15 parameters as XNJ.

Even though SMA is a relatively complex model (21 parameters), its performance was not satisfactory, and sometimes even poorer than the simplest, SMAR model partly because of to its non-standard unit hydrograph (unitgraph) used to convert the computed channel inflow into the catchment outflow. This unitgraph more or less matches the traditional unit hydrograph if surface runoff is the dominant runoff. The difference between the SMA unitgraph and the traditional unit hydrograph increases as the flow regime becomes more and more dominated by mixed flow instead of surface flow. Since the dry catchments have a short wet season and a long dry season, it is unlikely that one set of unit hydrograph ordinates will work well.

SMA's unitgraph ordinates were first derived through model calibration. Because of parameter interactions and the presence of two distinct seasons, it was not possible to get a realistic set of unitgraph ordinates from this approach. In a separate attempt, streamflow data (dependent variable) were regressed against rainfall data (independent variable). The performance of SMA based on the unitgraph ordinates derived from this approach turned out to be worse than before. Perhaps as a more plausible approach, two sets of unitgraphs, one for routing low flows and one for routing high flows, will improve SMA's performance. We believe SMA's performance on Sunkosi2 and Shiquan2 are comparable to other models partly because the unitgraph approximated the traditional unit hydrograph reasonably well in these two cases. Because complex models may have one or more components that sometimes do not function properly, such as SMA's unitgraph, complex models may perform poorer than simpler models (e.g., Loague and Freeze, 1985).

PTM that was initially developed for southern African catchments has three unique features compared to other four models. First, other than simulating direct and saturation overland flows, PTM is also designed to simulate Horton overland flow not considered in other models. Even then, PTM still has problems simulating high flows partly because data of time steps finer than daily (used in this study) to adequately reflect major storms undergoing significant changes within hours was not available.

Second, PTM breaks down daily rainfall depths to hourly increments according to a regression of the form, Duration (hours) = $\alpha + \beta$(rainfall) (mm), where α and β are regression parameters. Pitman derived α and β parameters for Pretoria, South Africa. This approach has two potential pitfalls. The relationship between storm duration and rainfall depths for a geographical location may not necessarily be linear and hourly rainfall data are needed to derive α and β for that location. Since no actual hourly rainfall data was available for the Great Usuthu or Ihimbu, the α and β values derived for Pretoria were also used for the Great Usuthu catchment. For Ihimbu, it was necessary to derive new α and β values through model calibration.

Third, PTM uses a variable recession constant to compute the baseflow while other models use fixed recession constants. Given the prolonged dry spell of dry catchments, the former is probably more realistic than the latter. Even with this additional feature, PTM could not do better than XNJ, or NAM in the low flows, perhaps partly the approach that PTM used to convert the daily to hourly rainfall was not applied properly since there was no hourly rainfall data to accurately determine α and β.

Since complex models can either do better or worse than a simple model, it seems that model complexity is less crucial than the model structure in modeling dry and wet catchments. On a whole, other than a few inappropriate features identified in this study, traditional CRR models with 9 to 15 parameters are generally applicable to dry and wet catchments if good calibration data are available, even though results from dry catchments are less predictable and are not expected to be as good as for temperate or wet catchments. Further tests on the dry Great Usuthu catchment were carried out.

5.2. Effects of Data Variability and Length based on the Great Usuthu

For the Great Usuthu, the five models were calibrated with different numbers of wet years and validated with dry years and vice versa, or calibrated and validated with mixed year data based on four sets of tests. In Test I all CRR were calibrated with five relatively wet years (1953 to 1958) and the calibrated models verified with five relatively dry years (1963 to 1968). Test II is the same as Test I except that two years of data were used instead of five. Test III, which was used to investigate the effect of dry versus wet calibration data, was a repetition of Test II except that the calibration was carried out using two relatively dry years of data (1963 to 1965). In Tests I and II, models calibrated from five and two years of data were also validated with 11 years of mixed (wet, dry and average) data (1963 - 1974). Finally, in Test IV, models calibrated with 10 years of mixed data (1953 - 1963) were validated with 11 years of data (1963 - 1974). These test strategies are summarized in Table 3.

For the calibration run in Test I, the E_f for PTM, XNJ, NAM, SMAR and SMA are about 66%, 79%, 76%, 66% and 69% respectively, while the Ef at the validation stage is 65%, 77%, 73%, 66% and 75% respectively. These E_f values show that the calibrations achieved are mainly moderate. For the calibration and validation runs in Test II, the results are generally better than that of Test I, but comparable to that in Test IV. Overall, the results obtained for Test III are also better than that for Tests I and IV. Tests II and III results are better than that of Tests I and IV partly because of shorter calibration and validation data sets used. Figure 2 shows that moving from the calibration to the validation runs, the drop in model performance (in terms of E_f) ranges from less than 1% to over 25%. Given that the Ef obtained can vary over a wide range for the Great Usuthu (and also for other two dry catchments), the success of a model calibration likely depends heavily on the calibration data and the model used. Apparently dry catchments are harder to calibrate than wet or temperate catchments (Gan and Biftu, 1996) and the outcome of calibrating dry catchments can be unpredictable.

In many calibration runs, XNJ generally performed slightly better than the other four models, especially at the calibration stage, even though XNJ was developed for humid catchments of China. NAM also did reasonably well, while SMA did badly in both calibration and validation stages, while SMAR and PTM are in-between.

Logically, we would expect a longer set of calibration data to achieve a better calibration because by going through a longer calibration experience, model parameters should be more accurately calibrated. This philosophy is generally not true, as shown by the results of Tests I, II and IV on the Great Usuthu, which used calibration data lengths of 2, 5, and 10 years respectively, and which were tested with a common, 11-year validation run (1963-74). For example, of the 2 and 5 wet years and a 10 mixed-year calibration cases, the PTM model produced an Ef of about 66%, 58%, and 71% for the 11 year validation runs respectively (Figure 2). The SMAR had Ef of about 69%, 58% and 61%, while XNJ's E_f was about 66, 64, and 69%. Therefore there is no obvious indication that model performance is related to the calibration data length. In some instances, models calibrated with two years of data could out-perform models calibrated with ten years of data. It seems that the data length is not that crucial, as long as it is not less than one hydrological year, and as long as the data used contain enough information for calibrating the parameters.

5.3 Effects of Model Structures and Data on Optimized Model Parameters based on SMA and PTM

As a mean to study the effects of different calibrating data and model structure on automatic CRR calibration, we used SCE-UA to derive the optimized model parameters of SMA (Table 4) for four different catchments, while that of PTM was based on only one catchment (Great Usuthu) but of different calibration data length and wetness (Table 5).

Among the four sets of optimized parameters obtained for SMA, it seems that upper zone parameters (UZTWM, UZFWM, and UZK) are more consistent than lower zone parameters (LZTWM, LZFSM, LZFPM, LZPK and LZSK) that tend to vary more widely. Apparently model structure exerts a larger influence than data on the model parameters optimized by SCE-UA. Further, for SMA, upper zone parameter are generally set to be of much smaller values than lower zone parameters, irrespective of what catchments (location) and the climatic conditions (dry or wet) we deal with.

Among SMA parameters, upper zone parameters are also more sensitive to the calibration data because their values are influenced largely by attempting to match recorded and simulated peak flows that can vary substantially over a short period of time. Also, with fewer parameters than the lower zone counterparts (3 versus 6), the formers are probably subjected to less parameter interactions, have less identifiability problems, and so their "optimum" parameters are either easier to detect by SCE-UA or more dependable.

Lower zone parameters rely mainly on the recession limbs of runoff hydrographs and most of these limbs decay slowly in an exponential manner. It may be surprising to obtain much higher lower zone free water values for Ihimbu (212.9 and 651.6 mm), a dry catchment, than Shiquan 2 (27.73 and 55.07mm), a semi-wet catchment, or another dry catchment, Bird Creek (60.89 and 52.10mm). However, being less sensitive parameters, driven primarily by low flows which do not change much for long periods of time, and with possibly strong parameter interactions, we could get lower zone parameters that vary a wide range, even though some such values may not adequately reflect the physical soil moisture capacities of the three catchments. Another example of an insensitive parameter is PFREE, for Sunkosi2, Shiquan2 and Ihimbu have almost identical PFREE. In other words, for SMA, it seems less sensitive parameters also suffer from more parameter interactions, and more parameter identifiability problem. Such problems are more of a result of model structure than climatic conditions, even though dry catchments are more prone to getting erroneous final parameters, such as Ihimbu.

Herein, to study the influence of data length and data content on CRR calibration, the four optimized parameter sets of PTM were derived by SCE-UA for the Great Usuthu

catchment based on four combinations of calibration data (Test I, II, III, and IV in Table 3). Except for ZMINN, which in a few instances took on the preset lower limits of 0.0, the four parameter sets are clearly different from each other, and the degree of differences vary from parameter to parameter (Table 5). Since different calibration data for the same catchment lead to different optimized parameters, it shows that optimized parameters and the calibration results (E_f, RMSE and BIAS) are data dependent. The E_f ranges from 63 to about 90%, which shows that some calibrated parameters are probably more realistic than others but most likely none of the parameter sets is of the global-optimum quality.

Ideally, if model parameters estimated are unique and realistic, the estimated parameters should be independent of the calibration data. In other words, if another calibration data set is used, the parameters estimated by the optimization method should be more or less the same (within the numerical accuracy and round-off errors). The concept of uniqueness used here is analogous to what Sorooshian and Gupta (1985) referred to as, "a model structure M parameterized by θ is *globally identifiable*, if and only if different parameter values of M give rise to different model output (streamflow)." In practice, because calibration involves adjusting the parameters until the difference between the simulated and observed streamflows is minimized, the final parameters are inevitably related to the calibration data. This data dependency feature is further complicated by the presence of insensitive parameters. If model parameters are insensitive or poorly identifiable, then different sets of parameters could essentially produce the same model output (Gan and Biftu, 1996). Sorooshian and Gupta (1985) attributed the parameter identifiability problems to model structure, which we have demonstrated through SMA (Section 5.3.1).

Table 4. Comparisons of optimized parameter values obtained by SCE-UA for SMA.

Parameters[*]	Sunkosi2	Shiquan2	Ihimbu	Bird Creek
UZTWM	6.267	19.25	5.074	37.84
UZFWM	29.02	14.68	12.60	13.46
UZK	0.659	0.424	0.608	0.973
ZPERC	10.20	31.43	42.68	54.64
REXP	1.010	1.233	2.038	2.93
LZTWM	54.22	153.1	153.5	190.5
LZFSM	150.4	27.73	212.9	60.89
LZFPM	310.5	55.07	651.6	52.10
LZSK	0.077	0.095	0.10	0.055
LZPK	0.097	0.006	0.007	0.014
PFREE	0.595	0.568	0.595	0.107
PXADJ	1.20	0.853	0.640	0.923
PEADJ	0.463	0.415	0.409	0.638

[*]LZ and UZ = Lower and Upper Zones; TWM and FWM = Tension and Free Water Maximum; UZK, LZSK and LZPK = Upper, Lower Secondary and Primary Zones runoff depletion coefficients; ZPERC and REXP = Maximum and shape of percolation curve; PFREE = % of free water that follows paths through cracks and faults; PXADJ and PEADJ = Precipitation and potential evaporation adjustment factors.

Table 5. Comparisons of optimized PTM parameters derived by SCE-UA for the Great Usuthu catchment using 4 sets of calibration data (Tests I, II, III and IV of Table 3).

Par	I	II	III	IV
ST	396.6	354.3	162.1	354.3
FT	0.690	1.000	0.460	0.934
ZMIN	0.788	0.055	0.00	0.00
XMAXN	12.46	13.66	13.30	13.01
TL	8.269	3.074	4.169	4.957
LAG	1.634	1.279	1.169	0.644
GL	0.056	0.285	17.830	0.039
DIV	0.842	0.444	0.997	0.561
OBSQ	2.500	3.116	1.341	0.504
E_f(Cal)	65.8	89.1	86.5	73.5
E_f(Val)	64.6/58.4[#]	81.1/ 66.2	63.1	70.8

[#] For Test I, 64.6% corresponds to the 5-year dry while 58.4% corresponds to the 11-year mixed data sets.

Comparisons of Tests II and III results (Figure 2) indicate that to model the Great Usuthu catchment, wet years are preferred over dry years as calibration data because dry years may not contain enough high flows to sufficiently activate model parameters responsible for simulating high flows during calibration. Wet years are more likely to contain both high and low flows and so they provide more ample information for calibration. However, when a model calibrated with wet years was validated with dry years, the chances of getting an over-estimated dry year flows tend to be higher than an under estimation. Test II shows that moving from calibration (2 wet years) to validation (2-dry years) runs, there is an increase in BIAS from -0.55 % to 20.4 %. In Test III, moving from calibration (2-dry years) to validation (2-wet years) runs, BIAS only increased from −0.79 to 3.31%. Since using wet years as calibration data tend to produce parameters that over-estimate streamflows at the validation stage or vice versa, this again shows the dependency of parameters on the data used for calibration.

5.4. Effects of Parameter Interactions – Xinanjiang Model (XNJ)

Besides data, the automatic calibration of parameters also depends on the model structure, that almost always suffers from parameter interactions and parameter identifiability problems, irrespective of what CRR we use. Some preliminary results on parameter interactions (indicated by cross-correlation, ρ) obtained for XNJ applied to three catchments of different degree of wetness are shown in Tables 6 to 9. The relatively small standard deviations found among the 100 sets of optimal parameters obtained for each test case (Table 9) shows that SCE-UA could still identify the optimal parameters fairly consistently despite of the variability introduced in the contaminated calibration data described in Section 4.3. However, the differences between each set of optimized parameters are substantial enough that the ρ between various parameters are generally small, except parameters suffering from interaction problems.

The results provide some indications on the degree of parameter interactions resulted mainly from the model structure, e.g., parameters explicitly inter-related or coupled in the model, and partly from the calibration data or some unknown reasons. For example, the upper and lower layer tension zones of XNJ, WUM and WLM expressed in terms of X [WUM/WM] and Y [WLM/((1-X)WM)] generally exhibit higher degree of correlation with each other partly because they are inter-related by WM = WUM + WLM + WDM, where WDM is the deep layer tension zone. There is considerably less correlation between these parameters, and

CI, CG, IM, C, KI, KG etc. because the latter represent different components of basin hydrology.

The next obvious factor affecting the degree of parameter interaction is the watershed's hydroclimatic condition embedded in the calibration data. It seems that wet or semi-wet catchments suffer from less parameter interactions than the dry catchment, Bird Creek. From a total of 105 parameter pairs examined in each case, the number of ρ exceeding 0.4 is 16 for Sunkosi2, 12 for Shiquan2 and 31 for the Bird Creek. Among all correlation $\rho > 0.4$, only 4 parameter pairs are common among all three catchments, which are WM-K, X-Y, Ex-Imp and KI-KG. From the perspective of XNJ's model structure, it is not surprising to find these 4 pairs of parameters showing consistent interactions in all three cases. For example, K (ratio of potential ET to pan evaporation) should be related to the areal mean tension water capacity (WM), tension water in the upper (WUM) and lower (WLM) layers should be coupled, and coefficients to interflow (KI) and groundwater (KG) are expected to be closely related, etc.

Beyond these 4 parameter pairs, there are 3 common pairs of $\rho > 0.4$ between Sunkosi2 and Bird Creek (WUM-Imp, WUM-C, and WLM-Imp) and 2 pairs of $\rho > 0.4$ between Shiquan2 and Bird Creek only (CI-CG and CI-KG). On the basis of model structure, we would expect the latter 2 pairs to be related but less likely the former 3 pairs of parameters. For the former, it is harder to understand why the upper and lower tension water should be related to the impervious area (Imp). Partly because of Bird Creek's noisy data, and possibly because of some unclear reasons, WUM of Bird Creek was strongly ($\rho > 0.4$) related to a total 6 parameters (Table 6).

For pairs of parameters with $\rho > 0.4$ that are unique in each test case, some are expected, but some are less explainable, especially those found for Bird Creek. A perusal of Table 8 shows that parameters related to evaporation, namely WM, X, K, and B show more interactions with other parameters, e.g., 6 cases of $\rho > 0.4$ for WM and X, 4 cases for K and 3 for B. This possibly reflects the influence of evaporation that dominated the water budget of Bird Creek, e.g., the mean potential ET is 3.58mm/day while the mean streamflow is only 0.61mm/day (Table 2). SM, is strongly related to 4 parameters on free water or reservoir. Some of the aforementioned parameters are supposed to be sensitive to model output but some are not. According to Zhao (1992), model parameters sensitive to model output are K, SM, KG, KI, CG, CS and L. Apparently parameter interactions could be the result of model structure, as well as calibration data, especially for dry catchments such as Bird Creek. More results are needed to confirm the above observations on parameter interactions, especially as to whether

Table 6. Cross Correlation Matric (ρ) Showing Parameter Interaction of Xinanjiang Model (XNJ) Applied to the Semi-Wet Shiquan River Basin of China.

	WM	X	Y	K	B	SM	Ex	CI	CG	IM	C	KI	KG	n	Nk
WM	1														
X	0.37	1													
Y	-0.38	-0.98	1												
K	-0.43	-0.94	0.89	1											
B	-0.52	-0.11	0.09	0.18	1										
SM	-0.05	-0.09	0.07	0.11	0.17	1									
Ex	-0.02	-0.03	0.01	-0.06	0.07	-0.12	1								
CI	-0.04	-0.04	0.03	0.03	0.09	0.02	0.01	1							
CG	0.04	0.05	-0.04	-0.04	-0.09	-0.04	-0.01	-1.0	1						
IM	-0.08	-0.01	0.03	0.14	-0.18	0.18	-0.65	-0.08	0.09	1					
C	0.06	0.11	-0.14	-0.08	0.09	0.04	0.04	0.07	-.07	-.07	1				
KI	0.06	-0.01	0.02	-0.03	-0.10	-0.04	0.06	-0.84	0.84	0.01	-.03	1			
KG	-0.08	-0.09	0.10	0.07	0.12	0.05	-0.02	0.83	-.83	-.10	0.04	-.63	1		
N	0.10	0.08	-0.08	-0.08	0.02	0.05	0.25	-0.07	0.07	-.25	0.12	0.0	-.09	1	
Nk	-0.03	-0.08	0.07	0.08	0.085	-0.09	-0.13	-0.06	0.06	0.16	-.01	0.06	-.06	-.28	1

Table 7. Cross Correlation Matrix (ρ) Showing Parameter Interaction of Xinanjiang Model (XNJ) Applied to the Wet, Sunkosi River Basin of Nepal.

	WM	X	Y	K	B	SM	Ex	CI	CG	IM	C	KI	KG	n	Nk
WM	1														
X	0.19	1													
Y	-0.15	-0.96	1												
K	-0.51	-0.09	-0.06	1											
B	0.01	0.05	0.03	-.58	1										
SM	0.30	0.24	-0.29	-.26	0.29	1									
Ex	0.26	0.29	-0.37	-.14	0.22	0.97	1								
CI	0.05	-0.08	-0.01	-.03	0.12	0.24	0.23	1							
CG	0.01	-0.01	0.05	0.05	-0.04	-0.15	-0.14	-0.26	1						
IM	-0.20	-0.44	0.60	-.18	-0.04	-0.60	-0.72	-0.16	0.11	1					
C	-0.20	0.74	-0.80	0.02	0.19	0.27	0.31	0.10	-.08	-.44	1				
KI	-0.20	-0.16	0.09	0.31	-0.12	0.00	0.01	-0.04	-.01	-.13	-.09	1			
KG	0.32	0.05	-.001	-.16	-0.12	0.00	0.01	0.01	-.07	0.11	-.11	-.56	1		
N	-0.43	-0.15	0.05	0.75	-0.49	-0.18	-0.08	0.01	0.15	-.05	-.11	0.36	-.15	1	
Nk	-0.25	-0.12	0.19	-0.01	-0.13	-0.25	-0.24	-0.09	0.14	0.44	-.14	-.35	0.20	0.04	1

dry catchments would generally lead to more parameter interactions than wet or semi-wet catchments.

6. CONCLUDING REMARKS

To study the effects of model complexity and structure, and parameter interactions on conceptual hydrologic modeling of dry and wet catchments, 37 sets of CRR model-data calibration and validation experiments were formulated out of 5 CRR models and 5 catchments of wet, semi-wet and dry climate. Conclusions to the results are:

(1) The parameters optimized by SCE-UA are data dependent, as confirmed by the case of PTM. Global optimum parameters are impossible to derive, given that CRR model structures suffer from parameter inter-action, identifiability problem, simplified and point representation of nature that is highly variable spatially, and data errors. However, realistic parameters are possible to derive from current CRR models (as confirmed by validation results), if adequate data is available for model calibration;

(2) Even though dry catchments undergo more complex and a wider range of hydrologic processes, it seems that a standard, good quality hydrologic data can still support modeling dry catchments with standard CRR models of 10 to 20 parameters. Model performance depends more on the model structure, data quality and a variety of information than model complexity or calibration data length. On a whole, more dependable results are generally expected for wet than for dry catchments;

Table 8. Cross Correlation Matrix (ρ) Showing Parameter Interaction of XNJ Applied to Bird Creek of USA.

	WM	X	Y	K	B	SM	Ex	CI	CG	IM	C	KI	KG	N	Nk
WM	1														
X	-0.36	1													
Y	-0.02	0.65	1												
K	0.61	-0.72	-0.21	1											
B	0.20	-0.09	-0.19	0.14	1										
SM	0.72	-0.41	-0.34	0.62	0.56	1									
Ex	-0.27	-0.09	-0.23	-0.15	-0.01	-0.27	1								
CI	-0.73	0.38	0.30	-0.53	-0.55	-0.84	0.07	1							
CG	-0.68	0.42	0.30	-0.58	-0.60	0.89	0.09	0.79	1						
IM	0.20	0.56	0.57	-0.03	0.04	0.23	-0.39	-0.14	-0.17	1					
C	0.50	-0.59	-0.24	0.46	0.10	0.45	0.13	-0.43	-0.46	-0.18	1				
KI	0.04	-0.19	-0.26	0.02	-0.15	-0.07	0.21	-0.21	0.22	-0.20	0.11	1			
KG	-0.37	0.03	0.06	-0.15	0.02	-0.34	0.17	0.46	0.07	-0.18	-.23	-.55	1		
N	0.10	0.06	0.15	0.21	0.33	0.16	0.09	-0.15	-0.09	-0.07	-.06	-.07	-0.06	1	
Nk	0.44	-0.07	-0.04	0.18	0.05	0.45	-0.28	-0.37	-0.52	0.36	0.19	-0.1	-0.19	-.46	1

Table 9. Three Sets of Average and Standard Deviations (Bracketed Values) for Over 100 Sets of Optimized Parameters Per Catchment for the Xinanjiang Model (XNJ).

	Parameter Definition	Sunkosi2	Shiquan2	Bird Creek
WM	Areal mean tension water capacity	114 (2.0)	287 (2.64)	162 (1.57)
X	Ratio of the upper (WUM) to WM storage capacity	0.25 (0.07)	0.237(0.05)	0.014 (.003)
Y	Ratio of lower (WLM) to (1-X)WM storage capacity	0.27 (0.2)	0.573(0.08)	0.80 (0.024)
K	Ratio of potential to pan evaporation	0.5 (0.002)	0.61(0.008)	0.99(0.005)
B	Exponent of tension water capacity curve	1.95 (0.07)	1.92(0.062)	0.73(0.014)
SM	Areal mean free water storage capacity	46.5 (1.7)	48.8 (0.22)	20.66(1.97)
Ex	Exponent of the free water capacity curve	1.49 (0.22)	0.12(0.005)	1.14 (0.187)
CI	Interflow reservoir constant of the sub-basin	0.88 (0.12)	0.06(0.047)	0.056(0.092)
CG	Groundwater reservoir constant of the sub-basin	0.54 (0.16)	0.04(0.047)	0.070(0.092)
IM	Impervious area of the sub-basin	0.12(0.003)	0.002(0.002)	0.003(0.003)
C	ET contribution (less than 1) from the deep layer	0.27 (0.04)	0.15 (0.066)	0.297(0.002)
KI	The interflow recession coefficient (0 to 1)	0.95(0.001)	0.58 (0.113)	0.72 (0.085)
KG	The groundwater recession coefficient (0 to 1)	0.95(0.006)	0.635(0.135)	0.71 (0.078)
N	Number of cascade linear reservoir for runoff routing	0.80(0.004)	9.77 (0.146)	7.86 (0.19)
Nk	Scale parameter of cascade linear reservoir	0.91(0.003)	1.47 (0.002)	1.50 (0.004)

(3) XNJ has been marginally (but consistently) doing better than other models in most catchments probably because it is the only model that considers the non-uniform distribution of runoff producing areas in simulating the runoff, which is especially crucial for dry catchments;

(4) Even though SMA is likely more complex than other CRR, it did relatively poorly especially partly because it uses only one set of unitgraph ordinates to route low and high flows together, and partly because model performance does not depend directly on model complexity. However, as the simplest model, SMAR may have a model structure that is a little too simple for dry catchments;

(5) The need for hourly rainfall data to determine two of its model parameters makes PTM inapplicable in places without hourly data. Although developed for southern African basins, PTM's performance on the Great Usuthu catchment is still relatively poor at the validation stage, perhaps because of the approach that PTM used to convert daily to hourly rainfall was not applied properly;

(6) From the optimized parameters obtained for SMA, it seems less sensitive parameters and parameters explicitly designed to inter-relate to a few other parameters suffer from more identifiability problem. Further, from the parameter interaction results obtained for XNJ, it is obvious that some significant parameter interactions ($\rho>0.4$) resulted directly from parameters explicitly coupled together in the model structure, but some also from the calibration data.

Lastly, we believe the framework of our study is fairly independent of the optimization method used. In other words, given the same sets of calibration data and the same CRR models, we cannot expect much improvement over what we have already achieved in our calibration experience using SCE-UA.

Acknowledgments. This work was partially supported by the Natural Science and Engineering Research Council of Canada. Enoch Dlamini did part of the work presented herein, and Oscar Kalinga plotted Figure 2. S. Sorooshian, Q. Duan, and V. K. Gupta kindly provided the computer software for the SCE-UA algorithm. W. Pitman of the University of Witwatersrand provided the Pitman model software. The Great Usuthu data were provided by the Computing Center for Water Research of South Africa, while the test data for the catchments in Tanzania and the US were obtained from the Hydrology Department, University College of Galway, Ireland.

REFERENCES

Burnash R. J., Ferral R. L., and Mcguire R. A., 1973. A generalized stream flow simulation system, Conceptual modeling for digital computers, NWS, Sacramento, California.

Danish Hydraulic Institute, 1982. NAM Model Documentation, 82-892, JCR/Skn.

Duan Q., Sorooshian S., and Gupta V. K., 1992. Effective and efficient global optimization for conceptual rainfall runoff models. *Water Resour. Res., 28*(4), p. 1015 - 1031.

Gan, T. Y., and Biftu, G. B., 1996. Automatic calibration of conceptual rainfall-runoff models: optimization algorithms, catchment conditions, and model structure, *Water Resour. Res., 32* (12), p. 3513-3524.

Holland S. H., 1975. Adaptation in natural and artificial systems. University of Michigan Press, Ann Arbor.

Jakeman A. J. and Hornberger G. M., 1993. How much complexity is warranted in a rainfall-runoff model?, *Water Resour. Res., 29*(8): p. 2637 - 2649.

Loague K. M. and Freeze R. A., 1985. A comparison of rainfall-runoff modeling techniques on small upland catchment, *Water Resour. Res., 21*, p. 229 - 248.

Nash, J. E., and J. V., Sutcliffe, 1970. River flow forecasting through conceptual models; Part 1 - A discussion of principles, *J. Hydro., 10*(3): 282-290.

Nelder J. A., and Mead R., 1965. A simplex method for functional minimization, *Computer Journal, 9*: 308 -313.

O'Connell, P.E., Nash, J.E., and Farrell, J.P., 1970. River flow forecasting through conceptual models: Pt 2, The Brosna catchment at Ferbane, *J. Hydro., 10*:317-329.

Pitman W. V., 1976. A mathematical model for generating daily stream flow from meteorological data in South Africa. Rep. 2/76, HRU, U. Witwatersrand, Johannesburg.

Sorooshian, S., and Gupta, V. K., 1985. The analysis of structural identifiability: Theory and application to conceptual rainfall-runoff models, *Water Resour. Res., 21*(4), 487-495.

Yapo, P. O., H. V. Gupta, and S. Sorooshian, 1998. Multi-objective global optimization for hydrologic models, *J. Hydro.*, 204, p. 83-97.

Zhao, R. J., 1992. The Xinanjiang model applied in China, *J. Hydrology, 135*, p. 371-381.

Thian Yew Gan and Getu Fana Biftu, Department of Civil and Environmental Engineering, University of Alberta, Edmonton, Alberta, T6G 2G7, CANADA

Parameter Sensitivity in Calibration and Validation of an Annualized Agricultural Non-Point Source Model

Barbara Baginska

NSW Environment Protection Authority, Sydney, Australia

William A. Milne-Home

University of Technology, Sydney, Australia

The capability of the new, continuous model Annualized AGricultural Non-Point Source (AnnAGNPS version 2) for simulating flow events, peak discharge, and generation of nitrogen and phosphorus loads was tested on extensive field data. The hydrologic and water quality data were obtained from an intensively monitored, small rural watershed within the Hawkesbury-Nepean river system of New South Wales, Australia. AnnAGNPS is a large environmental simulation model for which prediction uncertainty is inherent both in the model structure and in parameter identification depending on how well watershed heterogeneity is represented. In this study AnnAGNPS was coupled with the model independent, nonlinear parameter estimation code, PEST, for calibration and sensitivity testing. This approach provided insight into the sensitivities of output predictions with respect to the variation of parameters from a base value. The base values can be defined in relation to the calibrated model outputs and field measurements. As all AnnAGNPS input parameters represent measurable properties and conditions, ranges of the parameters need to be specified and violations of the range limits monitored to minimise prediction errors and problems of non-uniqueness in the parameter selection. PEST calibration and sensitivity routines can be used systematically within these constraints for parameter optimization and identifiability. The measured event streamflows were matched satisfactorily by AnnAGNPS/PEST, but modelling of daily generation of particulate nitrogen and phosphorus achieved only moderate accuracy. The latter may reflect factors inherent in watershed processes as well as their representation by the model.

1. INTRODUCTION

Persistent algal outbreaks, low dissolved oxygen and elevated levels of nutrients are just a few symptoms of excessive eutrophication resulting in deterioration of aquatic habitat and water quality in major watersheds throughout southeastern Australia. These problems can be attributed to land use change and increasing nutrient loads in runoff from rural and urban land. Assessing contributions of nitrogen and phosphorus from nonpoint sources presents a constant challenge to researchers and water quality managers. The major difficulty with quantifying nutrient loads in runoff can be attributed to the fact that runoff events are highly unpredictable and rainfall has been long recognised as one of the major factors controlling nutrient movement in Australian watersheds [*Eyre*, 1995; *McKee et al.*, 2000]. Furthermore, a simple aggregation of single sources distributed across the

Calibration of Watershed Models
Water Science and Application Volume 6
Copyright 2003 by the American Geophysical Union
10/1029/006WS24

watershed does not reflect a tributary load entering a water-way and rapidly changing land use patterns and management practices contribute even further to the complexity of the problem.

An intensive, field monitoring study was conducted in a small subwatershed of Currency Creek to quantify nitrogen and phosphorus contributions from nonpoint sources and to provide improved long-term estimates of nutrient runoff from agriculture. The study area is situated on the southern slopes of the creek valley, 90 kilometres northwest of Sydney, Australia (Figure 1). It is a subwatershed of an unnamed, ephemeral stream draining 255 hectares of intensively used agricultural and rural residential land. The monitored area represents approximately 7.5% of the Currency Creek watershed and 0.01% of the Hawkesbury-Nepean watershed which, with an area of approximately 22,000 km², is one of the largest and most diverse coastal watersheds in New South Wales. We extended the study by attempting to simulate the generation and transport of nitrogen and phosphorous through the Currency Creek watershed with the Annualized Agricultural Nonpoint Source Pollution (AnnAGNPS version 2) watershed modeling package.

AnnAGNPS package is a large, environmental simulation model, which can suffer from the problems of parameter identifiability and sensitivity common to such models. Those models might be more applicable to rural watersheds with limited monitoring data, as they, in principle, do not require calibration. However, they require extensive amounts of information on watershed characteristics, which may or may not be readily available. Furthermore, *Jamieson and Clausen* [1988] maintain that all models must be carefully calibrated or verified for site specific conditions even if no calibration is claimed to be necessary in general. Nevertheless, AGNPS has been used extensively to model nonpoint source pollution and to assist with the management of runoff, erosion and nutrient movement in rural landscapes [*Summer et al.*, 1990; *Tim and Jolly*, 1994]. In Australia, *Foerster and Milne-Home* [1995] described the application of AGNPS to simulations of nutrient generation and movement under different farming practices in northern New South Wales. It was necessary in this case to calibrate the model for simulating peak flows by adjusting the runoff curve numbers. The calibrated model was then capable of simulating the effect of proposed best management practices on nutrient movement in agricultural watersheds.

The conversion of AGNPS to the annualized runoff and nutrient simulator, AnnAGNPS, lifted the capability of the package from modeling individual storm events to continuous simulation. This allows for better representation of the processes involved in transport and deposition of the sediment generated by sheet and rill erosion. As part of the delivery process, the overland deposition of the eroded sediment rather than a complete delivery of the material to the stream system, is simulated. The generation of phosphorus (P) was improved in Version 2. Process based models such as AnnAGNPS can often be applied by simply adjusting parameters from the initial input of physically realistic values until an acceptable fit is obtained to the observed field data. An effect of this procedure is the non-uniqueness of parameter estimates resulting from the over parameterisation inherent in large complex models. This problem may be overcome partly through the sensitivity analysis of parameter values.

Sensitivity analysis has been approached previously on a large scale within the parameter space [*Hornberger and Spear*, 1981; *Spear*, 1997] or on a restricted scale within a more localised region [*Pastres et al.*, 1997]. *Brun and Reichert* [2001] point out that the best results are obtained from a combination of both methods in cases of a high dimensional parameter space, with local parameters being used to indicate those areas which result in the best fits among the model outputs. Our approach to the problem was to couple the model-independent Parameter ESTimation software, PEST, with AnnAGNPS. PEST allows for the optimization of an initial set of parameter values to obtain the best fit. A sensitivity analysis routine (SENSAN) is included in the package. Previous applications of PEST have been with MODFLOW and HSPF modelling packages [*Doherty and Johnston*, 2002]. Our linked use of AnnAGNPS version 2 and PEST appears to be the first attempt in Australia to test the performance and applicability of these linked modeling packages for simulation and prediction of nutrient transport.

2. MODEL STRUCTURE AND DATA INPUTS

AnnAGNPS [*Cronshey and Theurer*, 1998] is a daily time-step model for the continuous simulation of pollutant loading on the scale of a watershed. The watershed is divided into homogeneous areas (cells) on the basis of soils, climate and land use. Runoff, sediment and nutrients are routed through each cell via a network of channels to the watershed outlet. The movement of contaminants from within their cell of origin can be tracked through the channel network in the watershed so that the relative contribution of point and non-point sources can be estimated.

The key feature of the package is the Input Data Preparation Model into which the data required by the two input files, AnnAGNPS input and Daily Climate Data, are entered. Up to 33 sections of data may be needed including soil type, land use, crop characteristics, pesticide and fer-

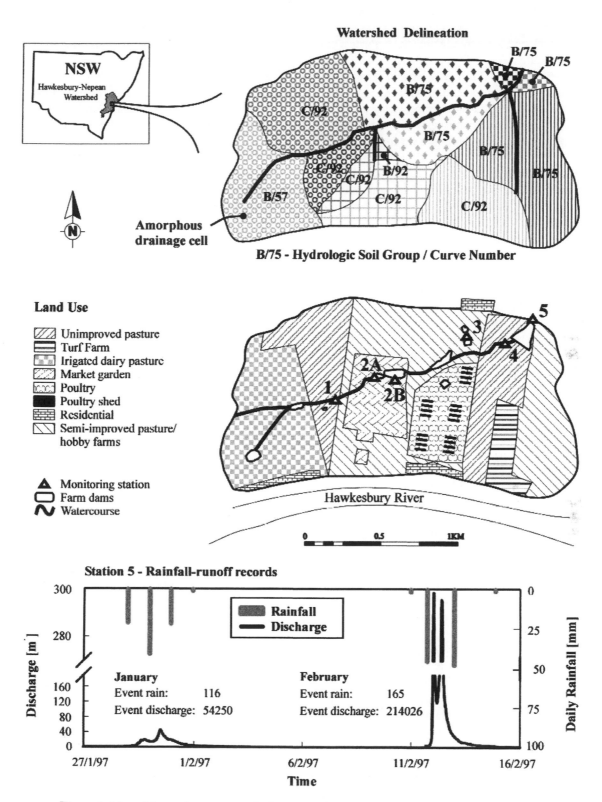

Figure 1. Map of the study area, watershed segmentation, land uses and typical rainfall-runoff records.

tiliser application, irrigation and land management practices. The terrain-based parameters are automatically generated by Flownet Generator Module which evaluates the topography and resultant drainage network of the watershed through the TopAGNPS, AGFlow and VBFlonet modules. Raster-type digital elevation model (DEM) data are required as input to TopAGNPS to delineate the modeled area into upstream and lateral subwatersheds and to set up the runoff and drainage channel network for the flow, sediment yield and pollutant simulations. Intuitively this flow driven discretization accounts better for spatial variability in hydrologic controls. AGFlow generates the reach and cell topographic characteristics which control the flow from the output of TopAGNPS. VBFlonet is a module for the graphics display of the generated networks. Output from all these modules are passed to the Pollutant Loading Model for the actual simulations followed by the Output Processor Model.

The surface runoff Pollutant Loading Model predicts nonpoint source pollutant generation and performs risk and cost/benefit analysis. It can simulate the chemical transport of particulate and soluble forms of phosphorus and nitrogen, organic carbon and pesticides using modified routines derived from the CREAMS model [Knisel, 1980].

2.1. Model Data Input

The boundaries of the modeled area and hydrologic segmentation of the watershed into amorphous cells contributing flow to channel links and the corresponding drainage divides required by AnnAGNPS were approximated through the analysis of the DEM obtained for this project from the Land and Property Information Centre, NSW. As a result of DEM data processing the modeled area of 264.9 hectares was discretized into 13 drainage areas (amorphous cells) and 6 reaches (Figure 1). Terrain-based geomorphic parameters such as slope, aspect, elevation and reach length were also determined as a result of DEM interpretation.

The AnnAGNPS Input Editor was used to develop and modify the input data to the pollutant-loading model. Most of the input parameters were sourced from measured data and where measured data were not available, the parameters were estimated based on the literature and the reference data provided with the modeling system. The simulation period for the Currency Creek watershed extends from 01/01/95 to 31/12/99. The key data inputs are the groups of parameters controlling rainfall, streamflow and related nutrient transport.

Rainfall dependent parameters, which reflect the ability of a storm to cause erosion, are expressed by average annual rainfall erosivity (R) and rainfall energy-intensity factor (EI30) for a 10-year average recurrence interval (ARI)

(Table 1). Spatial and temporal distribution of rainfall erosive power differs throughout Australia and during the year. In general, R increases during summer months when high intensity storms are most common. The average value of R for the study area was interpolated from a map showing the distribution of the R factor. A cumulative value of the R index based on a 15-day period formed part of the input data. The maximum rainfall intensity (I) for an event with a recurrence period of 10 years was determined from the IFD (intensity-frequency-duration) data for Richmond supplied by the Bureau of Meteorology, and the storm energy E (J/m2 mm) was estimated from the formula developed for eastern Australia by Rosewell, [1993]:

$$E = 29.0\ (1 - 0.596*10^{-0.04I})$$

where I is rainfall intensity (mm/h).

The TR-55 method [USDA, 1986] is used in AnnAGNPS to generate runoff, compute runoff volumes and peak discharges and to route the resulting excess precipitation through the watershed. The method applies the unit hydrograph theory and depends on travel time for peak discharge computation and watershed routing. Routing procedures describe the lagging and attenuation of water flow that occurs in the watershed. The simplified Manning's kinematic solution is used to compute travel time for generated sheet flow.

A 24-hour synthetic rainfall distribution provides means for estimation of peak discharges for a given watershed by specifying the length of the most intense rainfall duration contributing to the peak runoff. Each distribution is expressed as a mass curve of maximum rainfall intensities arranged in a sequence that is critical for producing runoff and is related to the time of concentration. The Type-II synthetic rainfall distribution was selected for the Currency Creek watershed. The selection was based on experimental studies [Browne, 1999] showing that it was the most representative hyetograph for areas where short-duration summer thunderstorms dominate.

Table 1. Selected Parameters for Runoff and Sediment Generation

Parameter	Value	Unit
Rainfall Erosivity (R)	2500	MJmm/ha-hr-annum
Energy Intensity (EI) (10-year ARI)	1888	MJmm/ha-hr
30-min rainfall intensity (10-year ARI)	65.2	mm/hr

ARI - average recurrence interval.

The runoff volumes are predicted using the SCS curve number (CN) method, which uses commonly available information such as soil type, cover and hydrologic conditions to estimate runoff. The method has been applied to a wide range of watersheds and climatic conditions for estimation of runoff volumes for ungauged areas in the United States [*Knisel*, 1980; *Rallison*, 1980]. The application of the method is aided by numerous tables and graphs giving examples of relevant curve numbers for different conditions including soil type, permeability, percent of impervious area, land cover, land use and vegetation.

The process of selecting runoff curve numbers for the purpose of the Currency Creek modeling is described here because simulated runoff volumes and nutrient transport proved to be sensitive to the values of these parameters.

A comprehensive evaluation of the applicability of the CN method for Australian conditions is provided by *Boughton* [1989]. *Dilshad and Peel* [1994] tested the performance of the CN method for Australian semi-arid tropics. Although the usefulness of the method is acknowledged, the Australian results show large variations in calculated runoff volumes and the importance of antecedent moisture conditions in determining the appropriate CN. Furthermore, the estimated runoff volumes are very sensitive to the selection of the curve number, such that a relatively small change of 15% to 20% in the selected CN may result in more than 100% difference in the estimated runoff volume.

In this study the initial curve numbers were selected using field measurements of rainfall and runoff. A method of CN curve fitting by graphical plotting of daily rainfall and runoff volumes was used [*Boughton*, 1989]. After con-structing the plot of rainfall (P) against direct runoff (Q), a visual comparison of plotted data with the USDA curve number plots was conducted to select the appropriate median curve number for the Currency Creek watershed (Figure 2). Although the curve numbers should be constant for a particular watershed, the comparison in Figure 2 shows considerable variations in the measured watershed responses. Three distinct groups of storm runoff curve numbers are noticeable, namely 50 – 55, 75 – 80 and 90 – 95 showing a very high runoff potential. The variations can be linked to soil characteristics and the high intensity and sporadic nature of the storm events recorded in the study watershed, which emphasises the importance of the soil moisture conditions to watershed responses. The CN plot represents solutions to the runoff equation for the average antecedent runoff conditions. Further adjustments to the CN are required to account for soil cover, land use and conditions preceding the storms, in order to fully describe Currency Creek watershed.

During analysis of the rainfall data for this study it became apparent that the highest daily precipitation recorded corresponded with the lowest CN of 52 (Figure 2). *Boughton* [1989] has noted that curve numbers have the tendency to decrease as the rainfall depth increases due to the empirical nature of the method and nonlinearity of the runoff equation. As a result, the remaining two groups of curve numbers which account for different hydrologic conditions in the watershed, were used in the calibration process. The initially selected curve numbers are documented in Table 2.

AnnAGNPS also requires the input of the terrain-based parameters for each cell derived from the DEM data together

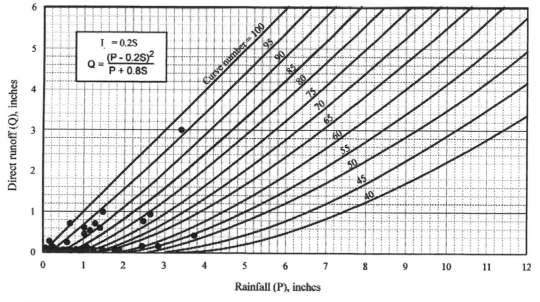

Figure 2. Graphical comparison of runoff for the Currency Creek watershed and the Curve Number plot.

Table 2. Selection of curve numbers

Land Use	CN Range	Hydrologic Soil Group	Moisture Conditions	Selected CN
Pasture	75 - 80	B – moderate infiltration rate	Average	75
Market Garden	90 - 95	C – slow infiltration rate, surface sealing	Wet	92

with detailed land management and fertiliser application data. Information on crop types and land use, soils and climate are also needed.

2.2. *Field Data*

Five monitoring stations were established in the Currency Creek watershed to measure runoff and water quality in order to assess the contributions of different agricultural land uses to water pollution. The event-based water quality monitoring was conducted from May 1995 to March 1997. Fully automated sampling and logging equipment was deployed to collect the data and runoff samples. In addition, a weather station was installed at the monitoring station St.3 (Figure 1) for validation of rainfall records and monitoring of climatic data. A continuous record of water level and stream flow and discrete water quality samples were collected during flow events.

The variable discharge-increment approach was selected as the most suitable for water quality sampling. This approach enabled the adjustable distribution of sampling through an event which prevented filling the autosampler bottles too soon. The datalogger continuously estimated streamflow and demonstrated the progressive increase in discharge increments from low to high flows. A good record of flow data together with the selective sampling approach allowed for a relatively high accuracy of assessment of watershed nutrient exports. Table 3 shows the hydrologic characteristics of the observed major events and corresponding measured loads of soluble and particulate nitrogen (N) and phosphorus (P) measured at the outlet of the watershed. The number of events sampled by each sta-

tion ranged from 3 to 13 depending on the duration of the monitoring period, the watershed area and performance of the sampling equipment. In general, only 13 out of 34 rainfall events were significant enough to result in runoff. Stations 1, 4 and 5 recorded the most comprehensive discharge and water quality data.

Altogether 420 water samples were collected over 23 months of monitoring and analysed for suspended solids, soluble and particulate forms of nitrogen and phosphorus as well as for total nitrogen (TN) and total phosphorus (TP). Discharge and concentration data were used to determine nutrient exports at each monitoring station for every event monitored. The period-weighted method was used to compute event nutrient loads which were then summed over 12 months to determine annual watershed loads. The pollutant load was computed as a product of average concentrations measured in samples taken at the beginning and at the end of an interval and the volume of water leaving the monitoring point during that interval. The interval length depended on the monitoring location and magnitude of the monitored storm and, in general, ranged from 5 minutes to 2 hours.

The detected concentrations of different species of N and P varied significantly between the monitoring sites and events. Nitrate was found as the dominant form of soluble N in irrigation runoff from different land uses in the watershed while the elevated concentrations of soluble P dominated in the uppermost part of the watershed where it could be attributed to grazing of improved pastures irrigated with dairy effluent. Typically a reduction in concentration and load of TP between sites 1 and 5 occurred despite the presence of vegetable farms contributing large amounts of phosphorus to runoff [*Baginska et al.*, 1998]. The concentrations

Table 3. Characteristics of the Events Monitored at the Outlet of Currency Creek

Event Duration	Rainfall (mm)	Discharge (m^3)	Peak Discharge (m^3/s)	Soluble N (kg)	Sediment N (kg)	Soluble P (kg)	Sediment P (kg)
24-25 / 09 / 1995	95.0	70225	2.82	554.3	36.6	29.3	10.0
30-31 / 08 / 1996	85.8	74572	1.69	598.7	84.6	86.7	37.1
29 / 09 / 1996	54.1	8773	0.29	30.2	16.2	6.9	5.1
28-31 / 01 / 1997	115.8	54246	0.75	361.6	57.2	41.5	29.2
11-12 / 02 / 1997	164.6	214026	5.15	798.5	162.3	257.8	101.7

of nitrogen usually increased during runoff. Also, nitrogen dominated the nutrient loads at the outlet of the watershed.

3. SENSITIVITY ANALYSIS AND PARAMETER ESTIMATION WITH PEST

Performing sensitivity analysis of AnnAGNPS responses to varying the initial input values is critical to identifying key model parameters and guiding the calibration process. It is particularly important in the case of the AnnAGNPS model, which has extensive data requirements and may often suffer from poor parameter identifiability in respect of available observation data. Despite the fact that all input parameters required by the model attempt to represent measurable properties and conditions, we are only able to assign parameter values within physically acceptable ranges based on usually sparse point measurements within a watershed. Initial runs of the model clearly indicated that modeled discharges and nutrient loads both displayed complex and often contradictory responses to changing input variables. We used PEST to examine the non-linear model responses, optimize selected model inputs and to assess model sensitivity over a large region of the parameter space.

PEST offers a unique combination of model-independent calibration tools. The tools include parameter optimization routines for typical model calibration, predictive analysis module and a sensitivity module (SENSAN). PEST uses the Gauss-Marquardt-Levenberg algorithm in the model calibration/parameter optimization process. This procedure combines the advantages of the inverse Hessian method and the steepest descent method [*Press et al.*, 1989]. It attempts to minimize the weighted sum of squared differences between the model-generated values and those measured in the field by varying user defined model inputs. The goodness of fit is apparent from the value of the optimized objective function and is also provided by a computed correlation coefficient. The coefficient is independent of the number of observations and levels of uncertainty associated with those observations, thus allowing for direct comparison of different parameter estimation runs. A user can observe the results of iterative runs in tabular and graphical forms while PEST optimization process is in progress and intervene in the execution of the model at any time. PEST offers many additional options, such as parameter scaling and weighting and freezing of sensitive parameters to support the identification of an optimal objective function and to avoid local minima [*Doherty*, 2001].

SENSAN is a command-line program, which facilitates sensitivity analysis by allowing a user to initiate numerous model runs and post-processing the results of those runs. It subsequently generates a range of formatted output files from user specified model outputs. This is particularly useful for interpreting large parameter space models such as AnnAGNPS as it allows testing of a large number of input variables at a time. The three sensitivity output files from SENSAN contain: 1) predictions read from the model outcome file for the range of values specified by the user, 2) relative differences between the model generated responses for a given/tested range of input parameters and a user defined base set of input data, and 3) model outcome sensitivities defined with respect to parameter variations from their base values. Sensitivity is calculated as the difference between the model outcome for a particular set of input variables and the pertinent outcomes for the base values, divided by the difference between the current parameter set and the base parameter set. For a simplified scenario, when only a single parameter p varies from the base set, the relative sensitivity is defined as:

$$S_r = (O - O_b)/(p - p_b)$$

where O_b and p_b are the model outcome and the parameter base values and O and p are the model outcome and the parameter values pertaining to a particular model run.

SENSAN uses the same model interface protocol as the PEST optimization routines and the same structure and format of the control files.

3.1. Initial Conditions for Sensitivity Testing and Parameter Optimization

Approximately 770 data fields had to be assigned in the model application to the study watershed. These fields ranged from simple vegetation codes or links between the model cells and data sections to detailed topographic, hydrologic, geomorphic and agronomic parameters. Topographic features of the watershed, such as slope, overland flow segment length, drainage dimensions and the subsequent time of concentration may substantially influence the magnitude and the dynamics of runoff. However, these parameters were not optimized as they were computed from the DEM data.

Sensitivity analysis was conducted only for those input variables, which exhibited large physical variations due to natural heterogeneity within the watershed, or variables for which data was not routinely collected. Due to the complexity of the model, sensitivity testing and parameter optimization were carried out in two steps. This approach minimized interactions between the calibrated parameters and enabled clear determination of the input parameters affecting model-simulated discharges and respective nutrient exports. Table 4 shows the groups of parameters included in the sensitivity analysis.

Better optimization of input parameters was necessary to compensate for the possible underperformance of AnnAGNPS because the degree of variation exhibited by the results of sensitivity analysis was quite large. Calibration was undertaken using PEST, as the usual stepwise adjustment of input parameters until model outputs matched observations was ineffective due to model nonlinearities and interdependencies between inputs.

Optimization focused on the input variables displaying the highest sensitivities. Parameters representing soil and fertilizer properties were optimized with PEST such that discrepancies between AnnAGNPS generated outputs and field measurements were minimized. The optimization runs were carried out separately for hydrologic responses and for daily nutrient loads of soluble and particulate N and P. Scaling of parameters and weighting of measured loads of N and P were used to minimize the impact of outliers on the computed objective function. The best initial values for all parameters and limits for all physically based parameters had to be supplied. Performance assessment of the optimization process and the subsequent adjustments of the parameters controlling calibration were based upon achieved reductions in the objective function and the values of correlation coefficient as defined by *Cooley* and *Naff* [1990]. The correlation coefficient, R, is calculated as follows:

$$R = \frac{\sum (w_i c_i - m)(w_i c_{oi} - m_o)}{[\sum (w_i c_i - m)(w_i c_i - m) \sum (w_i c_{oi} - m_o)(w_i c_{oi} - m_o)]^{1/2}}$$

For the range of tested parameters c_i represents i^{th} observed value, c_{oi} is the relevant model generated value, m is the mean of weighted observed values and m_o is the mean of weighted model-generated outcomes. R is independent from the number of observed model outputs included in the optimization process and from the absolute levels of uncertainty associated with those outputs. It therefore allows for direct comparison of different parameter estimation runs in the context of goodness of fit. Values of R above 0.9 indicate good agreement between the observed and simulated results.

Sensitivity of model responses was based on the analysis of outputs generated at the outlet of the Currency Creek watershed. A 4-day rainfall event observed in January 1997, plotted in Figure 1, was selected for comparison of simulated and observed values of daily flows and soluble and particulate loads of nitrogen and phosphorus. This event was used because it was observed at all sampling stations in the catchment. The subsequent optimization was first undertaken with reference to the January 1997 event, then all major events were included in the optimization process to detect temporal patterns in model responses.

3.2. Sensitivity of Model Inputs

Sensitivity testing showed a complex matrix of responses depending on the observed outputs and the range of the initially selected base parameter values in relation to which the relative sensitivities were calculated (Table 5). The analyzed patterns were further obscured by correlations between parameters and by inherent differences in processes governing generation and delivery of soluble and sediment associated forms of N and P. Nevertheless, sensitivity testing aided by SENSAN, enabled evaluation of a large parameter space and resulted in identifying the key model input parameters, and in determination of uncertainty and degree of influence of parameter perturbation on the model outcomes.

Table 5 shows a relative comparison of the sensitive model parameters and their implications for the modeled results. As can be expected, curve numbers (CN), soil moisture properties (FC1, FC2) and hydraulic properties (SC1, SC2) are the sole factors determining the capacity of the watershed to generate runoff. They also have a visible impact on the computed emissions of soluble N and P. Although calculated sensitivities for runoff and soluble N indicate gradual and steady change within the tested parameter space of the representative curve numbers (Figure 3), they result in the generation of opposing responses. That is, increasing surface runoff due to raising curve numbers results in higher flow velocities and less contact with soil, which, subsequently, reduces the amounts of soluble nutrients generated by the model.

The response of soluble P to parameter perturbation exhibits a large local variability in the computed sensitivities (Figure 3) and is the most susceptible to changes of pH

Table 4. Parameters Tested in the Sensitivity Analysis

Testing Scenarios	Selected Parameters
Discharge	Curve numbers (CN), Field capacity, Saturated conductivity
Nutrients	pH[a], Field capacity[a], Saturated conductivity[b], Organic and inorganic N ratio in soil[c], Organic and inorganic P ratio in soil[c] Fraction of organic and inorganic N and P in fertilizer[d], Annual mass root for pasture,

[a] Tested for two soil types and subsequent two soil layers
[b] Tested for two soil types, topsoil layers only
[c] Tested for the soil representative for the cropping area. Organic and inorganic ratios represent the initial amounts of nitrogen (N) and phosphorus (P) at the start of the simulation
[d] Fertilizer fraction which is organic N and P and mineralizable (inorganic) N and P

Table 5. Sensitive Parameters and the Expected Level of Change in Model Predictions

Parameter	CN	SC1	SC2	FC1	FC2	pH1	pH3	N_{org}	N_{inorg}	P_{inorg}	RM	FN_{inorg}
Discharge	L	L	L	L	H							
Particulate N			M		L						H	
Soluble N	L	L	L	M	L				H	H		M
Particulate P			M		L		L				H	
Soluble P			L		L	H	H			H		

CN – Curve Number; SC1,SC2 – saturated conductivity, FC1, FC2 – field capacity; N_{org}, N_{inorg}, P_{inorg},– organic and inorganic ratios of N and P in soil, RM – root mass, FN_{inorg}– inorganic ratio of N in fertilizers

L – Low change - up to 5% change in the output
M – Medium change - up to 25% change in the output
H – High change - often more than 25% change in the output

in the top layer of soil (Figure 4). Maximum changes induced by pH perturbations in the top soil layer can be two to three fold in magnitude. The increase of pH from 4 to 5 could result in reduction in particulate and soluble phosphorus generation by 12 – 25% and 9 – 34%, respectively. The importance of pH values for the model outcomes was examined by comparing the extent of differences between the observed and model generated daily phosphorus loads for the full parameter space of pH and field capacity. The results given in Figure 4 demonstrate very steep gradients resulting almost exclusively from the change in pH and identify two matching regions of sensitivity in the tested parameter space. For the given structure of the model for the Currency Creek watershed, the best match between the observed and modeled phosphorus loads can be achieved by applying two entirely different sets of pH values ranging from 4.5 to 5.2 or from 7 to 8. Large changes in pH measured in the Currency Creek watershed, which varied from 4.4 to 8.1, and the occurrence of identical regions in the objective function, may be the source of significant discrepancies in model predictions.

The intrinsic correlation between parameters and a local minimum is evident in the response surfaces generated for runoff (Figure 5). While attempting to minimize differences between the observed and model generated daily discharges it was noted that an incremental change in curve numbers from 70 to 75 could have a significant effect on the values of contributing parameters. Although the general shape of the response surface remained unchanged, the best match between the observed and predicted discharges was achieved for reversed values of the saturated conductivity and field capacity data.

The observed changes in soluble and particulate N and P due to variations in field capacity provide a snapshot of model sensitivities and indicate links to land management practices. Field capacity was tested for two soil groups and

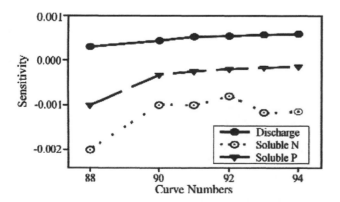

Figure 3. Sensitivities of discharge and soluble N and P to curve numbers.

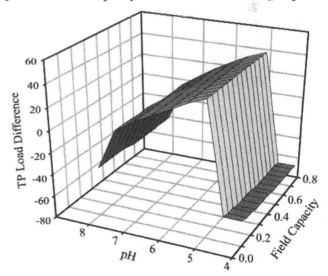

Figure 4. Difference between the observed and predicted daily loads of total phosphorus (TP) with unit change of pH in the topsoil.

for pertinent land uses representing predominantly unimproved pasture (FC1) and a mixture of intensive agricultural activities dominated by vegetable growing (FC2). The results indicated that under natural conditions only soluble N was susceptible to changes in field capacity, as the total available pool of nutrients would determine the outer limits of the extent of change in model predictions. In addition, sensitivity testing identified a distinct group of parameters, which exhibited exceptionally high sensitivities and for which no measured data were available. These parameters were fertilizer properties, as a proportion of inorganic nitrogen, land use reference data, such as annual root mass, and soil properties listed in Table 5. The selection of appropriate values for these parameters remains difficult and may introduce considerable uncertainty in the estimated nutrient loads. The difficulties stem from the likely heterogeneity of these parameters on the field scale as values may range across most of their recommended domain.

3.3. *Optimization Results*

It is sometimes claimed that agricultural nonpoint pollution models are not to be calibrated as they contain numerous interdependent variables resulting in complex interactions between them [*Shepherd and Geter*, 1995]. Furthermore, process based models, such as AnnAGNPS, are designed to characterize watershed processes well enough to enable the use of measurable properties and conditions and, therefore, they do not require formal calibra-

tion. However, *Ndiritu and Daniell* [1997] claim that calibration is likely to remain one of the most important steps in model application, including and especially for process based models. Moreover, for large domain models, manual stepwise calibration may often result in underestimating or even entirely omitting key parameters and, therefore, automatic optimization procedures are more likely to warrant better results with less effort.

In this study the aim of calibration was to optimize the model inputs so the differences between the simulated and observed data could be minimized and better accuracy of model predictions accomplished. In addition, calibration allowed for basic verification of the initial assumptions of watershed parameterization and provided means for assessment of how well the model input parameters described the relevant characteristics of the Currency Creek watershed. Detailed optimization of AnnAGNPS using PEST started with 4 parameters controlling daily runoff volumes (Table 6). This was followed by the optimization of 11 parameters having major impacts on the simulated soluble and particulate N and P, such as pH, soil moisture, annual root mass and ratios of soil and fertilizer N and P. The values of optimized parameters are shown in Table 6. Optimization of discharge related parameters for the January 1997 event required only 4 iterations and a very high correlation coefficient R, exceeding 0.9, was achieved. As expected, the addition of extra events added inherent variability in watershed responses, and subsequently caused alterations in the optimized parameter values mainly by increasing the curve numbers. Nevertheless, a relatively

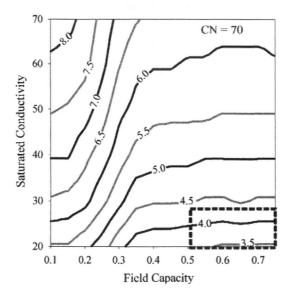
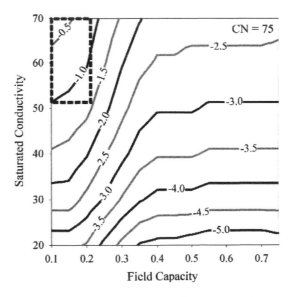

Figure 5. Contour graphs of differences between the observed and predicted runoff in the parameter space of saturated conductivity, field capacity and curve numbers. Contour values represent differences between observed and predicted daily runoff in cubic meters x1000. Regions outlined by a dotted line indicate the range of parameters for which the minimum discrepancy was observed.

high value of *R*, reaching 0.7, was maintained, and a good match between the observed and simulated daily discharge is clearly visible in Figure 6.

Despite good correlation achieved for simulation of event and daily discharges, in the subsequent optimization runs for nutrient emissions *R* values did not improve beyond 0.54 - 0.61. The objective function computed by PEST during the entire estimation process was dominated by contributions from soluble N and P. Sediment-associated P contributions exhibited two orders of magnitude lower impact (Figure 7) and showed insignificant variability despite the magnitude of the events tested. The main improvement in the optimization results came from assigning weights to the observations of particulate P and from scaling parameters, such as annual root mass for which the range of representative values was at least two orders of magnitude larger than for any other parameter used in the optimization. Although high sensitivities to pH were observed, in order to satisfy the criteria for minimizing the difference between the observed and simulated nutrient loads the optimized pH values had to remain close to the lower end of the pH range measured in field surveys (4.5 – 5.0). The very strong pH dependency and apparent inability of the model to adequately simulate daily variations in particulate N and P may introduce large uncertainties to model predictions for ungauged watersheds.

4. MODEL PREDICTIONS AND PERFORMANCE

Evaluation of the AnnAGNPS ability to predict flow and nutrient rates in the ephemeral stream subwatershed of Currency Creek involved sensitivity testing, systematic optimization of the key input parameters with PEST and verification of N and P loads generated by the model against

Table 6. Optimized Parameters

Parameter	Optimization Scenarios	
	January Event	All Events
Field capacity	0.17	0.1
Saturated conductivity	21.90	27.2
Curve Number 1	73	79
Curve Number 2	80	90

those measured in the field. This verification phase focused on how well the model could simulate runoff in the upland section of the watershed (Figure 1, St.1) and at the outlet of the watershed. The quality of predictions for nutrient generation was then tested by comparing simulated and observed loads on daily, event and annual basis.

Despite a few simplifying assumptions made in the process of runoff simulation, acceptable goodness of fit was achieved for runoff volumes. The level of calibration was quantified with the coefficient of efficiency (*E*) [*Nash and Sutcliffe*, 1970] and the mean, used as measures of degree of model accuracy and distribution of central tendency, respectively. Generally, acceptance criteria for rainfall-runoff modeling are still very much subjective and may vary significantly from application to application. The quality of AnnAGNPS hydrologic predictions was assessed with the criteria suggested by *Chiew et al.* [1993], which were based on 112 monthly streamflow simulations conducted throughout Australia. According to their findings flow estimates can be classified as acceptable if they have coefficient of efficiency (*E*) greater than 0.6 and mean simulated flow is always within 15% of mean recorded flow. The *E* criterion for event flows was met spatially for the study watershed, as the coefficient consistently exceeded 0.8 for the upper gaug-

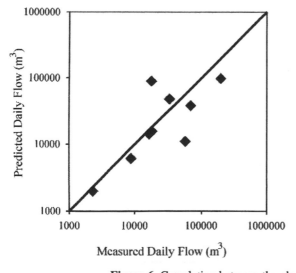

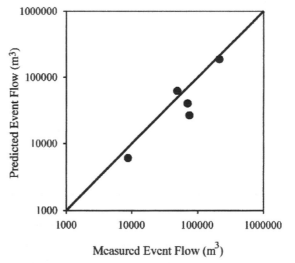

Figure 6. Correlation between the observed and predicted daily and event flows.

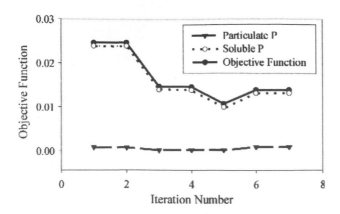

Figure 7. Objective function versus iteration number.

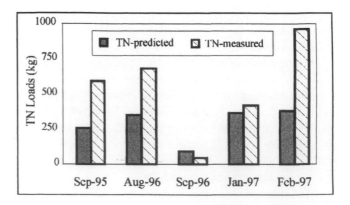

Figure 8. Measured and predicted event loads of total nitrogen at the watershed outlet.

ing site (St.1, Figure 1) and at the outlet of the watershed (Figure 6). However, the model did not perform equally well for daily and event-based assessments, representing the two temporal scales tested.

The simulated daily flows showed a wider scatter, which often caused the coefficient of efficiency (E) to be lower than the threshold limit for an acceptable simulation. For the majority of the events, the model predictions overestimated the recorded daily flows at the upper gauging site (St.1). The optimization of hydrologic parameters helped reduce previously observed significant variations around the 1:1 line for daily flows at the outlet of the watershed, causing low E values range between 0.26 and 0.44.

In order to determine the predictive capabilities of the model for simulation of nitrogen and phosphorus loads in semi-arid conditions, considerable effort was directed towards not only optimization of input parameters but also to revisions of the structure of the model. As recommended by *Novotny and Olem* [1994], in order to avoid temporal and spatial error propagation, hydrology and sediment need to be calibrated before any attempts to model water quality are made. Despite achieving acceptable levels of optimization for flow, predictions of nitrogen and phosphorus loads with the calibrated model still differed significantly from the observed values (Figure 8, 9).

The best fit between the observed and the simulated nitrogen load was achieved while comparing the results on an event basis. Despite the mean values for the predicted and measured loads matching closely, the calculated coefficient of efficiency (E) was usually negative indicating high deviations of the predicted nitrogen exports from the measured ones. Large deviations from the measured data were observed for daily simulations regardless of calibration efforts. Notwithstanding the uncertainty in absolute predictions of nutrient exports, relatively close patterns between the simulated and the observed data could be seen for total nitrogen and phosphorus (Figure 8, 9). The simulated total

nitrogen loads at the outlet of the watershed were mostly underestimated while the opposite occurred for the total phosphorus loads. It seems evident that the current model formulation underestimates particulate N, which subsequently results in lower than expected ability of the model to simulate TN loads.

The model failed in its ability to predict particulate phosphorus and nitrogen loads on all temporal and spatial levels tested. Those predictions were consistently low, exhibited small sensitivity to the event magnitude and underestimated the observed data by at least an order of magnitude. Despite extensive optimization and sensitivity testing the results of particulate N and P simulations were not changing much unless unrealistically high erosion rates were allowed. This indicates that the model cannot adequately simulate transport, immobilization and re-suspension of particle-bound pollutants in the Currency Creek watershed. The somewhat low overall quality of the model predictive capacity indicated by this study may result from a combination of factors and may be specific for the watershed and not necessarily for the model. Internal model deficiencies may be related to the representation of watershed processes and the selection of relevant assessment methods, while external problems may be related to the conditions within the modeled watershed and the quality of observed data.

The description of physical processes of runoff generation in the model may not be adequate. As mentioned earlier, calculation of runoff is based on the SCS Curve Number technique which was designed to predict the total channel flow at the watershed outlet, for which considerations of watershed flow paths and runoff generation areas were not essential [*Garen et al.*, 1999]. Although the method is still quite useful, its original design and applications are extended in the model to account for runoff occurring on the land surface and not in the stream channel. A high number of input parameters can also contribute to difficulties in calibration

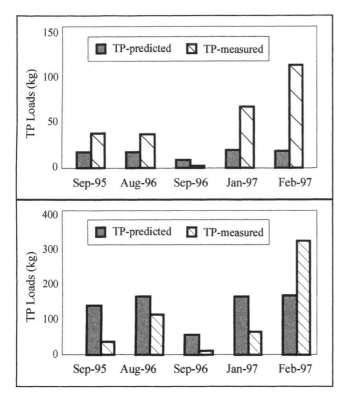

Figure 9. Measured and predicted event loads of total phosphorus at the gauging site St. 1 (top) and at the watershed outlet (bottom).

resulting from interdependence of various parameters and discontinuities in the response surface of the model [*Chiew et al.*, 1993]. In every model application, including this one, the assessment of the model's ability to simulate watershed responses is based on a fundamental assumption of absolute quality of the measured data. However, the data is error prone because of uncertainty usually associated with the estimation of event flow, in particular for short-lasting, high-magnitude events. Furthermore, simulation of stream flow and water quality in the Currency Creek watershed is complicated by the fact that flow occurs only as a result of significant rainfalls and watershed responses to a particular storm can be to a high degree modified by the antecedent moisture conditions which are highly unpredictable. The factors mentioned above can influence the predictive capacity of the AnnAGNPS model and reduce significantly its ability to simulate phosphorus emissions in ephemeral stream watersheds.

5. CONCLUSIONS

The objective of this study was to model nitrogen and phosphorus loadings from nonpoint sources for the subwatershed of Currency Creek in order to examine applicability, predictive power and implementation effort of the new continuous simulation AnnAGNPS model. The Currency

Creek subwatershed is characterized by a mixture of agricultural and rural-residential land uses and is representative of management practices and land use patterns common in the Hawkesbury-Nepean watershed. The modeled subwatershed also experiences widespread environmental and water quality problems often related to agriculture. These problems such as extensive soil erosion and high losses of nutrients in runoff, in turn, may contribute to the deterioration of water quality in the entire Hawkesbury-Nepean river system.

The quantification of nutrient loads from nonpoint sources is the primary focus of many watershed studies. Estimates of nutrient loads form the basic prerequisite for the subsequent assessment of how agriculture may influence the long-term quality of surface and ground water. On the other hand, direct measurements of nonpoint nutrient loads are always difficult, costly and even impractical in some management applications. Consequently, new generation water quality models capable of simulating erosion rates and water quality on a watershed scale are highly desirable in the light of increasing need for such models in land management planning and implementation of conservation measures.

Evaluation of the model predictions undertaken in this study demonstrates that AnnAGNPS produces results of satisfactory quality when simulating event flows but a high degree of uncertainty is associated with predictions of nutrient loadings. The ability of the model to adequately simulate phosphorus loads in watersheds with no permanent flow and multi-peak runoff events is, at this stage questionable.

This deficiency does not discredit the quality of model predictions other than particulate nitrogen and phosphorus loads and does not prevent a more successful use of the model in Australian conditions. There is indication that the model predictive capacity increases in perennial streams. An improvement in water quality predictions can be noticed at the watershed outlet where a stream channel is better defined and baseflow occurs due to limited groundwater recharge which makes the flow and soil moisture conditions more predictable.

In addition, the results suggest that the model may be better suited for studies on a larger regional scale than for small subwatersheds. Local conditions may prevail in the latter and over-parameterization is likely, causing adverse effects on model predictive capacity. It should also be noted that annual and event based predictions are better than those for shorter time increments, which is not uncommon in continuous simulation models. Despite the fact that all model inputs have physical meaning and can be measured in the field, calibration is always recommended as it allows tuning of the parameters controlling major delivery processes and, subsequently, may improve the quality of the results. Although the popularity of models similar to AnnAGNPS comes to some extent from the fact that they can be applied in data-poor watersheds, the need for calibration should be recognized, as it also helps to understand the uncertainty associated with the results. Nevertheless, as interdependence of model parameters is evident, calibration can be a difficult process. The issues of parameter identification and sensitivity can be addressed by including an optimization process as part of the modeling. This study has shown that coupling AnnAGNPS and PEST provided a semi-quantitative estimation of parameter sensitivity. The SENSAN feature of PEST allows the modeler to track and control the incremental change in parameter values.

Runoff generation and sediment predictions are simulated in the model with separate functions, but nitrogen and phosphorus transport is flow dependent. Therefore, particular attention is needed during the verification process so the predicted flow volumes match those at the gauging stations, if available. Otherwise, any inconsistencies originating from inadequate predictions of the flow volumes and event patterns are likely to be transferred and amplified in the water quality simulations which follow.

The model input requirements can be very extensive and a considerable amount of time should be allowed for assessment of the initial input data and watershed conceptualization. The structure of the model input file permits a reasonable level of flexibility in selection of data sections to represent the desired watershed complexity depending on the aim of the modeling and the expected prediction accuracy. A high level of empirical knowledge and, in particular, prior knowledge of the watershed, agricultural activities, soil and climatic conditions is a big advantage during all phases of modeling, from watershed discretization to optimization and interpretation of the results.

Acknowledgments. The authors would like to thank Prof. Peter Cornish from University of Western Sydney–Hawkesbury for making the Currency Creek data available for this project.

REFERENCES

Baginska, B., P.S. Cornish, E. Hollinger, G. Kuczera, and D. Jones, Nutrient export from rural land in the Hawkesbury-Nepean watershed, in *Proceedings 10th Australian Agronomy Conference*, July 1998. Wagga Wagga, 753-756, 1998.

Boughton, W.C., A review of the USDA SCS Curve Number method, *Aust. J. Soil Res.*, *27*, 511-523, 1989.

Browne, F.X., Stormwater management, in *Standard Handbook of Environmental Engineering*, edited by R.A. Corbitt, pp. 7.1-7.127, McGraw-Hill, New York; 1999.

Brun, R., and P. Reichert, Practical identifiability analysis of large environmental simulation models, *Water Resour. Res.*, *37*(4), 1015-1030, 2001.

Chiew, F.H.S., M.J. Stewardson, and T.A. McMahon, Comparison of six rainfall-runoff approaches, *J. Hydrol.*, *147*, 1-36, 1993.

Cooley, R.L., and R.L. Naff, Regression modeling of groundwater flow: US Geological Survey Techniques in Water Resources Investigations, Book 3, Chapter B4, pp232, 1990.

Cronshey, R.G., and F. D. Theurer, AnnAGNPS – Non-Point Pollutant Loading Model, in *Proceedings 1st Federal Interagency Modeling Conference*, 19-23 April, Las Vegas, NV, 1998.

Dilshad, M., and L.J. Peel, Evaluation of the USDA Curve Number Method for agricultural watersheds in the Australian semi-arid tropics. *Aust. J. Soil Res.*, *32*, 673-685, 1994.

Doherty, J., PEST-ASP User's Manual. Watermark Numerical Computing, 2001.

Doherty, J., and J.M. Johnston, Methodologies for calibration and predictive analysis of a watershed model, *J. Am. Water Resour. Assoc.*, in press 2002.

Eyre, B., A first-order nutrient budget for the tropical Moresby Estuary and watershed, North Queensland, Australia, *J. Coastal Res.*, *11*(33), 717-732, 1995.

Foerster, J., and W.A. Milne-Home, Application of AGNPS to model nutrient generation rates under different farming management practices at the Gunnedah Research Centre watershed, *Aust. J. Exp. Agric.*, *35*, 961-967, 1995.

Garen, D., D. Woodward, and F. Geter, A user agency's view of hydrologic, soil erosion and water quality modelling, *Catena*, *37*, 277-289, 1999.

Hornberger, G.M., and R.C. Spear, An approach to the preliminary analysis of environmental systems, *J. Environ. Manage.*, *12*, 7-18, 1981.

Jamieson, C.A., and J.C. Clausen, Test of the CREAMS model on agricultural fields in Vermont, *Water Resources Bulletin*, *24*(6), 1219-1226,1998.

Knisel, W.G.(ed.), A Field-Scale Model for Chemicals, Runoff and Erosion from Agricultural Management Systems. US Department of Agriculture. Conservation, Research Report No. 26, 640pp, Washington DC, 1980.

McKee, L., B. Eyre, and S. Hossian, Intra- and interannual export of nitrogen and phosphorus in the subtropical Richmond River catchment, Australia, *Hydrol. Process.*, *14*, 1787-1809, 2000.

Nash, J.E., and J.V. Sutcliffe, River flow forcasting through conceptual models: Part I. A discussion of principles, *J. Hydrol.*, *10*, 282-290, 1970.

Ndiritu, J.G., and T.M. Daniell, An improved genetic algorithm for rainfall-runoff model calibration and function optimization, in *Proceedings of the International Congress on Modelling and Simulation MODSIM'97*, 4, 1683-1688, 1997

Novotny, V., and H. Olem, *Water Quality: Prevention, Identification, and Management of Diffuse Pollution*, Van Nostrand Reinhold. New York, 1994.

Pastres, R., D.Franco, G. Pecenik, C. Solidoro, and C. Dejak, Local sensitivity analysis of a distributed parameters water quality model, *Reliab. Eng. Syst. Safety*, *57*(1), 21-30, 1997.

Press, W.H., B.P. Flannery, S.A. Teukolsky, and W.T. Vetterling, Numerical Recipes, 702pp, Cambridge University Press, Cambridge, 1989.

Rallison, R.E., Origin and evolution of the SCS runoff equation, *Am. Soc. Civ. Eng. Symposium on Watershed Management*, Boise, Idaho, USA, 1980.

Rosewell, C.J., SOILOSS 5.0 User's Manual, A Program to Assist in the Selection of Management Practices to Reduce Erosion, NSW Department of Conservation and Land Management. Gunnedah Research Centre, 1993.

Shepherd, R.G., and W.F. Geter, Verification, calibration, validation, simulation: Protocols in groundwater and AGNPS modeling, in *Proceedings of the International Symposium: Water Quality Modeling*, April 2-5, Orlando, Florida, 87-91, American Society of Agricultural Engineers, 1995.

Spear, R.C., Large simulation models: Calibration, uniqueness and goodness of fit, *Environ. Modell. Software*, 12(2-3), 219-228, 1997.

Spear, R. C., T.M. Grieb, and N. Shang, Parameter uncertainty and interaction in complex environmental models, *Water Resour. Res.*, 30(11), 3159-3169, 1994.

Summer, R.M., R.A. Alonso,. and R.A. Young, Modeling linked watershed and lake processes for water quality managemnt decisions, *J. Environ. Qual.*, 19(3), 421-427, 1990.

Tim, U.S., and R. Jolly, Evaluating agricultural nonpoint source pollution using integrated geographic information systems and hydrologic/water quality model, *J. Environ. Qual.*, 23, 25-35, 1994.

USDA, Urban Hydrology for Small Watersheds – TR-55, www.ncg.usda.gov/tech_tools.html, 1986.

B. Baginska, NSW EPA, Water Science Section, PO Box A290, Sydney South, NSW 1232, Australia

W. Milne-Home, National Centre for Groundwater Management, University of Technology Sydney, PO Box 123, Broadway, NSW 2007, Australia

Printed and bound by CPI Group (UK) Ltd, Croydon, CR0 4YY